Progress in Botany

Volume 77

Series Editors
Ulrich Lüttge, TU Darmstadt,
FB Biologie (10), Schnittspahnstraße 3–5,
64287 Darmstadt, Germany

Francisco M Cánovas
Universidad de Málaga, Depto. Biología
Molecular Y Campus de Teatinos 29071
Málaga, Spain

Rainer Matyssek
Technische Universität München
Hans-Carl-von-Carlowitz-Platz
Wissenschaftszentrum Weihenstephan
85354 Freising, Germany

More information about this series at http://www.springer.com/series/124

Ulrich Lüttge • Francisco M. Cánovas •
Rainer Matyssek
Editors

Progress in Botany 77

Springer

Editors
Ulrich Lüttge
TU Darmstadt FB Biologie
Darmstadt
Germany

Francisco M. Cánovas
Depto. Biología Molecular y
Universidad de Málaga
Málaga
Spain

Rainer Matyssek
Technische Universität München
Hans-Carl-von-Carlowitz-Platz
Wissenschaftszentrum Weihenstephan
Freising
Germany

ISSN 0340-4773 ISSN 2197-8492 (electronic)
Progress in Botany
ISBN 978-3-319-79827-1 ISBN 978-3-319-25688-7 (eBook)
DOI 10.1007/978-3-319-25688-7

Printed on acid-free paper

This Springer imprint is published by Springer Nature
The registered company is Springer International Publishing AG Switzerland

Contents

List of Contributors

Thierry Améglio INRA, UMR PIAF, Clermont-Ferrand, France

Clermont University, Université Blaise Pascal, UMR PIAF, Clermont-Ferrand, France

Suzana C. Bertolli Programa de Pós-graduação em Biologia Vegetal, Instituto de Biociências, Univ Estadual Paulista "Júlio de Mesquita Filho" (UNESP), Rio Claro, Brazil

Patricia Brito Department of Plant Biology, Universidad de La Laguna (ULL), La Laguna, Tenerife, Spain

M Luisa Buide Área de Botánica, Departamento de Biología Molecular e Ingeniería Bioquímica, Universidad Pablo de Olavide, Sevilla, Spain

Myriam Calonje Institute for Plant Biochemistry and Photosynthesis, Plant Development Unit, CSIC and Universidad de Sevilla, Seville, Spain

Inés Casimiro-Soriguer Área de Botánica, Departamento de Biología Molecular e Ingeniería Bioquímica, Universidad Pablo de Olavide, Sevilla, Spain

Área de Botánica, Departamento de Biología Vegetal y Ecología, Universidad de Sevilla, Sevilla, Spain

Wolfgang zu Castell Scientific Computing Research Unit, Helmholtz Zentrum München, German Research Center for Environmental Health GmbH, Neuherberg, Germany

Department of Mathematics, Technische Universität München, München, Germany

Matthias Fladung Thünen-Institute of Forest Genetics, Grosshansdorf, Germany

Frank Fleischmann Section Pathology of Woody Plants, Center of Life and Food Sciences Weihenstephan, Technische Universität München, München, Germany

Águeda Ma. González-Rodríguez Department of Plant Biology, Universidad de La Laguna (ULL), La Laguna, Tenerife, Spain

Scott Heckathorn Department of Environmental Sciences, University of Toledo, Toledo, OH, USA

Tina Heger Restoration Ecology, Center of Life and Food Sciences Weihenstephan, Technische Universität München, München, Germany

José Ignacio Hormaza Instituto de Hortofruticultura Subtropical y Mediterránea La Mayora (IHSM-UMA-CSIC), Estación Experimental La Mayora, Málaga, Spain

María S. Jiménez Department of Plant Biology, Universidad de La Laguna (ULL), La Laguna, Tenerife, Spain

Nerea Larrañaga Instituto de Hortofruticultura Subtropical y Mediterránea La Mayora (IHSM-UMA-CSIC), Estación Experimental La Mayora, Málaga, Spain

José R. Lorenzo Department of Plant Biology, Universidad de La Laguna (ULL), La Laguna, Tenerife, Spain

Eva Lucas-Reina Institute for Plant Biochemistry and Photosynthesis, Plant Development Unit, CSIC and Universidad de Sevilla, Seville, Spain

Ulrich Lüttge Department of Biology, Technical University of Darmstadt, Darmstadt, Germany

Ryo Matsuda Department of Biological and Environmental Engineering, Graduate School of Agricultural and Life Sciences, The University of Tokyo, Tokyo, Japan

Rainer Matyssek Technische Universität München, Hans-Carl-von-Carlowitz-Platz, Wissenschaftszentrum Weihenstephan, Freising, Germany

Stefan Mayr Institute of Botany, University of Innsbruck, Innsbruck, Austria

Sasmita Mishra Department of Environmental Sciences, University of Toledo, Toledo, OH, USA

Domingo Morales Department of Plant Biology, Universidad de La Laguna (ULL), La Laguna, Tenerife, Spain

Keach Murakami Department of Biological and Environmental Engineering, Graduate School of Agricultural and Life Sciences, The University of Tokyo, Tokyo, Japan

Eduardo Narbona Área de Botánica, Departamento de Biología Molecular e Ingeniería Bioquímica, Universidad Pablo de Olavide, Sevilla, Spain

M Isabel Ortiz-Marchena Institute for Plant Biochemistry and Photosynthesis, Plant Development Unit, CSIC and Universidad de Sevilla, Seville, Spain

José M. Romero Institute for Plant Biochemistry and Photosynthesis, Plant Development Unit, CSIC and Universidad de Sevilla, Seville, Spain

Francisco J. Romero-Campero Department of Computational Sciences and Artificial Intelligence, Research Group in Natural Computing, Universidad de Sevilla, Seville, Spain

Sanjay Singh Faculty of Agriculture, Department of Plant Sciences, University of Gondar, Gondar, Ethiopia

Gustavo M. Souza Laboratory of Plant Intelligence and Ecophysiology "Ulrich Lüttge", University of Western São Paulo (UNOESTE), Presidente Prudente, São Paulo, Brazil

Michel Thellier Emeritus of the University of Rouen, Nantes, France

Federico Valverde Institute for Plant Biochemistry and Photosynthesis, Plant Development Unit, CSIC and Universidad de Sevilla, Seville, Spain

Gerhard Wieser Department of Alpine Timberline Ecophysiology, Federal Research and Training Centre for Forests, Natural Hazards and Landscape (BFW), Innsbruck, Austria

Curriculum Vitae

Born 16 July 1936, Berlin
Family Status: Married, 3 children, 5 grandchildren

Academic and Professional Career

1954–1959 Studies of Biology and Chemistry, Ludwig-Maximilian-University, Munich

1960 Ph.D. (Dr. rerum naturalium), Technical University of Darmstadt

1960 Assistant, Technical University of Darmstadt

1964 Habilitation in Botany, Technical University of Darmstadt

1965 Dozent, Technical University of Darmstadt

Postdoctoral Years: 1965/1966 University of California, Los Angeles

1968/1969 Research School of Biological Sciences, Australian National University, Canberra

1970 o. Professor, Technical University of Darmstadt
2004 Professor Emeritus, Technical University of Darmstadt

Honours and Awarded Memberships
1982 Regents Lecturer, University of California at Riverside
1992 Member Academia Europaea
1996 Körber Price for the European Science
1996 Member Deutsche Akademie der Naturforscher Leopoldina
1999 Regents Lecturer, University of California at Davis
2002 Honorary Member Deutsche Botanische Gesellschaft
2004 Membre étranger Académie d'Agriculture de France, Paris

Scientific Functions
1976–1984 Elected Reviewer Deutsche Forschungsgemeinschaft (DFG)
1979–2000 Trustee Studienstiftung des Deutschen Volkes
1988–2002 Editor-in-Chief of Botanica Acta and Plant Biology
1992 to present Editor of Progress in Botany
1993–2000 Chairman Sonderforschungsbereich 199 of Deutsche Forschungsgemeinschaft (DFG) "Molecular Ecophysiology of Plants"
1994–2001 Selection Committee "Germans to Overseas" German Academic Exchange Service (DAAD)
1998–1999 Review Committee Biological Research, Vereniging van Universiteiten, The Netherlands
2000 to present Editor-in-chief of Trees: Structure and Function

Ulrich Lüttge

Part I
Review

Transport Processes: The Key Integrators in Plant Biology

Ulrich Lüttge

Contents

U. Lüttge (✉)
Department of Biology, Technical University of Darmstadt, Schnittspahn-Straße 3-5, 64287 Darmstadt, Germany
e-mail: luettge@bio.tu-darmstadt.de

U. Lüttge et al. (eds.), *Progress in Botany*, Progress in Botany 77,
DOI 10.1007/978-3-319-25688-7_1

Abstract In the 57 years of research reviewed in this essay, transport functions were studied in a variety of plant systems. Processes of membrane transport are essential in the operation of various glands, such as nectary glands, the glands of carnivorous plants, and the salt glands of halophytes. In the photosynthetic mode of crassulacean acid metabolism (CAM), a central feature is nocturnal accumulation of organic acids in the vacuoles. Thus, CAM poses a transport problem, which was resolved by the identification of the complement of an H^+-transporting ATPase, a malate channel, and a passive diffusion of non-dissociated malic acid at the tonoplast. The free running endogenous rhythm of CAM is operated by a biochemical-biophysical oscillator where the tonoplast acts as a master switch.

The paths of transport with apoplastic and symplastic transport and diffusion in the gas phase of aerenchymas couple and integrate cells within tissues. The energization of membrane transport is linked to the multicomponent network of energy metabolism. Transport in roots and leaves was investigated to show this.

All these features of plant biology indirectly or directly bear relations to physiological ecology of field performance. CAM is an ecophysiological adaptation to limited water supply which was studied intensively in the field in various tropical environments with respect to physiological autecology and ecosystem-relevant synecology.

Whole-plant physiology shows that transport is the basis of the functioning of entire plants. Transport is the pathway for interaction and integration creating plant's individuality as unitary organisms. The integration of modules via transport leads to the emergence of holistic systems across a large range of scalar levels from compartments within cells to cells and eventually the whole biosphere. Comprehending emergence of holism leads to understanding life beyond mechanistic modularity.

1 The Scene of Transport in Plants in the 1950s

Transport in plants: How did the landscape of research look in the 1950s? First, there were the membranes, barriers to, and paths of *short-distance transport*. Two alternative views were dominated by the lipid-permeability theory for permeation of lipid membranes by Overton (1899) and by the ultrafilter theory for permeation of porous membranes by Ruhland (1912; Ruhland and Hoffmann 1925), respectively. Both theories were combined within the lipid-filter theory (Wartiovaara and Collander 1960). Second, there were connections for *medium-distance transport* especially in the symplast of tissues. The symplastic transport was intensely studied by Arisz (1956, 1960). Third, there were the avenues of *long-distance transport* in the xylem and phloem. Two researchers covered this in the Institute of Forstbotanik (Forest Botany) of the Ludwig Maximilian University of Munich. Bruno Huber was advancing the cohesion–tension theory of xylem sap flow, applying the heat pulse method for measurements of velocity and derivation of flow. This theory was challenged time and again (Ziegler et al. 2009). Hubert Ziegler was dedicated to transport in the phloem (Ziegler 1956) based on the pressure flow theory of Münch (1930).

I started my experimental work in 1957 in this Institute of Forstbotanik. At that time laboratories studying transport in plants were not very numerous. Even in the 1960s, it proved easy to assemble all relevant books on the topic on one's personal small book shelve, namely, Briggs, Hope, and Robertson (1961), Sutcliffe (1962), Jennings (1963), and Robertson (1968). It is hard to imagine today that at the end of the 1970s, it appeared still possible to publish a book covering the entire scope of transport in plants (Lüttge and Higinbotham 1979). Since then the field exploded, currently identifying a vast number of families of membrane transporters at the molecular level, such as ATPases, carriers, channels, and porins. The last decades of the twentieth century and the first decades of the twenty-first century might be named an era of transport physiology.

In this essay, I shall try to develop the progress of my interests in transport physiology within context of knowledge and not stringently chronologically. Reviewing the various topics in such a way, I shall try to provide the links to the current states of research. However, space does not allow developing this in depth. Rather, the essay will essentially remain a historical treatise. The integrating power of transport overcoming borders and barriers allowed advancing from the platform of transport studies to physiological ecology of tropical plants and consideration of emergence at higher scalar levels. Much chance, serendipity, guidance by excellent teachers, stimulation by admirable peers and friends, and achievements of wonderful dedicated coworkers and students were involved. All of these were exceptional presents that now allow looking back with gratitude on what could be completed.

2 Glands, Salt Hairs, and Epidermal Bladders

2.1 Nectaries

The great question in Hubert Ziegler's reflections on phloem transport was the function of the companion cells. They were supposed to secrete sugars into the sieve tubes for their long-distance transport. However, inside the phloem tissue, the companion cells were not accessible for direct experimental studies (Ziegler 1956). As an analogy, Hubert Ziegler had made acquaintance with nectar glands, another example of sugar-secreting cells, during a stay in the laboratory of Albert Frey-Wyssling in Zürich in 1956/1957. Hence, when I joined him in 1957 as his first Ph. D. student, he asked me to work on nectaries.

I used all the available analytical techniques for analyzing the chemical composition of nectar from various species, with the profile of sugars mainly of sucrose, and glucose and fructose mostly in stoichiometrically equal amounts, and some oligosaccharides, carbonic acids, amino acids, mineral ions, protein, and vitamins (Lüttge 1961, 1962a; Ziegler et al. 1964). At that time our major interest in these analyses was to assess the array of solutes that are transported. Subsequently, others continued the chemical analyses revealing the rich chemical composition of nectar including secondary metabolites such as alkaloids (Kessler and Baldwin 2007; Manson et al. 2010). The sugar and amino acid spectra of nectars are of ecological relevance particularly with respect to mutualism of plants and the preferences of specific pollinators (Baker and Baker 1977; Corbet et al. 1979; Alm et al. 1990; Erhardt 1992; Schmidt-Lebhuhn et al. 2007; Nepi et al. 2012). Proteins and enzymology of nectars are studied to understand the metabolic activity of the secretion. A redox cycle has been shown to be active with redox compounds and reactive oxygen species such as H_2O_2 and ascorbate for the control of microbial contamination (Carter and Thornburg 2000, 2004a, b; Carter et al. 2007; Horner et al. 2007; González-Teuber et al. 2010; Hillwig et al. 2010, 2011; Escalante-Pérez and Heil 2012). Extrusion of vesicles ["granulocrinous" secretion *sensu* Fahn (1979) and Schnepf (Schnepf and Christ 1980)] might be the mechanism of secretion of macromolecules such as proteins.

For me the major challenge of studying gland functions remained that of transport physiology. The established task of the companion cells was to control the composition of the transported sieve tube sap. This would involve both secretion and reabsorption of compounds. Just like the phloem-sap, the chemical composition of nectar is rich in qualitative terms considering the diversity of compounds. However, quantitatively, the sugars are absolutely dominant. Thus, there is specificity in the secretion. I could show that the degree of specificity regarding the dominance of sugars and the low concentrations of noncarbohydrate solutes in the nectar was related to the degree of the anatomic specialization of the glandular tissue (Lüttge 1961). Transport processes could create the specificity by specific secretion or reabsorption. Active transport and specific transporters, explicitly carrier proteins, were thought to be involved, but at the time not any transporter had been characterized in plants, and the molecular basis remained pure speculation. Respiratory activity of nectaries is usually high (Lüttge and Schnepf 1976) and can drive energy-demanding transport processes. Many nectaries are green, and their photosynthetic capacity can provide energy as well as part of the carbohydrates secreted (Lüttge 2013a).

A central question was if nectaries do not only secrete but also reabsorb solutes and if different transport processes can operate simultaneously in opposite

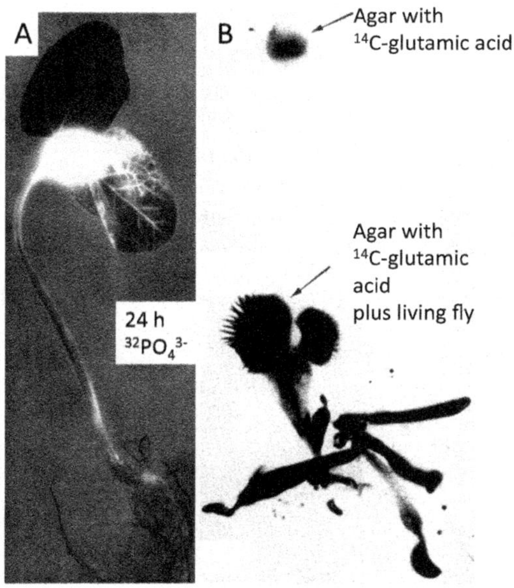

Fig. 1 Autoradiographs demonstrating resorption by glands. (**a**) A droplet of $^{32}PO_4^{3-}$ solution was placed on the extrafloral nectary of the left cotyledon of this seedling of *Ricinus communis* L. (Lüttge 1961). (**b**) A trap of *Dionaea muscipula* Ellis was fed (*arrows*) with ^{14}C-glutamic acid in dry agar (*top*) and with a living fly in addition to the labeled agar (*bottom*) (Lüttge 1963). Note that (**a**) is reproduced as a negative and (**b**) as a positive of the X-ray film, so that radioactivity appears *white* and *black*, respectively

directions. It proved useful to develop imaging techniques. The labeling technique of the time was that with radioactive isotopes. I applied radioactively labeled solutes to the nectaries and after some time of incubation pressed and freeze-dried the whole plants and covered them with X-ray film. In this way uptake of solutes by the nectaries and translocation in the whole plant could be demonstrated (Fig. 1a; Ziegler and Lüttge 1959; Lüttge 1962b). This contributed to explaining the specific chemical composition of nectar. At the end of the pollination period, entire surplus nectar may also be reabsorbed for economy of the plant's resources (Búrquez and Corbet 1991; Nepi and Stpiczyńska 2007). Thus, the studies on nectar secretion exerted some feedback on the reflections about companion cell functions in the phloem.

The refinement of autoradiography to make the anatomical level accessible then became a major occupation. It was essential to develop procedures preventing the artificial redistribution of water-soluble compounds during preparation for microscopic inspection. The task was tedious but highly rewarding for many further studies. The developed silver grains in the film of the micro-autoradiographs allowed localization at the cellular and subcellular level. Quantification was possible by densitometry or measurement of the reflection of incident illumination by the silver grains. The most precise way although extremely wearisome proved to be grid-based counting of silver grains under the microscope. It was proven that absorption indeed is by the nectary gland tissue (Lüttge 1962b). A book on methodology emerged from this work (Lüttge 1972). I met André Läuchli who had similar aims of cellular solute localizations and was engaged in X-ray microanalysis. A long friendship emerged, and André came in 1972 for 6 years as professor to our department in Darmstadt.

2.2 Carnivorous Plants

The glands of carnivorous plants fulfill a number of different functions. They secrete digestive enzymes and serve absorption of low molecular compounds from the digested prey. They secrete mucilage where this is involved in the capture of prey. In the pitchers of *Nepenthes*, they secrete chloride for the acidification of the pitcher fluid with hydrochloric acid (Fig. 2b; Lüttge 1966a, b). Hence, in

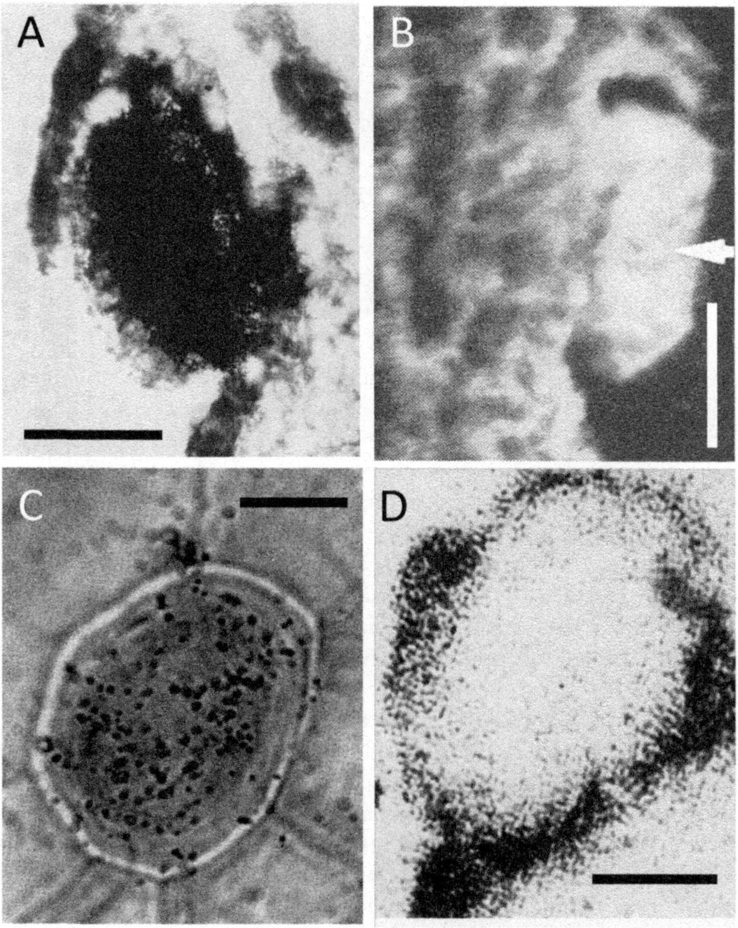

Fig. 2 Secretion and absorption by glands shown by micro-autoradiography. (**a**) Gland of *Nepenthes* where the pitcher was fed with $^{35}SO_4^{2-}$. Scale bar: 50 μm (Lüttge 1965). (**b**) Gland of *Nepenthes* secreting $^{36}Cl^-$ (*arrow*) fed to the pitcher wall tissue. *Scale bar*: 50 μm (Lüttge 1966b). (**c**) Uptake of $^{35}SO_4^{2-}$ by a hydropote gland of *Nymphaea*. *Scale bar*: 10 μm (Lüttge 1964b). (**d**) Stalk and bladder cells of *Atriplex spongiosa* James whose leaves were supplied with $^{35}SO_4^{2-}$. *Scale bar*: 25 μm (Osmond et al. 1969, copyright http://www.publish.csiro.au/nid/280/paper/BI9690797.htm). Note that in transmitted light, the silver grains of the micro-autoradiographs are *black* (**a**, **c**, **d**) and in incident light they are *bright white* (**b**)

carnivory, it appeared clear at the outset that glands would perform transport in two opposite directions.

The two lobes of the leaves of Venus' fly trap (*Dionaea muscipula* Ellis) close by a thigmonastic turgor movement when one of the six trigger hairs (three on each lobe) is touched twice or two different hairs are touched once. With each touching, an action potential is triggered and the two action potentials in series elicit trap closure. However, a chemical signal is required for a firm closure of the trap by a chemonastic growth movement (Fig. 37.7 in Lüttge et al. 2010) and for the induction of gland activity. Placing a dead fly, some dry cheese, or agar into the trap, this latter movement does not occur, and upon mechanic stimulus only the trap soon reopens. However, a living fly excreting fluid while struggling for life, a piece of moist meat, or a piece of moist cheese will elicit the chemonastic movement and also gland activity for absorption, e.g., of amino acids, as I could show by autoradiography (Fig. 1b; Lüttge 1963, 1965). Micro-autoradiography demonstrated the role of the pitcher glands of *Nepenthes* for absorption (Fig. 2a) which is specific. Proportional rates of absorption from 1 mM solutions in the pitcher of L-alanine/phosphate/sulfate were 1 : 0.4 : 0.1 (Lüttge 1965). Both specific transporters, such as channels and carriers, and vesicle endocytosis can be involved in absorption (Adlassnig et al. 2012).

The pitcher glands secrete digestive fluid before the lid of the pitcher opens. The fluid in the closed pitchers is microbiologically sterile, i.e., not contaminated. It contains proteinase activity which is of genuine origin of secretion by the plants (Lüttge 1964a). The proteinase processed to electrophoretic homogeneity has a pH optimum close to pH 2.2 (Steckelberg et al. 1967). The pH of open pitchers with prey was recorded at around pH 3.5 (Lüttge 1964a). The plasma membrane H^+-ATPase appears to be responsible for such acidification (An et al. 2001). The pitcher glands secrete chloride as shown by micro-autoradiography (Fig. 2b; Lüttge 1966b). Chloride secretion is inhibited by cyanide and arsenate. The chloride concentration of the fluid of still closed pitchers was about 30 mM on average, ranging up to about 65 mM (Lüttge 1966a). Hence, it is HCl that acidifies the pitcher fluid. The proteinase has biochemical properties akin to the pepsin of our own stomach. This was subsequently detailed by determination of amino acid sequences of the enzyme protein now named nepenthesin (Jentsch 1972; Athauda et al. 2004; Takahashi and Tanji 2007). Hydrochloric acid also acidifies the digestive fluid of *D. muscipula* (Rea 1983; Rea et al. 1983). Unfortunately space does not allow this essay to further pursue the fascinating research on carnivorous plants, "the most wonderful plants in the world" (Król et al. 2012).

2.3 Salt Glands and Salt Hairs

2.3.1 Salt-Excreting and Salt-Absorbing Glands

In many halophytes, the adaptation to salinity is based on the transport function of very specific structures, namely, salt glands excreting sodium chloride (Lüttge

1975). With the micro-autoradiographic imaging technique, chloride excretion by the salt glands of *Limonium vulgare* Miller was depicted (Ziegler and Lüttge 1967). For the hydropote glands on the lower, i.e., submerged, surface of the floating leaves of water lily (*Nymphaea*), this imaging showed that their function is salt uptake (Fig. 2c; Lüttge 1964b).

2.3.2 Salt Bladder Hairs of *Atriplex spongiosa* (James)

My major occupation with gland-type salt transport, however, arose from some serendipity. In 1965/1966 in the laboratory of George G. Laties at the University of California, Los Angeles, I had shared benches with C. Barry Osmond as postdocs. Barry spent another year with Tom ap Rees in Cambridge, UK, while I had returned to Darmstadt in September 1966. Barry visited me there for a few weeks on his way back to Australia (September–October 1967). Barry had been irreversibly infected with the C_4-fever after this new mode of photosynthesis had been discovered by Kortschak et al. (1965) and Hatch and Slack (1966). With this he also had fallen in love with *Atriplex spongiosa* James, not so much because of its nature as a halophyte, but since it was the C_4 partner of a couple together with *Atriplex hastata* L. serving as the C_3 partner for interspecific comparisons of the two modes of photosynthesis. *A. spongiosa* has epidermal salt hairs composed of a stalk cell and a large bladder cell. My interests in salt transport given, during Barry's visit in Darmstadt, we performed some micro-autoradiographs. The stalk cells are densely filled with cytoplasm, and like salt gland cells they function in secretion of salt into the large vacuoles of the bladder cells (Fig. 2d; Osmond et al. 1969).

Together with Ralph O. Slatyer, Barry invited me to come for an entire year to the just founded Research School of Biological Sciences at the Australian National University in Canberra (August 1968–July 1969). Our interests in photosynthesis and salt transport were combined to fathom the energetics of salt concentrating in the bladder cells of *A. spongiosa* (Osmond et al. 1969; Lüttge and Osmond 1970). Chloride accumulation in the bladder vacuoles is strongly stimulated by light. However, the bladder, stalk, and epidermis cells are not photosynthetically active. In plants grown on 0.25 M NaCl, the accumulation of NaCl in bladders is about 4–5 times higher than in the green lamina (Table 1). Hence, energy captured by photosynthetic light absorption in the lamina is able to energize active salt export by the distant stalk cells and accumulation in the bladder vacuoles (Lüttge and

Table 1 Light dependence of NaCl accumulation in salt bladders of *A. spongiosa* plants grown with 0.25 M NaCl solution (Osmond et al. 1969)

	Bladders	Green lamina
Photosynthetic CO_2 fixation mol $g_{FW}{}^{-1}$ h^{-1}	0.2	10.6
Light stimulation of Cl^- accumulation	4.7×	1.9×
Na^+ concentration (M)	1.02	0.28
Cl^- concentration (M)	0.72	0.13

Osmond 1970). In this way, the salt load on the metabolically active lamina tissue is highly reduced.

2.4 *Giant Epidermal Bladders:* Mesembryanthemum crystallinum *L. and* Capsicum annuum *L.*

The work on *A. spongiosa* led to a serendipity which marked the starting point of a significant revolution in plant stress physiology with now of a global dimension. In 1969 on the way back from Australia, I made a stopover in California, where I also was invited to visit the laboratory of Andy Benson in La Jolla. I assembled all my courage and all I knew about photosynthesis and lipids to meet the great man. However, at dinner in a Mexican restaurant, he asked me what I was really interested in, and I told him the *Atriplex* story. Andy immediately got excited, and in the darkness of this evening of 31 July 1969 he drove me to the beach and came up with collecting samples of *Mesembryanthemum crystallinum* L. There are really large huge epidermal bladder cells on the leaves and stems with a volume of up to 2 mm^3, and I should work with these. I shall continue telling the anecdote and its consequences below (Sect. 6.1.2) when it comes to talk about crassulacean acid metabolism.

Seeds of *M. crystallinum* were taken back to Darmstadt. Plants were grown, and a student was asked to check if the bladders accumulated salt. Although other authors later reported and believed that they did, we did not find NaCl concentrations in bladder cell sap higher than in the leaf mesophyll when plants were grown with watering by NaCl solutions of up to 500 mM (Lüttge et al. 1978). The bladder cells of *M. crystallinum* were already described by Haberlandt (1904) as peripheral water storage cells. In the leaves, they may sequester salt due to their sheer size, but they do not concentrate it. By contrast to the salt hairs of *Atriplex*, they are just inflated epidermal cells and there is no gland like cell underneath. The water storage capacity of the bladder cells of *M. crystallinum* is large. Biophysical studies showed how their cellular water transport dynamically integrates and stabilizes water relations of the leaves (Steudle et al. 1975, 1977; Lüttge et al. 1978; Rygol et al. 1986, 1989).

Giant cells with large cell sap vacuoles are an attribute of plant succulence. In a subepidermal cell layer of the inner pericarp wall of *Capsicum annuum* L., the cells have a similarly large volume as the bladder cells of *M. crystallinum*, namely, 0.5 mm^3 on average and up to 1.7 mm^3. The water relation parameters of the large *C. annuum* cells, i.e., higher hydraulic conductivity, L_p, and volumetric cell wall elasticity modulus, ε, and shorter half-life of water exchange as compared to the much smaller mesocarp parenchyma cells, led us to an ecophysiological interpretation of the biophysics of succulence in relation to water storage and remobilization (Rygol and Lüttge 1983).

3 Paths of Transport: Coupling and Integration Within Tissues

For the transport within tissues, there are three different types of pathways. Solutes in aqueous media can be transported in the apoplastic space of the cell walls and in the symplasm, i.e., outside the plasma membranes and within the cytoplasm, respectively. The third way is diffusion of gaseous substances in the gas phase within intercellular spaces, in particular, of leaf aerenchyma.

3.1 Apoplastic and Symplastic Transport

With respect to the entry of solutes into plant tissues, the apoplastic space was termed "apparent free space" (Briggs et al. 1961) because it appears to be freely accessible without metabolic control over the uptake into and the transport within this space. Conversely uptake across the plasma membrane is required before transport in the symplasm can occur. For his studies of symplastic transport, the authoritative investigator Arisz (1956, 1960) used the long band-like submerged leaves of *Vallisneria*.

Our micro-autoradiographic imaging contributed to visualizing the transport pathways. Frequently one sees that the cell borders are particularly labeled when radioactive-labeled solutes are transported. In some cases, the resolution was not high enough to distinguish apoplast versus cytoplasm as it was the case in the nectar glands of *Heracleum sphondylium* L. (Fig. 3a) and the root cortex of *Zea mays* L. (Fig. 3d). Transport in nectary tissues potentially occurs both in the apoplastic and the symplastic space (Sawidis 1991; Vassilyev 2010). In the bladder hairs of *A. spongiosa*, distinctively the dense cytoplasm of the stalk cells and the bladder cytoplasm were preferentially labeled (Fig. 2d). In the mesophyll and gland cells of *Nepenthes*, labeling of the cytoplasm and, hence, the symplastic pathway also became evident (Fig. 3b, c).

The transport from the external medium across the root can take place both apoplastically and symplastically in the cortex up to the endodermis which constitutes a barrier for apoplastic transport. Using micro-autoradiography, we found that entry of labeled sulfate into the stele was blocked at the endodermis by the metabolic inhibitor azide (Weigl and Lüttge 1962). This is the final point where transported solutes must undergo metabolically controlled transport from the apoplast into the symplast across the plasma membrane to move further towards the xylem vessels of the stele. In the tertiary endodermis, only the living passage cells allow transport between the cortex and stele of roots (Fig. 3e). Later Ernst Steudle and his collaborators devoted a large and impressive body of work to the elucidation of these pathways for the transport of ions and water, and they developed the "composite model" of transport across the root (Steudle 2011).

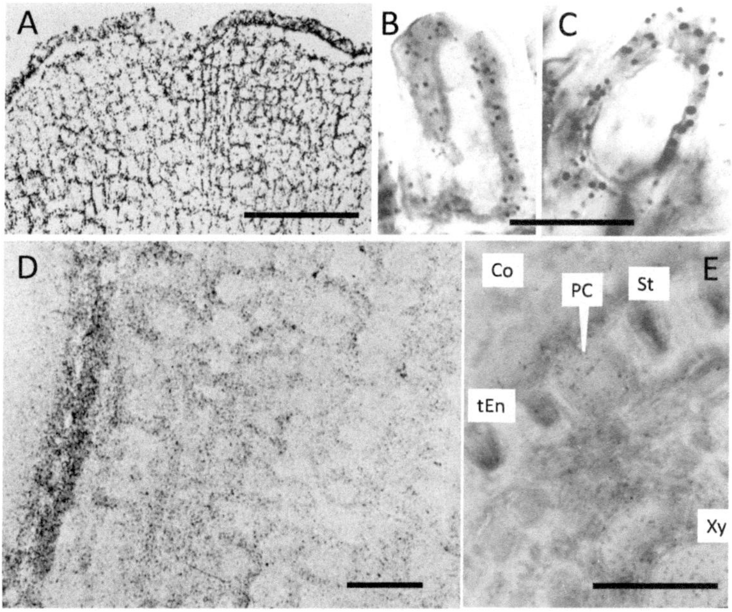

Fig. 3 Micro-autoradiographic imaging of transport in tissues. (**a**) Nectary of *Heracleum sphondylium* L. after resorption of $^{35}SO_4^{2-}$ (Lüttge 1962a). (**b**) Gland cell and (**c**) mesophyll cell of the pitcher wall of *Nepenthes* secreting $^{36}Cl^-$ (Lüttge 1966b). (**d**) Maize root cortex after uptake of $^{35}SO_4^{2-}$ (Weigl and Lüttge 1962; Lüttge and Weigl 1965). (**e**) Root of *Iris pumila* L. to which $^{35}SO_4^{2-}$ was administered via the transpiration stream in the stele. Co cortex, PC passage cell, tEn tertiary endodermis, St stele, Xy xylem (Ziegler et al. 1963). *Scale bars*: (**a**, **d**) 100 μm, (**b**, **c**), 25 μm, (**e**) 50 μm

3.2 *Cell Coupling*

Integration of cells in tissues requires their coupling. The essential structural basis of such coupling is the plasmodesmata, i.e., tubing-like bridges of endoplasmic-reticulum membranes, between the cytoplasm of adjacent cells (Spanswick 1976), resulting in a symplastic network of connected cells, i.e., the symplasm. By precipitating Cl^- ions with Ag^+ ions to obtain electron-dense AgCl particles, we visualized ion transport through plasmodesmata (Fig. 4).

The continuity of plasma membranes across the plasmodesmata allows electrical coupling. This was studied by Spanswick and collaborators (Spanswick 1976). Our own contributions made use of the phenomenon of transient changes of membrane electric potential in green cells when light is switched on or off (see below Sect. 4.3.2). This signal depends on photosynthesis and is not produced in nongreen cells. However, we could pick up such signals with microelectrodes in the nongreen bladder cells of *A. spongiosa* and *Chenopodium album* L. (Lüttge and Pallaghy 1969; Osmond et al. 1969; Pallaghy and Lüttge 1970) and in the nongreen cells of variegated leaves of *Oenothera* (Brinckmann and Lüttge 1974) demonstrating their electrotonic coupling to adjacent green mesophyll cells.

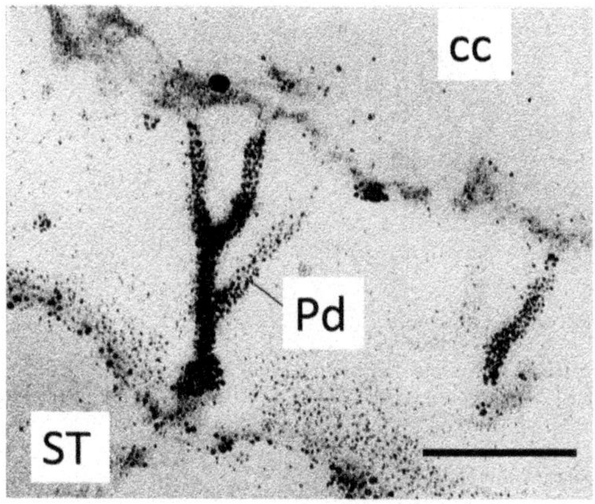

Fig. 4 Plasmodesmata (Pd) between a companion cell (CC) and a sieve tube (ST) in the phloem of a leaf of *Limonium vulgare* Miller transporting Cl^-. *Scale bar*: 250 nm. Cl^- was precipitated by addition of Ag-acetate to the electron-microscopic fixation medium; the *black dots* are precipitates of AgCl (Ziegler and Lüttge 1967)

3.3 The Gas Phase: Integrating Cells in Leaves with CAM

Our interests in the third pathway, i.e., lateral diffusion in the gas phase of leaves, arose later when we were involved in studies of crassulacean acid metabolism (CAM) (Sect. 6). The discussion of diffusion in the gas phase of leaves is determined by the heterobaric/homobaric leaf concept, i.e., whether the gas partial pressures are heterogeneous or homogeneous within the air space of whole leaves (Neger 1912, 1918; Terashima 1992; Pieruschka et al. 2005). Anatomical constraints leading to compartmentalization of the leaf air space are responsible for heterobaric conditions.

The friendship with Barry Osmond initiated a new methodological development in our laboratory. He had received the Alexander-von-Humboldt-Forschungspreis and arrived in our laboratory with a camera and a strong interest in chlorophyll fluorescence imaging. We studied nonuniform patchiness of photosynthetic activity in virus-infected leaves of *Abutilon* (Osmond et al. 1998). Heterobaric conditions in leaves are responsible for stomatal patchiness (Beyschlag and Eckstein 1997) because CO_2 is a signal molecule for stomatal movements (Lüttge and Hütt 2006). Nonuniform, i.e., patchy, photosynthetic activity in leaves is associated with this. Photosynthetic patchiness, however, can also arise independent of stomata as we found with Barry when studying wilting and recovery of leaves. Under such conditions, patchiness in chlorophyll fluorescence proved to arise from metabolic interference of drought stress (Osmond et al. 1999). The technique of imaging was then advanced in our laboratory by Uwe Rascher and later Heitor M. Duarte when they were Ph.D. students and the creative electronics expert Karl Schuller in the workshop of the Institute of Botany at Darmstadt. This enabled us to measure gas exchange, photorespiration, and spatially resolved photosynthetic electron transport synchronously and online.

The CAM plant *Kalanchoë daigremontiana* Hamet et Perrier has a rather uniform mesophyll of large succulent spherical cells, and therefore structurally the leaves would be homobaric. However, the cells are densely packed impeding lateral gas diffusion, and therefore the leaves become functionally heterobaric (Rascher et al. 2001; Maddess et al. 2002; Duarte et al. 2005b; Lüttge and Hütt 2006). Imaging shows that heterogeneity or patchiness of relative quantum use efficiency of photosystem II (Φ_{PSII}) is rather low in phase III of CAM during the light period. This is the phase when stomata are closed and CO_2 concentration in the gas phase of the leaves (p_i^{CO2}) is very high due to the CO_2 remobilization from organic acid, mainly malic acid, accumulated nocturnally in phase I of CAM (Lüttge 2002). High CO_2 concentration drives lateral diffusion and reduces patchiness. Conversely, patchiness of Φ_{PSII} is high in phases II and IV when stomata are open at the beginning and towards the end of the light period, respectively, at fairly low p_i^{CO2}. Patchiness is particularly high in the transition between phases III and IV, when stomata open after nocturnally accumulated organic acid is consumed. At that stage, p_i^{CO2} equilibria within the leaves are rearranged (Rascher et al. 2001; Maddess et al. 2002; Rascher and Lüttge 2002; Duarte et al. 2005b; Lüttge and Hütt 2006; Duarte and Lüttge 2007a). The transition is a highly dynamic spatiotemporal event. In phase III due to high p_i^{CO2}, Φ_{PSII} is high. During the transition between phases III and IV, low Φ_{PSII} patches develop from which waves of high Φ_{PSII} may run towards each other and extinguish each other when they meet (Fig. 5).

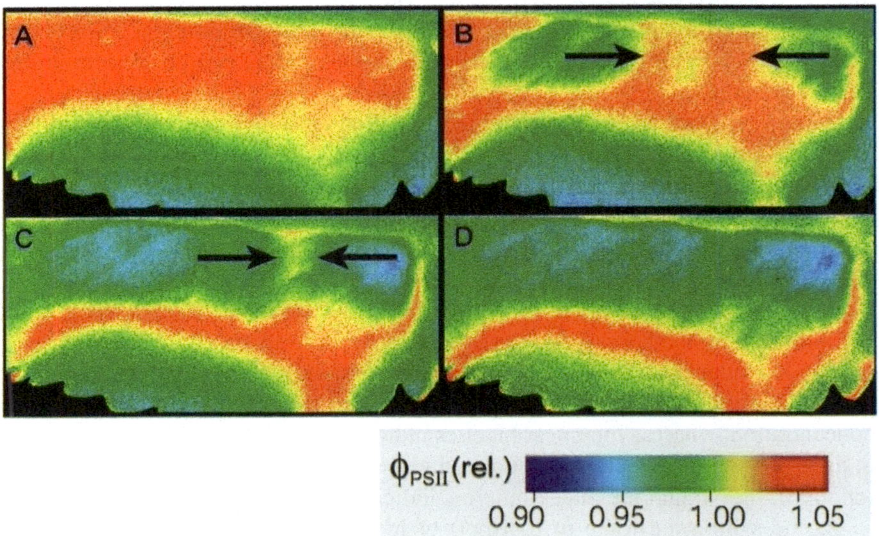

Fig. 5 Relative quantum use efficiency of photosystem II (Φ_{PSII}) in a leaf of *Kalanchoë daigremontiana* Hamet et Perrier during the transition between phases III and IV of CAM. Frames (**a**) to (**d**) were taken 20 min apart. Wavefronts of high Φ_{PSII} (*red*, see color code), which were initiated at different spots on the leaf, ran in opposite directions (*arrows*) and extinguished each other when they met (Rascher and Lüttge 2002, Copyright John Wiley and Sons)

4 Compartmentation: Transport at the Cellular Level

In assessing transport dynamics that govern cellular compartmentation, our work for some while focused on studies of transport kinetics (Lüttge 1968), namely, uptake kinetics (Sect. 4.1) and efflux kinetics (Sect. 4.2). In addition we used electrophysiology (Sect. 4.3). With this we were looking at the major cell compartments apoplast, cytoplasm, and vacuole.

4.1 Uptake Kinetics of Mineral Ions: Michaelis-Menten Hyperbolae

In the early 1950s, Emanuel Epstein had discovered that the concentration dependence of ion uptake by root tissues followed hyperbolic kinetics similar to the Michaelis-Menten kinetics of enzymes. With his group he elaborated this in a remarkable series of studies (Epstein and Hagen 1952; Epstein et al. 1963; Epstein and Rains 1965). They concluded that the enzyme kinetics of transport demonstrated the involvement of carrier molecules, which bind ions as substrates during the transport process across membranes. Moreover, they found that the kinetics were complex showing a dual hyperbolic isotherm, one saturating at low concentrations, i.e., with high affinity, and the other one saturating at high concentrations, i.e., with low affinity.

Epstein and his group had assumed that both the high and the low affinity systems were located at the same membrane, i.e., the plasma membrane, operating in parallel. When I came to George Laties' laboratory in October 1965, they had just tested an alternative view, namely, that the high affinity system was at the plasma membrane and the low affinity system was at the tonoplast (Torii and Laties 1966). They compared root tips, where the cells were not yet vacuolated having no tonoplasts, with vacuolated root tissue. In the non-vacuolated cells, they found the high affinity isotherm at low external ion concentrations and a linear concentration dependence of ion uptake at high concentrations. Their interpretation was that the high affinity system operated and limited transport into the cells at low concentration, whereas the linear kinetics indicated dominance of passive diffusion at high concentrations. In the vacuolated cells, having both plasma membrane and tonoplast, they found the dual isotherms indicating activity of both systems.

As the symplastic route of transport of ions from the medium across the root towards loading the xylem conduits does not involve vacuoles and tonoplasts (Sect. 3.1), George had postulated that it should follow the same kinetics as uptake by non-vacuolated cells, i.e., with the high affinity isotherm at low and a linear relationship at high concentrations. I got to check this, and indeed in a number of studies (Lüttge and Laties 1966, 1967a, b), the kinetic evidence corroborated these expectations (Fig. 6).

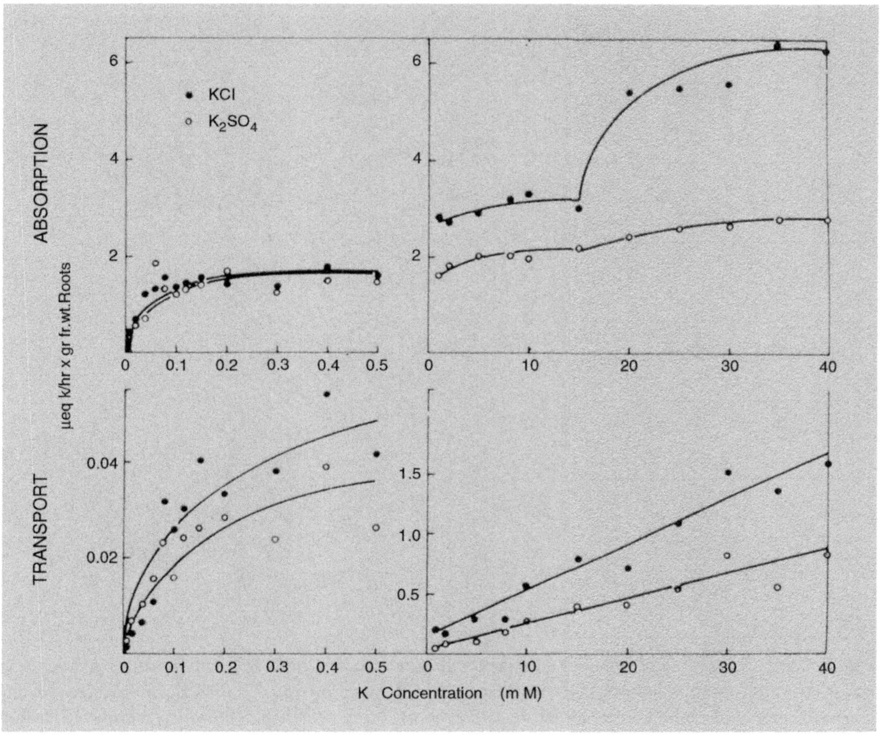

Fig. 6 Potassium absorption by the root tissue (*above*) and transport into xylem exudates (*below*) by maize seedlings from KCl (*closed circles*) and K_2SO_4 (*open circles*) solutions in a low (*left*) and high (*right*) concentration range (Lüttge and Laties 1966, www.plantphysiol.org, Copyright American Society of Plant Biologists)

For visualizing the uptake of ions into the cytoplasm and checking its isotherm, we returned to micro-autoradiography and used a trick. When one gently centrifuges tissues, one can precipitate the cytoplasm within cells without damaging the cells. The procedure enhances the resolution of imaging the cytoplasm. After uptake of labeled ions, the high affinity isotherm can be derived from quantifying the autoradiograms (Fig. 7), and similarly X-ray microanalysis can be used with the centrifuged material (Läuchli and Lüttge 1968).

Epstein's discovery of enzyme kinetics of ion uptake proved to be extraordinarily fruitful. From molecular characterizations in several laboratories, we now know that there are a plethora of transporters for various ions both with high and low affinity at both the plasma membrane and the tonoplast. On such grounds, the cellular localization of mechanisms which bring about the various Michaelis-Menten isotherms and compartmentation of the high and low affinity transport processes need new interpretation.

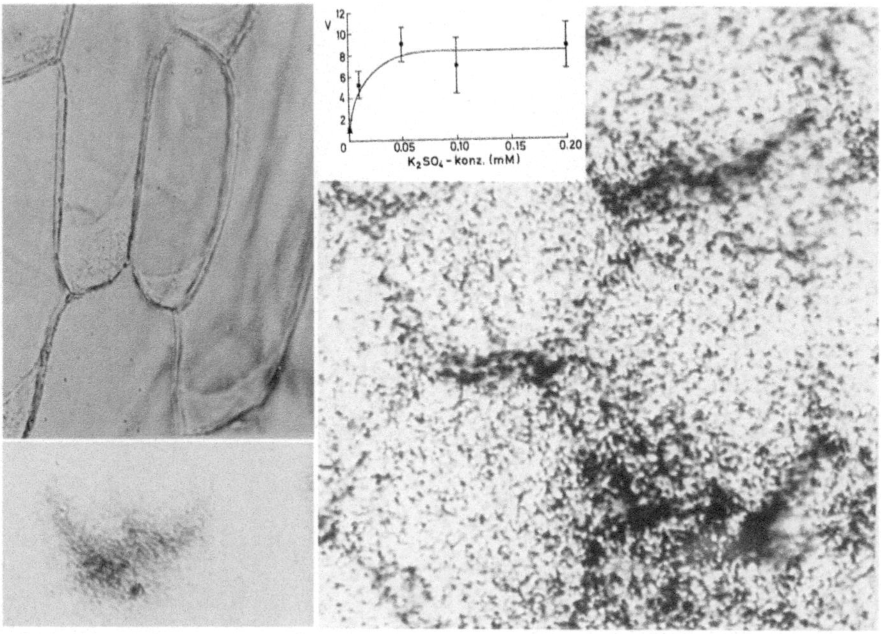

Fig. 7 *Left*: Cells of a centrifuged adaxial epidermis of onion bulb (*Allium cepa* L.) (*above*) with micro-autoradiograph (*below*) after uptake of $^{35}SO_4^{2-}$ for 3 h from 7.5 mM K_2SO_4. *Right*: Micro-autoradiograph of cells from a centrifuged leaflet of the moss *Plagiomnium cuspidatum* (Hedw.) T.J. Kop after uptake of $^{35}SO_4^{2-}$ for 7 h from 5 mM K_2SO_4, where the insert shows the relative rate of $^{35}SO_4^{2-}$ uptake derived from the density of the silver grains in micro-autoradiographs after uptake from increasing concentrations of K_2SO_4 in a low concentration range (Läuchli and Lüttge 1968)

4.2 *Efflux Kinetics*

While the uptake kinetics discussed in Sect. 4.1 are based on influx of labeled ions into the cells, for measuring efflux kinetics, the tissue is first labeled by extended incubation in the radioactive ion solution up to establishment of kinetic equilibrium of influx and efflux. The tissue is then transferred to a non-labeled solution with the same ion concentration as that of the labeling solution, and the rate of label efflux is measured. This essentially means that rates of isotope exchange are assessed. Basically efflux shows three linear phases: (1) a rather rapid one, (2) a slower one, and (3) a quite slow one extending over many hours. The phases arise from exchanges at the apoplast (1), the cytoplasm (2), and the vacuoles (3), represented by efflux of labeled and influx of non-labeled ions at efflux–influx equilibrium. A set of equations is developed relating fluxes and pools, i.e., for assessing the individual fluxes at the plasma membrane and tonoplast and the ion contents of cytoplasm and vacuole (Pitman 1963; Lüttge 1968). A shoulder of efflux in phase (2) indicated further sub-compartmentation of the cytoplasm (Pallaghy et al. 1970;

Lüttge and Pallaghy 1972). Concentration dependence of the fluxes calculated when efflux was measured at varied ion concentrations of the labeling and exchange solutions confirmed the pattern of plasma membrane and tonoplast transport kinetics found by Torii and Laties (1966, Sect. 4.1). However, it was already noted that molecular analyses were needed to actually identify the transporters involved (Lüttge and Bauer 1968). It is good to see now that, in the age of molecular "transportomics," where so many membrane transporter molecules are qualitatively identified and characterized, efflux kinetics are still used to quantify ionic relations (e.g., Kronzucker et al. 1999; Britto et al. 2004; Abbaspour et al. 2013).

4.3 Electrophysiology

4.3.1 Transcellular Electrical Profiles

In assessing gross compartmentation of plant cells, it is the aim to separately measure electrical membrane potentials of the plasma membrane and the tonoplast. In cells of higher plants, the plasma membrane potential is typically highly negative (cytoplasmic side negative), and the tonoplast potential is close to zero or somewhat positive (vacuolar side positive). In vacuolated higher plant cells with only a thin layer of cytoplasm along the cell wall, it is usually very difficult to place the tip of microelectrodes into the cytoplasm for measuring the plasma membrane potential separately from the tonoplast potential. Normally microelectrodes directly penetrate into the vacuoles, and in the measurements electrical potentials of both membranes are additive.

I had learned electrophysiology from Charles Pallaghy in Canberra in 1968/1969, and employing the approach received another boost when Noe Higinbotham visited us in Darmstadt twice and the second time for a whole year (July–September 1975, October 1977–September 1978). Noe was a pioneer and the first who had punched electrodes into cells of higher plants (Etherton and Higinbotham 1960; Higinbotham et al. 1964). We used our centrifugation technique (Sect. 4.1) to place electrode tips precisely into the cytoplasm (Lüttge and Zirke 1974; Fischer et al. 1976). More elegantly Jean-Pierre Rona in Paris, with whom we cooperated, produced transcellular electrical profiles of *Kalanchoë* cells showing the plasma membrane potential from −110 to −122 mV and the tonoplast potential from +20 to +25 mV (Rona et al. 1980, Fig. 8). The slightly positive tonoplast potential was also seen later, when we inserted electrodes into isolated vacuoles of *K. daigremontiana* (Jochem et al. 1984).

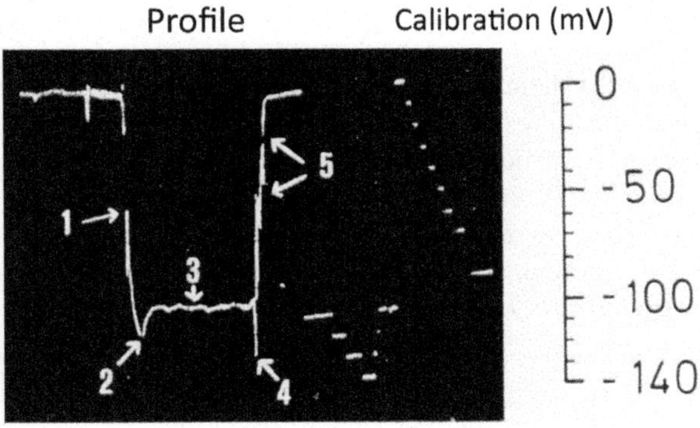

Fig. 8 Transcellular electrical profile obtained by pushing the tip of a microelectrode across a leaf cell of *Kalanchoë daigremontiana* Hamet et Perrier. From *left* to *right*: 1, 5 tip of the electrode in the cell wall; 2, 4 tip of the electrode in the cytoplasm measuring the plasma membrane potential; 3 tip of the electrode in the vacuole measuring the sum of the plasma membrane and the tonoplast potentials (Rona et al. 1980)

4.3.2 Light-Triggered Transient Photosynthesis-Dependent Membrane Potential Changes

We saw the transient membrane potential oscillations after light/dark changes first when studying the epidermal salt bladder hairs of Chenopodiaceae leaves (Sect. 3.2). While we were working on *Atriplex*, Andrianov et al. (1968) had also discovered such oscillations in the green cells of the charophyte alga *Nitella*. Turning the light off causes an initial hyperpolarization. Conversely, switching the light on incites a depolarization. The initial changes are followed by several oscillations of the membrane potential which last for several minutes until the potential returns close to the level of the resting potential observed before the light/dark changes (Fig. 9). Inhibitor studies (Andrianov et al. 1968; Lüttge and Pallaghy 1969) showed that switching on and off photosynthesis was responsible for the effect. Using light filters to distinguish between photosystems I and II, we demonstrated that it was due to the activity of PS II (Lüttge and Pallaghy 1969).

Our interpretation linked these observations to the compartmentation of the cytoplasm with the chloroplasts and their thylakoids. At the onset of photosynthetic electron flow in the thylakoid membrane, protons are transported into the thylakoid lumen. The idea was that this proton transport changes the equilibria at the chloroplast envelope and at the plasma membrane. Hence, an initially lowered H^+ concentration in the cytoplasm due to H^+ influx into the chloroplasts (light on) would explain the depolarization, whereas an increased H^+ concentration due to H^+ efflux from the chloroplasts (light off) would explain the hyperpolarization of the electrical potential at the plasma membrane. H^+ fluxes at the plasma membrane after some oscillations then reestablish equilibria and the resting membrane potential. By submerging green tissues in aqueous media, we supported this interpretation by showing H^+ transport between the tissues and the media which was not

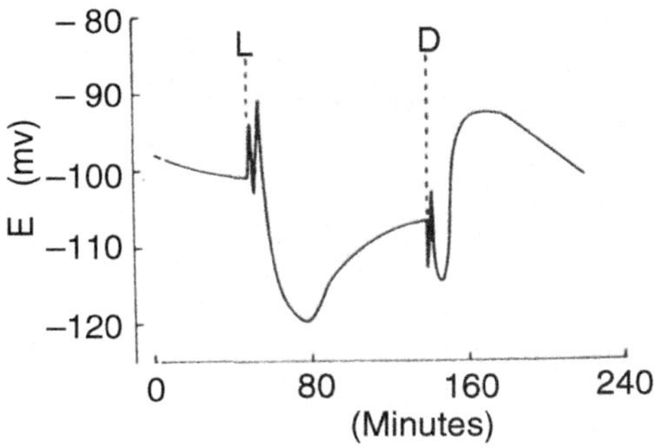

Fig. 9 Transient oscillations of membrane potential in *green* mesophyll cells of *Atriplex spongiosa* James triggered by switching light on (L) or off (D) (Pallaghy and Lüttge 1970)

solely due to photosynthetic CO_2 fixation after uptake of bicarbonate (Pallaghy and Lüttge 1970; Brinckmann and Lüttge 1972; Hope et al. 1972). Nevertheless, originally this was received with reservation because at the time the view in photosynthesis research was that the proton permeability of the chloroplast envelope was too low. Later and up until quite recently, the phenomenon was much studied by other authors, who basically confirmed our earlier interpretation (Bulychev and Turovetsky 1983; Vanselow et al. 1988, 1989; Bulychev and Kamzolkina 2006a, b).

5 Energization of Transport

5.1 Network of Energy Metabolism

A topic of major interest in the physiology of ion transport in the 1970s was the coupling of primary and secondary active transport at membranes to cellular energy sources (Lüttge and Higinbotham 1979). A hypothesis had been developed in the laboratory of Lundegårdh. They had discovered that ion accumulation was dependent on respiration. This was named salt respiration. The hypothesis derived was that ion accumulation was directly powered by mitochondrial electron transport (Lundegårdh and Burström 1933, 1935; Lundegårdh 1950, 1955). In analogy to that, from studies with green algal cells (*Nitella translucens* (Persoon) C. Agardh., *Hydrodictyon africanum* Yamanoudu), the idea was derived that photosynthetic electron transport via both photosystems I and II was driving anion (chloride) uptake without requiring ATP (MacRobbie 1965; Raven 1967). Conversely, the advanced understanding of the coupling of electron transport with ATP synthesis by oxidative and photophosphorylation, respectively, led to the view that ATP, the general energy currency of cells, would also drive ion transport at the membranes.

Our own interest arose from the studies of salt accumulation in the bladder cells of *Atriplex* (Sect. 2.3.2, Lüttge et al. 1970) and the photosynthesis-dependent membrane activities (Sect. 4.3.2). We got involved in studying energetics of ion transport in the aerial leaves of higher plants. We used thin 0.5 mm wide leaf slices submerged in aqueous solutions. Only in one case, during the visit of Alex Hope in our laboratory (June–September 1971), we recurred to unicellular green algae (*Scenedesmus*), of which we had obtained photosynthesis mutants (Table 2). We distinguished the various knots and edges of the complex cellular network of energy metabolism, such as respiratory and photosynthetic electron transport, oxidative phosphorylation, and noncyclic and cyclic photophosphorylation. For this purpose we used various inhibitors and applied special experimental conditions. We modified wavelengths of photosynthetically active irradiance, we established anaerobic conditions by saturating the solutions with nitrogen, and we chose variegated leaves, greening etiolated leaves, C_3 and C_4 leaves, and photosynthesis mutants. The general result emerging from all these treatments (Table 2) is that specific processes by themselves can drive ion transport, e.g., noncyclic photophosphorylation in *Scenedesmus obliquus* (Turpin) Kuetzing (Hope et al. 1974). However, any alternative paths of the network available by the manipulation of conditions are also effective. The basic energy state of the cells appears essential. Evidently as long as any meshes in the network can produce ATP, this can serve as energy source for transport (Johansen and Lüttge 1974, 1975; Lüttge and Ball 1976). With carrot tissue we also found that salt respiration did not directly power ion uptake. This excluded specific operation of a redox-type energization as postulated by the Lundegårdh hypothesis (Lüttge et al. 1971b).

5.2 Electrophysiology of H^+ Solute Cotransport

With electrophysiological experiments on the uptake of sugars, amino acids, and phosphate, predominantly using the water plant *Lemna gibba* L., we came closer to the actual mechanism of energy coupling of membrane transport with ATP. These studies were mainly performed when Anton Novacki visited Darmstadt many times including longer periods (August 1976–April 1977, September 1983–June 1984) and cooperated with Cornelia Ullrich-Eberius (Novacky et al. 1978a, b, 1980; Fischer and Lüttge 1980; Jung and Lüttge 1980; Jung et al. 1982; Lüttge et al. 1981a; Ullrich-Eberius et al. 1981). The negative electrical potential at the plasma membrane is built up due to active extrusion of protons by an H^+-pumping membrane ATPase. When a solute is added to the medium which is taken up into the cells via H^+ solute cotransport, the potential is depolarized with the onset of H^+ influx together with the cotransported solute. Then the ATPase reacts and begins to pump more strongly so that the previous resting potential is reattained. Conversely removal of the solute leads to a hyperpolarization as the pump is still working strongly, but then its activity decreases and again the resting potential is reached (Fig. 10). The amplitudes of the potential changes are concentration-dependent, and many insights about the physiology of transport can be gained. Effects of fusicoccin, a fungal toxin which specifically stimulates the plasma membrane

Table 2 Assessment of cellular energy sources for ion transport

Inhibitors conditions	Effects processes	Materials	References
CCCP (dichlorophenyl-carbonyl cyanide-phenylhydrazone)	Uncoupler	*Tradescantia albiflora* Kunth	Johansen and Lüttge (1974, 1975)
FCCP (p-CF_3O-carbonyl cyanide-phenylhydrazone)	Uncoupler	*Amaranthus caudatus* L. *Atriplex spongiosa* James *Atriplex hastata* L. *Oenothera albicans x hookeri* *Spinacia oleracea* L. *Tradescantia albiflora* Kunth *Zea mays* L.	Lüttge et al. (1970, 1971a), Lüttge and Ball (1971)
NaN_3 (acide)	Uncoupler	*Scenedesmus obliquus* (Turpin) Kuetzing	Hope et al. (1974)
DNP (di-nitro-phenol)	Uncoupler	*Tradescantia albiflora* Kunth	Johansen and Lüttge (1974, 1975)
Oligomycin A	Oxidative phosphorylation	*Tradescantia albiflora* Kunth	Johansen and Lüttge (1974, 1975)
DCMU (dichlorophenyl-dimethylurea)	Photosystem II and photosyn-thetic electron transport	*Atriplex spongiosa* James *Hordeum vulgare* L. *Scenedesmus obliquus* (Turpin) Kuetzing *Tradescantia albiflora* Kunth	Lüttge et al. (1970), Hope et al. (1974), Johansen and Lüttge (1974, 1975), Ullrich-Eberius et al. (1976)
Light filters	Photosystems I and II	*Atriplex spongiosa* James	Lüttge et al. (1970)
N_2-bubbled solutions	Anaerobiosis, respiration	*Hordeum vulgare* L. *Scenedesmus obliquus* (Turpin) Kuetzing *Tradescantia albiflora* Kunth	Hope et al. (1974), Johansen and Lüttge (1974, 1975), Lüttge and Ball (1976), Ullrich-Eberius et al. (1976)
Variegated leaves with green and mutated pale areas	Photosynthesis, respiration	*Oenothera albicans x hookeri* *Tradescantia albiflora* Kunth	Johansen and Lüttge (1974, 1975), Lüttge and Ball (1971)
Greening etio-lated leaves	Photosynthesis, respiration	*Hordeum vulgare* L.	Lüttge and Ball (1976)
Leaves of C_3 and C_4 plants	Photosynthesis	*Amaranthus caudatus* L. (C_4) *Atriplex spongiosa* James (C_4) *Atriplex hastata* L. (C_3) *Oenothera albicans x hookeri* (C_3) *Spinacia oleracea* L. (C_3) *Zea mays* L. (C_4)	Lüttge et al. (1971a)
Photosynthesis mutants	Photosynthesis	*Scenedesmus obliquus* (Turpin) Kuetzing	Hope et al. (1974)

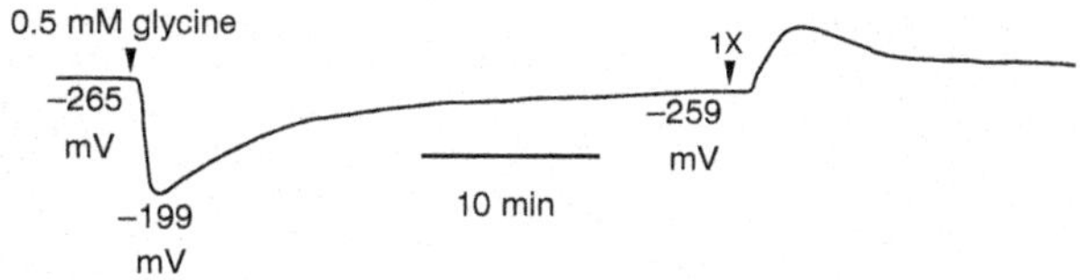

Fig. 10 Transient depolarization of the membrane potential of *Lemna gibba* L. by addition of 0.5 mM glycine and transient hyperpolarization after removal of the amino acid from the nutrient solution (1×) (Fischer and Lüttge 1980, www.plantphysiol.org, Copyright American Society of Plant Biologists)

ATPase, underline the involvement of proton pumping in energy coupling (Lüttge et al. 1981a).

6 CAM: A Problem of Transport

6.1 Dual Serendipity Leading to the Occupation with CAM

Two serendipities, which occurred around the same time at the beginning of the 1970s, brought us in contact with CAM which then remained my main research interest for the decades to come.

6.1.1 Feedback Inhibition of Vacuolar Ion Uptake, Osmotic Relations, and Turgor Pressure

We were engaged in the relations of vacuolar ion uptake and turgor pressure because we were studying the stoichiometry of K^+/H^+ exchange and the vacuolar potassium and malate accumulation responsible for turgor-driven elongation growth of *Avena* coleoptiles. Turgor pressure built up due to osmotically active K_2-malate drives cell elongation (Haschke and Lüttge 1973, 1975a, b, 1977).

At that time feedback of vacuolar ion contents on uptake of ions was much discussed as it was shown that NO_3^- and Cl^- accumulation in the vacuoles inhibited further uptake of anions (Cram 1973, 1976). Manfred Kluge had joined our department on the second chair of botany in 1974, and thus, CAM was introduced to Darmstadt. We thought that with just a few experiments comparing CAM leaves with high levels of malate in the morning and low levels in the evening we would be able to quickly check if vacuolar malate was to feedback inhibit ion uptake. It was a failure because there was no effect, but a lifelong occupation with CAM and a cooperation and friendship with Manfred arose from it. For the studies of ion uptake, we had incubated leaf slices of *Kalanchoë daigremontiana* Hamet et Perrier with KCl solutions of 0–50 mM. Erika Ball, the technician performing the experiments, noticed some strange changes of the color of the external medium during incubation, when KCl concentrations were low but not when they were high.

She did not miss the serendipity and stuck pH electrodes into the media. She found that at low KCl, they acidified. Analyses then proved that this was due to efflux of malic acid from the leaf slices. The media had changed reddish due to some anthocyanin also leaking from the vacuoles. Using mannitol we found the effect to be osmotic rather than caused by salt (Lüttge and Ball 1974a, b). The conclusion was that high turgor (in parallel to low KCl concentrations and low osmotic pressure of the medium) facilitated efflux of anthocyanin and malic acid from the vacuoles.

The principle of CAM is a switch from nocturnal malic acid accumulation (phase I) to daytime malic acid efflux (phase III) of the vacuoles (for CAM phases, see Sect. 3.3). Under natural situations, it is not a problem to explain the switch by the strongly changing conditions of dark and light periods, respectively. However, in the free running endogenous rhythm of CAM under constant external conditions (Sect. 7), this is not a trivial problem. Malic acid accumulation is osmotically active and brings about an increase of turgor (Smith and Lüttge 1985), so that turgor may be a trigger for the change from net influx to net efflux. We studied this hypothesis in detail also together with Hank Greenway when he was visiting our laboratory (January–December 1976) (Lüttge et al. 1975a, b, 1977). It followed that nocturnal malic acid accumulation contributes to osmotic acquisition of water by the CAM plants (Lüttge 1986). This was further confirmed during a visit of Park S. Nobel (August–September 1983) when we worked with the CAM plants *Cereus validus* Haw. and *Agave deserti* Engelm. and compared gas exchange, malate levels, osmotic pressure, and turgor. In these experiments, turgor was measured directly with a pressure probe in the vacuoles (Lüttge and Nobel 1984; Nobel et al. 1984). Turgor pressure proved to be an important module in the biophysical oscillator driving the overt output of the endogenous CAM rhythm (Sects. 7.4 and 7.6).

6.1.2 Salinity and the Induction of CAM in *Mesembryanthemum crystallinum* L., the "Common Ice Plant"

The other serendipity came from Andy Benson's seeds of *Mesembryanthemum crystallinum* L. (Sect. 2.4). We had grown plants and obtained our own seeds. After checking for salt accumulation in the epidermal bladders (Sect. 2.4), in the summer of 1971, the student had not cleaned up and left the plants in the greenhouse with their labels indicating irrigation with NaCl solutions of 0–500 mM. Around this time, Klaus Winter was a student in an advanced course and was supposed to perform gas exchange measurements. He had problems to allocate experimental plants until he found the *M. crystallinum* plants nobody was interested in anymore. The gas exchange curves he recorded appeared to be strange. Some plants showed C_3 and others CAM-like patterns. However, the labels resolved it, and Klaus Winter came up from the course experiments with his discovery that salinity induced CAM in *M. crystallinum* (Winter and von Willert 1972; Winter 1973a, b), which he subsequently subjected to in-depth study during the work for his Ph.D. thesis (Winter 1975).

Much work followed in our own laboratories in Darmstadt (Winter and Lüttge 1979; Heun et al. 1981 and see below). However, *M. crystallinum*, with its nickname "common ice plant" due to the crystal like epidermal bladder cells (Sect. 2.4), developed to a model plant of stress physiology and molecular stress biology with currently broad input of research activities by many renowned laboratories worldwide. One might wonder what had happened without the journey of *M. crystallinum* seeds from La Jolla to Darmstadt in 1969 and the serendipity arising from it. Would the C_3-CAM shift ever have been discovered by somebody else and under which circumstances? Would an incredible development of international plant biology have not occurred at all? In any manner, here we have another great gift of Andy Benson to photosynthesis, unwittingly in this case.

6.2 *Transport at the Tonoplast for the Performance of CAM*

The nocturnal vacuolar accumulation and daytime remobilization of large amounts of organic acids, mainly malic acid, in CAM imply massive transport processes of protons, organic acid anions, and also non-dissociated acid at the tonoplast. This makes CAM a problem of transport physiology.

6.2.1 Thermodynamics and Energetics

The first step was to ascertain the H^+/malate^{2-} stoichiometry of nocturnal accumulation. This is always $2H^+$:1 malate^{2-}, and it cannot be disturbed, e.g., by feeding the plants with a lot of K^+ (Lüttge et al. 1975a). Various other attempts to perturb this stoichiometry also failed (Lüttge and Ball 1979, 1980). Hence, the net effect is an accumulation of malic acid.

We then assessed thermodynamics comparing the free energy of ATP hydrolysis, ΔG_{ATP}, and the H^+ electrochemical gradient, $\Delta\mu_{H^+}$, at the tonoplast against which the acid needs to be accumulated. This showed that thermodynamically malic acid accumulation can be explained by an ATPase that transports $2H^+$ per one ATP with the malate anion following electrophoretically, i.e., passively. Moreover, measurements of dark respiration allowed calculating that the respiratory supply of ATP would just be sufficient to explain the observed malic acid accumulation (Lüttge et al. 1981b; Smith et al. 1982). Hence, this led to strong support of a $2H^+$-ATPase at the tonoplast (V-ATPase) being responsible for malic acid accumulation.

Considering vacuolar pH, dissociation constants, and dissociation equilibria, we concluded that remobilization from the vacuoles could occur by passive diffusion of the electrically neutral non-dissociated malic acid via the lipid phase of the membrane (Lüttge and Smith 1984).

6.2.2 Molecular Identification of the V-ATPase of CAM Plants

The thermodynamic evidence summarized in Sect. 6.2.1 is circumstantial. Thermodynamics describe the biophysical limits within which a process can operate but does not unveil underlying molecular mechanisms. Hence, the great challenge had arisen to provide a molecular identification of the V-ATPase. This challenge was exasperated because our friend Phibus Matile had published that they did not find such an ATPase in tonoplast membranes of the CAM plant *Kalanchoë daigremontiana* Hamet et Perrier, and therefore, our hypothesis appeared to be falsified (Buser-Suter et al. 1982; Matile 1982). However, we were convinced to be right. It was a long and hard effort to win the game. We owed the success to the never intimidated almost obstinate persistence of J. Andrew C. Smith and the expertise of Ernie Uribe he brought in when he joined us in Darmstadt (March–June 1982) in the endeavor.

The key for the success was to prepare tonoplast membranes fast enough and to measure ATPase activity quickly enough within hours, because the activity is rapidly lost during storage of the isolated membranes (Smith et al. 1984a, Fig. 11a). Such preparations then enabled us to describe the properties of the V-ATPase in some detail (Jochem et al. 1984; Smith et al. 1984b; Struve and Lüttge 1987; Haschke et al. 1988). Using detergents the ATPase was separated from the tonoplast membrane and reconstituted in liposomes (Behre et al. 1992; Bañuls et al. 1994). By getting hold of this V-ATPase, we detected and characterized the decisive building block of the tonoplast transport complement for the operation of CAM. It was also for the first time a V-ATPase had been observed in green mesophyll cells of higher plant leaves. That the V-ATPase activity is stimulated by malate may be relevant for CAM performance (Jochem and Lüttge 1987).

During handling of the large isolated vacuoles in suspension, they fragment into tiny tonoplast vesicles, which we used to visualize the ATPase. In freeze-fractured vesicles, the ATPase molecules appear as membrane particles (Fig. 11b). Negative staining shows the head and stalk structure (side views, Fig. 11c) and the hexameric structure of the head (top views, Fig. 11d). Subunit composition of the enzyme was analyzed (e.g., Ratajczak et al. 2003; Drobny et al. 2002). In cooperation with the laboratory of Bettina Böttcher and based on the group's expertise, a molecular model of the complex multi-subunit enzyme was obtained (Domgall et al. 2002).

6.2.3 Stress Responses of the V-ATPase

The induction of CAM in *M. crystallinum* is an ecophysiological stress reaction to salinity (Sect. 6.1.2), and the V-ATPase is a key element in the functioning of CAM (Sect. 6.2.2). Therefore we studied the responses of the ATPase to stress. In *M. crystallinum*, ATPase activity was increased when CAM was induced by salinity

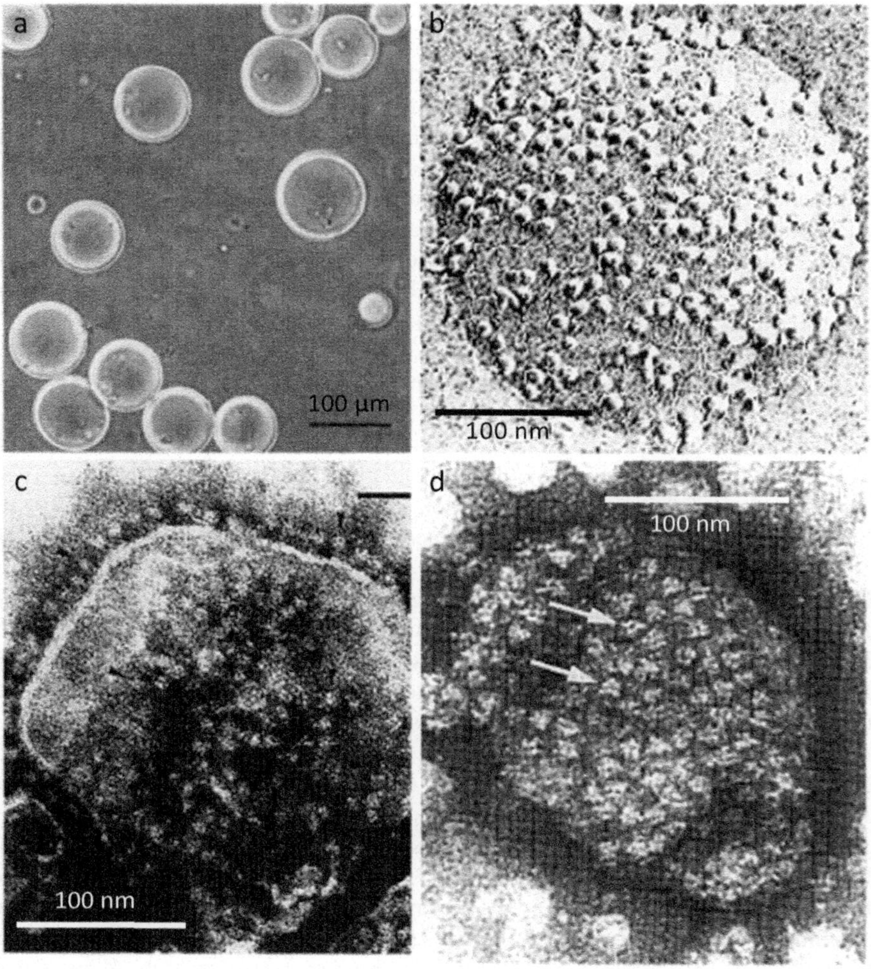

Fig. 11 (**a**) Isolated vacuoles of *Kalanchoë daigremontiana* Hamet et Perrier (Smith et al. 1984a). (**b**) Freeze-fractured tonoplast vesicle of *Mesembrynthemum crystallinum* L. with ATPase particles (Klink et al. 1990). (**c**) and (**d**) Negatively stained tonoplast vesicles of *M. crystallinum* with the ATPase in side view showing the head and stalk structure (**c**) and in top view showing the hexameric structure of the head (**d**) (Klink and Lüttge 1991, Copyright John Wiley and Sons)

(Struve et al. 1985; Struve and Lüttge 1988; Ratajczak et al. 1994b). Biochemical studies showed that this was not due to an increase of specific activity but to an increased amount of the V-ATPase in the tonoplast membrane (Bremberger and Lüttge 1992; Ratajczak et al. 1994b) and modifications of the enzyme's subunit composition (Bremberger and Lüttge 1992; Ratajczak et al. 1994b; Zhigang et al. 1996).

The biochemical results were supported by microscopic inspection. The density of the particles of freeze-fractured tonoplast vesicles increased after CAM induction in *M. crystallinum* (Klink and Lüttge 1992; Rockel et al. 1994). Specific immunogold labeling of the V-ATPase in negative staining of tonoplast vesicles also showed an increased density and confirmed that the particles (Fig. 11b) indeed represented the V-ATPase (Ratajczak et al. 1995). The diameter of the particles increased from 6.5 to 8.5 nm (Klink and Lüttge 1992). Subunit analysis suggested that the larger size in the CAM state was due to an increased number of subunit *c* forming the intramembranous H^+ channel of the V-ATPase (Rockel et al. 1994). The mRNA level for subunit *c* proved to be particularly sensitive to developmental and environmental changes (Löw et al. 1996; Rockel et al. 1998), which also includes nitrogen nutrition in *Nicotiana tabacum* L., where different isoforms of subunit *c* can be expressed (Fischer-Schliebs et al. 2000). Larger amounts of the V-ATPase and larger particle sizes in freeze-fractured tonoplast vesicles were also found in the C_3/CAM intermediate species *Kalanchoë blossfeldiana* Poellnitz cv. Tom Thumb when CAM was induced by short-day treatment (Struve et al. 1985; Mariaux et al. 1997). Baoshan Wang, our frequent visitor from Shandong in China (October 1988–May 1990, July 1998–October 1998, June 2000–May 2001), found that in the C_3 halophyte *Suaeda salsa* L., salinity increased the amount of the V-ATPase (Wang et al. 2001).

All this work in the 1990s was carried out in a wonderful cooperation with Rafael Ratajczak, who was then destined to a tragic premature death. Our work had shown many stress responses of the V-ATPase. The significance of the V-ATPase for the adaptation to stressful growth conditions was subsequently studied further extensively at the molecular and biochemical levels by other groups (Dietz et al. 2001). As a physiological, biochemical, and molecular basis of ecological adaptations, we suggested to call the V-ATPase an "eco-enzyme" (Lüttge et al. 1995b).

6.2.4 The V-PPase

Besides the V-ATPase, tonoplasts have a second H^+-transporting enzyme, namely, the V-PPase which uses the energy of inorganic pyrophosphate. We separated the two enzymes in CAM plants and purified and reconstituted the V-PPase in liposomes (Bremberger et al. 1988; Mariaux et al. 1994; Becker et al. 1995). In *M. crystallinum* the PPase activity did not increase and rather decreased under salinity inducing CAM (Bremberger and Lüttge 1992). In the literature, it was discussed that the two enzymes may have different functions under different environmental and developmental conditions, which was not confirmed by us, however, for all circumstances (Fischer-Schliebs et al. 1998). However, in *K. daigremontiana*, the V-PPase was shown to stimulate the V-ATPase suggesting a regulative role (Marquardt-Jarczyk and Lüttge 1990; Fischer-Schliebs et al. 1997). The general importance of the V-PPase in regulating metabolic processes has currently been revealed (Ferjani et al. 2013).

6.2.5 The Malate Channel of the Tonoplast

The task of the molecular identification of the tonoplast-malate transporter using our methods of protein biochemistry proved to be a nightmare for Ph.D. theses due to the extremely low amount of transporter protein in the membrane. The transporter protein could be partially purified (Ratajczak et al. 1994a). A protein fraction containing the transporter was reconstituted into liposomes and showed malate transport activity (Steiger et al. 1997). Malate transport into tonoplast vesicles of *M. crystallinum* was increased after the induction of CAM (Lüttge et al. 2000). The breakthrough came from collaboration with J. Andrew C. Smith, now in Oxford, and Gerhard Thiel, when patch clamp studies on isolated vacuoles of *Kalanchoë daigremontiana* (Fig. 11a) revealed that the malate transporter was an inward-rectifying anion channel (Hafke et al. 2003, Fig. 12).

This was the coronation of our work on CAM as a transport problem. Overall, in our laboratories, this let us get hold of the complete complement of the molecular building blocks or modules which mediate the transport of malic acid at the tonoplast during the CAM cycle, namely, (1) two proton pumps, the V-ATPase and the V-PPase, (2) the malate channel, and (3) passive efflux of non-charged non-dissociated malic acid.

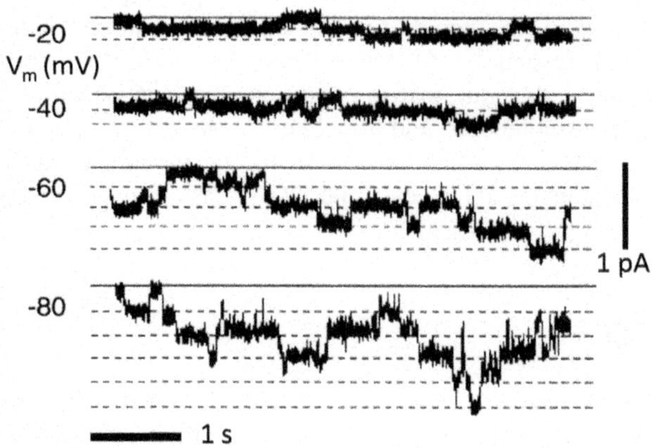

Fig. 12 Inward-rectifying malate currents: single-channel fluctuations in an isolated cytoplasmic-side-out tonoplast patch of *K. daigremontiana*. Currents were recorded at the voltages indicated on the *left side* in the presence of 100 mM malate on the cytoplasmic side and 10 mM malate on the vacuolar side. *Continuous horizontal lines* indicate the closed state of the channel, and *dashed lines* indicate the unitary channel amplitude (pA); time (s) runs from *left* to *right* (Hafke et al. 2003, Copyright John Wiley and Sons)

7 Transport and the Endogenous Rhythm of CAM

7.1 *The Magic Tripod: Experiment, Theory, and Model/Simulation*

It is a basic requirement of endogenous rhythms that they need a beat oscillator or hysteresis switch as pacemaker of the overt rhythmic output. Our studies of the transport of malate and water into and out of the vacuoles had suggested that in the endogenous rhythm of CAM, turgor pressure could be the basis of a biophysical hysteresis switch (Sect. 6.1.1). To study this under exactly controlled external conditions, I had to wait until the building of our little phytotron in Darmstadt was completed in 1988. Then I wound up with rapidly piling up long rolls of recorder paper with the traces of oscillations of net CO_2 exchange and transpiration. Attempting to analyze the structure of such time series, which showed a complex pattern, it became soon clear that theory was badly needed. I could readily convince my friend, the theoretical physicist Freder Beck, to join. In what followed we built up an interdisciplinary research team, and we received additional stimulating input when the theoretical physicist Marc-Thorsten Hütt joined us for many years as postdoctoral fellow (1999–2002) and junior professor (2002–2006). Our team philosophy was that any given problem should be worked on simultaneously by two Ph.D. students, an empiricist and a theorist, with very close day-to-day exchange and cooperation, for which the magic tripod became our metaphor. The ancient Greek goddess Pythia sat on a tripod when uttering the oracles of Delphi. The witches of the Middle Ages cooked their broths on a tripod. Our tripod had the three legs of experiment, theory, and model/simulation with continuous cross talk between them.

What Freder Beck first did to the time series was to submit them to fast Fourier transform (FFT) analysis to derive their power spectra. Then he reproduced the unexpected intriguing features observed (Sect. 7.2) in model simulations. Our first publication appeared in 1992 (Lüttge and Beck 1992).

7.2 *Structure of Time Series of Net CO_2 Exchange*

The power spectra of the oscillations in the endogenous rhythm of net CO_2 exchange under constant conditions and continuous illumination in the CAM plant *K. daigremontiana* show the basic frequency of the circadian rhythm as well as several harmonic overtones (Lüttge et al. 1996). On occasion of a public lecture, I also could exemplify them on a grand piano. There is a very sharp upper threshold of irradiance intensity and an upper and lower threshold of temperature above and below of which the rhythmic time series change to an arrhythmic behavior (Lüttge and Beck 1992; Grams et al. 1996, 1997a).

7.3 *Models for Simulations*

With mathematical modeling, we followed the philosophy of the "minimal model" (or "skeleton model") which approximates a biological system by taking into account only the relevant mechanisms for an observed dynamical behavior (Hütt and Lüttge 2007). The minimal model asks the question of which ingredients are really necessary for the model to reproduce the essential traits of the system's dynamics. The approach has an important heuristic value, because as long as one eliminates elements from the model without affecting the reproduction of actual experimental system behavior by the model simulation, one knows that these elements were not basically essential. In our theoretical CAM model, the essential ingredients were the three pools of internal CO_2, of malate in the cytoplasm, and of malate in the vacuole connected by flows between the pools, the state of order of lipid molecules in the tonoplast membrane, and the three external control parameters atmospheric CO_2, irradiance, and temperature (Blasius et al. 1997, 1998; Beck et al. 2001; Fig. 13). With this we could perfectly reproduce very many experimental observations on the endogenous circadian CAM rhythm in computer simulations. We started with a discrete hysteresis switch of the tonoplast built in the model, i.e., switching between two discrete states (Blasius et al. 1997, 1998), and then advanced to a dynamic hysteresis switch using continuous time differential equations (Blasius et al. 1999).

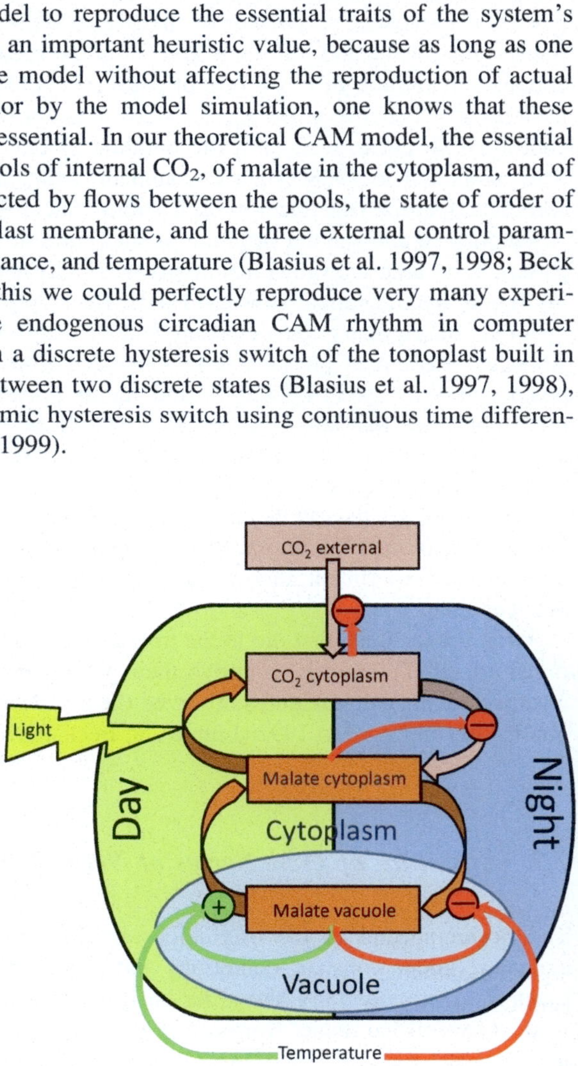

Fig. 13 The biochemical-biophysical oscillator of CAM. Modules and flows according to the minimal model described in Sect. 7.3; + and −: positive and negative feedback, respectively

7.4 Temperature and the Lipid Order of the Tonoplast

The absolute value of the upper temperature threshold (Sect. 7.2) below which rhythmicity can occur depends on the growth temperature of the plants (Grams et al. 1995). This can be explained by homeoviscous adaptation of membrane fluidity as we could learn from cooperation with the group of Manfred Kluge (Kliemchen et al. 1993), who showed that fluidity was reduced and increased in response to growth at high and low temperature, respectively.

This strengthened our idea that fluidity and the nature and order of membrane lipids governing permeability for passive efflux of malic acid from the vacuoles (see Sect. 6.2.1) were decisive parameters of the CAM rhythm. Experimental support came from observations with *K. daigremontiana*. The vacuoles had low malic acid levels and osmotic potential (π) when the rhythm stopped above the upper threshold of 29 °C, high membrane fluidity facilitating malic acid efflux. The vacuoles had high malic acid levels and π when rhythmicity stopped below the lower threshold of 8 °C, low membrane fluidity inhibiting malic acid efflux. The rhythm was reinitiated by increasing and decreasing, respectively, the temperature to be within the window of 8–29 °C. When coming from the high temperature, the rhythm started with increased activity of phospho*enol*pyruvate carboxylase (PEPC) and malate synthesis to refill the vacuole. Vice versa, when coming from the low temperature, the reinitiated rhythm began with emptying the vacuole (Grams et al. 1996, 1997a).

With temperature acting on passive efflux of malate from the vacuole, the theoretical model perfectly reproduced these observations (Grams et al. 1997a). The advanced model with the dynamic hysteresis switch (Sect. 7.3) showed fixed points with “empty” and “full” vacuole at high and low simulation temperatures, respectively, and limit cycle oscillations in between (Beck et al. 2001). We modeled the order state of the tonoplast with temperature determining the membrane surface area available for each lipid molecule and with that the osmotic consequences of malate accumulation. This showed a first-order phase transition with two coexisting phases of lipid order exhibiting hysteretic behavior (Neff et al. 1998). Taken together these experimental results and model simulations supported our view that the tonoplast membrane acts as hysteresis switch explaining the circadian period by the filling time of the vacuole (Blasius et al. 1999).

An intriguing result from modeling was the effect of noise. It is known that nonstructured so-called white noise can establish order, e.g., when noise lifts subthreshold peaks of oscillations above the threshold so that rhythmicity is seen by overt output at the peaks. This phenomenon is called stochastic resonance (Hütt and Lüttge 2002). Low noise intensity is not effective, high noise intensity overrides the oscillations, but intermediate noise intensity generates overt rhythmicity. The latter occurred in simulations by the CAM model (Beck et al. 2001). We also got close to demonstrating this experimentally, but sadly, unfortunate personal problems of a member of our team prevented us to bring this to maturity for publication.

7.5 Synchronization and Desynchronization

7.5.1 Oscillators of Individual Leaf Cells

Reinitiation of the net CO_2 exchange rhythm of *K. daigremontiana* by lowering the temperature when it had become arrhythmic above the upper threshold requires a strong temperature signal. Lowering the temperature in small steps failed to reinitiate the rhythm. This was also seen in the model simulations (Rascher et al. 1998; Beck et al. 2001). The explanation is that each leaf cell contains its own copy of the oscillator, with the strong temperature signal serving as synchronizer. This was confirmed by chlorophyll fluorescence imaging of the dynamic spatiotemporal patterns over single leaves (Rascher et al. 2001; Maddess et al. 2002; Rascher and Lüttge 2002). These observations also made the synchronizing power of high internal CO_2 concentrations, p_i^{CO2}, evident, i.e., during specific phases of the normal diurnal rhythm, especially in the daytime phase III with high p_i^{CO2}, and in corresponding phases of the endogenous circadian rhythm, as already mentioned in relation to mechanisms of cellular coupling (Sect. 3.3).

7.5.2 Oscillating Functions

In the minimal model the flows between pools fulfill three oscillating functions: (1) CO_2 uptake via the stomata linking the control parameter external CO_2 to internal CO_2, p_i^{CO2}, (2) CO_2 fixation via PEPC leading to synthesis of malate and its accumulation in the vacuole, and (3) CO_2 fixation via ribulose-bisphosphate carboxylase/oxygenase (RuBisCO) recapturing the CO_2 generated by decarboxylation of the malate after vacuolar efflux of malic acid. These oscillating functions can be coupled or uncoupled. If an external light rhythm is imposed on the plants with a period close to the endogenous circadian period, stomatal conductance and CO_2 assimilation are synchronous, but if the period length of the external rhythm is too short, synchrony between the two is lost and the overt output is arrhythmic (Bohn et al. 2001, 2003). In the obligate CAM plant *K. daigremontiana*, C_3 photosynthesis carboxylation via RuBisCO may take over from C_4 carboxylation via PEPC while the circadian rhythm is running without any change in the overt output of the rhythmic net CO_2 exchange, i.e., without affecting the stomatal rhythm (Wyka and Lüttge 2003; Wyka et al. 2004, 2005). C_3/CAM intermediate species, such as *M. crystallinum* and the tropical tree *Clusia minor* L., show endogenous rhythmicity in the state of both modes of photosynthesis (Boxall et al. 2005; Duarte and Lüttge 2007a). In *C. minor* we followed the oscillations of net CO_2 exchange; stomatal conductance, p_i^{CO2}; and effective quantum yield of photosystem II and their correlations. Like in *K. daigremontiana* we saw that during endogenous rhythmicity, the plant may change from the CAM mode to the C_3 mode. Hence, it is likely that a CAM oscillator based on tonoplast functions hands over to a C_3 oscillator possibly based on RuBisCO (Duarte and Lüttge 2007a).

By interruption of the gas stream of air with 21 % O_2 by pulses of air with only 1 % O_2, which causes non-photorespiratory conditions, with *C. minor* we also demonstrated for the first time endogenous oscillations of photorespiration (Duarte and Lüttge 2007b).

7.6 The Biochemical-Biophysical Oscillator of CAM

The molecular structure of the biological clock of plants is currently studied with great intensity in *Arabidopsis thaliana* (L.) Heynh. Basic clock genes were identified with *TOC1* (timing of chlorophyll a, b binding protein expression) being active in the evening and *CCA1/LHY* (circadian clock-associated 1/late elongated hypocotyl) active in the morning. There are other evening and morning genes and a plethora of downstream clock-controlled genes (Alabadí et al. 2001). *TOC1* and *CCA1/LHY* are also involved in the circadian rhythm of *M. crystallinum* where they change phases of their expression in C_3-CAM transitions (Boxall et al. 2005).

However, for the overt output of the CAM rhythm functions, a specific machinery is needed downstream of the basic molecular clock structure. The active night form of PEPC serving nocturnal CO_2 fixation in CAM is phosphorylated. Expression of the gene for the responsible enzyme, PEPC kinase, is under circadian control. This was also considered for some time as the core of the CAM oscillation machinery. However, the PEPC kinase gene expression turned out to be under metabolic control mainly by cytoplasmic pH and malate levels, i.e., "by treatments that affect the content and compartmentation of malate" (Borland et al. 1999, see Lüttge 2000 for detailed discussion and references). Malate compartmentation takes us back to the vacuole and the tonoplast. Therefore, with the modules of our minimal model, which so perfectly simulated many experimental observations (Sects. 7.3, 7.4, and 7.5.1), we built the biochemical-biophysical functional model of the CAM oscillator as shown in Fig. 13 in which the tonoplast acts as the master switch (Lüttge 2000).

8 Physiological Ecology in the Field

8.1 Trinidad: Our Gate to the Tropics

Much of the work described above directly or indirectly has ecological implications, such as the gland work on nectaries and carnivory, with the inherent implications of biotic interactions, and on the salt hairs and salt glands active under salinity stress (Sect. 2). Mineral ion uptake by roots (Sects. 4.1 and 4.2) and energization of transport (Sect. 5) are also related to ecological interests.

Particularly the work on CAM (Sect. 6) has ecological relevance because CAM is an ecophysiological adaptation to limited water supply. CAM saves water as it much reduces its loss in transpiration. Stomata are opened and CO_2 is taken up, fixed, and stored in the form of malic acid in the vacuoles in the nocturnal dark period when the evaporative demand driving transpiratory loss of water is low (phase I of CAM). Stomata are closed during the day when the driving force for transpiration, i.e., low atmospheric water potential, is high and CO_2 as it is remobilized from the malic acid is refixed internally via RuBisCO (phase III of CAM). Thus, the work on malate, π, turgor, and water relations (Sect. 6.1.1), the work on salinity inducing CAM in *M. crystallinum* (Sect. 6.1.2), and the work on the V-ATPase as a stress enzyme (Sect. 6.2.3) addressed physiological and biochemical bases of ecological adaptations. Since the biological clock is thought to bring about alertness to rhythmically changing external conditions, the work on endogenous rhythmicity of CAM (Sect. 7) may also be listed under the heading of ecophysiological implications. Under the chairmanship of Manfred Kluge, we even had a specially funded research group with several teams entitled "Biochemical bases of ecological adaptations".

However, at that stage, we did not go to the field where one needs to check the actual ecophysiological significance of mechanisms studied in the laboratory. This appeared particularly important regarding the various facets of CAM. It then developed, once again elicited by chance and serendipity. J. Andrew C. Smith had joined our laboratories as a postdoctoral fellow for 7 years (1979–1985) of extraordinarily fruitful, stimulating, and highly rewarding cooperation (see also Sect. 6.2.2). One day he came up with the request to get a leave for accepting the invitation to join an expedition of friends from zoology at the University of Dundee, Scotland, to Trinidad. Howard Griffiths also was a member of the party. But what should the two plant biologists do in Trinidad? As I was told, it was Barry Osmond, who happened to be in Dundee for a visit and who suggested that they might study the expression of CAM among the very many bromeliads on the island. On such grounds, we could even get a small travel grant for Andrew for the project. Andrew and Howard picked up Colin Pittendrigh's list of the bromeliads of Trinidad with the division into what he had called an exposure group, a sun group, and a shade-tolerant group (Pittendrigh 1948). The two plant biologists returned with a census of Trinidadian bromeliads and the relative occurrence of CAM among the three ecological groups (Griffiths and Smith 1983). The report to the granting agency (Deutsche Forschungsgemeinschaft) was well received with the suggestion to have more of such work in the wet tropics, where at that time little ecophysiological field work was performed. This was one aspect of our good chance. The other one arose when a panel of reviewers came to Darmstadt on 10 May 1982 to evaluate our research group. We had asked for some portable gas exchange equipment to take to the field. However, the requested equipment did not appear to be a very good choice. Otto L. Lange, chairman of the panel, knew much better instrumentation (Schulze et al. 1982), which, however, had not yet become public. During the evaluation, at a time when reviewers obviously still even could upgrade grant requests rather than only curtailing them, he came up with the suggestion or even demand that we should get the better equipment. Hence, we obtained number 4 of the manufactured series of a newly developed water vapor/CO_2 porometer of the firm of Heinz Walz (Effeltrich, Germany). In February/March 1983, we took the novel porometer to Trinidad, which became our gate to the tropics with many subsequent measuring campaigns many times in Venezuela and in Brazil, in the Virgin Islands, in Ethiopia, in French Guyana, and in Rajasthan, India. Several books emerged from these campaigns (e.g., Lüttge 1989, 2007a, 2008a).

8.2 *Photosynthetic Ecology of Bromeliads in Trinidad*

In Trinidad, we obtained on-site field data of the photosynthetic performance and water relations of 16 different bromeliad species including terrestrial and epiphytic ones and C_3 and CAM plants, and with five manuscripts we eventually filled a special issue of "Plant, Cell and Environment" (Plant, Cell Environment, 1986).

Regarding the CAM bromeliads, we found that the activity of CAM and the expression of CAM phases were strongly dependent on short-term and long-term water availability. Under dry conditions, the start of phase I CO_2 uptake was delayed in the night just as we had seen it before in the laboratory with *K. daigremontiana* (Smith and Lüttge 1985). Internal recycling of respiratory CO_2, an aspect of the so-called CAM idling, was increased. We also confirmed the osmotic consequences of nocturnal organic acid accumulation as we had worked it out in the laboratory (Sect. 6.1.1), with π increasing during acid accumulation. This drove osmotic uptake of water from dew late in the night period and also led to a reduced xylem tension.

Ecologically we related the epiphytic bromeliads to annual rainfall, altitude, and forest types. Very dry deciduous seasonal forest sustains low epiphytic biomass, and the small number of species occurring is CAM plants. With increasing altitude, the abundance of the epiphytic bromeliad species is highest in the evergreen seasonal forest and then decreases again in the lower montane rain forest. The relative contribution of CAM species among the epiphytic bromeliads decreases as forests become progressively wet with increasing altitude.

A distinction is made in the literature between synecology and autecology. The former addresses the synthetic emergence of ecosystem performance from the biodiversity of its species. Conversely, autecology covers the individual performance and adaptation of a plant species based on its traits, mainly physiological and biochemical ones characterized in the laboratory. Our work in the tropics led us to the expectation that a physiological synecology could and should also be developed (Lüttge and Scarano 2004, 2007; Lüttge 2005a). By characterizing the phytogeographical habitat and ecosystem distribution of the bromeliads based on traits of photosynthesis and water relations at the regional scale on the island of Trinidad, a first step was made towards such an aim.

On the peak of the Mount El Aripo in Trinidad, we also met *Clusia intertexta* Britton. We did not pay much attention to it not anticipating that the genus *Clusia* would later become a major occupation (Sect. 8.4).

8.3 Physiological Synecology in the Tropics

8.3.1 Alluvial Sand Plain at the Caribbean Coast of Northern Venezuela

On the way to Trinidad in February 1983 (Sect. 8.1), we had made a stopover in Caracas and visited Ernesto Medina at the Instituto Venezolano de Investigaciones Cientificas (IVIC). We undertook an excursion and saw various types of tropical vegetation including the alluvial sand plain at the Caribbean coast near Chichiriviche. This elicited a long cordial collaboration with five campaigns for ecophysiological measurements in the field in Venezuela in November/December 1985, March/April 1986, January/February 1989, February/March 1991, and February/March 1993 supported by substantial grants for investments by the Volkswagen Foundation. Our field work, excursions, and stimulating critical discussions on tropical plant ecology also led to conceiving a book on *Physiological Ecology of Tropical Plants* (now Lüttge 2008a), until it stalled exactly after 10 years with the last campaign in March 1993.

The alluvial plain at Chichiriviche was intriguing because it can be flooded knee-deep by freshwater in the rainy season and present a hypersaline substratum in the dry season, covered then by a centimeter-thick crust of salt. With a mosaic of saline and less-saline environments, it houses a great diversity of plant communities (Medina et al. 1989). Of particular interest are small vegetation islands on the otherwise bare sand plain (Fig. 14b, c). To understand vegetation-island dynamics under the highly contrasting seasonal changes, we decided to perform comparative studies in the wet and dry season, November/December 1985 and March/April 1986, respectively. From 34 species identified along a transect and on the islands (Medina et al. 1989), we chose ten for close ecophysiological inspection of photosynthesis, water, and mineral relations, representing different life forms and modes of photosynthesis (Medina et al. 1989, Table 3). In this way, we approached our aim of physiological synecology (Sect. 8.2).

The vegetation islands individually show spatiotemporal nonlinear dynamics of growth and decline (Medina et al. 1989). There is great physiological plasticity of the plants which respond differentially to the seasonal environmental challenges (Table 3). Among the perennial halophilic succulent herbs lining the fringe of the islands, *Sesuvium portulacastrum* is prostrate and more inhibited by the dry season conditions than the upright *Batis maritima*. *Portulacaria rubricaulis* is a C_4 plant shedding its leaves in the dry season.

All the CAM plants are salinity stress avoiders. *Tillandsia flexuosa* and *Schomburgkia humboldtiana* are epiphytes not touched by the hypersaline

Fig. 14 Coastal vegetation islands. (**a**) Coastal Atlantic restinga of Brazil. (**b**, **c**) Vegetation islands on the alluvial sand plain at the Caribbean coast near Chichiriviche, Venezuela in the dry season (**b**) and the wet season (**c**)

substratum in the dry season. They adapt using their plastic CAM options with variable expression of CAM phases (Sect. 3.3). *Bromelia humilis* has no functional soil roots and only adventitious absorptive roots inside its tanks. It simply lies on the sand plain and on rocks without firm contact. On ramets clonal phenotypes are produced with large green rosettes under the shade of deciduous woodland shrubs and small trees, medium-sized pale-green rosettes in semi-shade, and small lemon-yellow rosettes under full sun exposure on the open sand plain. Individuals

Table 3 Physiological synecology: species studied ecophysiologically at the Caribbean coastal alluvial plain in Venezuela

Species	Life form	Mode of photosynthesis	References
Batis maritima L. *Sesuvium portulacastrum* L. *Portulacaria rubricaulis* H.B.K.	Perennial halophilic succulent herbs	C_3 C_3 C_4	Lüttge et al. (1989b)
Bromelia humilis Jacq.	Terrestrial tank forming bromeliad	CAM	Lee et al. (1989)
Subpilosocereus ottonis Backeberg	Columnar cactus	CAM	Lüttge et al. (1989a)
Avicennia germinans (L.) Stern *Conocarpus erectus* L.	Mangrove shrubs	C_3	Smith et al. (1989)
Pereskia guamacho Weber	Leaf deciduous shrub	Leaves: C_3	Lüttge et al. (1989a)
Tillandsia flexuosa Sw. *Schomburgkia humboldtiana* Reichb.	Epiphytic, tank forming, leaf succulent Epiphytic, leaf succulent	CAM	Griffiths et al. (1989)

differentially use the CAM options. Especially the exposed yellow plants strongly depend on internal recycling of CO_2. Nocturnally, respiratory CO_2 is refixed into malate when stomata are more or less closed (phase I) and remobilized during the light period for fixation via RuBisCO (phase III). CAM here is coming close to the mode of CAM idling (see Sect. 8.2) with almost 100 % internal recycling of CO_2, i.e., 87 % during the dry period (see also Fetene et al. 1990; Fetene and Lüttge 1991). Columnar cacti, such as *Pilosocereus ottonis*, are salt excluders (see also Lüttge and Nobel 1984; Nobel et al. 1984). They sacrifice their absorptive fine roots in the dry season and rely on their water storage tissue to overcome by CAM idling, when stomata are closed night and day and respiratory CO_2 is recycled.

Among the mangrove shrubs, *Avicennia germinans* can work with NaCl excretion via salt glands of the leaves, while *Conocarpus erectus* can respond with leaf succulence. Restriction of performance in the dry season is more severe in the mangrove associate shrub *C. erectus* than in the true mangrove species *A. germinans*.

8.3.2 Richness of Tropical Environments and Major Shift of Vocation to Brazil

The richness of tropical environments where we performed ecophysiological studies included:

– The US Virgin Islands' St. John, where we studied *Clusia* together with Irvin Ting (Sect. 8.4)

- Ethiopia, where we investigated the performance of natural formations and plantations of trees and effects of facilitation and nurse plants in montane forests (Fetene et al. 1997; Rascher et al. 2000; Lüttge et al. 2001, 2003) and the giant rosette plant *Lobelia rhynchopetalum* Hemsl. in the tropical alpine regions (Fetene et al. 1997) under the guidance and hospitality of my former Ph.D. student Masresha Fetene (in Darmstadt April 1988–December 1990) and in collaboration with Erwin Beck
- India, where we measured the ecophysiological plasticity of *Butea monosperma* Taub. in SE Rajasthan as guests of Vinay Sharma at the University of Banasthali (Kumari et al. 2005; Mikosch et al. 2012)
- French Guiana, where in addition to Venezuela we studied the desiccation tolerance and performance of cyanobacterial crusts on the surface of granitic inselberg rocks (Büdel et al. 1994; Lüttge et al. 1995a; Ziegler and Lüttge 1998; Rascher et al. 2003; Dojani et al. 2007)

However, after 1993, the major vocation for tropical ecophysiology shifted to Brazil. Augusto C. Franco, who had spent 2 years in our laboratory in Darmstadt (1989–1991), initiated the cooperation and put it on many shoulders with great friendship also arising to Eduardo A. de Mattos, Geraldo Fernandes Wilson, and Fabio R. Scarano. Seven visits with teams for measuring campaigns resulted between 1993 and 2004. We measured the desiccation tolerance of mosses on an inselberg Pedra Grande in São Paulo (Lüttge et al. 2008), the performance of *Mimosa naguirei* Barneby subject to the parasitism by *Pilostyles ingae* (Karst.) Hook. F. (Fernandes et al. 1998), and the photosynthesis of *Vellozia* species (Lüttge et al. 2007) on rupestrian fields of the Serra do Cipó in Minas Gerais, the photosynthesis of *Araucaria angustifolia* (Bertol.) Kuntze in the mountains of the Itatiaia massive (Franco et al. 2005), and the photosynthesis of savanna species in the Cerrados (Franco and Lüttge 2002). The major occupation regarded, however, the coastal restingas (Sect. 8.3.3) and other ecosystems of the Atlantic forest periphery (Sect. 8.3.4) and monographic studies of the genus *Clusia* (Sect. 8.4).

8.3.3 Coastal Atlantic Restinga of Brazil

The restinga ecosystems of Brazil occupy sandy coastal plains. They are geologically young, i.e., of Quaternary origin (3,000–120,000 years BP). These ecosystems belong to the periphery of the Brazilian Atlantic rain forest (Sect. 8.3.4), from where most of their vegetation originated due to migration during the relatively short geological history. The coastal plains can be dry or wet and harbor a mosaic of plant communities ranging from open formations to forest ecosystems. The phytosociologically dominant tree is the CAM species *Clusia hilariana* Schltdl (Table 4). It forms the so-called *Clusia* scrub, which is the characteristic physiognomy in the restingas at the northern coast of the State of Rio de Janeiro (Pimentel et al. 2007). For ecophysiological synecology of the restingas, we studied the plants listed in Table 4. Most notably like on the alluvial coastal sand plain in Venezuela

Table 4 Physiological synecology: species studied ecophysiologically in the coastal Atlantic restingas of Brazil

Species	Mode of photosynthesis	References
Allagoptera arenaria (Gomes) O. Ktze.	C_3	Scarano et al. (2001), Gessler et al. (2008)
Andira legalis (Vell. Conc.) Toledo	C_3	de Mattos et al. (1997), Scarano et al. (2001), Gessler et al. (2005a, 2008)
Clusia fluminensis Planch. et Triana	CAM	Scarano et al. (2001, 2005b)
Clusia hilariana Schltdl.	CAM	Franco et al. (1996, 1999), de Mattos et al. (1997), Herzog et al. (1999b), Liebig et al. (2001), Berg et al. (2004), Scarano et al. (2005b), Gessler et al. (2008)
Clusia parviflora Saldanha et Engl.	C_3/CAM	Herzog et al. (1999b)
Mollugo verticillata L.	C_3	Scarano et al. (2001)
Myrsine parvifolia A. DC.	C_3	Scarano et al. (2001)
Neoregelia cruenta (Grah.) L.B. Smith	CAM	Scarano et al. (2001)
Panicum trinii Kunth	C_3	Scarano et al. (2001)
Philodendron corcovadense Kunth	C_3	Scarano et al. (2001)
Protium icicariba (DC) March	C_3	de Mattos et al. (1997)
Psittacanthus dichroos Mart.	C_3	Scarano et al. (2001)
Rheedia brasiliensis (Mart.) Planch. et Triana	C_3	Scarano et al. (2001)
Vriesea neoglutinosa Mart. ex Schult f.	cf	Scarano et al. (2001)

(Sect. 8.3.1), the open restinga consists of vegetation islands of various sizes surrounded by white sand (Fig. 14a). Creating these vegetation islands, *C. hilariana* and also *Allagoptera arenaria* (Gomes) O. Ktze. are pioneer species functioning as nurse species for the establishment of other plants. These nurse species give illustrative examples for the spatiotemporal dynamics of the interactions between facilitation and competition (de Araujo and Scarano 2007; Lüttge et al. 2012; more references in Table 4).

8.3.4 Ecosystems of the Atlantic Forest Periphery of Brazil

Many of our field studies in Brazil addressed the various ecosystems at the periphery of and influenced by the Atlantic rain forest, such as the dry and wet restingas whose vegetation mostly originated from the Atlantic forest (Sect. 8.3.3).

Table 5 Studies in ecosystems peripheral to the Atlantic rain forest of Brazil

Ecosystem	References
Restinga (dry, wet)	Sect. 8.3.3, Table 4
Dry dune forest	Gessler et al. (2005a), Scarano et al. (2005b)
Semideciduous dry forest	Duarte et al. (2005a), Gessler et al. (2005b), Scarano et al. (2005b)
Swamp forest	Scarano et al. (1999), Duarte et al. (2005a)
Inselberg	de Mattos et al. (1997), Duarte et al. (2005a)

A list of these ecosystems is given in Table 5. They form an array of interrelated systems at a high scalar ecological level (Scarano et al. 2001, 2005a; Duarte et al. 2005a). Using our data, especially photosynthesis-saturating irradiance and maximum photosynthetic electron transport rates based on light saturation curves (Rascher et al. 2000), we tried to produce matrices of synecological fingerprinting. Although preliminary, these show that it is possible to separate generalists and specialists as the former show superior performance in wet-dry gradients and the latter occupy specific niches (Lüttge and Scarano 2007).

8.4 Clusia, *the Only Dicotyledonous Genus with Trees Performing CAM: A Monographic Treatise*

8.4.1 CAM in Dicotyledonous Trees of *Clusia*

In 1985 Ernesto Medina had brought with him to Chichiriviche the paper by the Mexicans Tinoco Ojanguren and Vazquez-Yanes (1983) written in Spanish language, where they described the performance of CAM by trees of *Clusia*. Although some measurements of Alexander von Humboldt in Venezuela in 1800 had already unveiled some characteristic physiological features of CAM photosynthesis in *Clusia rosea* Jacq., especially the absence of gas exchange in the light period (Faak 2000; Lüttge 2007b), this had never been adequately interpreted and put in actual context of CAM. Therefore, the paper of the Mexicans was a great revelation to us, as so far no real dicotyledonous trees were known to perform CAM. Ting et al. (1985) picked it up in a publication in 1985, and our first respective publication appeared in 1987 (Popp et al. 1987). An occupation with *Clusia* arose, unbroken ever since. From cuttings collected over the years, I built up a large life collection of more than 20 *Clusia* species in the Botanical Garden of the Technical University of Darmstadt which could be copied via cuttings and was established in several other botanical gardens. We performed laboratory work in growth chambers and made measurements in the field with a continuous ping-pong-like feedback between the two, in which Helen J. S. Lee and Augusto C. Franco became most active when they were postdocs in Darmstadt (1986–1988 and 1989–1991, respectively), and a great collaboration developed with other peers around the world, namely, Miriam Diaz and Elizabeth Olivares in Venezuela, Eduardo A. de Mattos and Fabio R. Scarano in Brazil, Irvin P. Ting in California, Marianne Popp in Austria, Zbigniew Miszalski and Andrej Kornas in Poland, and Arthur Geßler, Manfred Kluge, and Heinz Rennenberg in Germany. The monograph book (Lüttge 2007a) and a number of reviews (e.g., Lüttge 1999, 2006, 2008b) cover this work which I shall try to summarize in the following subsections.

8.4.2 Plasticity

About 300–400 species exist of *Clusia*. Leaves are morphologically and anatomically very similar, always entire, leathery, and somewhat succulent so that they constitute one typical morphotype. Diversity arises as many species express different life forms, i.e., terrestrial, epiphytic, and hemiepiphytic stranglers (Fig. 15).

However, in physiological and biochemical terms, pants of *Clusia* are characterized by extraordinary plasticity (Lüttge 2007c). There are C_3, CAM, and C_3/CAM intermediate species. Expression of the different CAM phases (Sect. 3.3) can be very versatile in response to various environmental factors. Among the C_3/CAM-*Clusias, C. minor* L. proved to be the most astonishing plant I ever had in my hands. Switches between to two modes of photosynthesis are rapidly reversible. If one puts the two opposite leaves at one given node in a moist and a dry atmosphere, they can simultaneously perform C_3 photosynthesis and CAM, respectively (Schmitt et al. 1988). Besides malic acid, *Clusia* species can also use citric acid for the nocturnal acid accumulation of CAM. Modifying the three parameters intensity of irradiance, night temperature, and day temperature, Angela Haag-Kerwer was able to elicit any kind of behavior in *C. minor*, viz., C_3 photosynthesis and any combinations of malate and citrate accumulation in the CAM mode (Haag-Kerwer et al. 1992).

8.4.3 Photorespiration and Oxidative Stress

In *C. minor* we also demonstrated photorespiration during CAM performance including phase III (Duarte and Lüttge 2007a). The reason for this is that during intensive photosynthetic CO_2 assimilation in phase III behind closed stomata, there is also intensive O_2 evolution. This causes oxidative stress, and notwithstanding high internal CO_2 concentrations photorespiration is not suppressed due to simultaneously high internal O_2 concentrations (Lüttge 2010).

We studied oxidative stress in *Clusia* in cooperation with Z. Miszalski and A. Kornas in Kraków. Species of *Clusia* do get under photoinhibition (Lüttge 2007d; Kornas et al. 2009). Increased irradiance stress alone, i.e., without concomitant drought stress, does not stimulate CAM activity, and the xanthophyll cycle reactions for dissipation of surplus excitation energy are involved in photoprotection (Kornas et al. 2010). In addition mitochondrial processes are playing a role. At low light mitochondria contribute to energization, but in the

Fig. 15 (**a**, **b**) *Clusia rosea* Jacq., St. John, US Virgin Islands, (**a**) hemiepiphytically, (**b**) strangler. (**c**) *Clusia multiflora* H.B.K. (*large leaves*) and *Clusia minor* L. (*small leaves*) sympatrical, growing terrestrially in a secondary savanna in Venezuela

CAM state the tricarboxylic acid cycle in the dark period is downregulated to prevent breakdown of organic acids (Miszalski et al. 2007, 2013).

Table 6 Ecosystems with *Clusia* species studied and their mode of photosynthesis (after Lüttge 2007d)

Ecosystem	Location	Mode of photosynthesis
Restingas	Brazil	C_3, CAM
Coastal rocks	Virgin Islands	CAM
Savanna/cerrado	Venezuela, Brazil	C_3, C_3/CAM
Gallery forest—cerrado ecotone	Brazil	Weak inducible CAM
Semideciduous dry low land forest	Brazil	CAM, C_3/CAM
Secondary shrub forest	Venezuela	C_3, C_3/CAM
Dry montane karstic limestone forest	Venezuela	C_3, CAM, C_3/CAM
Montane (rain) forest	Virgin Islands	C_3, CAM, C_3/CAM
Atlantic rain forest	Brazil	C_3, CAM
Cloud forest/fog forest/elfin forest	Venezuela	C_3
Inselberg	Venezuela, Brazil, French Guiana	C_3, weak inducible CAM

8.4.4 Ecological Amplitude

The ecological amplitude of the genus *Clusia* covers an enormous range of tropical ecosystems (Table 6, Lüttge 2007d; Lüttge et al. 2015). Even individual species get established in many different ecosystems. The C_3 species *C. multiflora* is found in six, the obligate CAM species *C. rosea* in four, and the C_3/CAM intermediate species *C. minor* in five of the ecosystems listed in Table 6. *C. hilariana* migrated from the Atlantic forest to the sandy restingas in Brazil, where it functions as a pioneer and nurse plant (Sect. 8.3.3).

8.4.5 Niche Width and Speciation

There is uncertainty about the numbers of existing *Clusia* species quoted to vary between 300 and 400 (Sect. 8.4.2). The uncertainty arises from the many synonyms found in herbarium collections and in the literature and from the expectation that many species in the tropics have just not been discovered. This also led to the view that the rate of ongoing speciation in the genus is high, which is supported by molecular studies (Gustafsson et al. 2007; Vaasen et al. 2007). The high plasticity of *Clusia*s may play a role in speciation by allowing increased width of niche occupation. In fact it is seen in the field that the C_3/CAM intermediate species *C. minor* does occur in the semi-shade of deciduous forest as well as under full sun exposure in savannas nearby, whereas the obligate C_3 species *C. multiflora* only occupies the open savanna (Grams et al. 1997b). Phytotron studies showed that *C. multiflora* adapts to the high irradiance conditions during its growth and development, while *C. minor* can rapidly acclimate to exposed conditions using its CAM option (Herzog et al. 1999a). *C. minor* can intrude into the sites occupied by *C. multiflora* where both species occur sympatrically (Fig. 15c). Hence, the

C_3-CAM options enable *C. minor* to cover a larger niche width. Such performance, if accompanied by the formation of ecotypes prior to segregation and separation, can drive speciation (Lüttge 2005b, 2007d).

9 Integration and Emergence

9.1 Transport Creating Integration at Many Scalar Levels

Dialectically a general philosophical view of any kind of separation by barriers or borders is that on the one hand they are absolutely required to allow distinctions and compartmentalization and that on the other hand there must be ways to overcome them for controlled interaction to avoid obstruction that would lead to impoverishment. In such a sense life must separate itself from its environment, and at the same time it must always remain an open system through which a continuous flow of matter and energy is maintained. Such characteristics of life are dictated by thermodynamics. Transport processes are ways to overcome borders and to achieve systemic integration. Integration is required for the creation of new systems from modules composing them. Integration is a prerequisite for hierarchically higher systems to emerge, originating from systemic components as they assemble towards new functional entity. This guided me to consider my work as "Transport Processes: The Key Integrators in Plant Biology". My essay documents transport as an integrative phenomenon at different scalar levels in plants, such as multi-subunit molecules (Sect. 6.2.2), membranes (Sects. 6.2, 7.4, and 7.6), cells (Sects. 2 and 4–7), tissues and organs (Sects. 3 and 7.5), and whole plants and ecosystems (Sect. 8). I have covered the integrative power of transport in various books (Lüttge 1969, 1973; Lüttge and Pitman 1976a, b; Lüttge and Higinbotham 1979). Moreover, naturally it would not have been possible to write textbooks overarching the entire field of plant biology (Lüttge et al. 2010; Lüttge and Kluge 2012) without considering it as a whole.

9.2 Modules and Emergence

In the literature on plant biology, we find two irritatingly contrasting views about the life of plants (Lüttge 2012a, b). One of them is purely mechanistic modularity. Modules are structural and functional building blocks. Plants are considered as modular organisms whose performance is nothing more than the sum of the module properties. In this view "a tree is not a tightly integrated organism but a by-product of its parts" (Haukioja 1991). The other extreme of views advocates far reaching anthropomorphic homologies of plant intelligence, with individuality, communication, learning, foresight, and intentions (Trewavas 2003).

There is a deep crevasse between these two extremes of views which we should bridge. Interaction and integration of modules at all scalar levels (Sect. 9.1) create plants' individuality as unitary organisms. Without doubt, signal perception exists together with internal management of information including storage and recall functions of learning and memorizing (Lüttge 2012b; Thellier and Lüttge 2013). I am grateful to my friend Michel Thellier for letting me participate in developing thoughts about the general implications of his lifelong work on plant memory. Related functions are based on a molecular inventory found in all organisms. Such functions, therefore, are basic features of life and, thus, also operative in plants. Conversely, although there is signal transduction in plants mediated by chemical, i.e., hormonal, and hydraulic signals and including electrical action potentials (Lüttge 2012b), plants evidently do not have nervous systems. There are no neurons as specialized cells transmitting nerve impulses in plants. Conceiving a "plant neurobiology" with inference of foresight and intentions in plants surrenders to the menace of crossing the border between natural science and speculative philosophical extrapolation.

The rescue from the controversy of the two contrasting views of mechanistic modularity and plant neurobiology, respectively, and casting the bridge between them comes from the concept of emergence (Lüttge 2012a). Emergence is self-organization from the modules where the self-organizing unitary entity is more than the sum of its compounds. The new emergent systems have completely new innovative properties as compared to those of the modules from which they are built up. This occurs at all scalar levels, including the inorganic world of the physical laws (Laughlin 2005) as well as in life (Lüttge 2012a). Emergence is based on interaction and integration as I have pursued it in this essay on transport processes as key integrators. That transport creates integration becomes explicit in the whole-plant perspective, which we had begun to adhere to, stimulated by the collaboration with Michael Pitman in the 1970s and 1980s. Michel had visited us in Darmstadt from July 1972 to January 1973, and a long wonderful friendship and the joint edition of two Encyclopedia volumes on transport arose (Lüttge and Pitman 1976a, b). Michael was one of the outstanding advocates of whole-plant physiology (Pitman et al. 1974a, b; Läuchli et al. 1978; Lüttge 2012b).

A stimulating idea conceived by viewing emergence from integrated modules is the holobiont concept. For the development of thinking about it, I am much indebted to the exchange with my friend Rainer Matyssek and the members of his group. A holobiont is a host organism with all its associated microorganisms forming an entity for selection and evolution. In a broader sense, holobionts may be any regular organisms, regarding both their internal functionality and their interactions among each other (Matyssek and Lüttge 2013). With this broader understanding of holobionts, we can scale up integration and emergence. We can start with endosymbiosis that gave rise to eukaryotic cells and advance to various symbioses of unicellular and pluricellular organisms and to micro- and macro-ecosystems from soil crusts to forests and further to biomes. Finally we arrive at the entire biosphere or Gaia *sensu* Lovelock (1979) as one supraorganism, the ultimate emergent system of life (Matyssek and Lüttge 2013). A new writing cooperation and friendship with Gustavo M. Souza highlight the importance of the term "hierarchy." The integration over the spatiotemporal levels creates hierarchies, which is an essential aspect of biological systems (Souza et al. 2016). In space and time, hierarchies operate both as top-down and bottom-up hierarchies. From interactions based on plasticity, diversity, and complexity of integrated systems, robustness and system persistence apparently emerge (this series, Souza and Lüttge 2014).

Mankind is part of this supraorganism Gaia, if conceived within the scope of natural sciences (*sensu* Lovelock 1979). In such terms, the Gaia concept is relevant to the question if sustained self-organization and self-maintenance (i.e., as based on intrinsic repair functions) can keep the biosphere fit for accommodating an increasing human population on Earth (Lüttge et al. 2012; Lüttge 2013b; Lüttge 2016).

9.3 The Biology of Plants and the Power of Apprehending Life

After 60 years of studying plant biology, it is extraordinarily disturbing to be confronted with the opinion that biology might not contribute to the understanding of life. This opinion was expressed by Viktor von Weizsäcker in lectures presented in 1919/1920, where he said that with each successful step forward towards understanding mechanisms, biology diverts more and more from understanding life and that the mechanistic way of biology unavoidably diverts from life rather than approaching it, which is its tragedy as a science. ("*Denn jeder Schritt, den diese Biologie tut, und jeder Erfolg, den sie hat, ist ein Nagel zum Sarge des Lebens,*" and "*Dass der mechanistische Weg der Biologie zwangsläufig vom Leben fort statt zu ihm hin führt; daran liegt der Urwiderspruch und eine Tragödie dieser Wissenschaft*" (pages 73 and 67, respectively, in von Weizsäcker 1954). As long as this view is resulting from a transcendental perception of life, the scientific biologist can be reconciled. With a strict methodological dualism of approaches, philosophy of transcendence and natural science consider conceptually completely different aspects and reveal qualitatively different types of "truth" (Lüttge and Mayer 2012). If, however, the argument that current scientific biology fails to understand life comes from within natural science, as in the writings of Weber (2010a, b), the disturbance is not relieved.

It is intriguing then that Weber (2010b) says that biology may be at the threshold of a change of paradigm and that it can provide decisive impulses towards attaining a holistic attitude. On the one hand, this is neglecting the fact that biology needs characterization of the modules. The modules always must be understood for understanding emergence. Moreover, what is an emergent system at one scalar level can become a module at the next higher scalar level in the emergence of systems over all the scalar levels up to Gaia. On the other hand, there is no need of a change in paradigm. Weber's "new paradigm" is already realized with the understanding of emergence in the biological sciences. Even within the methodological self-constraints of natural science, this advances us from purely mechanistic and materialistic modularity to holistic comprehension.

I hope that with the consideration of my work under the auspices of transport as a key integrator, in this essay it is seen that the biological science of plants contributes to such holistic comprehension of life, even though plants are nonconscious organisms. Emergence of holism casts the bridge and resolves the intrinsic contradiction within biology as assumed by von Weizsäcker (1954). I am grateful that destiny allowed me to perform all this work together with so many coworkers and friends. I think that after all, this work led to an at least humble contribution to the understanding of life, without losing sight of the different spiritual qualities of life separate from natural science (Lüttge and Mayer 2012). A rose is both a complex emergent biological system and a wonderful flower of overwhelming beauty.

Acknowledgment I thank Manfred Kluge, Rainer Matyssek, and C. Barry Osmond for the critical reading of my text before publication.

References

Abbaspour N, Kaiser B, Tyerman S (2013) Chloride transport and compartmentation within main and lateral roots of two grapevine rootstocks differing in salt tolerance. Trees 27:1317–1325

Adlassnig W, Koller-Peroutka M, Bauer S, Koshkin E, Lendl T, Lichtscheidl IK (2012) Endocytotic uptake of nutrients in carnivorous plants. Plant J 71:303–313

Alabadí D, Oyama T, Yanovsky MJ, Harmon FG, Más P, Kay SA (2001) Reciprocal regulation between *TOC1* and *LHY/CCA1* within the *Arabidopsis* circadian clock. Science 293:880–883

Alm Y, Ohnmeiss TE, Lanza Y, Vriesenga L (1990) Preference of cabbage white butterflies and honey bees for nectar that contains amino acids. Oecologia 84:53–57

An C-I, Fukusaki E-I, Kobayashi A (2001) Plasma-membrane H^+-ATPases are expressed in pitchers of the carnivorous plant *Nepenthes alata* Blanco. Planta 212:547–555

Andrianov VK, Kurella GA, Litvin EF (1968) Influence of light on the electrical activity of *Nitella* cells. Abh Dtsch Akad Wiss Berlin 4a:187–196

de Araujo DSD, Scarano FR (2007) Biogeographic features of *Clusia*, with emphasis on South American and especially Brazilian species. In: Lüttge U (ed) *Clusia*. A woody neotropical genus of remarkable plasticity and diversity, vol 194, Ecological studies. Springer, Berlin, pp 31–54

Arisz WH (1956) Significance of the symplasm theory for transport across the root. Protoplasma 46:5–62

Arisz WH (1960) Symplasmatischer Transport in *Vallisneria*-Blättern. Protoplasma 52:309–343

Athauda SB, Matsumoto K, Rajapakshe S, Kuribayashi M, Kojima M, Kubomura-Yoshida N, Iwamatsu A, Shibata C, Inoue H, Takahashi K (2004) Enzyme and structural characterization of nepenthesin, a unique member of a novel subfamily of aspartic proteinases. Biochem J 381:295–306

Baker HG, Baker I (1977) Intraspecific constancy of floral nectar amino acid complements. Bot Gaz 138:183–191

Bañuls J, Ratajczak R, Lüttge U (1994) Solubilization and functional reconstitution of the tonoplast H^+-ATPase from *Citrus* in liposomes. J Plant Physiol 144:74–79

Beck F, Blasius B, Lüttge U, Neff R, Rascher U (2001) Stochastic noise interferes coherently with a model biological clock and produces dynamic bahaviour. Proc R Soc B 268:1307–1313

Becker A, Canut H, Lüttge U, Maeshima M, Marigo G, Ratajzcak R (1995) Purification and immunological comparison of the tonoplast H^+-pyrophosphatase from cells of *Catharanthus roseus* and leaves from *Mesembryanthemum crystallinum* performing C_3-photosynthesis and the obligate CAM plant *Kalanchoë daigremontiana*. J Plant Physiol 146:88–94

Behre B, Ratajczak R, Lüttge U (1992) Selective reconstitution of the tonoplast H^+-ATPase of the crassulacean-acid metabolism plant *Kalanchoë daigremontiana*. Bot Acta 105:260–265

Berg A, Orthen B, de Mattos EA, Duarte HM, Lüttge U (2004) Expression of crassulacean acid metabolism in *Clusia hilariana* Schlechtendal in different stages of development in the field. Trees 18:553–558

Beyschlag W, Eckstein J (1997) Stomatal patchiness. Prog Bot 52:283–298

Blasius B, Beck F, Lüttge U (1997) A model for photosynthetic oscillations in crassulacean acid metabolism (CAM). J Theor Biol 184:345–351

Blasius B, Beck F, Lüttge U (1998) Oscillatory model of crassulacean acid metabolism: structural analysis and stability boundaries with a discrete hysteresis switch. Plant Cell Environ 21:775–784

Blasius B, Neff R, Beck F, Lüttge U (1999) Oscillatory model of crassulacean acid metabolism with a dynamic hysteresis switch. Proc R Soc B 266:93–101

Bohn A, Geist A, Rascher U, Lüttge U (2001) Responses to different external light rhythms by the circadian rhythm of crassulacean acid metabolism in *Kalanchoë daigremontiana*. Plant Cell Environ 24:811–820

Bohn A, Hinderlich S, Hütt M-T, Kaiser F, Lüttge U (2003) Identification of rhythmic subsystems in the circadian cycle of crassulacean acid metabolism under thermoperiodic perturbations. Biol Chem 384:721–728

Borland AM, Hartwell J, Jenkins GI, Wilkins MB, Nimmo HG (1999) Metabolite control overrides circadian regulation in phosphoenolpyruvate carboxylase kinase and CO_2 fixation in crassulacean acid metabolism. Plant Physiol 121:889–896

Boxall SF, Foster JM, Bohnert HJ, Cushman JC, Nimmo HG, Hartwell J (2005) Conservation and divergence of circadian clock operation in a stress-inducible crassulacean acid metabolism species reveals clock compensation against stress. Plant Physiol 137:969–982

Bremberger C, Lüttge U (1992) Dynamics of tonoplast proton pumps and other tonoplast proteins of *Mesembryanthemum crystallinum* during the induction of crassulacean acid metabolism. Planta 188:575–580

Bremberger C, Haschke H-P, Lüttge U (1988) Separation and purification of the tonoplast ATPase and pyrophosphatase from plants with constitutive and inducible CAM. Planta 175:465–470

Briggs GE, Hope AB, Robertson RN (1961) Electrolytes and plant cells. Blackwell, Oxford

Brinckmann E, Lüttge U (1972) Vorübergehende pH-Änderungen im umgebenden Medium intakter grüner Zellen bei Beleuchtungswechsel. Z Naturf 27b:277–284

Brinckmann E, Lüttge U (1974) Lichtabhängige Membranpotentialschwankungen und deren interzelluläre Weiterleitung bei panaschierten Photosynthese-Mutanten von *Oenothera*. Planta 119:47–57

Britto DT, Ruth TJ, Lapis S, Kronzucker HJ (2004) Cellular and whole-plant chloride dynamics in barley: insights into chloride-nitrogen interactions and salinity responses. Planta 218:615–622

Büdel B, Lüttge U, Stelzer R, Huber O, Medina E (1994) Cyanobacteria of rocks and soils of the Orinoco Lowlands and the Guayana Uplands, Venezuela. Bot Acta 107:422–431

Bulychev AA, Kamzolkina NA (2006a) Differential effects of plasma membrane electric excitation on H^+ fluxes and photosynthesis in characean cells. Bioelectrochemistry 69:209–215

Bulychev AA, Kamzolkina NA (2006b) Effect of action potential on photosynthesis and spatially distributed H^+ fluxes in cells and chloroplasts of *Chara corallina*. Russ J Plant Physiol 53:1–9

Bulychev AA, Turovetsky VB (1983) Light-triggered changes of membrane potential in cells of *Anthoceros punctatus* and their relation to activation of chloroplast ATPase. J Exp Bot 34:1181–1188

Búrquez A, Corbet SA (1991) Do flowers reabsorb nectar? Funct Ecol 5:369–379

Buser-Suter C, Wiemken A, Matile P (1982) A malic acid permease in isolated vacuoles of a crassulacean acid metabolism plant. Plant Physiol 69:456–459
Carter C, Thornburg RW (2000) Tobacco nectarin I. Purification and characterization as a germin-like manganese superoxide dismutase implicated in the defense of floral reproductive tissues. J Biol Chem 275:36726–36733
Carter C, Thornburg RW (2004a) Is the nectar redox cycle a floral defense against microbial attack? Trends Plant Sci 9:320–324
Carter C, Thornburg RW (2004b) Tobacco nectarin III is a bifunctional enzyme with monodehydroascorbate reductase and carbonic anhydrase activities. Plant Mol Biol 54:415–425
Carter C, Healy R, O'Tool NM, Saqlan Naqvi SM, Ren G, Park S, Beattie GA, Horner HT, Thornburg RW (2007) Tobacco nectaries express a novel NADPH oxidase implicated in the defense of floral reproductive tissues against microorganisms. Plant Physiol 143:389–399
Corbet SA, Willmer PG, Beamet JWL, Unwin DM, Prŷs-Jones OE (1979) Post-secretory determinants of sugar concentration in nectar. Plant Cell Environ 2:293–298
Cram WJ (1973) Internal factors regulating nitrate and chloride influx in plant cells. J Exp Bot 24:328–341
Cram WJ (1976) Negative feedback regulation of transport in cells. The maintenance of turgor, volume and nutrient supply. Encycl Plant Physiol 2A: 284–316
Dietz KJ, Tavakoli N, Kluge C, Mimura T, Sharma SS, Harris GC, Chardonnes AN, Golldack D (2001) Significance of the V-type ATPase for the adaptation to stressful growth conditions and its regulation on the molecular and biochemical level. J Exp Bot 52:1969–1980
Dojani S, Lakatos M, Rascher U, Wanek W, Lüttge U, Büdel B (2007) Nitrogen input by cyanobacterial biofilms of an inselberg into a tropical rainforest in French Guiana. Flora 202:521–529
Domgall I, Venzke D, Lüttge U, Ratajczak R, Böttcher B (2002) Three dimensional model of a plant V-ATPase based on electron microscopy. J Biol Chem 277:13115–13121
Drobny M, Schnölzer M, Fiedler S, Lüttge U, Fischer-Schliebs E, Christian A-L, Ratajczak R (2002) Phenotypic subunit composition of the tobacco (*Nicotiana tabacum* L.) vacuolar-type H^+-translocating ATPase. Biochim Biophys Acta Biomembr 1564:243–255
Duarte H, Lüttge U (2007a) Correlation between photorespiration, CO_2-assimilation and spatio-temporal dynamics of photosynthesis in leaves of the C_3-photosynthesis/crassulacean acid metabolism-intermediate species *Clusia minor* L. (Clusiaceae). Trees 21:531–540
Duarte H, Lüttge U (2007b) Circadian rhythmicity. In: Lüttge U (ed) *Clusia*: A woody neotropical genus of remarkable plasticity and diversity, vol 194, Ecological studies. Springer, Berlin, pp 245–256
Duarte HM, Gessler A, Scarano FR, Franco AC, de Mattos EA, Nahm M, Rennenberg H, Rodrigues PJFP, Zaluar HLT, Lüttge U (2005a) Ecophysiology of six selected shrub species in different plant communities at the periphery of the Atlantic Forest of SE-Brazil. Flora 200:456–476
Duarte H, Jakovljevic I, Kaiser F, Lüttge U (2005b) Lateral diffusion of CO_2 in leaves of the crassulacean acid metabolism plant *Kalanchoë daigremontiana* Hamet et Perrier. Planta 220:809–816
Epstein E, Hagen CE (1952) A kinetic study of the absorption of alkali cations by barley roots. Plant Physiol 27:457–474
Epstein E, Rains DW (1965) Carrier mediated cation transport by barley roots: kinetic evidence for a spectrum of active sites. Proc Natl Acad Sci USA 53:1320–1324
Epstein E, Rains DW, Elzam OE (1963) Resolution of dual mechanisms of potassium absorption by barley roots. Proc Natl Acad Sci USA 49:684–692
Erhardt A (1992) Preferences and non-preferences for nectar constituents in *Ornithoptera priamus poseidon* (Lepidoptera, Papilionidae). Oecologia 90:581–585
Escalante-Pérez M, Heil M (2012) The production and protection of nectars. Prog Bot 74:239–261
Etherton B, Higinbotham N (1960) Transmembrane potential measurements of cells of higher plants as related to salt uptake. Science 131:409–410

Faak M (ed) (2000) Alexander von Humboldt. Reise durch Venezuela. Auswahl aus den amerikanischen Reisetagebüchern. Beiträge zur Alexander von Humboldt-Forschung 12, Akademie Verlag, Berlin, p 193

Fahn A (1979) Ultrastructure of nectaries in relation to nectar secretion. Am J Bot 66:977–985

Ferjani A, Segami S, Asaoka M, Maeshima M (2013) Regulation of PPi levels trough the vacuolar membrane H^+-pyrophosphatase. Prog Bot 75:145–165

Fernandes GW, de Mattos EA, Franco AC, Lüttge U, Ziegler H (1998) Influence of the parasite *Pilostyles ingae* (Rafflesiaceae) on some physiological parameters of the host plant, *Mimosa naguirei* (Mimosaceae). Bot Acta 111:51–54

Fetene M, Lüttge U (1991) Environmental influences on carbon recycling in a terrestrial CAM bromeliad, *Bromelia humilis* Jacq. J Exp Bot 42:25–31

Fetene M, Lee HSJ, Lüttge U (1990) Photosynthetic acclimation in a terrestrial CAM bromeliad: *Bromelia humilis* Jacq. New Phytol 114:399–406

Fetene M, Nauke P, Lüttge U, Beck E (1997) Photosynthesis and photoinhibition in a tropical alpine giant rosette plant, *Lobelia rhynchopetalum*. New Phytol 137:453–461

Fischer E, Lüttge U (1980) Membrane potential changes related to active transport of glycine in *Lemna gibba* G1. Plant Physiol 65:1004–1008

Fischer E, Lüttge U, Higinbotham N (1976) Effect of cyanide on the plasmalemma potential of *Mnium*. Plant Physiol 58:240–241

Fischer-Schliebs E, Mariaux J-B, Lüttge U (1997) Stimulation of H^+-transport activity of vacuolar H^+-ATPase by activation of H^+-PPase in *Kalanchoë blossfeldiana*. Biol Plant 39:169–177

Fischer-Schliebs E, Ratajczak R, Weber P, Tavakoli N, Ullrich CI, Lüttge U (1998) Concordant time-dependent patterns of activities and enzyme-protein amounts of V-PPase and V-ATPase in stimulated and non-stimulated plant tissues. Bot Acta 111:130–136

Fischer-Schliebs E, Drobny M, Ball E, Ratajczak R, Lüttge U (2000) Variation in nitrate nutrition leads to changes in the performance of the V-ATPase and immunological differences of proteolipid subunit *c* in tobacco (*Nicotiana tabacum* L.) leaves. Aust J Plant Physiol 27:639–648

Franco AC, Lüttge U (2002) Midday depression in savanna trees: coordinated adjustments in photochemical efficiency, photorespiration, CO_2 assimilation and water use efficiency. Oecologia 131:356–365

Franco AC, Haag-Kerwer A, Herzog B, Grams TEE, Ball E, de Mattos EA, Scaranos FR, Barreto S, Garcia MA, Manotvani A, Lüttge U (1996) The effect of light levels on daily patterns of chlorophyll fluorescence and organic acid accumulation in the tropical CAM tree *Clusia hilariana*. Trees 10:359–365

Franco AC, Herzog B, Hübner C, de Mattos EA, Scarano FR, Ball E, Lüttge U (1999) Diurnal changes in chlorophyll *a* fluorescence, CO_2-exchange and organic acid decarboxylation in the tropical CAM tree *Clusia hilariana*. Tree Physiol 19:635–644

Franco AC, Duarte HM, Gessler A, de Mattos EA, Nahm M, Rennenberg H, Ribeiro KT, Scarano FR, Lüttge U (2005) In situ measurements of carbon and nitrogen distribution and composition, photochemical efficiency and stable isotope ratios in *Araucaria angustifolia*. Trees 19:422–430

Gessler A, Duarte HM, Franco AC, Lüttge U, de Mattos EA, Nahm M, Scarano FR, Zaluar HLT, Rennenberg H (2005a) Ecophysiology of selected tree species in different plant communities at the periphery of the Atlantic Forest of SE-Brazil. II. Spatial and ontogenetic dynamics in *Andira legalis*, a deciduous tree. Trees 19:510–522

Gessler A, Duarte HM, Franco AC, Lüttge U, de Mattos EA, Nahm M, Rodrigues PJFP, Scarano FR, Rennenberg H (2005b) Ecophysiology of selected tree species in different plant communities at the periphery of the Atlantic Forest of SE-Brazil. III. Three legume trees in a semideciduous forest. Trees 19:523–530

Gessler A, Nitschke R, de Mattos EA, Zaluar HTL, Scarano FR, Rennenberg H, Lüttge U (2008) Comparison of the performance of three different ecophysiological life forms in a sandy coastal restinga ecosystem of SE-Brazil: a nodulated N_2-fixing C_3-shrub (*Andira legalis*

(Vell.) Toledo), a CAM-shrub (*Clusia hilariana* Schltdl.) and a tap root C_3-hemicryptophyte (*Allagoptera arenaria* (Gomes) O. Ktze.). Trees 22:105–119

González-Teuber M, Pozo MJ, Muck A, Svatos A, Adame-A'lvarez RM, Heil M (2010) Glucanases and chitinases as causal agents in the protection of *Acacia* extrafloral nectar from infestation by phytopathogens. Plant Physiol 152:1705–1715

Grams TEE, Kluge M, Lüttge U (1995) High temperature adapted plants of *Kalanchoë daigremontiana* show changes in temperature dependence of the endogenous CAM rhythm. J Exp Bot 46:1927–1929

Grams TEE, Beck F, Lüttge U (1996) Generation of rhythmic and arrhythmic behaviour of crassulacean acid metabolism in *Kalanchoë daigremontiana* under continuous light by varying the irradiance or temperature: measurements in vivo and model simulations. Planta 198:110–117

Grams TEE, Borland AM, Roberts A, Griffiths H, Beck F, Lüttge U (1997a) On the mechanism of reinitiation of endogenous crassulacean acid metabolism rhythm by temperature changes. Plant Physiol 113:1309–1317

Grams TEE, Haag-Kerwer A, Olivares E, Ball E, Arndt S, Popp M, Medina E, Lüttge U (1997b) Comparative measurements of chlorophyll a fluorescence, acid accumulation and gas exchange in exposed and shaded plants of *Clusia minor* L. and *Clusia multiflora* H.B.K. in the field. Trees 11:240–247

Griffiths H, Smith JAC (1983) Photosynthetic pathways in the Bromeliaceae of Trinidad: relations between life-forms, habitat preference and the occurrence of CAM. Oecologia 60:176–184

Griffiths H, Smith JAC, Lüttge U, Popp M, Cram WJ, Diaz M, Lee HSJ, Medina E, Schäfer C, Stimmel K-H (1989) Ecophysiology of xerophytic and halophytic vegetation of a coastal alluvial plain in northern Venezuela. IV. *Tillandsia flexuosa* SW. and *Schomburgkia humbodltiana* Reichb., epiphytic CAM plants. New Phytol 111:273–282

Gustafsson MHG, Winter K, Bittrich V (2007) Diversity, phylogeny and classification of *Clusia*. In: Lüttge U (ed) *Clusia*. A woody neotropical genus of remarkable plasticity and diversity, vol 194, Ecological studies. Springer, Berlin, pp 95–116

Haag-Kerwer A, Franco AC, Lüttge U (1992) The effect of temperature and light on gas exchange and acid accumulation in the C_3-CAM plant *Clusia minor* L. J Exp Bot 43:345–352

Haberlandt G (1904) Physiologische Pflanzenanatomie, 3. Aufl. W. Engelmann, Leipzig

Hafke JB, Hafke Y, Smith JAC, Lüttge U, Thiel G (2003) Vacuolar malate uptake is mediated by an anion-selective inward rectifier. Plant J 35:116–128

Haschke H-P, Lüttge U (1973) ß-Indolylessigsäure-(IES)-abhängbiger K^+-H^+-Austauschmechanismus und Streckungswachstum bei *Avena*-Koleoptilen. Z Naturf 28c:555–558

Haschke H-P, Lüttge U (1975a) Interactions between IAA, potassium, and malate accumulation, and growth in *Avena* coleoptiles segments. Z Pflanzenphysiol 76:450–455

Haschke H-P, Lüttge U (1975b) Stoichiometric correlation of malate accumulation with auxin-dependent K^+-H^+-exchange and growth in *Avena* coleoptile segments. Plant Physiol 56:696–698

Haschke H-P, Lüttge U (1977) Action of auxin on CO_2 dark fixation in *Avena* coleoptile segments as related to elongation growth. Plant Sci Lett 8:53–58

Haschke H-P, Grötsch S, Lüttge U (1988) Proton transporting enzymes at the tonoplast of leaf cells of the CAM plant *Kalanchoë daigremontiana*. III. Regulation of the ATPase. J Plant Physiol 132:604–607

Hatch MD, Slack CR (1966) Photosynthesis by sugarcane leaves. Biochem J 101:103–111

Haukioja E (1991) The influence of grazing on the evolution, morphology and physiology of plants as modular organisms. Philos Trans R Soc Lond B Biol Sci 333:241–247

Herzog B, Hoffmann S, Hartung W, Lüttge U (1999a) Comparison of photosynthetic responses of the sympatric tropical C_3-species *Clusia multiflora* H.B.K. and the C_3-CAM intermediate species *Clusia minor* L. to irradiance and drought stress in a phytotron. Plant Biol 1:460–470

Herzog B, Hübner C, Ball E, Bastos RD, Franco AC, Scarano FR, Lüttge U (1999b) Comparative study of the C_3/CAM-intermediate species *Clusia parviflora* Saldanha et Engl. and the obligate

CAM-Species *Clusia hilariana* Schlecht. growing sympatrically exposed and shaded in the coastal restinga of Brazil. Plant Biol 1:453–459

Heun A-M, Gorham J, Lüttge U, WynJones RG (1981) Changes of water-relation characteristics and levels of organic cytoplasmic solutes during salinity induced transition of *Mesembryanthemum crystallinum* from C_3-photosynthesis to crassulacean acid metabolism. Oecologia 50:66–72

Higinbotham N, Etherton B, Foster RJ (1964) Effects of external K, NH_4, Na, Ca, Mg, and H ions on the cell transmembrane electropotential of *Avena* coleoptile. Plant Physiol 39:196–203

Hillwig MS, Liu X, Liu G, Thornburg RW, MacIntosh GC (2010) Petunia nectar proteins have ribonuclease activity. J Exp Bot 61:2951–2965

Hillwig MS, Kanobe C, Thornburg RW, MacIntosh GC (2011) Identification of S-RNase and peroxidase in petunia nectar. J Plant Physiol 168:734–738

Hope AB, Lüttge U, Ball E (1972) Photosynthesis and apparent proton fluxes in *Elodea canadensis*. Z Pflanzenphysiol 68:73–81

Hope AB, Lüttge U, Ball E (1974) Chloride uptake in strains of *Scenedesmus obliquus*. Z Pflanzenphysiol 72:1–10

Horner HT, Healy RA, Ren G, Fritz D, Klyne A, Seames C, Thornburg RW (2007) Amyolplast to chromoplast conversion in developing ornamental tobacco floral nectaries provides sugar for nectar and antioxidants for protection. Am J Bot 94:12–24

Hütt M-T, Lüttge U (2002) Nonlinear dynamics as a tool for data analysis and modeling in plant physiology. Plant Biol 4:281–297

Hütt M-T, Lüttge U (2007) Noise-induced phenomena and complex rhythms: theoretical considerations, modeling and experimental evidence. In: Mancuso S, Shabala S (eds) Rhythms in plants: phenomenology, mechanisms, and adaptive significance. Springer, Berlin, pp 313–339

Jennings DH (1963) The absorption of solutes by plant cells. Oliver and Boyd, Edinburgh

Jentsch J (1972) Enzymes from carnivorous plans (*Nepenthes*). Isolation of the protease nepenthacin. FEBS Lett 21:273–276

Jochem P, Lüttge U (1987) Proton transporting enzymes at the tonoplast of leaf cells of the CAM plant *Kalanchoë daigremontiana*. I. The ATPase. J Plant Physiol 129:251–268

Jochem P, Rona J-P, Smith JAC, Lüttge U (1984) Anion-sensitive ATPase activity and proton transport in isolated vacuoles of species of the CAM genus *Kalanchoë*. Physiol Plant 62:410–415

Johansen C, Lüttge U (1974) Respiration and photosynthesis as alternative energy sources for chloride uptake by *Tradescantia albiflora* leaf cells. Z Pflnzenphysiol 71:189–199

Johansen C, Lüttge U (1975) A comparison of potassium and chloride uptake by *Tradescantia albiflora* leaf cells at different KCl concentrations. Aust J Plant Physiol 2:471–479

Jung K-D, Lüttge U (1980) Amino acid uptake by *Lemna gibba* by a mechanism with affinity to neutral L- and D-amino acids. Planta 150:230–235

Jung K-D, Lüttge U, Fischer E (1982) Uptake of neutral and acidic amino acids by *Lemna gibba* correlated with the H^+-electrochemical gradient at the plasmalemma. Physiol Plant 55:351–355

Kessler D, Baldwin IT (2007) Making sense of nectar scents: the effects of nectar secondary metabolites on floral visitors of *Nicotiana attenuata*. Plant J 49:840–854

Kliemchen A, Schomburg M, Galla H-J, Lüttge U, Kluge M (1993) Phenotypic changes in the fluidity of the tonoplast membrane of crassulacean-acid metabolism plants in response to temperature and salinity stress. Planta 189:403–409

Klink R, Lüttge U (1991) Electron-microscopic demonstration of a "head and stalk" structure of the leaf vacuolar ATPase in *Mesembryanthemum crystallinum*. Bot Acta 104:122–131

Klink R, Lüttge U (1992) Quantification of visible structural changes of the V_0V_1-ATPase in the leaf tonoplast of *Mesembryanthemum crystallinum* by freeze-fracture replicas prepared during the C_3-photosynthesis to CAM transition. Bot Acta 105:414–420

Klink R, Haschke H-P, Kramer D, Lüttge U (1990) Membrane particles, proteins and ATPase activity of tonoplast vesicles of *Mesembryanthemum crystallinum* in the C_3 and CAM state. Bot Acta 103:24–31

Kornas A, Fischer-Schliebs E, Lüttge U, Miszalski Z (2009) Adaptation of the obligate CAM plant *Clusia alata* to light stress: metabolic responses. J Plant Physiol 166:1914–1922

Kornas A, Miszalski Z, Surówka E, Fischer-Schliebs E, Lüttge U (2010) Light stress is not effective to enhance crassulacean acid metabolism. Z Naturf 65c:79–86

Kortschak HP, Hartt CE, Burr GO (1965) Carbon dioxide fixation in sugarcane leaves. Plant Physiol 40:209–213

Król E, Płancho BJ, Adamec L, Stolarz M, Dziubińska H, Trębacz K (2012) Quite a few reasons for calling carnivores 'the most wonderful plants in the world'. Ann Bot 109:47–64

Kronzucker HJ, Siddiqi MY, Glass ADM, Kirk GJD (1999) Nitrate-ammonium synergism in rice. A subcellular flux analysis. Plant Physiol 119:1041–1045

Kumari N, Sharma V, Mikosch M, Unfried C, Geßler A, Fischer-Schliebs E, Lüttge U (2005) Seasonal photosynthetic performance and nutrient relations of *Butea monosperma* Taub. in comparison to two other woody species of a seasonal deciduous forest in SE-Rajasthan and to planted trees in the area. Indian J For 28:116–126

Läuchli A, Lüttge U (1968) Untersuchung der Kinetik der Ionenaufnahme in das Cytoplasma von *Mnium*-Blattzellen mit Hilfe der Mikroautoradiographie und der Röntgen-Mikrosonde. Planta 83:80–98

Läuchli A, Pitman MG, Lüttge U, Kramer D, Ball E (1978) Are developing xylem vessels the site of ion exudation from root to shoot? Plant Cell Environ 1:217–223

Laughlin RB (2005) A different universe—reinventing physics from the bottom down. Basic Books, New York, NY

Lee HSJ, Lüttge U, Medina E, Smith JAC, Cram WJ, Diaz M, Griffiths H, Popp M, Schäfer C, Stimmel K-H, Thonke B (1989) Ecophysiology of xerophytic and halophytic vegetation of a coastal alluvial plain in northern Venezuela. III: *Bromelia humilis* Jacq., a terrestrial CAM bromeliad. New Phytol 111:253–271

Liebig M, Scarano FR, de Mattos EA, Zaluar HLT, Lüttge U (2001) Ecophysiological and floristic implications of sex expression in the dioecious neotropical CAM tree *Clusia hilariana* Schltdl. Trees 15:278–288

Lovelock J (1979) Gaia. A new look at life on earth. Oxford University Press, Oxford

Löw R, Rockel B, Kirsch M, Ratajczak R, Hörtensteiner S, Martinoia E, Lüttge U, Rausch T (1996) Early salt stress effects on the differential expression of vacuolar H^+-ATPase genes in roots and leaves of *Mesembryanthemum crystallinum*. Plant Physiol 110:259–265

Lundegårdh H (1950) The translocation of salts and water through wheat roots. Plant Physiol 2:103–151

Lundegårdh H (1955) Mechanisms of absorption, transport, accumulation, and secretion of ions. Annu Rev Plant Physiol 6:1–24

Lundegårdh H, Burström H (1933) Untersuchungen über die Salzaufnahme der Pflanzen. III. Quantitative Beziehungen zwischen Atmung und Anionenaufnahme. Biochem Z 261:235–251

Lundegårdh H, Burström H (1935) Untersuchungen über die Atmungsvorgänge in Pflanzenwurzeln. Biochem Z 277:223–249

Lüttge U (1961) Über die Zusammensetzung des Nektars und den Mechanismus seiner Sekretion. I. Planta 56:189–212

Lüttge U (1962a) Über die Zusammensetzung des Nektars und den Mechanismus seiner Sekretion. II. Der Kationengehalt des Nektars und die Bedeutung des Verhältnisses Mg^{++}/Ca^{++} im Drüsengewebe für die Sekretion. Planta 59:108–114

Lüttge U (1962b) Über die Zusammensetzung des Nektars und den Mechanismus seiner Sekretion. III. Die Rolle der Rückresorption und der spezifischen Zuckersekretion. Planta 59:175–194

Lüttge U (1963) Die Bedeutung des chemischen Reizes bei der Resorption von ^{14}C-Glutaminsäure, $^{35}SO_4^{--}$ und $^{45}Ca^{++}$ durch *Dionaea muscipula*. Naturwissenschaften 50:22

Lüttge U (1964a) Untersuchungen zur Physiologie der Carnivoren-Drüsen. I. Mitteilung. Die an den Verdauungsvorgängen beteiligten Enzyme. Planta 63:103–117
Lüttge U (1964b) Mikroautoradiographische Untersuchungen über die Funktion der Hydropoten von *Nymphaea*. Protoplasma 54:157–162
Lüttge U (1965) Untersuchungen zur Physiologie der Carnivoren-Drüsen. II. Mitteilung. Über die Resorption verschiedener Substanzen. Planta 66:331–344
Lüttge U (1966a) Untersuchungen zur Physiologie der Carnivoren-Drüsen. IV. Mitteilung. Die Kinetik der Chloridsekretion durch das Drüsengewebe von Nepenthes. Planta 68:44–56
Lüttge U (1966b) Untersuchungen zur Physiologie der Carnivoren-Drüsen. V. Mitteilung. Mikroautoradiographische Untersuchung der Chloridsekretion durch das Drüsengewebe von Nepenthes. Planta 68:269–285
Lüttge U (1968) Die Kinetik von Parenchymtransporten. Vortr Gesamtgeb Botanik NF 2:66–78
Lüttge U (1969) Aktiver Transport. Kurzstreckenstransport bei Pflanzen. Protoplasmatologia, Handb der Protoplasmaforschung, Viii/7b, Springer, Wien
Lüttge U (ed) (1972) Micro-autoradiography and electron probe analysis. Springer, Berlin
Lüttge U (1973) Stofftransport der Pflanzen. Heidelberger Taschenbücher. Springer, Berlin
Lüttge U (1975) Salt glands. In: Baker DA, Hall JL (eds) In transport in plant cells and tissues. North Holland, Amsterdam, pp 335–376
Lüttge U (1986) Nocturnal water storage in plants having crassulacean acid metabolism. Planta 168:287–289
Lüttge U (ed) (1989) Vascular plants as epiphytes. Evolution and ecophysiology, vol 76, Ecological studies. Springer, Berlin
Lüttge U (1999) One morphotype three physiotypes: sympatric species of *Clusia* with obligate C_3-photosynthesis, obligate CAM and C_3-CAM intermediate behavior. Plant Biol 1:138–148
Lüttge U (2000) The tonoplast functioning as the master switch for circadian regulation of crassulacean acid metabolism. Planta 211:761–769
Lüttge U (2002) CO_2-concentrating: consequences in crassulacean acid metabolism. J Exp Bot 53:2131–2141
Lüttge U (2005a) Physiologische Ökologie der Photosynthese—autökologische und synökologische Aspekte anhand von δ^{13}C- und δ^{18}-Daten. Rundgespr Komm Ökologie Bay Akad Wiss 30: 69–82, Auf Spurensuche in der Natur, München Verlag Dr. F. Pfeil
Lüttge U (2005b) Genotypes–phenotypes–ecotypes: relations to crassulacean acid metabolism. Nova Acta Leopoldina NF 92(342):177–193
Lüttge U (2006) Photosynthetic flexibility and ecophysiological plasticity: questions and lessons from *Clusia*, the only CAM tree, in the neotropics. New Phytol 171:7–25
Lüttge U (ed) (2007a) *Clusia*. A woody neotropical genus of remarkable plasticity and diversity, vol 194, Ecological studies. Springer, Berlin
Lüttge U (2007b) Historical recollections. In: Lüttge U (ed) *Clusia*. A woody neotropical genus of remarkable plasticity and diversity, vol 194, Ecological studies. Springer, Berlin, pp 3–9
Lüttge U (2007c) Photosynthesis. In: Lüttge U (ed) *Clusia*. A woody neotropical genus of remarkable plasticity and diversity, vol 194, Ecological studies. Springer, Berlin, pp 135–186
Lüttge U (2007d) Physiological ecology. In: Lüttge U (ed) *Clusia*. A woody neotropical genus of remarkable plasticity and diversity, vol 194, Ecological studies. Springer, Berlin, pp 187–234
Lüttge U (2008a) Physiological ecology of tropical plants, 2nd edn. Springer, Berlin
Lüttge U (2008b) *Clusia*: holy grail and enigma. J Exp Bot 59:1503–1514
Lüttge U (2010) Photorespiration in phase III of Crassulacean acid metabolism: evolutionary and ecophysiological implications. Prog Bot 72:371–384
Lüttge U (2012a) Modularity and emergence: biology's challenge in understanding life. Plant Biol 14:865–871
Lüttge U (2012b) Whole-plant physiology: synergistic emergence rather than modularity. Prog Bot 74:165–190
Lüttge U (2013a) Green nectaries: the role of photosynthesis in secretion. Bot J Linn Soc 173:1–11

Lüttge U (2013b) The planet Earth: can it feed nine billion people? In: Matyssek R, Lüttge U, Rennenberg H (eds) The alternatives growth and defense: resource allocation at multiple scales in plants. Nova Acta Leopold NF 114(391): 345–364

Lüttge U (2016) Plants shape the terrestrial environment on Earth: Challenges of management for sustainability. Prog Bot 77:187–217

Lüttge U, Ball E (1971) Light-independent uncoupler-sensitive ion uptake by green and by pale cells of variegated leaves of higher plants in relation to protein content and chloroplast integrity. Z Naturf 26b:158–161

Lüttge U, Ball E (1974a) Proton and malate fluxes in cells of *Bryophyllum daigremontianum* leaf slices in relation to potential osmotic pressure of the medium. Z Pflanzenphysiol 73:326–338

Lüttge U, Ball E (1974b) Mineral ion fluxes in slices of acidified and de-acidified leaves of the CAM plant *Bryophyllum daigremontianum*. Z Pflanzenphysiol 73:339–348

Lüttge U, Ball E (1976) ATP levels and energy requirements of ion transport in cells of slices of greening barley leaves. Z Pflanzenphysiol 80:50–59

Lüttge U, Ball E (1979) Electrochemical investigation of active malic acid transport at the tonoplast into the vacuoles of the CAM plant *Kalanchoë daigremontiana*. J Membr Biol 47:401–422

Lüttge U, Ball E (1980) $2H^+$: 1 $malate^{2-}$ stoichiometry during Crassulacean Acid Metabolism is unaffected by lipophilic cations. Plant Cell Environ 3:195–200

Lüttge U, Bauer K (1968) Evaluation of ion uptake isotherms and analysis of individual fluxes of ions. Planta 80:52–64

Lüttge U, Beck F (1992) Endogenous rhythms and chaos in crassulacean acid metabolism. Planta 188:28–38

Lüttge U, Higinbotham N (1979) Transport in plants. Springer, New York, NY

Lüttge U, Hütt M-T (2006) Spatiotemporal patterns and distributed computation—a formal link between CO_2 signalling, diffusion and stomatal regulation. Prog Bot 68:242–260

Lüttge U, Kluge M (2012) Botanik. Die einführende Biologie der Pflanzen, 6th edn. Wiley-VCH, Weinheim

Lüttge U, Laties GG (1966) Dual mechanism of ion absorption in relation to long distance transport in plants. Plant Physiol 41:1531–1539

Lüttge U, Laties GG (1967a) Absorption and long distance transport by isolated stele of maize roots in relation to the dual mechanisms of ion absorption. Planta 74:173–187

Lüttge U, Laties GG (1967b) Selective inhibition of absorption and long distance transport in relation to the dual mechanisms of ion absorption in maize seedlings. Plant Physiol 42:181–185

Lüttge U, Mayer E (2012) Natur und Geist. Konfliktgeschichte und Kooperationsmöglichkeit. Evangelische Zentralstelle für Weltanschauungsfragen, Berlin – Texte 217

Lüttge U, Nobel PS (1984) Day-night variations in malate concentration, osmotic pressure, and hydrostatic pressure in *Cereus validus*. Plant Physiol 75:804–807

Lüttge U, Osmond CB (1970) Ion absorption in *Atriplex* leaf tissue. III. Site of metabolic control of light dependent chloride secretion to epidermal bladders. Aust J Biol Sci 23:17–25

Lüttge U, Pallaghy CK (1969) Light triggered transient changes of membrane potentials in green cells in relation to photosynthetic electron transport. Z Pflanzenphysiol 61:58–67

Lüttge U, Pallaghy CK (1972) Unerwartete Kinetik des Efflux'und der Aufnahme von Ionen bei verschiedenen Pflanzengeweben. Z Pflanzenphysiol 67:359–366

Lüttge U, Pitman MG (eds) (1976a) Transport in plants II, part A cells, vol 2, Encyclopedia of plant physiology new series. Springer, Berlin

Lüttge U, Pitman MG (eds) (1976b) Transport in plants II, part B tissues and organs, vol 2, Encyclopedia of plant physiology new series. Springer, Berlin

Lüttge U, Scarano FR (2004) Ecophysiology. Rev Bras Bot 27:1–10

Lüttge U, Scarano FR (2007) Synecological comparisons sustained by ecophysiological fingerprinting of intrinsic photosynthetic capacity of plants as assessed by measurements of light response curves. Braz J Bot 30:355–364

Lüttge U, Schnepf E (1976) Organic substances. In: Lüttge U, Pitman MG (eds) Transport in plants II. Part B tissues and organs, vol 2, Encyclopedia of plant physiology new series. Springer, Berlin, pp 244–277
Lüttge U, Smith JAC (1984) Mechanism of passive malic-acid efflux from vacuoles of the CAM plant *Kalanchoë daigremontiana*. J Membr Biol 81:149–158
Lüttge U, Weigl J (1965) Zur Mikroautoradiographie wasserlöslicher Substanzen. Planta 64:28–36
Lüttge U, Zirke G (1974) Attempts to measure plasmalemma and tonoplast electropotentials in small cells of the moss *Mnium* using centrifugation techniques. J Membr Biol 18:305–314
Lüttge U, Pallaghy CK, Osmond CB (1970) Coupling of ion transport in green cells of *Atriplex spongiosa* leaves to energy sources in the light and in the dark. J Membr Biol 2:17–30
Lüttge U, Ball E, von Willert K (1971a) A comparative study of the coupling of ion uptake to light reactions in leaves of higher plant species having the C_3- and C_4-pathway of photosynthesis. Z Pflanzenphysiol 65:336–350
Lüttge U, Cram WJ, Laties GG (1971b) The relationship of salt stimulated respiration to localised ion transport in carrot tissue. Z Pflanzenphysiol 64:418–426
Lüttge U, Ball E, Tromballa H-W (1975a) Potassium independence of osmoregulated oscillations of malate^{2-} levels in the cells of CAM-leaves. Biochem Physiol Pflanzen 167:267–283
Lüttge U, Kluge M, Ball E (1975b) Effects of osmotic gradients on vacuolar malic acid storage. A basic principle in oscillatory behavior of crassulacean acid metabolism. Plant Physiol 56:613–616
Lüttge U, Ball E, Greenway H (1977) Effects of water and turgor potential on malate efflux from leaf slices of *Kalanchoë daigremontiana*. Plant Physiol 60:521–523
Lüttge U, Fischer E, Steudle E (1978) Membrane potentials and salt distribution in epidermal bladders and photosynthetic tissue of *Mesembryanthemum crystallinum* L. Plant Cell Environ 1:121–129
Lüttge U, Jung K-D, Ullrich-Eberius CI (1981a) Evidence for amino acid-H^+ cotransport in *Lemna gibba* by effects of fusicoccin. Z Pflanzenphysiol 102:117–125
Lüttge U, Smith JAC, Marigo G, Osmond CB (1981b) Energetics of malate accumulation in the vacuoles of *Kalanchoë tubiflora* cells. FEBS Lett 126:81–84
Lüttge U, Medina E, Cram WJ, Lee HSJ, Popp M, Smith JAC (1989a) Ecophysiology of xerophytic and halophytic vegetation of a coastal alluvial plain in northern Venezuela. II. Cactaceae. New Phytol 111:245–251
Lüttge U, Popp M, Medina E, Cram WJ, Diaz M, Griffiths H, Lee HSJ, Schäfer C, Smith JAC, Stimmel K-H (1989b) Ecophysiology of xerophytic and halophytic vegetation of a coastal alluvial plain in northern Venezuela. V. The *Batis maritima—Sesuvium portulacastrum* vegetation unit. New Phytol 111:283–291
Lüttge U, Büdel B, Ball E, Strube F, Weber P (1995a) Photosynthesis of terrestrial cyanobacteria under light and desiccation stress as expressed by chlorophyll fluorescence and gas exchange. J Exp Bot 46:309–319
Lüttge U, Ratajczak R, Rausch T, Rockel B (1995b) Stress responses of tonoplast proteins: an example for molecular ecophysiology and the search for eco-enzymes. Acta Bot Neerl 44:343–362
Lüttge U, Grams TEE, Hechler B, Blasius B, Beck F (1996) Frequency resonances of the circadian rhythm of CAM under external temperature rhythms of varied period lengths in continuous light. Bot Acta 109:422–426
Lüttge U, Pfeifer T, Fischer-Schliebs E, Ratajczak R (2000) The role of vacuolar malate-transport capacity in crassulacean acid metabolism and nitrate nutrition. Higher malate-transport capacity in the ice plant after crassulacean acid metabolism-induction and in tobacco under nitrate nutrition. Plant Physiol 124:1335–1347
Lüttge U, Fetene M, Liebig M, Rascher U, Beck E (2001) Ecophysiology of niche occupation by two giant rosette plants, *Lobelia gibberoa* Hemsl and *Solanecio gigas* (Vatke) C. Jeffrey, in an afromontane forest valley. Ann Bot 88:267–278

Lüttge U, Berg A, Fetene M, Nauke P, Peter D, Beck E (2003) Comparative characterization of photosynthetic performance and water relations of native trees and exotic plantation trees in an Ethiopian forest. Trees 17:40–50

Lüttge U, Duarte HM, Scarano FR, de Mattos EA, Cavalin PO, Franco AC, Fernandes GW (2007) Physiological ecology of photosynthesis of five sympatric species of Velloziaceae in the rupestrian fields of Serra do Cipó, Minas Gerais, Brazil. Flora 202:637–646

Lüttge U, Meirelles ST, de Mattos EA (2008) Strong quenching of chlorophyll fluorescence in the desiccated state in three poikilohydric and homoiochlorophyllous moss species indicates photo-oxidative protection on highly light-exposed rocks of a tropical inselberg. J Plant Physiol 165:172–181

Lüttge U, Kluge M, Thiel G (2010) Botanik. Die umfassende Biologie der Pflanzen. Wiley-VCH, Weinheim

Lüttge U, Garbin ML, Scarano FR (2012) Evo–devo–eco and ecological stem species: potential repair systems in the planetary biosphere crisis. Prog Bot 74:191–212

Lüttge U, Scarano FR, de Mattos EA, Franco AC, Broetto F, Dias ATC, Duarte HM, Uehlein N, Wendt T (2015) Does ecophysiological behaviour explain habitat occupation of sympatric *Clusia* species in a Brazilian Atlantic rainforest? Trees 29:1973–1988

MacRobbie EAC (1965) The nature of the coupling between light energy and active ion transport in *Nitella translucens*. Biochim Biophys Acta 94:64–73

Maddess T, Rascher U, Siebke K, Lüttge U, Osmond CB (2002) Definition and evaluation of the spatio-temporal variations in chlorophyll fluorescence during the phases of CAM and during endogenous rhythms in continuous light, in thick leaves of *Kalanchoë daigremontiana*. Plant Biol 4:446–455

Manson JS, Ottersatter MC, Thomson JD (2010) Consumption of a nectar alkaloid reduces pathogen load in bumble bees. Oecologia 162:81–89

Mariaux J-B, Becker A, Kemna I, Ratajczak R, Fischer-Schliebs E, Kramer D, Lüttge U, Marigo G (1994) Visualization by freeze-fracture electron microscopy of intramembraneous particles corresponding to the tonoplast H^+-pyrophosphatase and H^+-ATPase of *Kalanchoë daigremontiana* Hamet et Perrier de la Bâthie. Bot Acta 107:321–327

Mariaux J-B, Fischer-Schliebs E, Lüttge U, Ratajczak R (1997) Dynamics of activity and structure of the tonoplast V-type H^+-ATPase in plants with different expression of CAM and in C_3 plant under salt stress. Protoplasma 196:181–189

Marquardt-Jarczyk G, Lüttge U (1990) PP_iase-activated ATP-dependent H^+ transport at the tonoplast of mesophyll cells of the CAM plant *Kalanchoë daigremontiana*. Bot Acta 103:203–213

Matile P (1982) Vacuoles come of age. Physiol Vég 20:303–310

de Mattos EA, Grams TEE, Ball E, Franco AC, Haag-Kerwer A, Herzog B, Scarano FR, Lüttge U (1997) Diurnal patterns of chlorophyll a fluorescence and stomatal conductance in species of two types of coastal tree vegetation in southeastern Brazil. Trees 11:363–369

Matyssek R, Lüttge U (2013) Gaia: the planet holobiont. In: Matyssek R, Lüttge U, Rennenberg H (eds) The alternatives growth and defense: resource allocation at multiple scales in plants. Nova Acta Leopold NF 114(391): 325–344

Medina E, Cram WJ, Lee HSJ, Lüttge U, Popp M, Smith JAC, Diaz M (1989) Ecophysiology of xerophytic and halophytic vegetation of a coastal alluvial plain in northern Venezuela. I. Site description and plant communities. New Phytol 111:233–243

Mikosch M, Kumari N, Sharma T, Sharma V, Gessler A, Fischer-Schliebs E, Lüttge U (2012) Plasticity of photosynthetic performance of the Indian tree *Butea monosperma* TAUB. at three sites with different microclimates. Photos Res 113:287–295

Miszalski Z, Kornas A, Gawronska K, Ślesak I, Niewiadomska E, Kruk J, Christian AL, Fischer-Schliebs E, Krisch R, Lüttge U (2007) Superoxide dismutase activity in C_3 and C_3/CAM intermediate species of *Clusia*. Biol Plant 51:86–92

Miszalski Z, Kornas A, Rozpądek P, Fischer-Schliebs E, Lüttge U (2013) Independent fluctuations of malate and citrate in the CAM species *Clusia hilariana* Schltdl. under low light and high light in relation to photoprotection. J Plant Physiol 170:453–458

Münch E (1930) Die Stoffbewegungen in der Pflanze. Gutstav Fischer, Jena

Neff R, Blasius B, Beck F, Lüttge U (1998) Thermodynamics and energetics of the tonoplast membrane operating as a hysteresis switch in an oscillatory model of crassulacean acid metabolism. J Membr Biol 165:37–43

Neger FW (1912) Spaltöffnungsschluss und künstliche Turgorsteigerung. Ber Dt Bot Ges 30:179–194

Neger FW (1918) Die Wegsamkeit der Laubblätter für Gase. Flora 111:152–161

Nepi M, Stpiczyńska M (2007) Nectar resorption and translocation in *Cucurbita pepo* L. and *Platanthera chlorantha* Custer (Rchb.). Plant Biol 9:93–100

Nepi M, Soligo C, Nocentini D, Abate M, Guarnieri M, Cai G, Bini L, Puglia M, Bianchi L, Pacini E (2012) Amino acids and protein profile in floral nectar: much more than a simple reward. Flora 207:475–481

Nobel PS, Lüttge U, Heuer S, Ball E (1984) Influence of applied NaCl on crassulacean acid metabolism and ionic levels in a cactus, *Cereus validus*. Plant Physiol 75:799–803

Novacky A, Fischer E, Ullrich-Eberius CI, Lüttge U, Ullrich WR (1978a) Membrane potential changes during transport of glycine as a neutral amino acid and nitrate in *Lemna gibba* G1. FEBS Lett 88:264–267

Novacky A, Ullrich-Eberius CI, Lüttge U (1978b) Membrane potential changes during transport of hexoses in *Lemna gibba* G1. Planta 138:263–270

Novacky A, Ullrich-Eberius CI, Lüttge U (1980) pH and membrane-potential changes during glucose uptake in *Lemna gibba* G1 and their response to light. Planta 149:321–326

Osmond CB, Lüttge U, West KR, Pallaghy CK, Shacher-Hill B (1969) Ion absorption in *Atriplex* leaf tissue. II. Secretion of ions to epidermal bladders. Aust J Biol Sci 22:797–814

Osmond CB, Daley PF, Badger MR, Lüttge U (1998) Chlorophyll fluorescence quenching during photosynthetic induction in leaves of *Abutilon striatum* Dicks. Infected with *Abutilon* mosaic virus, observed with a field-portable imaging system. Bot Acta 111:390–397

Osmond CB, Kramer D, Lüttge U (1999) Reversible water stress-induced non-uniform chlorophyll fluorescence quenching in wilting leaves of *Potentilla reptans* may not be due to patchy stomatal responses. Plant Biol 1:618–624

Overton E (1899) Über die allgemeinen osmotischen Eigenschaften der Zelle, ihre vermutlichen Ursachen und ihre Bedeutung für die Physiologie. Vierteljahrschr Naturf Ges Zürich 44:88–135

Pallaghy CK, Lüttge U (1970) Light-induced H^+-ion fluxes and bioelectric phenomena in mesophyll cells of *Atriplex spongiosa*. Z Pflanzenphysiol 62:417–425

Pallaghy CK, Lüttge U, von Willert K (1970) Cytoplasmic compartmentation and parallel pathways of ion uptake in plant root cells. Z Pflanzenphysiol 62:51–57

Pieruschka R, Schurr U, Jahnke S (2005) Lateral gas diffusion inside leaves. J Exp Bot 56:857–864

Pimentel MCP, Barros MJ, Cirne P, de Mattos EA, Oliveira RC, Pereira MCA, Scarano FR, Zaluar HLT, Araujo DSD (2007) Spatial variation in the structural and floristic composition of "restinga" vegetation in southeastern Brazil. Rev Bras Bot 30:543–551

Pitman MG (1963) The determination of the salt relations of the cytoplasmic phase in cells of beetroot tissue. Aust J Biol Sci 16:647–668

Pitman MG, Lüttge U, Läuchli A, Ball E (1974a) Effect of previous water stress on ion uptake and transport in barley seedlings. Aust J Plant Physiol 1:377–385

Pitman MG, Lüttge U, Läuchli A, Ball E (1974b) Action of abscisic acid on ion transport as affected by root temperature and nutrient status. J Exp Bot 25:147–155

Pittendrigh CS (1948) The bromeliad-*Anopheles*-malaria complex in Trinidad. I. The bromeliad flora. Evolution 2:58–89

Plant Cell Environment (1986) Special issue: bromeliad ecophysiology. Plant Cell Environ 9:359–419

Popp M, Kramer D, Lee H, Diaz M, Ziegler H, Lüttge U (1987) Crassulacean acid metabolism in tropical dicotyledonous trees of the genus *Clusia*. Trees 1:238–247
Rascher U, Lüttge U (2002) High-resolution chlorophyll fluorescence imaging serves as a non-invasive indicator to monitor the spatio-temporal variations of metabolism during the day-night cycle and during the endogenous rhythm in continuous light in the CAM plant *Kalanchoë daigremontiana*. Plant Biol 4:671–681
Rascher U, Blasius B, Beck F, Lüttge U (1998) Temperature profiles for the expression of endogenous rhythmicity and arrhythmicity of CO_2 exchange in the CAM plant *Kalanchoë daigremontiana* can be shifted by slow temperature changes. Planta 207:76–82
Rascher U, Liebig M, Lüttge U (2000) Evaluation of instant light-response curves of chlorophyll fluorescence parameters obtained with a portable chlorophyll fluorometer on site in the field. Plant Cell Environ 23:1397–1405
Rascher U, Hütt M-T, Siebke K, Osmond CB, Beck F, Lüttge U (2001) Spatiotemporal variation of metabolism in a plant circadian rhythm: the biological clock as an assembly of coupled individual oscillators. Proc Natl Acad Sci USA 98:11801–11805
Rascher U, Lakatos M, Büdel B, Lüttge U (2003) Photosynthetic field capacity of cyanobacteria of a tropical inselberg of the Guiana highlands. Eur J Phycol 38:247–256
Ratajczak R, Kemna I, Lüttge U (1994a) Characteristics, partial purification and reconstitution of the vacuolar malate transporter of the CAM plant *Kalanchoë daigremontiana* Hamet et Perrier de la Bâthie. Planta 195:226–236
Ratajczak R, Richter J, Lüttge U (1994b) Adaptation of the tonoplast V-type H^+-ATPase of *Mesembryanthemum crystallinum* to salt stress, C_3-CAM transition and plant age. Plant Cell Environ 17:1101–1112
Ratajczak R, Hille A, Mariaux J-B, Lüttge U (1995) Quantitative stress responses of the V_0V_1-ATPase of higher plants detected by immuno-electron microscopy. Bot Acta 108:505–513
Ratajczak R, Pfeifer T, Drobny M, Schnölzer M, Lüttge U (2003) Molecular evidence for the occurrence of H^+-transporting V-ATPase subunit D and two different forms of subunit E in leaves of the obligate CAM species *Kalanchoë daigremontiana* Hamet et Perrier. Biol Plant 46:13–21
Raven JA (1967) Light stimulation of active transport in *Hydrodictyon africanum*. J Gen Physiol 50:1627–1640
Rea PA (1983) The dynamics of H^+ efflux from the trap lobes of *Dionaea muscipula* Ellis (Venus' flytrap). Plant Cell Environ 6:125–134
Rea PA, Joel DM, Juniper BE (1983) Secretion and redistribution of chloride in the digestive glands of *Dionaea muscipula* Ellis (Venus' flytrap) upon secretion stimulation. New Phytol 94:359–366
Robertson RN (1968) Protons, electrons, phosphorylation and active transport. Cambridge University Press, Cambridge
Rockel B, Ratajczak R, Becker A, Lüttge U (1994) Changed densities and diameters of intramembrane tonoplast particles of *Mesembryanthemum crystallinum* in correlation with NaCl-induced CAM. J Plant Physiol 143:318–324
Rockel B, Lüttge U, Ratajczak R (1998) Changes of message amount of V-ATPase subunits during salt-stress induced C_3-CAM transition in *Mesembryanthemum crystallinum*. Plant Physiol Biochem 36:567–573
Rona J-P, Pitman MG, Lüttge U, Ball E (1980) Electrochemical data on compartmentation into cell wall, cytoplasm, and vacuole of leaf cells in the CAM genus *Kalanchoë*. J Membr Biol 57:25–35
Ruhland W (1912) Studien über die Aufnahme von Kolloiden durch die pflanzliche Plasmahaut. Jb Wiss Bot 51:376–431
Ruhland W, Hoffmann C (1925) Die Permeabilität von *Beggiatoa mirabilis*. Planta 1:1–83
Rygol J, Lüttge U (1983) Water-relation parameters of giant and normal cells of *Capsicum annuum* pericarp. Plant Cell Environ 6:545–553

Rygol J, Büchner K-H, Winter K, Zimmermann U (1986) Day/night variations in turgor pressure in individual cells of *Mesembryanthemum crystallinum* L. Oecologia 69:171–175
Rygol J, Zimmermann U, Balling A (1989) Water relations of individual cells of *Mesembryanthemum crystallinum* plants grown at low and high salinity. J Membr Biol 107:203–212
Sawidis T (1991) A histochemical study of nectaries of *Hibiscus rosa-sinensis*. J Exp Bot 42:1477–1487
Scarano FR, de Mattos EA, Franco AC, Herzog B, Ball E, Grams TEE, Mantovani A, Barreto S, Haag-Kerwer A, Lüttge U (1999) Habitat segregation of C_3 and CAM species of *Nidularium* (Bromeliaceae) in response to different light regimes in the understory of a swamp forest in southeastern Brazil. Flora 194:281–288
Scarano FR, Duarte HM, Ribeiro KT, Rodrigues PJFP, Barcellos EMB, Franco AC, Brulfert J, Deléens E, Lüttge U (2001) Four sites with contrasting environmental stress in southeastern Brazil: relations of species, life form diversity, and geographic distribution to ecophysiological parameters. Bot J Linn Soc 136:345–364
Scarano FR, Duarte HM, Franco AC, Gessler A, de Mattos EA, Rennenberg H, Lüttge U (2005a) Physiological synecology of tree species in relation to geographic distribution and ecophysiological parameters at the Atlantic forest periphery in Brazil: an overview. Trees 19:493–496
Scarano FR, Duarte HM, Franco AC, Gessler A, de Mattos EA, Nahm M, Rennenberg H, Zaluar HTL, Lüttge U (2005b) Ecophysiology of selected tree species in different plant communities at the periphery of the Atlantic Forest of SE Brazil. I. Performance of three different species of *Clusia* in an array of plant communities. Trees 19:497–509
Schmidt-Lebhuhn AN, Schwerdtfeger M, Kessler M, Lohaus G (2007) Phylogenetic constraints vs. ecology in the nectar composition of Acanthaceae. Flora 202:62–69
Schmitt AK, Lee HSJ, Lüttge U (1988) The response of the C_3-CAM tree *Clusia rosea*, to light and water stress. I. Gas exchange characteristics. J Exp Bot 39:1581–1590
Schnepf E, Christ P (1980) Unusual transfer cells in the epithelium of the nectaries of *Asclepias curassavica* L. Protoplasma 105:135–148
Schulze E-D, Hall AE, Lange OL, Walz H (1982) A portable steady-state porometer for measuring the carbon dioxide and water vapour exchanges of leaves under natural conditions. Oecologia 53:141–145
Smith JAC, Lüttge U (1985) Day-night changes in leaf water relations associated with the rhythm of crassulacean acid metabolism in *Kalanchoë daigremontiana*. Planta 163:272–282
Smith JAC, Marigo G, Lüttge U, Ball E (1982) Adenine-nucleotide levels during crassulacean acid metabolism and the energetics of malate accumulation in *Kalanchoë tubiflora*. Plant Sci Lett 26:13–21
Smith JAC, Uribe EG, Ball E, Lüttge U (1984a) ATPase activity associated with isolated vacuoles of the crassulacean acid metabolism plant *Kalanchoë daigremontiana*. Planta 162:299–304
Smith JAC, Uribe EG, Ball E, Heuer S, Lüttge U (1984b) Characterization of the vacuolar ATPase activity of the crassulacean-acid metabolism plant *Kalanchoë daigremontiana*. Eur J Biochem 141:415–420
Smith JAC, Popp M, Lüttge U, Cram WJ, Diaz M, Griffiths H, Lee HSJ, Medina E, Schäfer C, Stimmel K-H, Thonke B (1989) Ecophysiology of xerophytic and halophytic vegetation of a coastal alluvial plain in northern Venezuela. VI. Water relations and gas exchange of mangroves. New Phytol 111:293–307
Souza GM, Lüttge U (2014) Stability as a phenomenon emergent from plasticity—complexity—diversity in ecophysiology. Progr Bot 76:211–239
Souza GM, Bertolli SC, Lüttge U (2016) Hierarchy and information in a system approach to plant biology: explaining the irreducibility in plant eco-physiology. Prog Bot 77:167–186
Spanswick RM (1976) Symplasmic transport in tissues. Enc Plant Physiol 2B:35–56
Steckelberg R, Lüttge U, Weigl J (1967) Reinigung der Proteinase aus *Nepenthes*-Kannensaft. Planta 76:238–241
Steiger S, Pfeifer T, Ratajczak R, Martinoia E, Lüttge U (1997) The vacuolar malate transporter of *Kalanchoë daigremontiana*: A 32 kDa polypeptide? J Plant Physiol 151:137–141

Steudle E (2011) Hydraulic architecture of vascular plants. In: Lüttge U, Beck E, Bartels D (eds) Plant desiccation tolerance, vol 215, Ecological studies. Springer, Berlin, pp 185–207

Steudle E, Zimmermann U, Lüttge U (1975) Water relations of the epidermal bladder cells of the halophytic species *Mesembryanthemum crystallinum*: direct measurements of hydrostatic pressure and hydraulic conductivity. Planta 126:229–246

Steudle E, Zimmermann U, Lüttge U (1977) Effect of turgor pressure and cell size on the wall elasticity of plant cells. Plant Physiol 59:285–289

Struve I, Lüttge U (1987) Characteristics of $MgATP^{2-}$-dependent electrogenic proton transport in tonoplast vesicles of the facultative crassulacean-acid-metabolism plant *Mesembryanthemum crystallinum* L. Planta 170:111–120

Struve I, Lüttge U (1988) Biochemical and immunological properties of solubilized tonoplast ATPase of the facultative CAM plant *Mesembryanthemum crystallinum* in the C_3- and CAM state. Bot Acta 101:39–44

Struve I, Weber A, Lüttge U, Ball E, Smith JAC (1985) Increased vacuolar ATPase activity correlated with CAM induction in *Mesembryanthemum crystallinum* and *Kalanchoë blossfeldiana* cv. Tom Thumb. J Plant Physiol 117:451–468

Sutcliffe JF (1962) Mineral absorption in plants. Pergamon, Oxford

Takahashi K, Tanji M (2007) Variation in the content and isozymic composition of nepenthesin in the pitcher fluids among *Nepenthes* species. Carnivorous Plant Newsl 36:73–76

Terashima I (1992) Anatomy of non-uniform leaf photosynthesis. Photos Res 31:195–212

Thellier M, Lüttge U (2013) Plant memory: a tentative model. Plant Biol 15:1–12

Ting IP, Lord EM, Lda Sternberg S, DeNiro MJ (1985) Crassulacean acid metabolism in the strangler *Clusia rosea* Jacq. Science 229:969–971

Tinoco Ojanguren C, Vázquez-Yanes C (1983) Especies CAM en la selva húmeda tropical de los tuxtlas, Veracruz. Bol Soc Mex 45:150–153

Torii K, Laties GG (1966) Mechanisms of ion uptake in relation to vacuolation of corn roots. Plant Physiol 41:863–870

Trewavas A (2003) Aspects of plant intelligence. Ann Bot 92:1–20

Ullrich-Eberius CI, Lüttge U, Neher L (1976) Energy relations of phosphate uptake and distribution in barley leaf slices as affected by cutting and adaptive ageing. Z Pflanzenphysiol 79:347–359

Ullrich-Eberius CI, Novacky A, Fischer E, Lüttge U (1981) Relationship between energy-dependent phosphate uptake and the electrical membrane potential in *Lemna gibba* G1. Plant Physiol 67:797–801

Vaasen A, Scarano FR, Hampp R (2007) Population biology of different *Clusia* species in the state of Rio de Janeiro. In: Lüttge U (ed) *Clusia*. A woody neotropical genus of remarkable plasticity and diversity, vol 194, Ecological studies. Springer, Berlin, pp 117–127

Vanselow KH, Dau H, Hansen UP (1988) Indication of transthylakoid proton-fluxes in *Aegopodium podagraria* L. by light-induced changes of plasmalemma potential, chlorophyll fluorescence and light-scattering. Planta 176:351–361

Vanselow KH, Kolbowski Y, Hansen UP (1989) Further evidence for the relationship between light-induced changes of plasmalemma transport and transthylakoid proton uptake. J Exp Bot 40:239–245

Vassilyev AE (2010) On the mechanism of nectar secretion revisited. Ann Bot 105:349–354

von Weizsäcker V (1954) Am Anfang schuf Gott Himmel und Erde, 5th edn. Vandenhoeck and Ruprecht, Göttingen

Wang B, Lüttge U, Ratajczak R (2001) Effects of salt treatment and osmotic stress on V-ATPase and V-PPase in leaves of the halophyte *Suaeda salsa*. J Exp Bot 52:2355–2365

Wartiovaara V, Collander T (1960) Permeabilitätstheorien. Protoplasmatologia, vol II, C8d. Springer, Wien

Weber A (2010a) Zwischen Biomaschine, Artenkollaps und Wachstumswahn: was ist der Irrtum in unserem Bild vom Leben? In:Wobus AM,Wobus U, Parthier B (eds) Der Begriff der Natur. Wandlungen unseres Naturversta¨ndnisses und seine Folge. Nova Acta Leopoldina NF 109 (376):25–43

Weber A (2010b) Biokapital. Die Versöhnung von Ökonomie, Natur und Menschlichkeit. Berliner Taschenbuchverlag, Berlin

Weigl J, Lüttge U (1962) Mikroautoradiographische Untersuchungen über die Aufnahme von $^{35}SO_4^{2-}$ durch die Wurzeln von *Zea mays* L. Die Funktion der primären Endodermis. Planta 59:15–28

Winter K (1973a) CO_2-Fixierungsreaktionen bei der Salzpflanze *Mesembrynthemum crystallinum* unter variierten Außenbedingungen. Planta 114:75–85

Winter K (1973b) Zum Problem der Ausbildung des Crassulaceensäurestoffwechsels bei *Mesembryanthemum crystallinum* unter NaCl-Einfluß. Planta 109:135–145

Winter K (1975) Die Rolle des Crassulaceen-Säurestoffwechsels als biochemische Grundlage zur Anpassung von Halophyten an Standorte hoher Salinität. PhD-thesis, Darmstadt

Winter K, Lüttge U (1979) C_3-Photosynthese und Crassulaceen-Säurestoffwechsel bei *Mesembryanthemum crystallinum* L. Ber Dt Bot Ges 92:117–132

Winter K, von Willert DJ (1972) NaCl-induzierter Crassulaceensäurestoffwechsel bei *Mesembryanthemum crystallinum*. Z Pflanzenphysiol 67:166–170

Wyka TP, Lüttge U (2003) Contribution of C_3 carboxylation to the circadian rhythm of carbon dioxide uptake in a crassulacean acid metabolism plant *Kalanchoë daigremontiana*. J Exp Bot 54:1471–1479

Wyka TP, Bohn A, Duarte HM, Kaiser F, Lüttge U (2004) Perturbations of malate accumulation and the endogenous rhythms of gas exchange in the crassulacean acid metabolism plant *Kalanchoë daigremontiana*: testing the tonoplast oscillator model. Planta 219:705–713

Wyka TP, Duarte HM, Lüttge U (2005) Redundancy of stomatal control for the circadian photosynthesis rhythm in *Kalanchoë daigremontiana* Hamet et Perr. Plant Biol 7:176–181

Zhigang A, Löw R, Rausch T, Lüttge U, Ratajczak R (1996) The 32 kDa tonoplast polypeptide D_i associated with the V-type H^+-ATPase of *Mesembryanthemum crystallinum* L. in the CAM state: a proteolytically processed subunit B? FEBS Lett 398:314–318

Ziegler H (1956) Untersuchungen über die Leitung und Sekretion der Assimilate. Planta 47:447–500

Ziegler H, Lüttge U (1959) Über die Resorption von ^{14}C-Glutaminsäure durch sezernierende Nektarien. Naturwissenschaften 46:176–177

Ziegler H, Lüttge U (1967) Die Salzdrüsen von *Limonium vulgare*. II. Mitteilung. Die Lokalisierung des Chlorids. Planta 74:1–17

Ziegler H, Lüttge U (1998) Carbon isotope discrimination in cyanobacteria of inselbergs and soils of savannas in the neotropics. Bot Acta 111:212–215

Ziegler H, Weigl J, Lüttge U (1963) Mikroautoradiographischer Nachweis der Wanderung von $^{35}SO_4^{2-}$ durch die Tertiärendodermis der *Iris*-Wurzel. Protoplasma 56:362–370

Ziegler H, Lüttge U, Lüttge U (1964) Die wasserlöslichen Vitamine des Nektars. Flora 154:215–222

Ziegler H, Weber J, Lüttge U (2009) Thermal dissipation probe measurements of sap flow in the xylem of trees documenting dynamic relations to variable transpiration given by instantaneous weather changes and the activities of a mistletoe parasite. Trees 23:441–450

Part II
Revisiting Principles of Plant Life – Integration of Whole-Plant Functionality Under Ecological and Evolutionary Perspective

The conceptual perception of plant life has gained new impetus towards functionally understanding the inextricable "eco–evo" interrelationships (Müller 2007; Gilbert and Epel 2009). For Higher Plants, which are sessile and cannot escape stress, survival at their sites is a big challenge. They need high plasticity in effective stress response, higher than in mobile animals. Indeed, response plasticity turns out in plants to represent a functional regularity on its own (Matyssek et al. 2012; Souza and Lüttge 2014). High plasticity inherently requires high degrees of internal functional integration of the plant as a whole, i.e. multi-scale systemic ability of control. The control is challenged externally and determined by biotic interactions, i.e. through competition, parasitism, and mutualistic relationships. Multi-organismic associations prove to drive resource allocation and, as a consequence, the diverse functions of plants (Matyssek et al. 2012).

Evolution teaches us that plants master that challenge at their firm rooting sites quite successfully. Plant science has been caught in a tradition that tended to address plants (i) under a more "static" and "modular" perspective than animals regarding internal functional organization, and (ii) in the field of plant (eco-) physiological research, in aut- rather than syn-ecological terms. The aut-ecological emphasis is owed to the analytical research with its reductionist tendency of reducing degrees of freedom while neglecting the systemic synthesis of evidence within and across spatiotemporal scales. To take the latter step, one needs to comprehensively embrace whole-plant functioning in ecologically meaningful ways, both *per se* and as part of environmental contexts. Overcoming deficits of (i) and (ii) requires merging reductionist and systemic approaches. The new impetus of conceptual perception of plant life has gained momentum recently in breaking (i) and (ii) related restrictions in understanding whole-plant performance under ecological and evolutionary perspectives. This present volume 77 of "Progress in Botany" gathers ten contributions reflecting this recent momentum.

Regarding (i), the contributions by Lüttge (2016a) and Lüttge and Thellier (2016) each elucidate the highly integrated functional interrelationships at the level of whole plants, enabling for rapid, variant-rich, and highly differentiated performance within and responsiveness to the environment. One central means of functional whole-plant integration are transport processes at various organizational levels within plants from membranes and cells to organs in the whole plant (Lüttge 2016a). The vitally essential water fluxes at membranes, cells, and tissues and the pathways of long-distance transport of not only the xylem but also the phloem are integrated in the functional network of the phenomenon of root pressure (Singh 2016). These transports carry the distribution of nutrients and assimilates within plants and with that the resource-supply-based information for whole-plant performance. For fine-tuned regulation, the transports are the intrinsic platform of phytohormonal plant-level signalling, beyond electrical and hydraulic signalling.

How is this plethora of signalling evaluated and "distilled" into highly coordinated whole-plant response, as plants do not have a nervous system specialized for signal pulsing? Neither a central processing unit of information exists. Instead of a "central processing unit", a diffuse information-processing network of cells may serve whole-plant integration. For example, the systemic regulation of photosynthesis in response to light and CO_2 operates with the integration of chloroplasts and green cells at the whole-plant level (Matsuda and Murakami 2016). Hence, the evolutionary approach would contrast with that of animals but may be similarly effective, perhaps differentially "tuned" during evolution to the particular ecological needs of the life form of plants. Doubtlessly, information storage exists in plants, as demonstrated by Lüttge and Thellier (2016), and the highly specific and differential memory of internal and external signalling is controlled by multiple metabolic mechanisms, including biological clocks that rely on regular matching with phenological and ontogenetic stimuli. Such kind of memory remarkably is systemic, mirroring the evolutionary approach in plants of diffuse information processing.

Regarding (ii), evidence suggests that individual-centred views of aut-ecology fall short if trying to unveil ecological and evolutionary mechanisms crucial in promoting adaptability and niching as prerequisites of genotype evolution and persistence. This has been shown in research on biological invasions (Heger et al. 2013), in restoration ecology (Zaplata et al. 2013), and during biodiversity experiments (Allan et al. 2011; Scherber et al. 2010). Rather, evolutionary significance arises from multi-organismic associations as the ecologically relevant entities which apparently possess emergent functional features that are conducive to niche formation as an intrinsic aspect of fitness. Such biological systems are hierarchically organized. Their irreducibility is explained by the flow of information across the various hierarchical levels, where we must abandon the conception of a one-way hierarchical order and realize that hierarchy works not only top down but also bottom up (Souza et al. 2016).

Multi-organismic associations comprising a highly evolved host and microorganisms (MOs) have been termed "holobionts"; however, such notion is much more introduced and functionally understood by means of animal (including humans) than plant systems (zu Castell et al. 2016). Such holobionts are represented through their "hologenome", responding as a functional entity to environmental impact. The ecological and evolutionary strength of such systems is anchored in the hologenome responding, via MOs, rapidly and effectively to stress as compensating for the sluggishness in host responsiveness. The holobiontic performance is concluded to promote acclimation and adaptation to changing environmental conditions, with the holobiont being the actual platform of selection and (co-) evolution. Environmental information can be stored and inherited in holobionts not only in the conventional genetic way, but also through structural and functional changes, e.g. in the multi-organismic assembly.

Holobiontic principles in plant systems begin to gain attention in ecological research (Vandenkoornhuyse et al. 2015) and provide the functional grounds of plasticity in stress response, highlighted above as a crucial feature of plant persistence.

Souza and Lüttge (2014, last vol.) elucidate functional stability at the plant and ecosystem level as a phenomenon that emerges—as a new synergistic quality of system functionality—from the holobiont plasticity in combination with the complexity and diversity of the underlying ecophysiological responsiveness. Lüttge (2016b) demonstrates such principles as the functional grounds of the natural self-management of ecosystems and as guidance for stand management in agronomy and forestry. Such considerations lead us to spatiotemporal scales hierarchically higher than that of holobionts *sensu stricto*.

The question arises, if holobiont-like principles are realized in scale-invariant ways. If so, the holobiont *sensu stricto* would represent just one variant of holobiont-like systems (HLS) and its principles of biotic interactions would represent an intrinsic and generic characteristic of any biological system. Such considerations are further propagated by zu Castell et al. (2016). Respective consistencies appear to be keys to understanding the systems' self-organization, pseudo-steady states, and self-maintenance, perhaps mediated through particular settings of driving forces, feedback or feed-forward mechanisms, and organismic control components. Clearly, we have left the scale now of holobionts *sensu stricto*, and we may be inclined of thinking in terms of the conventional hierarchically vertical spatio-temporal scaling of interaction principles (Souza et al. 2016). However, the hypothesized scale invariance of HLS principles also—and in particular—demands for functional cross-linking within spatiotemporal scales. The demand reflects adaptive cycles that are nested one within the other within scales and extending beyond.

For such kind of interrelationships, the term "panarchy" was coined by Holling (2001) and Gunderson and Holling (2002). The "panarchical" view becomes compelling in recognizing HLS principles and potential scale invariance to complement and ultimately replace the hierarchical perspective (zu Castell et al. 2016). To the extent that such overarching principles may substantiate through upcoming research, a quality of evidence would become available casting new bridges in the understanding of biotic interaction in the interrelated "eco–evo" research field, i.e. unifying explorations of ecology and evolution (Müller 2007; Lüttge et al. 2012). Hence, the new impetus to plant science may turn into a fundamental driver of theory building about plant life.

A prerequisite towards such goal is a new understanding, however, of "systems biology" for becoming "syn-ecological" ecosystem biology. For this to be achieved, "systems biology" needs to embed molecular biology into ecophysiological, multi-organismic, and ecosystem-level networking. This means that genomic biology must also advance to exploring different aspects of plant biology, such as elaborating the role of genetic diversity (Larrañaga and Hormaza 2016) and its functional analysis (Fladung 2016) in relation to evolutionary aspects, where the evolution of flowering has eminent eco–evo implications (Lucas-Reina et al. 2016). We wish the addressed collection of contributions to this present volume 77 of "Progress in Botany" to provide a stimulating input towards reaching the outlined goal.

References

Allan E, Weisser W, Weigelt A, et al (2011) More diverse plant communities have higher functioning over time due to turnover in complementary dominant species. Proc Nat Acad Sci USA 108:17034–17039

Fladung M (2016) Transposon activation tagging in plants for gene function discovery. Progr Bot 77:265–289

Gilbert SF, Epel D (2009) Ecological development biology: integrating epigenetics, medicine and evolution. Sinauer Ass., Palgrave-MacMillan, New York

Gunderson L, Holling CS (2002) Panarchy: understanding transformations in human and natural systems. Island Press, Washington (DC)

Heger T, Pahl AT, Botta-Dukat Z, Gherardi F, Hoppe C, Hoste I, Jax K, Lindström L, Boets P, Haider S, Kollmann J, Wittmann MJ, Jeschke JM (2013) Conceptual frameworks and methods for advancing invasion-ecology. AMBIO 42:527–540

Holling CS (2001) Understanding the complexity of economic, ecological, and social systems. Ecosystems 4:390–405

Larrañaga N, Hormaza I (2016) Advances in genetic diversity analysis in fruit tree crops. Progr Bot 77:245–264

Lucas-Reina E, Ortíz-Marchena MI, Romero-Campero FJ, Calonje M, Romero JM and Valverde F (2016) Evolution of the flowering pathways. Progr Bot 77:291–329

Lüttge U (2016a) Transport processes—The key integrators in plant biology. Progr Bot 77:3–65

Lüttge U (2016b) Plants shape the terrestrial environment on Earth: Challenges of management for sustainability. Progr Bot 77:187–217

Lüttge U, Thellier M (2016) Roles of memory and the circadian clock in the ecophysiological performance of plants. Progr Bot 77:73–104

Lüttge U, Garbin ML, Scarano FR (2012) Evo-devo-eco and ecological stem species: Potential repair systems in the planetary biosphere crisis. Progr Bot 74:191–212

Matsuda R, Murakami K (2016) Light- and CO_2-dependent systemic regulation of photosynthesis. Progr Bot 77:151–166

Matyssek R, Gayler S, zu Castell W, Oßwald W, Ernst D, Pretzsch H, Schnyder H, Munch JC (2012) Predictability of plant resource allocation: New theory needed? In Matyssek R, Schnyder H, Oßwald W, Ernst D, Munch JC, Pretzsch H (eds) Growth and defence in plants—Resource allocation at multiple scales. Ecol Studies 220, Springer, Heidelberg, pp 433–449

Müller GB (2007) Evo-devo extending the evolutionary synthesis. Nat Rev Genet 8:939–949

Scherber C, Eisenhauer N, Weisser WW, et al (2010) Bottom-up effects of plant diversity on multitrophic interactions in a biodiversity experiment. Nature 468:553–556

Singh S (2016) Root pressure: getting to the root of pressure. Progr Bot 77:105–150

Souza GM, Lüttge U (2014) Stability as a phenomenon emergent from plasticity—complexity—diversity in ecophysiology. Progr Bot 76:211–239

Souza GM, Bertolli SC, Lüttge U (2016) Hierarchy and information in a system approach to plant biology: explaining the irreducibility in plant ecophysiology. Progr Bot 77:167–186
Vandenkoornhuyse P, Quaiser A, Duhamel M, Le Van A, Dufresne A (2015) The importance of the microbiome of the plant holobiont. New Phytol doi: 10.1111/nph.13312
Zaplata MK, Winter S, Fischer A, et al. (2013) Species-driven phases and increasing structure in early-successional plant communities. Am Nat 181:E17–E27
zu Castell W, Fleischmann F, Heger T, Matyssek R (2016) Shaping theoretic foundations of holobiont-like systems. Progr Bot 77:219–244

Roles of Memory and Circadian Clock in the Ecophysiological Performance of Plants

Ulrich Lüttge and Michel Thellier

Contents

Abstract Adaptation and acclimation of metabolism and development to environmental conditions at the site of rooting requires nonmobile plants to memorize information introduced by external signals. These act at various spatiotemporal levels of structure and function and ecophysiological performance. There are different types of memory, among which are priming memory, store/recall memory (STO/RCL), where both the storage and the recall function as well as their combination have ecophysiological significance, and epigenetic memory. Timing is

U. Lüttge (✉)
Department of Biology, Technical University of Darmstadt, Schnittspahnstr. 3-5, 64287 Darmstadt, Germany
e-mail: luettge@bio.tu-darmstadt.de

M. Thellier
Emeritus of the University of Rouen, 29 bis rue de la Chézine, 44100 Nantes, France

U. Lüttge et al. (eds.), *Progress in Botany*, Progress in Botany 77,
DOI 10.1007/978-3-319-25688-7_2

important. Therefore, ultradian, circadian and annual rhythms are underlying memory functions, where the circadian clock may represent a prominent component. Memorization associated with adaptation and acclimation needs implementation of memory as backbone. A plethora of ecological impacts require memory, some of which will be exemplified and critically examined, namely, molecular aspects of membrane transport, fitness, photosynthesis, osmotic stress and salinity, pollution events and priming by volatile organic compounds and by vibrations. Memory is not an occasional episode but a fundamental property of general importance in the life of plants.

1 Introduction

Memory is not a straightforward concept. In the basic meaning, the term "memory" applies to animals (especially higher animals and humans). It is defined as an "ability to retain and recall information, ideas, images and thoughts" (Sinclair et al. 1987), and it is based on the activity and interactions of neurones, especially in the central nervous system (Dudai 2004; Lesburguères et al. 2011). However, the meaning has now broadened in two ways: (1) it is employed not only to animals but also to practically any sort of living organism including plants (Thellier et al. 1982) and prokaryotes (Thellier and Lüttge 2013). (2) It is not necessarily based on neuronal activity so that one may speak of genetic and epigenetic memories in living beings, the memory of an instrument (such as a computer or a pocket calculator) or even the memory of anything involving processes with a hysteretic behaviour (see the chapter "Hysteresis" in Wikipedia, the free encyclopaedia). The semantic difficulty is that one single word, "memory", stands for all these different aspects. To elude this difficulty, one might say "memory sensu stricto", i.e. when routinely speaking, for instance, of human memory, and "memory sensu lato", i.e. for memories of any kind within more recent contexts of accepted understanding. However, such terminology would be quite cumbersome, so that we shall rather continue to use "memory", as everybody does, although staying aware of the many different meanings this one notion may have.

A tentative model of the network modules incorporated in the system of operations of memory is shown in Fig. 1. Its various features shall be revealed as we go along in this assay, where we aim at unravelling the role of memory in the autecological and synecological performance of plants. Autecological performance is determined by adaptation and/or acclimation of given species or individual plants to their environment, while synecological behaviour is characterized by the interactions of species or individuals among each other and with the environment at the community level. In two sections we shall first consider the different types of memory (Sect. 2) and relations to biorhythmicity and the biological clock (Sect. 3) with an ecological perspective in mind. Then, we shall assess concrete ecological functions which require memorization and, hence, implementation of

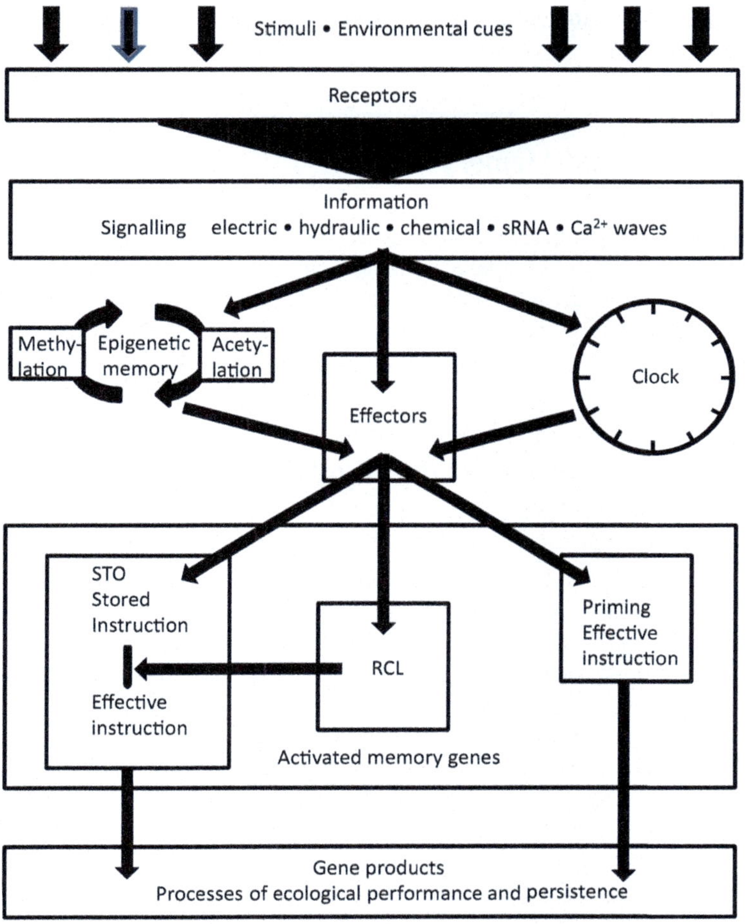

Fig. 1 Tentative model of the network of modules incorporated in the system of operations of memory for ecophysiological performance of plants. Stimuli of environmental cues are received by molecular receptors, and their information is translated into signalling of various forms, such as electric, hydraulic, phytohormonal/chemical and small RNA (sRNA) signals and Ca^{2+} waves. These signals generate molecular effectors either directly or indirectly via the epigenetic memory or the biological clock. The effectors activate memory genes of priming and of store (STO) and recall (RCL) functions, where the STO and RCL boxes are independent of each other. With the operation of the activated priming genes, the instruction inherent in the original stimuli is effective directly. With the operation of the activated STO genes, the instruction is stored but becomes effective only via the operation of the activated RCL genes. The effective instruction leads to gene products of ecophysiological relevance

memory and the biological clock as supporting grounds (Sect. 4). Eventually, we shall explore the ecophysiological potential of the priming and store/recall forms of plant memory (Sect. 5). Finally, we shall conclude considering plant memory in relation to that of other organisms as an essential means for persistence in the environment.

2 Types of Plant Memory

We will only briefly summarize those aspects of memory that are mainly studied in physics or engineering (computer's memory and hysteresis). We shall develop in more detail the biological aspects with particular reference to plants (genetic and epigenetic memory, priming and store/recall forms of plant memory and the relations of memory to developmental phenological phases in plants).

2.1 Aspects of Memory Mainly Studied in Physics and Engineering

Briefly, "*a computer's memory is the part of the computer where information is stored, especially for a short time, before it is transferred to magnetic tapes or disks*" (Sinclair et al. 1987). "*Hysteresis is the dependence of the output of a system not only on its current input, but also on its history of past inputs; the dependence arises because the history affects the value of an internal state; to predict its future outputs, either its internal state or its history must be known; if a given input alternately increases and decreases, a typical mark of hysteresis is that the output forms a loop*" (see the chapter "Hysteresis" in Wikipedia, the free encyclopaedia; for a possible application to plant systems, see, for instance, Sect. 4.2.4 or Desbiez et al. 1994).

2.2 Aspects of Memory Common to All Living Beings

2.2.1 Genetic Memory

Genomes reflect the history of organisms because the genotypes are selected in evolution. In a very broad and general sense, we may therefore think of this genetic information as being a kind of memory of past events that have affected evolutionary selection. In a similar vein we may think of the genomes of sexually produced organisms, being memory of mother and father having contributed their sets of chromosomes via the gametes. Reading the genetic information for transcription followed by translation and then regulating metabolic and physiological responses could be considered in recalling the genetic information. The processes of the control of reading genetic information will lead to more specific considerations of which we shall give examples later, for instance, the activation and inactivation of operons (Sect. 4.2.1). A dynamic process in reading the genetic information is epigenetics, which will lead us to more concrete memory functions in the next section.

2.2.2 Epigenetic Memory

Epigenetic modifications of the genome are induced by internal and external signals. They can be stored and affect gene expression beyond cell cycles and even generations. Therefore, we can speak of an "epigenetic memory" (Thellier and Lüttge 2013; Kinoshita and Seki 2014). Epigenetic modifications are currently the best understood molecular mechanism of memory.

Molecular epigenetics is a system of reading the genetic information of DNA where the structure and conformation properties of chromatin are modulated by acetylation and methylation, respectively, of DNA and nucleosomal histones. In the DNA methyl or acetyl groups are attached to the cytosine groups. In the histone proteins the lysine and arginine residues are post-translationally modified (Yaish et al. 2011), i.e. by acetylation/methylation (Grunstein 1997; Zhang and Reinberg 2001), ADP-ribosylation (Tanigawa et al. 1984), glycosylation (Cervantes-Laurean et al. 1996), phosphorylation (Lo et al. 2001) and ubiquitination (Sridhar et al. 2007). Acetylation allows access of regulator molecules of gene activation or deactivation due to the larger size of the acetyl group as compared to the smaller methyl group. Methylation leads to repression of gene transcription, and the genetic information is silenced (Chinnusamy and Zhu 2009). Under the perspective of ecological performance of plants, it is important that epigenetic modifications are triggered by environmental cues (Jablonka and Lamb 1989; Boyko and Kovalchuk 2008; Alvarez et al. 2010; Chen et al. 2010; Yaish et al. 2011; Kinoshita and Seki 2014). Both histone and chromatin methylation patterns are strongly modified by environmental stress (Molinier et al. 2006; Bond and Finnegan 2007; Chinnusamy and Zhu 2009; Adams 2010; Daxinger and Whitelaw 2010; Verhoeven et al. 2010).

The methylation status is not necessarily reset when the stress is relieved (Chinnusamy and Zhu 2009). In fact, it can even be transferred through cell divisions both mitotically and meiotically (Molinier et al. 2006). This means that the epigenetic memory is retained in somatic cell lines. With this it provides a more short-term epigenetic stress memory within a given organism. However, epigenetic memory can even last over generations. For a rather long-term trans-generational stress memory, it is remarkable that stress-induced methylation changes are not reset through the germ line and are mostly heritable, so that epigenetic information relative to stresses received by plants can be transferred through several subsequent generations (Jablonka and Lamb 1989; Bird 2002; Kakutani 2002; Molinier et al. 2006; Bond and Finnegan 2007; Saze 2008; Verhoeven et al. 2010). In brief, the epigenetic memory remains stable when stress is not continuous but occurs in episodes, and this can last for generations.

2.3 *Memory Capacities in Plants*

2.3.1 Plant Sensitivity to Stimuli and Types of Subsequent Response

Plants are sensitive to a variety of stimuli such as wind, rain, touch, drought, cold shock, heat shock, wounds inflicted by herbivorous animals, attack by fungi, bacteria or viruses and even electromagnetic irradiation in the approximate range 1–100 GHz (Tafforeau et al. 2002, 2004; Roux et al. 2006; Vian et al. 2006). There are basically two contrasting types of possible responses to environmental cues acting as signals or stimuli, namely, a direct immediate response and responses involving memory.

Usually, plants react almost immediately to a stimulus by generating a "calcium wave", i.e. a transient invasion of the cytosol by calcium originating from Ca^{2+}-rich internal and external pools (Knight et al. 1991; Trewavas 1999). This calcium wave triggers a chain of events, including the opening of ionic channels, the phosphorylation of existing proteins and changes in the genome expression (Dolmetsch et al. 1997; McAinsh and Hetherington 1998).

All of that eventually results in a final response that can be a modification of growth and/or metabolism and sometimes a macroscopic movement (*Dionaea muscipula*, *Mimosa pudica*). The response may be stereotyped and direct, i.e. independent of the previous history of the plant and involving no more delay than necessary for the intermediate events required to occur between the perception of the stimulus and the final response to this stimulus. The rapidity is advantageous for reacting to rare or unknown stimulations, especially those involving an attack by an herbivore or a pest.

However, if plants made such a direct, stereotyped response to each individual stimulus which they perceive, and if they responded with similar intensity to innocuous and harmful stimuli, erratic metabolic and growth behaviour would emerge, being unnecessarily costly in energy. Therefore, an apparent requirement exists for a mechanism that permits plants to adjust their response to the entirety of stimulations and their dynamics experienced in the past. This is achieved by means that functionally resemble animal memories, although the underlying mechanisms are very different (especially since plants neither have neurons nor anything comparable to a central nervous system).

At the beginning of the 1980s (see Thellier et al. 1982), it was discovered that plants possess memory capacities, which to some extent mimic our human "memory". Since then, a number of publications have been devoted to the occurrence and characteristics of that memory (for reviews, see, e.g. Thellier et al. 2000, 2013; Trewavas 2003; Ripoll et al. 2009). It has also been recognized (Trewavas 2003) that two different kinds of plant memory can be distinguished, namely, "priming", which resembles the animal "training" (Bailey and Chen 1983), and "store/recall (STO/RCL) memory" (resembling the animal "memorization/evocation").

2.3.2 Priming Memory

In the priming memory, the first stimulus, or sequence of stimuli, changes the transduction of subsequent stimuli, thus tending to either diminish or enhance the intensity of the plant response (observations carried out at the level of the final response or as early as the generation of the calcium wave).

For instance, in *Nicotiana plumbaginifolia* seedlings, a wind stimulus causes cytosolic calcium to rapidly increase, but repeated wind stimuli within very short periods of time make the plant cells refractory to further calcium signalling for approximately 1 min (Knight et al. 1992). In *Arabidopsis thaliana*, cold pretreatments attenuate the increase of cytosolic calcium due to cold shock (Plieth et al. 1999). Again in *Arabidopsis*, a hyperosmotic-stress pretreatment increases the elevation of cytosolic calcium due to hyperosmosis (mimicking drought), while an oxidative-stress pretreatment reduces it (Knight et al. 1998).

2.3.3 Store/Recall Memory

In the STO/RCL memory, the perception of a stimulus is responsible for storage (STO) of information within the plant; then, that information may be recalled (RCL) at a later time. During the lapse of time between storage and recall (memorization time), the stored information remains latent, i.e. without any apparent effect on the plant behaviour. When an appropriate stimulus or change in internal or environmental conditions causes the RCL function to be switched from "off" to "on", the plant is enabled to recall the stored information and to make it effective in the control of the its metabolism and growth.

Three experimental systems, which shall be termed here SR1 (Desbiez et al. 1983, 1987), SR2 (Desbiez et al. 1991) and SR3 (Verdus et al. 1997), have been mainly used in the basic original studies of STO/RCL memory (Table 1; for reviews, see, e.g. Thellier et al. 2000, 2013; Trewavas 2003; Ripoll et al. 2009). With system SR1, *Bidens* seedlings were stimulated by pricking one or both cotyledons, which caused the storage of "reduction of hypocotyl growth" information, acting as a kind of instruction governing the control of hypocotyl growth. However, it is only when the plants were grown on a very diluted medium that they were enabled to recall the stored information/instruction and let it take effect in reducing hypocotyl growth. (When the pricked and non-pricked plants were grown in a conventional nutrient solution, the growth of their hypocotyls was not significantly different). For brevity, see Table 1 for the description of the experiments with SR1 to SR3. These experiments were designed at the outset of memory investigations for testing and proving the very existence of memory under strictly controlled laboratory conditions. Before extrapolating from ecological observations to the involvement of memory functions in a framework of environmental conditions, the ground laying operation of STO/RCL functions had to be shown in readily reproducible experimental approaches. These experiments clearly revealed the

Table 1 Summarizing the experimental features concerning the three systems with store/recall memory

System	Seedling	Typical stimulus-inducing information storage	Information stored	Means of induction of the plant ability/inability to recall stored information
SR1	*Bidens*	Cotyledon pricking	Hypocotyl elongation inhibition	Diluted/nutrient medium
SR2	*Bidens*	Cotyledon pricking	Specification of bud dominance[a]	Various[b]
SR3	*Linum*	Manipulation[c]	Meristem production	Transient Ca^{2+} depletion/excess

[a]Specify which of the two cotyledonary buds will be the first to start to grow after removal of the seedling apex
[b]Time of the day when plant decapitation is carried out, pricking, thermal treatment, etc.
[c]"Manipulation" consists of transferring seedlings from the germination box to a grid that covers the vessels containing the growth medium

sequence of events relevant for any kind of ecological responses as exemplified in Sect. 4, i.e. > external stimulus > process of information/instruction storage > state of information/instruction being stored > triggering induction to put stored information/instruction into action (Table 1). The main results obtained from the compilation of data yielded with SR1 to SR3 are as follows.

The Storage Function

With SR1, it has been observed that a signal migrates from the stimulated area (here the pricked cotyledons) to the reactive area (here the hypocotyl) where information storage finally occurs. The rate of signal migration is of the order of one to a few tenths of a millimetre per second (Desbiez et al. 1983). Electric depolarization signals in phloem cells are involved in signal migration, but the mechanisms in action are different from those in animal nerves. At SR3, the application of pharmaceutical agents blocking calcium movements, during and shortly after the occurrence of the calcium wave, prevents information storage (Verdus et al. 2007). This means that these agents have blocked information storage either directly or indirectly by blocking the migration of the signal from the stimulated to the reactive area. In any case, the information induced by the initial stimulus becomes firmly stored in the responding tissue after a few minutes at the most.

When a stimulus has been perceived, the shape, amplitude and duration of the calcium wave (Dolmetsch et al. 1997; McAinsh and Hetherington 1998; Knight et al. 1998) and the early and transient modifications of existing proteins or of genome expression (Tafforeau et al. 2006) are specific of the stimulus perceived. However, the memory of the stimulus is finally lost and what is memorized is mere instruction. More precisely, it is a sort of instruction, which addresses the final response that has to be performed in reaction to the specific stimulus (SR1, SR2 and SR3).

Comparing SR1 and SR2, it appears that the application of the same stimulus (pricking one of the two plant cotyledons) stores two different pieces of instruction in the hypocotyl and in the cotyledonary buds, i.e. percentage of reduction of hypocotyl growth and specification of bud dominance (measured by the percentage of dominant buds at the axil of the non-pricked cotyledon), respectively. There is an apparent discrepancy in the storage behaviour in SR1 and SR2 because the percentage of reduction of hypocotyl elongation in SR1 is quasi proportional to the number of pricks, whereas, in SR2, the percentage of dominant buds at the axil of the non-pricked cotyledon is independent of the number of pricks. It is likely that this discrepancy can be explained by assuming that, in SR2, the application of a single prick suffices to saturate the storage capacity of the system, and therefore delivering one or several pricks has exactly the same effect. The reason is that, when taking into account much weaker stimuli, such as the small gradients of temperature or light that inevitably exist in the culture rooms, the behaviour in SR2 is fairly similar to that in SR1. In brief, as long as there is no saturation effect, after a stimulus the intensity of the stored instruction depends on the intensity of the stimulus (Thellier 2015).

Once a first stimulus has been perceived and an instruction for a response has been stored accordingly, subsequent stimuli can modulate quantitatively (i.e. in its intensity) this programmed response (SR3). Though a direct experimental test is still lacking, it may be reasonably inferred from the preceding paragraph that the instruction stored after a first stimulus can also be modulated qualitatively as a consequence of the perception of subsequent stimuli. This would mean that the very nature of the information for performance of a response may be modified.

The Recall Function

The recall function can usually be switched "off/on" or "on/off" reversibly, thus enabling/disabling the plant to recall instruction stored after the perception of a stimulus. However, cases exist when recall can be blocked in status "on" or in that of "off", thus always permitting or preventing, respectively, the plant to recall stored instruction. There is no universal way to enable/disable a plant to recall stored instruction: with the three systems studied, enabling/disabling plants to recall stored instruction was accomplished by (1) using a dilute/normal growth medium (SR1), (2) decapitating the seedlings at the onset/middle of daylight (SR2) or (3) imposing a transient Ca^{2+} depletion/excess (SR3). Stored instruction can be repetitively recalled (at least twice in SR2 and SR3). Recalling stored instruction, whether once or at several times, does not seem to alter the stored information. The functioning of the RCL box is apparently independent of the functioning of the STO box.

Hypothetical Mechanism of Functioning

It may be that various substances play a part in the memorization process in plants. Such substances are "memory metabolites" (Ueda and Nakamura 2006), molecules involved in the control of the cell cycle (Desbiez et al. 1998) or small bundles of messenger RNA termed "stress granules" (Alain Vian, personal communication, Davies et al. 2012). However, it is possible to account for the main facts observed by interpreting plant memory (especially the "store/recall" type) via an interaction, involving epigenetics, between a few genes (Thellier 2015). The perception of a stimulus would modify the histone and/or chromatin methylation patterns (see Sect. 2.2.2), thus unlocking a few locked genes (and/or locking a few unlocked genes): this is the storage function, and the genes involved are termed "STO genes". However, the unlocked genes would remain silent, until being activated by an appropriate ligand. Other genes would be unlocked and activated (on perception of an appropriate stimulus, after an appropriate treatment and/or depending on the external conditions), and their products would be the activators of the unlocked STO genes: this is the recall function, and the genes involved are termed "RCL genes" (for details see Thellier 2015). Hence, only the unlocked and activated STO genes would be functional, thus permitting the corresponding metabolic pathways to function, while the metabolic pathways depending on the locked STO gene would remain non-functional. Thanks to this "store/recall memory", the plant would be able to adjust its metabolism to the external conditions and stimuli.

In brief, a change of methylation/acetylation equilibrium means storage (STO) of stimulus-information. A changed access of transcription factors is modulating recall (RCL). These are mechanisms of the epigenetic memory (Sect. 2.2.2; Fig. 2 in Thellier and Lüttge 2013), which is one of the possible pathways of signal transduction in the priming and in the STO/RCL memory (Fig. 1; Fig. 3 in Thellier and Lüttge 2013; Hütt et al. 2015; Thellier 2015). Moreover, it is likely that "priming memory" can be interpreted using a similar conceptual model of functioning (Thellier 2015).

2.3.4 Memory and Developmental Phenological Phases

In rhythmic phenomena which normally are oscillations, a phase is a rhythmically occurring specific point or state be it in developmental cycles or any other shorter type of oscillation. Phenology relates developmental phases of plants to the times when they are expressed. Stress treatments in particular phenological phases during earlier stages of development can become effective at later stages or phases with transition periods in between such stages. This means that development is not simply a cumulative expression of genetically preprogrammed events (Amzallag 2002, 2005). Complex links exist between development and adaptation. Evidently memory is involved. An example is the memory-regulated meristem formation in flax seedlings particularly active during April through June (Verdus et al. 1997; see also section "Annual Fitness"). A fascinating challenge is posed for a thorough assessment of the role of memory in development as modified by the impact of environmental signals. Clearly the experience of stress in an early developmental phenological phase is stored. The signal for recalling it is the transition into a later developmental phenological phase. The phenomenon has been particularly studied with NaCl salinity as the stress signal and is of eminent ecological significance (Sect. 4.2.5).

3 Rhythmicity and Memory

3.1 Ultradian, Circadian and Annual Rhythmicity

Rhythms of plants as of other organisms including man in time can cover a vast range of period lengths. Basically we define ultradian rhythmicity having period lengths shorter than the 24 h of the day, diurnal rhythmicity with the period length of a day and infradian rhythmicity with period lengths longer than a day. When diurnal rhythms run endogenously, i.e. under constant conditions independent of external rhythms of environmental parameters, they normally do not have an exact 24 h period length but just a little shorter or longer than that. Therefore, these rhythms are called circadian rhythms (from Latin circa and dies = day).

In the SR2 example of pricking cotyledons of *Bidens* in Table 1 and looking at specific bud dominance, an inherent ultradian rhythmicity can be observed. When a plant is subjected to two successive stimuli (one of which is dissymmetrical, consisting of pricking only one of the two cotyledons), the RCL function exhibits a damped oscillation "on/off" according to whether the delay between the two stimuli is close to 1 h or 8.5 h or larger than 14 h (RCL "on") or close to 3 h or 12 h (RCL "off"; Desbiez et al. 1991; and see Hütt et al. 2015 for a theoretical approach). Again with SR2 under the experimental conditions used, the RCL function is "on" or "off" according to whether plant decapitation has been carried out in the morning or in the middle of the day (Desbiez et al. 1986, 1991; Thellier and Lüttge 2013). Finally, with SR3 (Verdus et al. 1997), when using plants all subjected to transient Ca^{2+} depletion, a very significant increase of the production of meristems was observed to take place in the period of April to June whether these plants were stimulated or non-stimulated. However, the number of meristems produced remained at least 5–10 times larger in the stimulated than in the non-stimulated plants. By contrast, when using plants nonsubjected to transient Ca^{2+} depletion, the number of meristems produced was always close to zero, whether the plants were stimulated or not and whatever the period of the year. From all these data, it may be inferred that the RCL function is linked (1) with an ultradian rhythm of the plant, reset by the dissymmetric stimulus, (2) with a circadian rhythm and (3) with an infradian annual rhythm (Fig. 2).

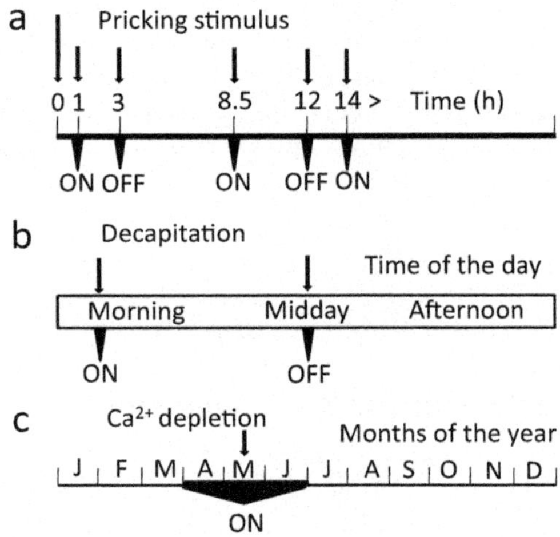

Fig. 2 Schematic overview of OFF and ON responses of RCL in SR2 and SR3 experiments of Table 1. Relations (**a**) to short ultradian periods, (**b**) to the time of the day and (**c**) to the months of the year. Experiments are (**a**) SR2 and pricking stimuli, (**b**) SR2 and decapitation and (**c**) SR3 and Ca^{2+} depletion

3.2 Circadian Clock and Memory

Memory is related to functions of timing. Therefore, it should be expected that the biological clock is part of mechanisms of memory. What are the relations between clocks and memory? A clock allows measuring the flow of time. However, as such this has nothing to do with memory. However, a clock becomes part of the structure and function of memory if it contains specific points set at a certain time at which recall functions are alerted. A familiar example of such set points is an alarm clock set on a specific point in time. It is a reminder because it causes us to remember. In the biological circadian clocks reflecting the natural day–night rhythms with period lengths close to 24 h when running free under constant environmental conditions (Lüttge 2003), genes are involved that label set points. Among the master genes of the clock *CIRCADIAN CLOCK ASSOCIATED (CCA1)* and *LATE ELONGATED HYPOCOTYL (LHY)* are expressed in the morning (morning genes) and *TIMING OF CHLOROPHYLL a/b BINDING (TOC1)* in the evening (evening genes). There are further morning and evening elements functioning as transcription factors (e.g. Kikis et al. 2005; Harmer and Kay 2005; McClung 2006; Nakamichi 2011). Downstream of these genes, a vast number of other genes are controlled by the clock, the so-called clock-controlled genes (CCGs). A plethora of plant functions are under the regime of the clock including complex processes such as growth (Farré 2012). Thus, there is a complex machinery of set points determining phases in the rhythmic oscillations of the clock.

The set points are affected by storage in the memory. Setting the clock is a result of entrainment (Dixon et al. 2014) by environmental ecologically relevant factors. The most obvious example is the exact operation of clock-dependent functions in the entrainment by the natural 24 h rhythm of the day. When the phases of external environmental rhythms change, new entrainment will change the set points. The recall function of the STO/RCL memory breaking the symmetry of bud growth after stimulation of cotyledons is dependent on diurnal timing (Table 1).

The molecular basis has to be assessed at the level of genes involved in the phase relations of resetting the clock. Genes giving phase information (Michael and McClung 2002) and phase mutants (Onai et al. 2004) have been identified. In *Arabidopsis* an *OUT OF PHASE 1* gene has been characterized (Salomé et al. 2002). Phase variations in populations are studied (Darrah et al. 2006). However, much more work is needed, precisely addressing the links between memory, clock and environmental responses at the molecular level.

The most relevant signals eliciting phase shifting, entrainment and resetting the circadian clock in plants are light, temperature and for rhythms of photosynthesis also CO_2 (Lüttge 2003). The responses to the signals in resetting are dependent on the phase of the rhythms, i.e. it matters at which time during oscillations the phase shifting signals are applied. This means that there is the so-called gating, i.e. that the input pathways to the clock are not constantly open over the entire 24-hour period of the day for the reception of the environmental signals (Edmunds and Tamponnet 1990; Rikin 1991; Johnson 1992; Millar and Kay 1996; Millar 1999; McWatters et al. 2000; Covington et al. 2001). The predominant light signals are blue and red light with cryptochrome and phytochrome, respectively, as their photoreceptors (see Table 2 in Lüttge 2003 with references; Devlin 2002; Wenden et al. 2011) and green light in *Chlamydomonas reinhardtii* (Forbes-Stovall et al. 2014). Marking new set points for the memory by clock resetting under the influence of these environmental control parameters demonstrates close relations between clock and memory in ecophysiological performance.

4 Ecophysiological Functions That Require Memory and Clock

4.1 Adaptation, Acclimation and Memory Functions

In conditioning for ecological and ecophysiological performance of plants, we distinguish adaptation and acclimation (Wilhelm and Wirth 2015). Adaptation is a long-term process. It builds up during evolution and is secured in the form of genetically stored information. Acclimation is based on expression of present instruction and modulation of phenotype via transcription, translation and the control by metabolites. This requires induction by perception of signals. If we accept that acclimation in general includes memory, of course, this leads us to consider memory as a fundamental quality in ecology. Such a general relation implies that there will be a plethora of examples for memory in the ecological performance of plants.

Induction can result in immediate responses by direct activation. However, induction can also involve the priming and the STO/RCL memory. This evidently is a matter of temporal dynamics. An example illustrating this is given by responses of defence. They can be activated directly upon attack. Then, obviously no memory functions are involved. Conversely, activation of defence can follow induction and priming indirectly after a certain lapse of time when an attack of predators, herbivores or parasites is a later event, but plants appear prepared or acclimatized to it (Conrath et al. 2001; Conrath 2009, 2011). An example is the cell-content-feeder *Tupiocoris notatus*, which by its feeding on tobacco plants elicits greater mortality of attacking hornworm (*Manduca sexta*) (Kessler and Baldwin 2004; Voelckel and Baldwin 2004; Gális et al. 2009). Similarly, chronically enhanced tropospheric ozone impact can prime against intermittent pathogenic interference, delaying or reducing infestation under controlled and field conditions in woody plants, irrespective of ontogenetic stage (Bahnweg et al. 2005; Luedemann et al. 2005, 2009; Olbrich et al. 2010), as both kinds of stress act through oxidant release (Matyssek and Sandermann 2003; Matyssek et al. 2008). In such context authors have also spoken of "immunity" and "vaccination" as a certain kind of memory.

4.2 Examples of Ecophysiological Performance Requiring Memory and Clock

In this section we address some specific ecophysiological examples at various scales beginning with the molecular level. Circadian clock genes act in stress responses (Kant et al. 2008). The epigenetic memory (Sect. 2.2.2) is explicitly involved in many ecophysiological functions. While we have seen that memory and clock can be intimately related (Sect. 3.2), it is noteworthy also to outline their participation in conveying fitness. Fitness is much more than Darwinian reproductive success. Among many aspects of plant performance, it also requires competitiveness given by growth (see zu Castell et al. 2016), where timing is essential on the levels of both diurnal and annual entrainment. Furthermore, other selected phenomena of memory functions in ecological performance will be addressed.

4.2.1 Examples for the Molecular Level: Induction of Mechanisms of Membrane Transport

Transport across membranes is of great ecophysiological importance, e.g. for the acquisition of substrates. A well-known example worked out at the molecular level is the regulator-operator model of François Jacob and Jacques Monod (Nobel Prize 1965; see textbooks, e.g. Lüttge et al. 2010). When cells of the bacterium *Escherichia coli* are grown in the absence of β-galactosides, e.g. lactose, a capacity to use these as substrate for growth is not expressed. When lactose is added, a membrane transporter lactose-permease and a β-galactosidase that splits the lactose in its hexose moieties glucose and galactose for further metabolism are induced. This occurs because lactose inactivates a repressor of the operator blocking the transcription of the genes of the operon coding the permease and the β-galactosidase. When the substrate lactose is removed, the cells will remember that they can metabolize lactose for a certain time as seen when the substrate is added again. The repressor is coded by a regulator gene. In the absence of lactose after some time, this gene will lead again to the production of active repressor, so that the previous experience of lactose will be forgotten. Hence, even prokaryotes have memory. In the eukaryotic cells of *Chlorella vulgaris*, a memory of glucose uptake was demonstrated (Tanner 1969; Tanner et al. 1970). The capacity for uptake is not expressed in the absence of glucose. When glucose was added to the medium, an uptake was induced within 20 min. When glucose was subsequently removed, 10 h later glucose-uptake capacity was still active. However, the memory did not last for much longer. After 13 h glucose-uptake capacity had been forgotten and needed to be induced again.

Both examples illustrate a molecular memory important for acclimation to the use of substrates available in the environment. Expression of products of transcription (mRNA), translation (proteins, enzymes) and regulatory metabolites constitutes store functions of memory. As long as the respective products are present and active, the events which led to their production will be remembered. When turnover results in their disappearance, the functions of induction, priming and storage will be lost with them.

4.2.2 Examples of Epigenetic Changes in Response to Environmental Cues and Their Inheritance

Epigenetic modifications of methylation patterns of histone and chromatin and, hence, the epigenetic memory have been shown to be involved in responses to stresses by salt (Wang et al. 2010), drought (Bruce et al. 2007; Baek et al. 2011; Ding et al. 2012; Kinoshita and Seki 2014), heat (Kinoshita and Seki 2014; Li et al. 2014), nutritional limitation, e.g. nitrogen deficiency (Kou et al. 2011), as well as herbivores and pathogens each inducing biochemical defences (Gális et al. 2009; Verhoeven et al. 2010), and virus infection (Kathiria et al. 2010).

The regulation of methylation patterns following stress reception involves chemical signals such as by phytohormones, electrical signals and calcium waves (Trewavas 2003; Thellier et al. 2013). These signals may transcribe into particular RNA signals by stress-induced expression of microRNAs, for example, under salinity, drought, low relative air humidity, cold and herbivore stress (Matzke et al. 2001, 2007; Sunkar and Zhu 2004; Gális et al. 2009; Shen et al. 2010; Yaish et al. 2011; Kinoshita and Seki 2014). Small interfering RNAs (siRNAs) of a length of 24–26 nucleotides direct DNA methylation and histone modifications (Richards 2006; Zhang et al. 2006; Bond and Finnegan 2007; Saze 2008; Zhang 2008; Chinnusamy and Zhu 2009). Small RNAs are mobile in the symplast via plasmodesmata and in the phloem. They can be transmitted within plants and function as systemic signals produced by stress (Saze 2008). Rasmann et al. (2012) demonstrated that *Arabidopsis* and tomato plants that experienced herbivory were more resistant to subsequent attack in the next generation. The induction of the defence process is dependent on the phytohormone jasmonic acid and the biogenesis of siRNA priming progeny plants for enhanced resistance.

4.2.3 Fitness

Diurnal Fitness

Logical common sense takes endogenous circadian rhythmicity as being essential for fitness because it provides preparedness or alertness for regularly changing conditions in the day–night rhythm. Some studies support this argument. Plants take advantage from circadian control of photosynthesis and physiological performance in general (Dodd et al. 2005; Hotta et al. 2007; Yerushalmi and Green 2009). The advantage of fitness to possess the suitable endogenous period length of rhythmicity matching with external rhythmicity was suggested by some work using period mutants showing entrainment of competitive fitness. Golden and collaborators obtained mutants of the cyanobacterium *Synechococcus elongatus* PCC7942 having different circadian periods of their endogenous clocks. They grew them in coculture under different external light-dark rhythms. The mutants having the correct endogenous period, i.e. closest to the imposed external light-dark rhythms, outcompeted the others during growth in the cultures. This competitive advantage disappeared in constant environments where the selective pressure of external light-dark rhythms was removed (Ouyang et al. 1998; Johnson and Golden 1999; Woelfle et al. 2004). Yerushalmi et al. (2011) crossed *Arabidopsis thaliana* mutants with different circadian period lengths and studied the F2 and F3 generations which they subjected to the selective pressure of altered external light-dark-cycle periods. Endogenous circadian rhythms that resonated with the environmental ones were positively selected. Nevertheless, a match of internal circadian period with external rhythmicity does not appear to be sufficient for guaranteeing positive clock effects on growth and competitiveness. The relationships are much more complex, which results from an interplay between the clock and metabolism. Metabolites can affect resetting of the clock where especially sucrose plays a dual role as signal and metabolite. In a feedback loop the circadian clock controls metabolism and is controlled by metabolites (Müller et al. 2014).

Annual Fitness

At latitudes north and south from the equator, the duration of the light period (or photoperiod) during the 24 h day changes over the year; the phases of the photoperiod change between short and long days or long and short nights. Changing entrainment in response of the transition between short days and long days allows flexibility of responses to phases (Dixon et al. 2014). For obtaining annual fitness plant growth and development, frost hardiness, flowering and seed production are subject to regulation by the photoperiod. Phase adjustment of the biological clock is an essential mechanism in photoperiod perception, where phytochrome acts as the photoreceptor (Frankhauser and Staiger 2002; Roden et al. 2002; Love et al. 2004; Ogudi et al. 2004; Fujiwara et al. 2008; Lüttge and Hertel 2009; Niwa et al. 2009; Ibáñez et al. 2010). The overwhelming evidence for the absolute necessity of the biological clock for plant fitness is provided by a huge volume of literature on photoperiodism and phenology. In experiments on epidermal meristem induction in flax seedlings, it was shown that the STO/RCL memory is involved. A stimulus induces the storage of meristem-production information (STO function), and a transient depletion of calcium enables the seedlings to recall stored information and let it take effect in the promotion of meristem production (RCL function). This is subject to seasonal modulations. The memory-controlled meristem formation is particularly active in April to June (Verdus et al. 1997; Sect. 3.1; Fig. 2), suggesting annual rhythmicity to be involved and ecologically relevant for growth and competitiveness. Hence, overall we can conclude that seasonal phenological memory is stored in the clock.

4.2.4 Photosynthesis

Photosynthesis is the foremost ecophysiological function of plants under the influence of primarily the impact of photosynthetically active radiation (PAR) and secondarily a broad array of almost all other prevailing environmental cues. In the context of memory in ecology, it is worthwhile to explore the involvement of memory functions in the ecophysiology of photosynthesis. Photosynthetic memory, as far as we can see, has not been addressed in the literature, but we can refer to some phenomena in which priming or storage and recall of information must obviously be involved.

In general activation–deactivation equilibria of metabolic activities, e.g. of enzymes, may imply memorization. This is the case if there is turnover with on and off states, respectively, being stable for some time in the absence of activation–deactivation elicitors (see also examples of membrane transporters in Sect. 4.2.1). In photosynthesis ribulose-bisphosphate carboxylase/oxygenase (RUBISCO) may be an interesting example for exploration in memory studies. The enzyme is active in the form of RUBISCO-carbamate-Mg^{2+} and needs carbamylation by binding of CO_2 plus Mg^{2+} as catalysed by RUBISCO activase (Buchanan et al. 2000; Portis 2003).

When we take a photosynthetically active plant sample from darkness and subject it to gradually increasing PAR, we can record hyperbolic light saturation curves of photosynthesis saturating at a certain level of high PAR. When we then decrease PAR again, we may observe hysteresis where the rates of photosynthesis are lower during the descent of PAR than at any given PAR during the ascent. The plant memorizes that it has been at these PARs before but responds in a different way. The reason is that it has become subject to photoinhibition at high PAR. This elicited protective molecular changes within the photosynthetic apparatus (Osmond and Grace 1995), which may be stable for some time at low PAR and in darkness reminding to the high PAR experienced before.

Photoinhibition can be acute, i.e. reversible within short periods of time up to the length of a nocturnal dark period, or chronic, i.e. irreversible or only reversible after long periods of repair. Chronic photoinhibition at high irradiance is due to photodestruction. However, not always photoinhibitory reactions are right out destructive. By contrast, there is a cascade of mechanisms leading to acute photoinhibition but functioning in protection of the photosynthetic apparatus under high irradiance (Figs. 10–14 in Lüttge et al. 2010). Some of the photoprotective mechanisms are based on conformational changes of the photosynthetic apparatus, providing acclimation which is subject to turnover. Thus, acute photoinhibition appears as an instructive case for being viewed under the perspective of memory, demanding for further unravelling priming and STO/RCL functions:

- Spill over of excitation from photosystem II (PSII) to photosystem I (PSI) is a protective mechanism based on the reversible separation of the light-harvesting complex LHII from PSII and its transfer from the grana thylakoids to the stroma thylakoids towards PSI. The reaction is mediated by a kinase phosphorylating LHII. Reversibility and turnover are given by dephosphorylation catalysed by a phosphatase (Jennings et al. 1986).
- Another mechanism with turnover is aggregation/dissociation of the LHII complex. This is based on the binding of the xanthophyll zeaxanthin replacing lutein, under involvement of the thylakoid protein PsbS. Conformational changes modify the structure of LHII so that it switches from a light harvesting to a heat dissipation state dispersing harmful excess of excitation energy (Bilger and Björkman 1994; Horton et al. 1994, 1996; Gilmore and Yamasaki 1998; Gilmore et al. 1996, 1998; Gilmore 1997; Gilmore and Govindjee 1999; Holt et al. 2005; Horton and Ruban 2005; Goss and Lepetit 2015).
- The D1 protein of LHCII is involved in transferring excitation to plastoquinone. The protein is damaged and destroyed under harmfully high irradiance. The repair requires protein synthesis. The protein is under continuous turnover, although only at low irradiance destruction and repair are balanced. At high irradiance destruction exceeds repair (Critchley and Russell 1994; Tyystjärvi and Aro 1996; He and Chow 2003).

A conspicuous example of memory in photosynthesis is the acclimation to light flecks on the floor of forests. Under the trees the solar irradiance reaching the forest floor is only a few percent of the irradiance at the upper canopy. However, light flecks occur when movements of leaves in the wind or the changing angle of irradiance allow direct light penetration through gaps in the canopy cover for intermittent periods of time. In tropical rain forests, light flecks may provide up to 80 % of the total irradiation reaching the forest floor, and their intensity varies between 10 % and 70 % of that of full sunlight (Lüttge 2008). Ecologically they are highly important for photosynthesis of the forest floor vegetation. When a sudden light fleck arrives after plants were at low-background irradiance, the photosynthetic apparatus must be activated and stomata must open for CO_2 uptake. Excitation of the photosynthetic apparatus is immediate, and activation of the electron transport reactions occurs within seconds to minutes. However, activations of stomatal opening for CO_2 uptake, of the reactions of CO_2 assimilation through RUBISCO and of Calvin-cycle enzymes, and the filling of pools of intermediates are sluggish with time constants of 10–30 min (Sassenrath-Cole and Pearcy 1992). The advantage of the slower induction processes is that they also are subject to slower decay, and this is the mechanism of memory in this case. The use of the energy from light flecks by photosynthesis accelerates with time when short light flecks alternate with low-background irradiance. Conditioned or induced leaves have considerably higher light use efficiency than non-induced ones (Pearcy et al. 1985; Sassenrath-Cole and Pearcy 1992; Valladares et al. 1997). Such memorizing of previous light flecks under the dynamic light environment on the forest floor is essential for the persistence of plants under closed canopies.

4.2.5 Tolerance of Osmotic Stress and Salinity

The clock and memory have been shown to be involved in plant responses to osmotic stress and salinity. Osmotic stress at the level of barley roots up-regulated expression of clock genes which control the expression of stress response genes (Habte et al. 2014). When young seedlings of different species, such as tomato and *Sorghum bicolor*, are subjected to sublethal salinity, stress response will be modulated during subsequent phenophases in development. Amzallag and coworkers (Amzallag et al. 1993, Amzallag 2005) found that in *S. bicolor*, there was a developmental window during the 5th and the 10th day following germination. After being treated during this particular period with mild salinity of 150 mM NaCl, the plants remembered this at later stages by proving to be resistant to 300 mM NaCl. This also changed other aspects of development, e.g. perturbing reproductive development. Signal and response occurring during a critical period may be adaptive for some aspects of development and disturbing for others.

4.2.6 Memory of Pollution Events

Some perennial plants like conifers have been shown to memorize events of pollution such as the effects of acid rain and ozone on photosynthesis. When seedlings of loblolly pine (*Pinus taeda* L.) were exposed to such stress, their ozone memory was expressed in the following season before they experienced any further stress (Sasek et al. 1991). Similar memorizing of ozone exposure was recorded in *Pinus sylvestris* (L.) and *Picea abies* ((L.) Karst.). The new flush appearing during spring after ozone exposure in the previous growing season showed visible stress symptoms including premature shedding of needles (Langebartels et al. 1998).

4.2.7 Signalling by Volatile Organic Compounds

Signalling triggers induction. Both are elements in the functioning of memory. Among the signals eliciting induction are volatile organic compounds (VOCs) of plants. There is a large variety of such compounds, including gaseous phytohormones (pheromones) like ethylene and methyl jasmonate, as recently surveyed in a special issue of "Plant, Cell and Environment" (Loreto et al. 2014). VOCs have signalling functions between different plant species, between individuals of given species and even within given individual plants avoiding vascular constraints of signal transport (Conrath 2009; Gols 2014; Karban et al. 2014). VOC signals play a prominent role in plant defence. Herbivore-triggered plant VOCs lead to induced defences (Kessler and Baldwin 2002; Conrath 2009; Dicke 2009; Pierik et al. 2014). Inducible defences are of eminent relevance in ecological contexts. In cases of induction of defence genes and "increased accumulation, and/or post-translational modification of inactive cellular signalling proteins", a molecular basis of memory in ecology becomes accessible (Conrath 2009). "Dormant signalling proteins" thus are memory molecules.

4.2.8 Signalling by Vibrations

The VOC systems of long-distance signalling pathways can be complemented by other signals, e.g. vibrations caused by caterpillar feeding. In *Arabidopsis thaliana* such signalling induced the production of increased levels of anti-herbivore compounds such as glucosinolate and anthocyanin when subsequently attacked by caterpillars. The plants can even distinguish between different specific vibration signals and discriminate between the chewing vibrations and wind or insect song (Appel and Cocroft 2014).

5 Ecophysiological Potential of Plant Priming and Store/Recall Memory

5.1 Potential of the "Priming" Form of Plant Memory

Plants are subject to many signals in their environment that occur repeatedly. Some of these signals are relatively innocuous, and reactions to them would be quite unnecessary and detrimental as wasting resources and energy. Thus, an ecological memory function is to remember irrelevant signals for filtering them out. Priming memory is suited to that since the repetition of a stimulus can change the transduction of subsequent stimuli of the same type in a way tending to *diminish* the intensity of the plant response (see the examples of wind stimulus and cold shock in Sect. 2.3.2). By analogy with the animal case, such a plant behaviour may be termed "familiarization" or "habituation". As an example, the leaf-folding behaviour of *Mimosa pudica* elicited by the mechanic signals from trampling herbivores is thought to be a defensive reflex of the plants to avoid being seen and readily exposed to the herbivores. However, it is costly because in the folded state, photosynthesis is drastically reduced. Thus, there is a trade-off between protection and productivity at low and limiting photosynthetically active radiation (PAR). At high PAR this may be less inflictive than at low PAR. To test this, Gagliano et al. (2014) compared the response of plants kept under low- and high-PAR environments, respectively. They found that the low-PAR plants learned faster to ignore the stimulus and retained the memory of this longer, i.e. for up to a month when undisturbed, than the high-PAR plants. For the low-PAR plants, the response to the stimulus leading to a reduction in photosynthesis is more severely disadvantageous and non-adaptive than for the high-PAR plants.

Conversely, memory is important for remembering grave signals to react more violently to them (Lodish et al. 2000). Again, priming memory is suited since the repetition of a stimulus can change the transduction of subsequent stimuli of the same type in a way tending to *increase* the intensity of the plant response. See the example with *Arabidopsis*, in which hyperosmotic-stress pretreatment amplifies the increase of cytosolic calcium due to hyperosmosis (Sect. 2.3.2). By analogy with the animal case, such a plant behaviour may be termed "sensitization". Many examples come from priming in defence physiology (Conrath 2009; Pastor et al. 2013) where the signals are mechanical and chemical injury associated with herbivory and phytopathology (Baluška and Ninkovic 2010; Bruce 2010; Heil 2010; van Hulten et al. 2010). Defence priming involves enhancement of molecular mechanisms having many analogies with immunology (Conrath 2011; Rasmann et al. 2012).

Signals can also become grave when the repetition of stimuli changes the response to other types of stimuli. The intensity of a reaction to new types of stimuli can be altered by previous stimuli of another nature. An example is when an oxidative-stress pretreatment reduces the elevation of cytosolic Ca^{2+} due to hyperosmosis (see Sect. 2.3.2).

Thus, the ecophysiological advantage for a plant to possess priming memory is evident. When such kind of memory is involved, the response is rapid, similar to a direct response (see Sect. 2.3.1), although negative or positive modulation is possible. Familiarization-like effects permit the plant to ignore harmless stimuli and thus to economize the cost of elaborating a full defence response to a non-dangerous stimulation. Conversely, sensitization-like effects can produce increasingly rapid and violent responses to harmful stimuli. More sophisticated effects occur when the perception of a first stimulus modulates the intensity of the response to a subsequent stimulus of different nature.

5.2 Potential of the "Store/Recall" Memory Functions

The ecophysiological advantage for a plant to possess store/recall memory lays in the individual properties of the two functions STO and RCL and their combined effect.

5.2.1 Ecophysiological Significance of the Storage Function

Concerning the STO box, two facts are of primary importance: (1) that what is stored after the perception of a first stimulus is a sort of instruction for the response to that stimulus and (2) that subsequent stimuli will modulate the intensity, and perhaps also the nature, of the programmed response. The STO/RCL memory can thus progressively elaborate an integrated, updated programme of response to the ensemble of stimuli that the plant has perceived during the preceding course of time. Such integration will be much more efficient than responding separately to each individual stimulus. It will efficaciously contribute to plant acclimation to the climatic conditions prevailing at the place where it has rooted.

However, the ecological advantage of the programmed responses to stimuli is sometimes questionable in systems such as SR1 to SR3 (cf. Table 1). For instance, what can be the advantage of specifying the dominance between the cotyledonary buds in response to cotyledon pricking or of responding to manipulation stimulus by producing meristems in the hypocotyl? The reason is that at the outset of plant-memory research, simple and strictly controllable conditions had to be used to be able to prove the very existence of memory and describe its basic functions in view of the great reservations of many people to accept such a property of plants. Plants thus have made unexpected responses to unusual environmental conditions, i.e. differing from natural site scenarios.

Ecologists have stressed the importance for a plant to adjust the allocation of its (usually limited) resources to its main living activities in order to optimize its probability of survival and reproduction (Herms and Mattson 1992; Gayler 2010; Gayler et al. 2006, 2008; Matyssek et al. 2012). We may assume as a working hypothesis that the stored information, permitting the elaboration of an integrated, updated response to the variety of stimuli perceived by the plant in the course of time, is in fact an instruction for an optimized allocation of the plant resources to its principal living activities. It will be rewarding to explore if stored information vanishes after each generation or if cases of trans-generational information conservation exist.

5.2.2 Ecophysiological Significance of the Recall Function

Thanks to the RCL function, a stored instruction of response will not be expressed at any arbitrary time but only when an appropriate stimulus or external conditions have enabled the plant to recall the stored instruction and let it take effect. The RCL function can thus synchronize the release of the memorized response with the occurrence of a particular environmental event. The RCL function being linked with ultradian, circadian and annual rhythms of the plant (Sect. 3.1) means that it can similarly synchronize the release of the memorized response with the occurrence of a particular internal event. Since a plant can repeatedly recall a stored instruction of response, the RCL function is able to synchronize the release of the memorized response with different external and/or internal events occurring at different times.

5.2.3 Ecophysiological Significance of the Combined Effect of the Storage and Recall Functions

We have seen above (section "The Storage Function") that (1) information storage induced by cotyledon pricking was corresponding in fact to the storage of a sort of instruction of growth reduction in the hypocotyl of very young *Bidens* seedlings and to the storage of an instruction of dominance specification in the cotyledonary buds of slightly older *Bidens* seedlings and (2) the plants were enabled to recall the stored instructions by being subjected to different external events or conditions. Hence, same types of stimulus can induce the storage of different instructions of response in the different plant tissues, so that plants are enabled to recall these stored instructions of responses at different times in different tissues. A store/recall memory thus supplies an extreme variety of possibilities to a plant to adjust its response to stimuli. Hence, the plant can keep track of the progress in internal and external processes possibly occurring at different times across different tissues.

5.3 Ecophysiological Significance of the Combined Effect of Priming and Store/Recall Memory

Although published data usually deal with the study of either the priming or store/ recall forms of memory, it is likely that plants operate both forms simultaneously. In that case, with the priming memory plants would be able to adjust their response intensity to the dangerousness of the signals perceived. Simultaneously with the store/recall memory, they could elaborate an integrated, updated response to the entirety of the environmental stimuli (especially the climatic stimuli) and synchronize the response to any particular internal or external event including ultradian, circadian or annual rhythms of the plant. Instead of appearing as a sort of cock-and-bull story, the equipment of plants with priming and store/recall memories could thus play a major part in optimizing plant response to aggressions and climatic hazards. Experimental clarification is required for strengthening such interpretation of plant memory. However, a straight explanation is offered already right now of the fact that plants can manage to survive under possibly awkward conditions that may prevail at rooting sites.

6 Conclusions

Originally, researchers have been mainly interested in proving that memory capacities exist in plants and in unravelling the main characteristics of plant memory (Thellier et al. 1982). Incidentally, with this work it has been observed early that these properties of the plant memory enabled them to produce a sophisticated and efficacious mode of adapting to the conditions at the places of rooting. Now, time has come to assess specific ecological advantages that the possession of a memory can confer to a plant for the adaptation and acclimation to the environment. Exploring such dimensions was the aim of this assay. Plants have a variety of ways to store ecologically relevant information for memorization. Turnover of pools of metabolites, of conformation states of individual enzymes or multi-protein-subunit enzymes and of transcript levels provides mechanisms of memory. This particularly holds true for the priming memory. The complex STO/RCL memory involves STO and RCL genes as incorporated in our model of memory (Thellier and Lüttge 2013), and epigenetic modifications are substantial mechanisms of memory (Fig. 1). The more we survey adaptations and acclimations to environmental signals and stress which are often stable over non-stressful intervals and even generations, the more examples we find of memory functions which are controlling and regulating ecological performance of plants. It is clear that memory of plants is not an occasional episode but a fundamental property of plant life governing ecological comportment. This is also a challenge for future research on specific examples of physiological and biochemical ecology for unravelling molecular mechanisms of memory. The ultimate aim is the understanding of functional linkages resulting in theory building and computer simulations of plant internal networks with the functions of memory and the biological clock (Hütt et al. 2015).

Natural selection has equipped animals and plants with memory. The underlying mechanisms are completely different, based on the existence and lack of a nervous system, respectively. Strategies of information management are different. Animals are able to change their location to feed, escape predators and, more generally, look for environmental conditions suited to their metabolism and mode of life: their memory helps them to orientate in space and time during migratory explorations. Evolution yielded sophisticated results as in the case of higher species and humans. By contrast, plant strategies consist of adapting their metabolism and development to the environmental conditions at rooting sites. To do that plants must be able (1) to perceive the stimuli and stresses issued from their environment and (2) to create a response optimizing probabilities of survival and reproduction, which requires memorization at various spatiotemporal levels of structure and function. Many processes at the genetic and molecular, the metabolic and the physiological levels are involved in plant memorization and make up parts of the hardware structure as well as the operational software functioning of memory.

All living beings that have biomembranes, polynucleotides and proteins have the molecular basis to develop memory, and this includes prokaryotes (Thellier and Lüttge 2013). Memory is a fundamental property of all life. Therefore, we should not be surprised that being equipped with memory is essential for the ecological performance and persistence of plants in their environment.

Acknowledgement We thank Professor Dr. Rainer Matyssek for carefully reading the manuscript and very many useful suggestions and stimulating comments.

References

Adams KL (2010) Dandelions 'remember' stress: heritable stress-induced methylation patterns. New Phytol 185:867–868

Alvarez ME, Nota F, Cambiagno DA (2010) Epigenetic control of plant immunity. Mol Plant Pathol 11:563–576

Amzallag GN (2002) The adaptive potential of plant development: evidence from the response to salinity. In: Läuchli A, Lüttge U (eds) Salinity: environment—plants—molecules. Kluver, Dordrecht, pp 291–312

Amzallag GN (2005) Perturbed reproductive development in salt-treated *Sorghum bicolor*: a consequence of modifications in regulation networks? J Exp Bot 56:2821–2829

Amzallag GN, Seligmann H, Lerner HR (1993) A developmental window for salt-adaptation in *Sorghum bicolor*. J Exp Bot 44:645–652

Appel HM, Cocroft RB (2014) Plants respond to leaf vibrations caused by insect herbivore chewing. Oecologia 175:1257–1266

Baek D, Jiang J, Chung JS, Wang B, Chen J, Xin Z, Shi H (2011) Regulated AtHKT1 gene expression by a distal enhancer and DNA methylation in the promoter plays an important role in salt tolerance. Plant Cell Physiol 52:149–161

Bahnweg G, Heller W, Stich S, Knappe C, Betz G, Heerdt C, Kehr RD, Ernst D, Langebartels C, Nunn AJ, Rothenburger J, Schubert R, Müller-Starck G, Werner H, Matyssek R, Sandermann H Jr (2005) Beech leaf colonization by the endophyte *Apiognomonia errabunda* dramatically depends on light exposure and climatic conditions. Plant Biol 7:659–669

Bailey C, Chen M (1983) Morphological basis of long-term habituation and sensitization in *Aplysia*. Science 220:91–93

Baluška F, Ninkovic V (2010) Plant communication from an ecological perspective. Springer, Berlin, p 252

Bilger W, Björkman O (1994) Relationships among violaxanthin deepoxidation, thylakoid membrane conformation, and non-photochemical chlorophyll fluorescence quenching in leaves of cotton (*Gossypium hirsutum* L.). Planta 193:238–246

Bird A (2002) DNA methylation patterns and epigenetic memory. Genes Dev 16:6–21

Bond DM, Finnegan EJ (2007) Passing the message on: inheritance of epigenetic traits. Trends Plant Sci 12:211–216

Boyko A, Kovalchuk I (2008) Epigenetic control of plant stress response. Environ Mol Mutagen 49:61–72

Bruce TJA (2010) Exploiting plant signals in sustainable agriculture. In: Baluška F, Ninkovic V (eds) Plant communication from an ecological perspective. Springer, Berlin, pp 215–227

Bruce TJA, Matthes MC, Napier JA, Pickett JA (2007) Stressful 'memories' of plants: evidence and possible mechanisms. Plant Sci 173:603–608

Buchanan RB, Gruissem W, Jones RL (2000) Biochemistry and molecular biology of plants. American Society Plant Physiologists, Rockville, MD, 1367p
Cervantes-Laurean D, Jacobson EL, Jacobson MK (1996) Glycation and glycoxidation of histones by ADP-ribose. J Biol Chem 271:10461–10469
Chen M, Lv S, Meng Y (2010) Epigenetic performers in plants. Dev Growth Differ 52:555–566
Chinnusamy V, Zhu J-K (2009) Epigenetic regulation of stress responses in plants. Curr Opin Plant Biol 12:133–139
Conrath U (2009) Priming of induced plant defense responses. Adv Bot Res 51:361–395
Conrath U (2011) Molecular aspects of defence priming. Trends Plant Sci 16:524–531
Conrath U, Thulke O, Katz V, Schwindling S, Kohler A (2001) Priming as a mechanism in induced systemic resistance of plants. Eur J Plant Pathol 107:113–119
Covington MF, Panda S, Liu XL, Strayer CA, Wagner DR, Kay SA (2001) ELF3 modulates resetting of the circadian clock in *Arabidopsis*. Plant Cell 13:1305–1315
Critchley C, Russell AW (1994) Photoinhibition of photosynthesis in vivo: the role of protein turnover in photosystem II. Physiol Plant 92:188–196
Darrah C, Taylor BL, Edwards KD, Brown PE, Hall A, McWatters HG (2006) Analysis of phase of *LUCIFERASE* expression reveals novel circadian quantitative trait loci in *Arabidopsis*. Plant Physiol 140:1464–1474
Davies E, Stankovic B, Vian A, Wood AJ (2012) Where has all the message gone? Plant Sci 185–186:23–32
Daxinger L, Whitelaw E (2010) Transgenerational epigenetic inheritance: more questions than answers. Genome Res 20:1623–1628
Desbiez MO, Champagnat P, Boyer N, Frachisse JM, Gaspar T, Thellier M (1983) Inhibition correlative de la croissance de l'hypocotyle de *Bidens pilosus* L. par des traumatismes cotylédonaires légers. Bull Soc Bot Fr (Actual Bot) 130:67–77
Desbiez MO, Champagnat P, Thellier M (1986) Mécanismes de mise en mémoire et de rappel de mémoire chez *Bidens pilosus*. CR Acad Sci Paris 302:573–578
Desbiez MO, Gaspar T, Crouzillat D, Frachisse JM, Thellier M (1987) Effect of cotyledonary prickings on growth, ethylene metabolism and peroxidase activity in *Bidens pilosus*. Plant Physiol Biochem 25:137–143
Desbiez MO, Tort M, Thellier M (1991) Control of a symmetry-breaking process in the course of the morphogenesis of plantlets of *Bidens pilosa* L. Planta 184:397–402
Desbiez MO, Mikulecky D, Thellier M (1994) Growth messages in plants: principle of a possible modeling and further experimental characteristics. J Biol Syst 2:127–136
Desbiez MO, Tort M, Monnier C, Thellier M (1998) Asymmetrical triggering of the cell cycle in opposite buds of a young plant, after a slight cotyledonary wound. CR Acad Sci Paris (Sciences de la Vie/Life Sciences) 321:403–407
Devlin PF (2002) Signs of the time: environmental input to the circadian clock. J Exp Bot 53:1535–1550
Dicke M (2009) Behavioural and community ecology of plants that cry for help. Plant Cell Environ 32:654–665
Ding Y, Fromm M, Avramona Z (2012) Multiple exposures to drought 'train' transcriptional responses in *Arabidopsis*. Nat Commun 3:740
Dixon LE, Hodge SK, van Ooijen G, Troein C, Akman OE, Millar AJ (2014) Light and circadian regulation of clock components aids flexible responses to environmental signals. New Phytol 203:568–577
Dodd AN, Salathia N, Hall A, Kevei E, Toth R, Nagy F, Hibberd JM, Millar AJ, Webb AAR (2005) Plant circadian clocks increase photosynthesis, growth, survival and competitive advantage. Science 309:630–633
Dolmetsch RE, Lewis RS, Goodnow CC, Healy JJ (1997) Differential activation of transcription factors induced by Ca^{2+} response amplitude and duration. Nature 386:855–858
Dudai Y (2004) The neurobiology of consolidations, or, how stable is the engram? Annu Rev Psychol 55:51–86

Edmunds LN, Tamponnet C (1990) Oscillator control of cell division cycles in *Euglena*: role of calcium in circadian time-keeping. In: O'Day DH (ed) Calcium as an intracellular messenger in eucaryotic microbes. American Society for Microbiology, Washington, DC, pp 97–123
Farré EM (2012) The regulation of plant growth by the circadian clock. Plant Biol 14:401–410
Forbes-Stovall J, Howton J, Young M, Davis G, Chandler T, Kessler B, Rinehart CA, Jacobshagen S (2014) *Chlamydomonas reinhardtii* strain CC-124 is highly sensitive to blue light in addition to green and red light in resetting its circadian clock, with the blue-light photoreceptor plant cryptochrome likely acting as negative modulator. Plant Cell Physiol 75:14–23
Frankhauser C, Staiger D (2002) Photoreceptors in *Arabidopsis thaliana*: light perception, signal transduction and entrainment of the endogenous clock. Planta 216:1–16
Fujiwara S, Oda A, Yoshida R, Niinuma K, Miyata K, Tomozoe Y, Tajima T, Nakagawa M, Hayashi K, Coupland G, Mizoguchi T (2008) Circadian clock proteins LHY and CCA1 regulate SVP protein accumulation to control flowering in *Arabidopsis*. Plant Cell 20:2960–2971
Gagliano M, Renton M, Depczynski M, Mancuso S (2014) Experience teaches plants to learn faster and forget slower in environments where it matters. Oecologia 175:63–72
Gális I, Gaquerel E, Pandey SP, Baldwin IT (2009) Molecular mechanisms underlying plant memory in JA-mediated defense responses. Plant Cell Environ 32:617–627
Gayler S (2010) Modélisation de l'effet de facteurs de l'environnement sur la répartition des ressources dans un système végétal mixte. CR Acad Agric France 96:89–90
Gayler S, Grams TEE, Kozovits A, Luedemann G, Winkler JB, Priesack E (2006) Analysis of competition effects in mono- and mixed cultures of juvenile beech and spruce by means of the plant growth simulation model PLATHO. Plant Biol 8:503–514
Gayler S, Grams TEE, Heller W, Treutter D, Priesack E (2008) A dynamic model of environmental effects on allocation to carbon-based secondary compounds in juvenile trees. Ann Bot 101:1089–1098
Gilmore AM (1997) Mechanistic aspects of xanthophyll cycle-dependent photoprotection in higher plant chloroplasts and leaves. Physiol Plant 99:197–209
Gilmore AM, Govindjee (1999) How higher plants respond to excess light: energy dissipation in photosystem II. In: Singhal GS, Renger G, Sopory SK, Irrgang KD, Govindjee (eds) Concepts in photobiology: photosynthesis and photomorphogenesis. Narosa Publishing House, New Delhi, pp 513–548
Gilmore AM, Yamasaki H (1998) 9-Aminoacridine and dibucaine exhibit competitive interactions and complicated inhibitory effects that interfere with measurements of ΔpH and xanthophyll cycle-dependent photosystem II energy dissipation. Photosynth Res 57:159–174
Gilmore AM, Hazlett TL, Debrunner PG, Govindjee (1996) Comparative time-resolved photosystem II chlorophyll *a* fluorescence analyses reveal distinctive differences between photoinhibitory reaction center damage and xanthophyll cycle-dependent energy dissipation. Photochem Photobiol 64:552–563
Gilmore AM, Shinkarev VP, Hazlett TL, Govindjee (1998) Quantitative analysis of the effects of intrathylakoid pH and xanthophyll cycle pigments on chlorophyll *a* fluorescence lifetime distributions and intensity in thylakoids. Biochemistry 73:13582–13593
Gols R (2014) Direct and indirect chemical defenses against insects in a multitrophic framework. Plant Cell Environ 37:1741–1752
Goss R, Lepetit B (2015) Biodiversity of NPQ. J Plant Physiol 172:13–32
Grunstein M (1997) Histone acetylation in chromatin structure and transcription. Nature 389:349–352
Habte E, Müller LM, Shtaya M, Davis SJ, von Korff M (2014) Osmotic stress at the barley root affects expression of circadian clock genes in the shoot. Plant Cell Environ 37:1321–1337
Harmer SL, Kay SA (2005) Positive and negative factors confer phase-specific circadian regulation of transcription in *Arabidopsis*. Plant Cell 17:1926–1940
He J, Chow WS (2003) The rate coefficient of repair of photosystem II after photoinactivation. Physiol Plant 118:297–304

Heil M (2010) Within-plant signaling by volatiles triggers systemic defences. In: Baluška F, Ninkovic V (eds) Plant communication from an ecological perspective. Springer, Berlin, pp 99–112
Herms DA, Mattson WJ (1992) The dilemma of plants: to grow or defend. Q Rev Biol 67:283–335
Holt NE, Zigmantas D, Valkunas L, Li X-P, Niyogi KK, Fleming GR (2005) Carotenoid cation formation and the regulation of photosynthetic light harvesting. Science 307:433–436
Horton P, Ruban A (2005) Molecular design of the photosystem II light-harvesting antenna: photosynthesis and photoprotection. J Exp Bot 56:365–373
Horton P, Ruban AV, Walters RG (1994) Regulation of light harvesting in green plants. Indication by nonphotochemical quenching of chlorophyll fluorescence. Plant Physiol 106:415–420
Horton P, Ruban AV, Walters RG (1996) Regulation of light harvesting in green plants. Annu Rev Plant Physiol Plant Mol Biol 47:655–684
Hotta CT, Gardner MJ, Hubbard KE, Baek SJ, Dalchau N, Suhita D, Dodd AN, Webb AAR (2007) Modulation of environmental responses of plants by circadian clocks. Plant Cell Environ 30:333–349
Hütt M-T, Lüttge U, Thellier M (2015) Noise-induced phenomena and complex rhythms: a test scenario for plant systems biology. In: Mancuso S, Shabala S (eds) Rhythms in plants, 2nd edn. Springer, Berlin, pp 279–321
Ibáñez C, Kozarewa I, Johansson M, Ögren E, Rohde A, Eriksson ME (2010) Circadian clock components regulate entry and affect exit of seasonal dormancy as well as winter hardiness in *Populus* trees. Plant Physiol 153:1823–1833
Jablonka E, Lamb MJ (1989) The inheritance of acquired epigenetic variation. J Theor Biol 139:69–83
Jennings RC, Islam K, Zucchelli G (1986) Spinach-thylakoid phosphorylation: studies on the kinetics of changes in photosystem antenna size, spill-over and phosphorylation of light-harvesting chlorophyll a/b protein. Biochim Biophys Acta [Bioenergetics] 850:483–489
Johnson CH (1992) Phase response curves: what can they tell us about circadian clocks? In: Hiroshige T, Honma K (eds) Circadian clocks from cell to human. Hokkaido University Press, Sapporo, Japan, pp 209–249
Johnson CH, Golden SS (1999) Circadian programs in cyanobacteria: adaptiveness and mechanism. Annu Rev Microbiol 53:389–409
Kakutani T (2002) Epi-alleles in plants: inheritance of epigenetic information over generations. Plant Cell Physiol 43:1106–1111
Kant P, Gordon M, Kant S, Zolla G, Davydov O, Heimer YM, Chalifa-Caspi V, Shaked R, Barak S (2008) Functional-genomics-based identification of genes that regulate *Arabidopsis* responses to multiple abiotic stresses. Plant Cell Environ 31:697–714
Karban R, Wetzel WC, Shiojiri K, Ishizaki S, Ramirez SR, Blande JD (2014) Deciphering the language of plant communication: volatile chemotypes of sagebrush. New Phytol 204:380–385
Kathiria P, Sidler C, Golubov A, Kalischuk M, Kawchuk LM, Kovalchuk I (2010) Tobacco mosaic virus infection results in an increase in recombination frequency and resistance to viral, bacterial, and fungal pathogens in the progeny of infected tobacco plants. Plant Physiol 153:1859–1870
Kessler A, Baldwin IT (2002) Plant responses to insect herbivory: the emerging molecular analysis. Annu Rev Plant Biol 53:299–328
Kessler A, Baldwin IT (2004) Herbivore-induced plant vaccination. Part I. The orchestration of plant defenses in nature and their fitness consequences in wild tobacco *Nicotiana attenuata*. Plant J 38:639–649
Kikis EA, Khanna R, Quail PH (2005) ELF4 is a phytochrome-regulated component of a negative-feedback loop involving the central oscillator components CCA1 and LHY. Plant J 44:300–313
Kinoshita T, Seki M (2014) Epigenetic memory for stress response and adaptation in plants. Plant Cell Physiol 55:1859–1863
Knight MR, Campbell AK, Smith SM, Trewavas AJ (1991) Transgenic plant aequorin reports the effect of touch and cold-shock and elicitors on cytoplasmic calcium. Nature 352:524–526

Knight MR, Smith SM, Trewavas AJ (1992) Wind-induced plant motion immediately increases cytosolic calcium. Proc Natl Acad Sci USA 89:4967–4971

Knight H, Brandt S, Knight MR (1998) A history of stress alters drought calcium signaling pathways in *Arabidopsis*. Plant J 16:681–687

Kou HP, Li Y, Song XX, Ou XF, Xing SC, Ma J, von Wettstein D, Liu B (2011) Heritable alteration in DNA methylation induced by nitrogen-deficiency stress accompanies enhanced tolerance by progenies to stress in rice (*Oryza sativa* L.). J Plant Physiol 168:1685–1693

Langebartels C, Heller W, Führer G, Lippert M, Simons S, Sandermann H (1998) Memory effects in the action of ozone on conifers. Ecotoxicol Environ Saf 41:62–72

Lesburguères E, Gobbo OL, Alaux-Cantin S, Hambucken A, Trifilieff P, Bontempi B (2011) Early tagging of cortical networks is required for the formation of enduring associative memory. Science 331:924–928

Li S, Liu J, Liu Z, Li X, Wu F, He Y (2014) *HEAT-INDUCED TAS1 TARGETG1* mediates thermotolerance via HEAT STRESS TRANSCRIPTION FACTOR A1-directed pathways in *Arabidopsis*. Plant Cell 26:1764–1780

Lo WS, Duggan L, Emre NCT, Belotserkovskya R, Lane WS, Shiekhattar R, Berger SL (2001) Snf1—a histone kinase that works in concert with the histone acetyltransferase Gcn5 to regulate transcription. Science 293:1142–1146

Lodish H, Berk A, Zipursky SL, Matsudaira P, Baltimore D, Darnell J (2000) Molecular cell biology, 4th edn. W. H. Freeman, New York, NY, Section 21-7: Learning and memory

Loreto F, Dicke M, Schnitzler J-P, Turlings TCJ (2014) Plant volatiles and the environment. Plant Cell Environ 37:1905–1908

Love J, Dodd AN, Webb AAR (2004) Circadian and diurnal calcium oscillations encode photoperiodic information in Arabidopsis. Plant Cell 16:956–966

Luedemann G, Matyssek R, Fleischmann F, Grams TEE (2005) Acclimation to ozone affects host/pathogen interaction and competitiveness for nitrogen in juvenile *Fagus sylvatica* and *Picea abies* trees infected with *Phytophthora citricola*. Plant Biol 7:640–649

Luedemann G, Matyssek R, Winkler JB, Grams TEE (2009) Contrasting ozone × pathogen interaction as mediated through competition between juvenile European beech (*Fagus sylvatica*) and Norway spruce (*Picea abies*). Plant Soil 323:47–60

Lüttge U (2003) Circadian rhythmicity: is the "biological clock" hardware or software. Progr Bot 64:277–319

Lüttge U (2008) Physiological ecology of tropical plants, 2nd edn. Springer, Berlin, 458p

Lüttge U, Hertel B (2009) Diurnal and annual rhythms in trees. Trees 23:683–700

Lüttge U, Kluge M, Thiel G (2010) Botanik. Die umfassende Biologie der Pflanzen. Wiley-VCH, Weinheim, 1215p

Matyssek R, Sandermann H (2003) Impact of ozone on trees: an ecophysiological perspective. Prog Bot 64:349–404

Matyssek R, Sandermann H, Wieser G, Booker F, Cieslik S, Musselman R, Ernst D (2008) The challenge of making ozone risk assessment for forest trees more mechanistic. Environ Pollut 156:567–582

Matyssek R, Koricheva J, Schnyder H, Ernst D, Munch JC, Osswald W, Pretzsch H (2012) The balance between resource sequestration and retention: a challenge in plant science. In: Matyssek R, Schnyder H, Osswald W, Ernst D, Munch JC, Pretzsch H (eds) Growth and defence in plants—resource allocation at multiple scales, Ecological studies 220. Springer, Heidelberg, pp 3–24

Matzke M, Matzke AJ, Pruss GJ, Vance VB (2001) RNA-based silencing strategies in plants. Curr Opin Genet Dev 11:221–227

Matzke M, Kanno T, Huettel B, Daxinger L, Matzke AJ (2007) Targets of RNA directed DNA methylation. Curr Opin Plant Biol 10:512–519

McAinsh MR, Hetherington AM (1998) Encoding specificity in Ca^{2+} signalling systems. Trends Plant Sci 3:32–36

McClung CR (2006) Plant circadian rhythms. Plant Cell 18:792–803

McWatters HG, Bastow RM, Hall A, Millar AJ (2000) The *ELF3 zeitnehmer* regulates light signaling to the circadian clock. Nature 408:716–720
Michael TP, McClung CR (2002) Phase-specific circadian clock regulatory elements in *Arabidopsis*. Plant Physiol 130:627–638
Millar AJ (1999) Biological clocks in *Arabidopsis thaliana*. New Phytol 141:175–197
Millar AJ, Kay SA (1996) Integration of circadian and phototransduction pathways in the network controlling *CAB* gene transcription in *Arabidopsis*. Proc Natl Acad Sci USA 93:15491–15496
Molinier J, Ries G, Zipfel C, Hohn B (2006) Transgeneration memory of stress in plants. Nature 442:1046–1049
Müller LM, von Korff M, Davis SJ (2014) Connections between circadian clocks and carbon metabolism reveal species-specific effects on growth control. J Exp Bot 65:2915–2923
Nakamichi N (2011) Molecular mechanisms underlying the *Arabidopsis* circadian clock. Plant Cell Physiol 52:1709–1718
Niwa Y, Yamashino T, Mizuno T (2009) The circadian clock regulates the photoperiodic response of hypocotyl elongation through a coincidence mechanism in *Arabidopsis thaliana*. Plant Cell Physiol 50:838–854
Ogudi T, Sage-Ono K, Kamada H, Ono M (2004) Characterization of transcriptional oscillation of an *Arabidopsis* homolog of *PnC401* related to photoperiodic induction of flowering in *Pharbitis nil*. Plant Cell Physiol 45:232–235
Olbrich M, Knappe C, Wenig M, Gerstner E, Häberle K-H, Kitao M, Matyssek R, Stich S, Leuchner M, Werner H, Schlink K, Müller-Starck G, Welzl G, Scherb H, Ernst D, Heller W, Bahnweg G (2010) Ozone Fumigation (twice ambient) reduces leaf infestation following natural and artificial inoculation by the endophytic fungus *Apiognomonia errabunda* of adult European beech trees. Environ Pollut 158:1043–1050
Onai K, Okamoto K, Nishimoto H, Morioka C, Hirano M, Kami-Ike N, Ishiura M (2004) Large-scale screening of *Arabidopsis* circadian clock mutants by a high-throughput real-time bioluminescence monitoring system. Plant J 40:1–11
Osmond CB, Grace SC (1995) Perspectives on photoinhibition and photorespiration in the field: quintessential inefficiencies of the light and dark reactions of photosynthesis? J Exp Bot 46:1351–1362
Ouyang Y, Andersson CR, Kondo T, Golden SS, Johnson CH (1998) Resonating circadian clocks enhance fitness in cyanobacteria. Proc Nat Acad Sc USA 95:8660–8664
Pastor V, Luna E, Mauch-Mani B, Ton J, Flors V (2013) Primed plants do not forget. Environ Exp Bot 94:46–56
Pearcy RW, Osteryoung K, Calkin HW (1985) Photosynthetic responses to dynamic light environments by Hawaiian trees. Plant Physiol 79:896–902
Pierik R, Ballaré CL, Dicke M (2014) Ecology of plant volatiles: taking a plant community perspective. Plant Cell Environ 37:1845–1853
Plieth C, Hansen UP, Knight H, Knight MR (1999) Temperature sensing by plants: the primary characteristics of signal perception and calcium response. Plant J 18:491–497
Portis AR (2003) Rubisco activase—Rubisco's catalytic chaperone. Photosyn Res 75:11–27
Rasmann S, de Vos M, Casteel CL, Tian D, Halitschke R, Sun JY, Agrawal AA, Felton GW, Jander G (2012) Herbivory in the previous generation primes plants for enhanced insect resistance. Plant Physiol 158:854–863
Richards EJ (2006) Inherited epigenetic variation—revisiting soft inheritance. Nat Rev Genet 76:395–401
Rikin A (1991) Temperature-induced phase shifting of circadian rhythms in cotton seedlings as related to variations in chilling resistance. Planta 185:407–414
Ripoll C, Le Sceller L, Verdus M-C, Norris V, Tafforeau M, Thellier M (2009) Memorization of abiotic stimuli in plants: a complex role for calcium. In: Baluska F (ed) Plant-environment interactions. Springer, Berlin, pp 267–283

Roden LC, Song H-R, Jackson S, Morris K, Carré IA (2002) Floral responses to photoperiod are correlated with the timing of rhythmic expression relative to dawn and dusk in *Arabidopsis*. Proc Natl Acad Sci USA 99:13313–13318

Roux D, Vian A, Girard S, Bonnet P, Paladian F, Davies E, Ledoigt G (2006) Electromagnetic fields (900 MHz) evoke consistent molecular responses in tomato plants. Physiol Plant 128:283–288

Salomé PA, Michael TP, Kearns EV, Fett-Neto AG, Sharrock RA, McClung CR (2002) The *out of phase 1* mutant defines a role for PHYB in circadian phase control in *Arabidopsis*. Plant Physiol 129:1674–1685

Sasek TW, Richardson CJ, Fendick EA, Bevington SR, Kress LW (1991) Carryover effects of acid rain and ozone on the physiology of multiple flushes of loblolly pine seedlings. For Sci 37:1078–1098

Sassenrath-Cole GF, Pearcy RW (1992) The role of ribulose-bisphosphate regeneration in the induction requirement of photosynthetic CO_2 exchange under transient light conditions. Plant Physiol 99:227–234

Saze H (2008) Epigenetic memory transmission through mitosis and meiosis in plants. Semin Cell Dev Biol 19:527–536

Shen J, Xie K, Xiong L (2010) Global expression profiling of rice microRNAs by one-tube stem-loop reverse transcription quantitative PCR revealed important roles of microRNAs in abiotic stress responses. Mol Gen Genet 284:477–488

Sinclair J, Hanks P, Fox G, Moon R, Stock P (1987) Collins Cobuild English dictionary. Collins, London, 1703p

Sridhar VV, Kapoor A, Zhang K, Zhu J, Zhou T, Hasegawa PM, Bressan RA, Zhu JK (2007) Control of DNA methylation and heterochromatic silencing by histone H2B deubiquitination. Nature 447:735–738

Sunkar R, Zhu JK (2004) Novel and stress-regulated microRNAs and other small RNAs from Arabidopsis. Plant Cell 16:2001–2019

Tafforeau M, Verdus M-C, Norris V, White G, Demarty M, Thellier M, Ripoll C (2002) SIMS study of the calcium-deprivation step related to epidermal meristem production induced in flax by cold shock or radiation from a GSM telephone. J Trace Microprobe Techn 20:611–623

Tafforeau M, Verdus MC, Norris V, White GJ, Cole M, Demarty M, Thellier M, Ripoll C (2004) Plant sensitivity to low intensity 105 GHz electromagnetic radiation. Bioelectromagnetics 25:403–407

Tafforeau M, Verdus M-C, Norris V, Ripoll C, Thellier M (2006) Memory processes in the response of plants to environmental signals. Plant Signal Behav 1:9–14

Tanigawa Y, Tsuchiya M, Imai Y, Shimoyama M (1984) ADP-ribosyltransferase from hen liver nuclei. Purification and characterization. J Biol Chem 259:2022–2029

Tanner W (1969) Light-driven active uptake of 3-O-methylglucose via an inducible hexose uptake system of *Chlorella*. Biochem Biophys Res Comm 36:278–283

Tanner W, Grünes R, Kandler O (1970) Spezifität und Turnover des induzierbaren Hexose-Aufnahmesystems von *Chlorella*. Z Pflanzenphysiol 62:376–386

Thellier M (2015) Les plantes ont-elles de la mémoire? Editions Quae, Versailles, 111p

Thellier M, Lüttge U (2013) Plant memory: a tentative model. Plant Biol 15:1–12

Thellier M, Desbiez MO, Champagnat P, Kergosien Y (1982) Do memory processes also occur in plants? Physiol Plant 56:281–284

Thellier M, Le Sceller L, Norris V, Verdus M-C, Ripoll C (2000) Long-distance transport, storage and recall of morphogenetic information in plants: the existence of a primitive plant "memory". CR Acad Sci Paris (Sciences de la Vie/Life Science) 323:81–91

Thellier M, Ripoll C, Norris V (2013) Memory processes in the control of plant growth and metabolism. Nova Acta Leopoldina NF 114(391):21–42

Trewavas A (1999) Le calcium c'est la vie: calcium waves. Plant Physiol 120:1–6

Trewavas A (2003) Aspects of plant intelligence. Ann Bot 92:1–20

Tyystjärvi E, Aro E-M (1996) The rate constant of photoinhibition measured in lincomycin-treated leaves is directly proportional to light intensity. Proc Natl Acad Sci USA 93:2213–2218
Ueda M, Nakamura Y (2006) Metabolites involved in plant movement and "memory": nyctinasty of legumes and trap movement in the Venus flytrap. Nat Prod Rep 23:548–557
Valladares F, Allen MT, Pearcy RW (1997) Photosynthetic responses to dynamic light under field conditions in six tropical rainforest shrubs along a light gradient. Oecologia 111:505–514
van Hulten M, Ton J, Pieterse CMJ, van Wees SCM (2010) Plant defense signaling from the underground primes aboveground defenses to confer enhanced resistance in a cost-efficient manner. In: Baluška F, Ninkovic V (eds) Plant communication from an ecological perspective. Springer, Berlin, pp 43–60
Verdus M-C, Thellier M, Ripoll C (1997) Storage of environmental signals in flax: their morphogenetic effect as enabled by a transient depletion of calcium. Plant J 12:1399–1410
Verdus M-C, Le Sceller L, Norris V, Thellier M, Ripoll C (2007) Pharmacological evidence for calcium involvement in the long-term processing of abiotic stimuli in plants. Plant Signal Behav 2:212–220
Verhoeven KJF, Jansen JJ, van Dijk PJ, Biere A (2010) Stress-induced DNA methylation changes and their heritability in asexual dandelions. New Phytol 185:1108–1118
Vian A, Roux D, Girard S, Bonnet P, Paladian F, Davies E, Ledoigt G (2006) Microwave irradiation affects gene expression in plants. Plant Signal Behav 1:67–70
Voelckel C, Baldwin IT (2004) Herbivore-induced plant vaccination. Part II. Array-studies reveal the transience of herbivore-specific transcriptional imprints and a distinct imprint from stress combinations. Plant J 38:650–663
Wang WS, Pan YJ, Zhao XQ, Dwivedi D, Zhu LH, Ali J, Fu BY, Li ZK (2010) Drought-induced site-specific DNA methylation and its association with drought tolerance in rice (*Oryza sativa* L.). J Exp Bot 62:1951–1960
Wenden B, Kozma-Bognár L, Edwards KD, Hall AJW, Locke JCW, Millar AJ (2011) Light inputs shape the *Arabidopsis* circadian system. Plant J 66:480–491
Wilhelm C, Wirth C (2015) Physiodiversity—New tools allow physiologist to embrace biodiversity and reconstruct the evolution of 'physiologies'? J Plant Physiol 172:1–3
Woelfle MA, Ouyang Y, Phanvijhitsiri K, Johnson CH (2004) The adaptive value of circadian clocks: an experimental assessment in cyanobacteria. Curr Biol 14:1481–1486
Yaish MW, Colasanti J, Rothstein SJ (2011) The role of epigenetic processes in controlling flowering time in plants exposed to stress. J Exp Bot 62:3727–3735
Yerushalmi S, Green RM (2009) Evidence for the adaptive significance of circadian rhythms. Ecol Lett 12:970–981
Yerushalmi S, Yakir E, Green RM (2011) Circadian clocks and adaptation in *Arabidopsis*. Mol Ecol 20:1155–1165
Zhang X (2008) The epigenetic landscape of plants. Science 320:489–492
Zhang Y, Reinberg D (2001) Transcription regulation by histone methylation: interplay between different covalent modifications of the core histone tails. Genes Dev 15:2343–2360
Zhang X, Yazaki J, Sundaresan A, Cokus S, Chan SW-L, Chen H, Henderson IR, Shinn P, Pellegrini M, Jacobsen SE, Ecker JR (2006) Genome-wide high resolution mapping and functional analysis of DNA methylation in *Arabidopsis*. Cell 126:1189–1201
zu Castell W, Fleischmann F, Heger T, Matyssek R (2016) Shaping theoretic foundations of holobiont-like systems. In: Cánovas FM, Lüttge U, Matyssek R (eds) Progress in botany, vol 77. Springer, Heidelberg

Root Pressure: Getting to the Root of Pressure

Sanjay Singh

Contents

S. Singh (✉)
Department of Plant Sciences, College of Agriculture & Rural Transformation, University of Gondar, Gondar City, Ethiopia
e-mail: sanju80gon@gmail.com

U. Lüttge et al. (eds.), *Progress in Botany*, Progress in Botany 77,
DOI 10.1007/978-3-319-25688-7_3

Abstract Water and solutes are essential ingredients of root pressure. For root pressure to develop, the entry into root cells including xylem tissues of these ingredients is necessary. Therefore, in the light of latest research findings, the integrated synthesis of the combined "osmotic" and "energetically driven uphill water transport" in plants against water potential gradient and the gravity has been presented with the hope to halt or at least dilute for some time to come the prevalent legacy of "riddle" or "enigma" of root pressure. Further, various techniques, both invasive and noninvasive including new ones, have been described focusing on the factors affecting and consequences and implications of root pressure for agriculture, horticulture, and forestry.

1 Introduction

The evolution and progression of unicellular to multicellular aquatic plants and to the present-day giant terrestrial tallest known living tree measuring 115.61 m (379.3 ft) (Preston 2007) on the Earth essentially call for an interdependency of their various organs deploying the principle of division of labor to drive their complex life activities through the integration of different structurally specialized functions for survival and growth under continually click-by-click changing environment since the birth of this planet about 4.5 billion years ago (Darwin 1859; Chamberlin 1916; Singh et al. 2009a). Obviously and quite understandably, in this long route of evolutionary journey of progression, the higher plants strategically through adaptation opted for the formation of two distinct organs, i.e., root and shoot, to perform interdependent functions translating them into growth and development through their mutual negotiations. These negotiations include, in the main, the supply of water and nutrients by root to the shoot and, in return, the supply of manufactured food and energy by shoot to the root making the existence of plants themselves as well as animals and humans possible. However, not surprisingly, we hear and read more of shoot functionings than those of root—the supplier of basic ingredients such as water and nutrients, for photosynthesis—the basis of life on this planet. This discrepancy for root victimization seems to have resulted from the general principle of human behavior under the rule of a well-known dictum—out of sight, out of mind—roots being invisible to human eyes due to their hidden habitat beneath the soil. However, of late, the increasing demand for food to feed the ever-rising population, both the diminishing availability of water and the restrictions imposed on the increasing use of fertilizers to provide nutrients to agricultural crops in the main, has attracted the attention of most plant biologists including physiologists, agronomists, geneticists, breeders, biotechnologists, and biometricians toward root research vis-a-vis shoot research worldwide (Barber and Bouldin 1984; Kirk 1994; Baluska 1995; Singh and Singh 1999, 2000; Singh et al. 1999; Henry 2013; Eshel and Beekchman 2013; Ahmed et al. 2014). Among root studies, the investigations on the physiology of uptake and transport of water and nutrients by roots have been in the past and the present time too the topic of great interest among plant physiologists on account of their economic and ecological significance for achieving global ecosystem sustainability (Jackson et al. 2000; Javot et al. 2003; Draye et al. 2010; Maurel et al. 2010; Aroca et al. 2012; Lobet et al. 2013). In this context, however, it has long been recognized that water and solute absorption by roots may either result from forces arising up in the leaves due to transpiration pull and transmitted to the roots and nearing soil solution (Steudle 2000, 2001; Tyree 2003a) or from forces developed down in the roots themselves called *root pressure* visibly evidenced by stem stump exudation or guttation or bleeding (Renner 1915, 1925; Kramer 1945; Singh and Singh 2013; Singh 2013, 2014a, b).

Pertinent to the present topic, there appear no variant opinions that it was Stephen Hales who first discovered the phenomenon of liquid exudation from the heavily pruned grapevines (*Vitis vinifera*) as early as 1727 and ascribed this event to some kind of force arising in the root which propelled the liquid up to exude from the cut surfaces terming this force as *root pressure*. Since then, it has engaged the mind of experimental botanists and plant physiologists who began to investigate this curious, complex, and intriguing physiological phenomenon, albeit with long intervals between investigations (Pfeffer 1881; Sachs 1887). It was only in the first quarter of the twentieth century that some definitive experimental data were provided to explain this phenomenon (Renner 1915; Priestley 1920; Overton 1921; Bose 1923; Kramer 1932) which followed, with the passage of time drawing the twentieth century to close, its detailed and accredited accounts written by a number of workers (White 1938; White et al. 1958; Lundegardh 1944; Kramer and Currier 1950; Stocking 1956; Davis 1961; Barrs 1966; O'Leary 1966; O'Leary and Kramer 1964; Singh and Singh 1989; Zholkevich 1991) leaving much scope for further work at the experimental front to explain the unresolved phenomenon of *root pressure* for this and perhaps, as it appears, the next century as well. Accordingly, the dawn of twenty-first century witnessed some illuminating studies advancing the concepts of "two-compartments," "three-interphases" (Pickard 2003a, b), "plant heart theory" (Kundt and Gruber 2006), "pushing water upward-like mechanism" (Singh et al. 2009a), and very recently proposed energetically driven uphill "water co-transport theory" (Wegner 2014) to explain the mechanism of root pressure. The problem of how plants manage to lift their water from the soil to their leaves has been with biologists at least since 1726, when Stephen Hales discussed it in the first edition of the *Proceedings of the Royal Society* of London. So much has been thought about it that present-day scientists often believe that the problem has been solved long ago or that there was no problem at all with mentions of osmotic suction, capillary forces, transpiration, and coherence. For plants, this supply problem is not answered in the biological textbooks, at least not satisfactorily so leaving the mechanism unresolved (Pickard 2003a, b; Kundt and Gruber 2006; Zholkevich et al. 2007; Wegner 2014).

I, therefore, in efforts to getting as close to the bottom of this physiological phenomenon as possible, have attempted to organize this chapter in various heads and subheads in the light of new and novel research findings which have led to the formulation of modern thermodynamically sound physical, physiological, metabolic, and molecular concepts and explanations based on the art-of-the-state technology of the techniques and instrumentation, to account for "pushing" the sap up in plants under "positive pressure" of roots in the absence or reduced transpiration which otherwise "pulls" the water up under "negative pressure" generated by cohesion–tension on the water present in the leaves at the cost of solar energy.

2 Definition of Root Pressure

The hydrostatic pressure developed in the roots of plants, causing exudation of sap from cut stems and/or guttation of water from uninjured leaves, is known as *root pressure*. This phenomenon is also known as "root exudation pressure" or "stump exudation pressure" or simply "sap pressure." It is inclusive of "radial cell pressure" as well as "axial xylem pressure." The pressure is assumed to be generated in combination by osmotically and energetically driven uphill transport of water across the plasma membrane of xylem parenchyma cells possibly taking advantage of the free energy gradients of ions and sugars (Wegner 2014).

3 Taxonomic Distribution of Root Pressure

Root pressures in temperate climate most frequently develop during warm nights though most of water transport occurs during daytime. Significant and consistent root pressures are developed in a wide range of plant species. Moreover, root pressure develops not only in the herbaceous species but in deciduous trees as well. Both dicotyledonous and monocotyledonous species have been reported to exhibit root pressure (Zachary 2009). Among dicots, prominent examples include several important agricultural plants such as tomato (*Solanum lycopersicum*) (White 1938), grapes (*Vitis vinifera*) (Sperry et al. 1987), sugar maple (*Acer saccharum*) (Sperry et al. 1988), birch tree (*Betula cordifolia*) (Sperry 1993), oak tree (*Quercus robur* and *Q. petraea*) (Steudle and Meshcheryakov 1996), walnut tree (*Juglans regia*) (Ameglio et al. 2001), sunflower (*Helianthus annuus*) (Dustmamatov et al. 2004), kiwifruit tree (*Actinidia* spp.) (Clearwater et al. 2007), and *Betula lenta* L. and *B. populifolia* Marsh. (Miller-Rushing and Primack 2008). However, in many of the above dicots, the observed root pressure was seasonal, being synced with the onset of spring. Such pressures are demonstrable in the spring before the buds open, but once the leaves have expanded and rapid water movement through the plant begins, root pressure can no longer be detected.

Further, a group of recent studies found root pressure in 61 of 109 tropical vine-like species despite the lack of freezing temperatures, much higher percentage of species with positive xylem pressures than those reported earlier by Ewers et al. (1997), suggesting this to be a regular, if not daily, occurrence (Fisher et al. 1997), both monocotyledonous and dicotyledonous vines evinced the phenomenon. These studies imply that root pressures are less common in lianas than in more herbaceous climbers and that certain families may have a strong tendency for root pressures, including the monocotyledonous family Araceae (ten of ten species) and the dicotyledonous families Vitaceae (ten of ten species) and Dilleniaceae (three of five species) (Fisher et al. 1997). In contrast to dicotyledonous lianas, the climbing fern *Lygodium venustum* and the viny bamboo *Rhipidocladum racemiflorum* also exhibit root pressures that might be adequate to refill embolized vessels throughout the entire plant. In addition, significant daily root pressure has been observed in herbaceous dicots (Milburn and McLaughlin 1974; Kramer and Kozlowski 1979), palms (Davis 1961), and banana (Davis 1961; Lu et al. 2002). Grasses in particular have widely demonstrated a daily pattern of root pressure, including Rhodes grass (*Chloris gayana* Kunth) (Ogata et al. 1985), corn (*Zea mays*) (Miller 1985; Tyree et al. 1986), sugarcane (*Saccharum* spp.) (Tyree et al. 1986; Neufeld et al. 1992), the vine-like bamboo (*R. racemiflorum*) (Cochard et al. 1994), rice (*Oryza sativa*) (Stiller et al. 2003), and several others (*Phleum pratense* and *Festuca pratensis*) (Macduff and Bakken 2003).

Davis (1961) reported positive root pressures in ten species of palm trees and in banana (*Musa sapientum*) but a lack of root pressure in five dicotyledons as trees and in a cycad. Root pressure, therefore, appears to be intimately linked to the hydraulics of plants, and it was suggested that monocotyledons, and other taxa with reduced or absent secondary growth for the production of new vascular tissue in their stems, may often depend on root pressures for water transport. Thus, far from a rare phenomenon, root pressure appears to be a widespread, if not common, factor in sap flow. It is, however, generally agreed that root pressures are not directly important for the high-volume daytime transpiration in vascular plants, as was implied by Davis (1961) for monocotyledons though Kundt and Gruber (2006) argue for a bigger role of root pressure in upward transport of water both in tall and short plants.

4 Quantification of Root Pressure

Root pressure is one of several important plant activities which serves a number of purposes required for growth and development. It is, however, affected by chemical, physical, environmental, edaphic, and genetic factors. Its quantification and determination of sap composition are essential for ecological, physiological, genetic, and metabolic studies on the one hand and for agricultural, horticultural, and forestry research on the other. It can be measured both invasively and noninvasively directly and computed and quantified indirectly as well.

4.1 *Direct Measurements of Root Pressure*

4.1.1 Invasive Techniques

Manometric Method

Stem Stump Method

Root pressure may be demonstrated and measured by attaching a mercury manometer by vinyl tubing to the cut end of the stem called stump. Within a short time, the level of mercury rises up indicating upward movement of sap due to root pressure. Bubble manometers can also be used to measure xylem pressure by attaching them to branch stumps (Ewers et al. 1997). The manometers made from glass tubes (1-mm internal diameter) are sealed at the distal end by flame. The distal half is filled with air, while water fills the basal half. The base is connected to the stump by a tight fitting of vinyl tube and hose clamps. Prior to attachment, the freshly cut stump is shaved with a new razor blade to permit unobstructed fluid flow between the stump and manometer. Each evening the cut stumps are reshaved and the manometers reattached. After allowing for equilibration overnight, the bubble length (L_{pd}) in the manometer is measured at predawn. The vinyl tubing is then cut and the bubble length (L_{atm}) immediately remeasured at atmospheric pressure. The xylem water pressure (P_x in kPa) is calculated from a relation derived from the ideal gas law:

$$P_x = 100\left[\left(L_{atm}/L_{pd}\right) - 1\right],$$

when L_{pd} is $>L_{atm}$ (including situations where all the water from the manometer is absorbed by the shoot), the P_x is recorded as negative, though manometers generally are not able to accurately measure the negative pressures that can occur in plants. The bubble manometers give P_x values near those often recorded by electronic pressure transducers (Cochard et al. 1994) and are used as such because they cost much less for especially surveys involving many species. Observations may be repeated three consecutive mornings at predawn.

Excised Root Method

Root exudation of an individual root may be quantified by excising the root at the base under water and inserting it 10 mm into a glass capillary (diameter = 0.55 mm) and sealing the capillary with a drop of super glue. The root is bathed in nutrient solution from the pot in which the plant had grown. The rise of xylem sap in the capillary may be recorded at time intervals (arbitrary) of 5 min for 1 h, and osmotic flow rate (Q_{ros}) may be calculated from the linear part of the flow vs time plot. At the end, the exudate in the capillary may be collected with a hypodermic needle attached to a syringe. The exudate may be analyzed for osmolality using picoliter osmometry separately.

Root Pressure Probe Techniques

Stem Stump Method

For measuring the root pressure by this technique which can be deployed on stump after cutting off the stem just below the first leaves, the remaining stump is fixed to the probe, a miniature pressure sensor "root pressure probe" with a rubber seal (Steudle and Jeschke 1983; Steudle et al. 1993). The pot with the root system is supplied continuously with nutrient solution or water as the case may be. With some root systems, root pressure could be recorded for several days. Curves may be digitized using a digitizing tablet.

Excised Root Method

The individual seminal roots are excised at their base under water. The root is fixed to the root pressure probe using a cylindrical silicon seal made from liquid silicon material (Knipfer and Fricke 2010). Seals are tightened using a screw, which compresses the seal and the root cortex. The root is then bathed in the same medium in which the plant had been grown. The medium is circulated with the aid of a peristaltic pump to ensure the uniformity of bathing medium. When a stable root/ xylem pressure is ensured to have been reached (after 0.5–2 h), pressure relaxations are induced. Hydrostatic pressure relaxations are produced by turning the metal rod of the probe rapidly to induce a hydrostatic pressure pulse (0.05 MPa) in the root xylem. Osmotic relaxations are induced by adding 20 or 25 mM NaCl to the root medium. Root hydraulic conductivity is calculated from the halftimes of overall hydrostatic ($T_{1/2hy}$) and osmotic pressure ($T_{1/2os}$) relaxations. Root radial hydraulic conductivity (*Lp*HYP and *Lp*HOP, in m s^{-1} MPa^{-1}) is related to hydrostatic and osmotic $T_{1/2}$ as follows:

$$\mathrm{Log_e} 2/T_{1/2} = L_\mathrm{P} \cdot A_\mathrm{r} \cdot \beta$$

The elastic modulus of the measuring system (β, 5.1×10^{-9} MPa m^{-3}) is determined by inducing step changes of volume and measuring the resulting changes in root pressure. The total root surface area (A_r) is calculated from root length and diameter including lateral roots. The reflection coefficient (r) for NaCl during osmotic pressure relaxations is calculated by relating the measured maximum decrease in root pressure caused by addition of NaCl to the expected decrease in pressure (20 or 25 mM NaCl; 0.1 or 0.125 MPa).

Root Vacuum Perfusion Method

Individual roots are excised under water. The excised root is inserted at its base into a 10-mm-long glass capillary (diameter = 0.55 mm), which is fixed to the bottom of a 100-mm-long water-filled glass capillary (diameter = 0.4 mm). The root is pushed through the small capillary piece until the root base protruded into the water-filled capillary. The root is sealed with a drop of super glue on the outer surface between the root and the capillary. The water-filled capillary is connected to a vacuum pump via rigid silicon tubing and is held by a clamp attached to a stand in such a way that the weight of the root is supported by the stand and does not contribute to the weight of nutrient solution in which the root is bathed. The beaker containing nutrient solution is placed on a balance, and root water uptake is measured gravimetrically as the decrease in weight with time. Before the application of hydrostatic pressure gradients, the osmotic flow rate Q_{ros} is measured from the linear part of the water uptake vs time plot. The driving force for this osmotic water flow is the difference in osmotic pressure (D_p, MPa) between root xylem and medium multiplied by the corresponding reflection coefficient for solutes (a value of 1.0 is used). Water loss from the beaker as a result of evaporation (Q_{eva}) is generally negligible and is corrected accordingly. To measure the hydrostatic flow rate Q_{rhy}, a partial vacuum of 20, 30, 40, 60, and 80 kPa is applied at the open end of the capillary in succession for 10 min each. Q_{rhy} is determined from the linear part of the water uptake vs time plot after Q_{eva} and Q_{ros} had been subtracted (Knipfer and Fricke 2010).

Externally Applied Pneumatic Pressure Method

This method also involves cutting off stem with a sharp razor blade above the ground level and allowing it to exude freely for some time. Then the cut end of the stump is attached to a vinyl or tygon thick-walled tubing sealing with silicon glue or parafilm properly as described earlier on the one end and to a pressure applying system equipped with a pressure gauge on the other. Subsequently, the pressure is gently applied to stop the exudation confirming it by viewing through a hand lens. At this point of time, the magnitude of pressure applied is noted which is interpreted to be equal to the root pressure.

4.1.2 Noninvasive Techniques

The minimally invasive measurements in the xylem of trunks remain the greatest challenge. Knowledge of the forces and flows in this compartment are crucial to unearthing the truth about water lifting because branches, twigs, and petioles may be segmented and separated from the main stream in the trunk. However, independent of the outcome of such experiments, in the light of the evidences reviewed here, it is obvious that nature has developed a broad spectrum of complementary strategies for water transport against gravity to cope with various water deficiency situations without the necessity of developing incredibly negative pressures (Zimmermann et al. 1995). Various noninvasive techniques are briefly described hereunder.

Isopiestic Method

This is a noninvasive technique for measuring root pressure. By adapting this technique, the root pressure can be measured isopiestically in intact plants by applying solutions of different concentrations separately of non-penetrable osmoticum such as polyethylene glycol (MW $\geq$ 4,000) to the root medium with the observations on the guttation process (Klepper and Kaufmann 1966; Steudle and Jeschke 1983; Zhu et al. 1995). The magnitude of the osmotic potential (negative) of the solution that stops the guttation is considered equal to the root pressure (positive).

Xylem Pressure Probe Method

The xylem pressure probe method allows the direct measurement of diurnal and seasonal changes in xylem pressure, xylem flow, and solute composition in intact plants (Balling and Zimmermann 1990). Briefly, the xylem pressure probe is filled with denucleated water and incorporates a water-wettable pressure transducer. The microcapillary of the probe is advanced through the tissue (Fig. 1).

Penetration is stopped immediately when the transducer registers a pressure below atmospheric. In order to verify that the probe is placed in a vessel, the microcapillary is loaded with a dye/water mixture prior to insertion. This results in staining of the penetrated vessel. Furthermore, injection of volume pulses into a probed vessel by appropriate displacement of the metal rod results in a rapid dissipation of the pressure provided that the tip is unobstructed and placed in the lumen of a large vessel. The probe can accurately read both the negative and positive pressures. Clearwater et al. (2007) measured root pressure using pressure transducers that were installed inside the xylem of kiwi roots. To my knowledge, this is the only existing nondestructive continuous method to measure root pressure

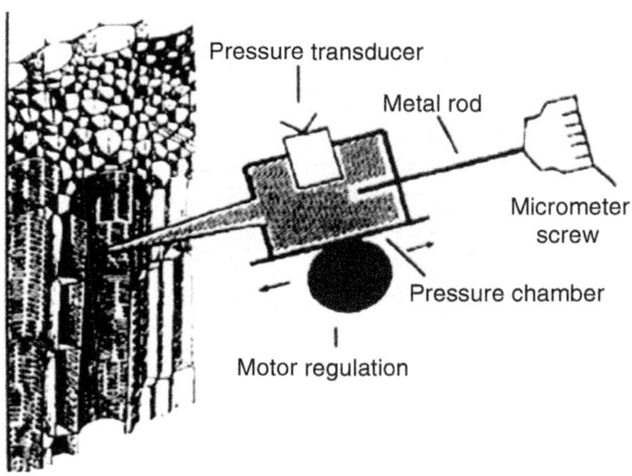

Fig. 1 A schematic diagram of the xylem pressure probe [*Source*: Balling and Zimmermann (1990)]

in a herbaceous plant. However, the application of this method on herbaceous plants with limited root and stem diameters such as tomato may be difficult, if not impossible, because this method requires installation of the system 10–15 mm inside the xylem.

Cell Pressure Probe Method

For this purpose, a cell pressure probe is used to measure hydrostatic pressure of the intact root cortical cells (Azaizeh et al. 1992). An oil-filled capillary (outer tip diameter, 4 ± 0.7 mm) is attached to a pressure chamber which contains an electronic transducer (silicon chip). The probe is fixed on a micromanipulator which allows insertion of the tip into individual cells. Cells may be probed at distances between 30 and 100 mm from the apex of roots grown in any growth medium. Thus, by using the pressure probe, the hydrostatic pressure of individual cells may be measured directly. However, the primary limitation of this method is that some cells are too small to measure. Furthermore, some cells tend to leak after being stabbed with the capillary, and others plug up the tip of the capillary, thereby preventing valid measurements. By measuring the depth of insertion of the tip in the cortex, the location of punctured cells could be measured. Prior to insertion into root tissue, the pressure probe is completely filled with silicon oil. When a cell is punctured, a meniscus is formed between oil and cell sap, and this meniscus is kept at a certain position. The pressure transducer converts the pressure signal into a proportional voltage. Pressure/time curves are recorded on a chart recorder. Happily, other hydraulic traits such as turgor, hydraulic conductivity, and elastic modulus can also be quantified by this method. For processing the data on such traits, recorder strips are digitized using a digitizing tablet. Cell elastic moduli are evaluated from changes in cell volume which caused changes in cell turgor pressure.

Axial and Radial Root Growth Confinement Method

A method was developed by Misra et al. (1986) for estimating radial root growth pressure of intact seedlings. Under this method, each root of desired seedling is confined both in the axial and radial directions in a cylindrical chalk sample at a constant water potential. In doing so, the root exerts radial stress which causes tensile failure in a proportion of the chalks. The measurement of tensile strength of duplicate chalks enables estimation of the maximum radial pressure exerted by the roots. The axial and radial root growth pressures measured in this manner were, for example, of comparable magnitudes registering 497, 289, and 238 kPa for pea, cotton, and sunflower seedlings of similar size, respectively (Misra et al. 1986).

4.2 *Indirect Measurements of Root Pressure in Intact Plants*

4.2.1 Root Pressure Computation by Difference in Leaf Thickness Under Different Relative Humidities

Recently, a technique was developed to measure the root pressure in which measurements of leaf thickness are compared with predictions of it using a mechanistic model (De Swaef and Bleyaert 2012). This model predicts diurnal variations in leaf thickness based on variations in transpiration and a concept of growth in relation to turgor. The principle is that when measurement of leaf thickness exceeds the predictions during the night, this difference could be attributed to root pressure. If this difference is not present, the diurnal differences in leaf thickness are entirely explained by the transpiration and growth concepts present in the model, and root pressure is expected not to be present. Thus, the difference between measured and predicted leaf thickness could mathematically be translated into an absolute value for root pressure. The leaf patch clamp pressure probes can also be used for this purpose (Zimmermann et al. 2008; Lee et al. 2012). The practical implication of these methods is that these can be used to relate the occurrence of tip burn in lettuce head to the occurrence of root pressure by manipulating nighttime relative humidity around the plant (De Swaef and Bleyaert 2012).

4.2.2 Root Pressure Computation by Using Sap Flow and Stem Diameter

Very recently, a nondestructive estimation of root pressure using sap flow, stem diameter measurements, and mechanistic modeling was developed by De Swaef et al. (2013). The magnitude of the root pressure destructively measured using a manometer installed on excised tomato stems has been found to agree well with the model-based estimations made by these authors. Under this technique, however, destructively measured root pressure showed a decrease during the night, presumably as a result of decreasing substrate temperature, whereas estimated root pressure in intact leafed plants showed an increasing pattern toward the end of the night. It is therefore hard to relate diurnal dynamics of destructively measured root pressure to diurnal dynamics of transpiring plants, because the excised stems did not transpire during the day; root pressure enhanced during the day because of the higher temperature in the greenhouse, whereas root pressure is not allowed to develop in transpiring plants. With critical considerations concurrent with the root pressure-induced nighttime increase in diameter, it might be expected that water flow must increase. This was, however, not always visible in the experiments conducted by the above referred authors but could be clarified by the model calculations. The measured nighttime diameter increase corresponds approximately to a calculated maximum water mass inflow of approximately 250 mg h^{-1} for an 8-m-long tomato stem, which is 400 times smaller compared with daytime water flow rates. Because of the low vapor pressure deficit at the end of the nights, plant water loss via transpiration could be neglected in such cases. It is, therefore, hypothesized that nearly all of the upward pushed water flow in the xylem is stored in the plant itself and thus results in the increase in diameter. However, the sensitivity of the heat balance sensors may not always allow detection of these low amounts of water flow (van Bavel and van Bavel 1990). The approach described above was validated in an extra experiment on tomato plants in which relative humidity of the air was manipulated to be high during the night. The same model was used for the forward simulation using sap flow as input variable and shows comparison between measured and simulated diameter for the forward simulation. The resulting estimates of root pressure were then compared with actual measurements of root pressure on excised shoots.

From the foregoing discussion, it is clear that no single method is available which is applicable to all situations. Therefore, depending upon the plant species and their age, different strategies are required for the measurement of root pressure which need to be noninvasive, simple, easy to use, cheap, and dependable.

5 Magnitude of Root Pressure

The exudation of liquid water from passive hydathodes, such as in *Colocasia*, and the secretion of water and solutes from wounds show that the roots of many plants under certain conditions can develop considerable pressure displaying a daily rhythm and a yearly rhythm: the mechanisms seem to be working only when required in an unspectacular way (Kramer and Kozlowski 1979; Kundt and Gruber 2006). The magnitude of pressures also depends upon the techniques used. For example, pressures measured by root pressure probe were higher (in the range of 0–0.5 MPa) than the values produced by the pressure chamber, although new experiments have recently been conducted with the pressure probe (Wei et al. 1999) and were found to agree with the pressure chamber. Over seven decades ago, White (1938) recorded pressures of 600–700 kPa in excised tomato roots. A pressure of 700 kPa, though capable of causing a flow of water in the xylem of tall trees, in view of existence of resistance to flow, cavitation, etc., may not be sufficient to push the water in most of the tallest trees. However, such magnitudes of root pressure would certainly be no problem for either agricultural and horticultural crops (Singh and Singh 1989; Tanner and Beevers 1999, 2001; Singh et al. 2009a) or woody as well as vine-like lianas (Fisher et al. 1997) and bamboos (Zachary 2009) or deciduous forest trees (Feild et al. 2005; Feild and Arens 2007) acting as supplementary device to the cohesion–tension mechanism (Steudle 2001) for upward movement of sap. A root pressure in the root xylem/stele has been typically found within the range 0.1–0.4 MPa (Steudle et al. 1987; Knipfer et al. 2007) and attributable to active solute uptake and subsequent passive water inflow. Actually, plants depending upon their habitat require an excess pressure inside a plant, not to be established by suction, which can amount to 0.6 MPa in the tomato, 1 MPa in grass stalks, or even 6 MPa in certain desert plants studied since many decades (White et al. 1958; Kundt and Gruber 2006). Positive xylem pressures in the stems of plants usually are attributed to "root pressure," i.e., the osmotic water uptake caused by solute uptake into roots (Tyree et al. 1994; Ewers et al. 1997; Fisher et al. 1997). The water flow caused by root pressure is normally much less than that caused by transpiration pull. Apparently, when transpiration is high, the osmotic force causing root pressure tends to disappear but not nonexistent because solutes are diluted by an influx of soil water in the xylem. Thus, root pressure is highest when transpiration is minimal, such as predawn and during rainstorms (Cochard et al. 1994).

Data for *Tetracera* recorded by Cochard et al. (1994) are similar to results of Scholander et al. (1957) for this genus at the same site, where they found xylem pressures of 10–80 kPa. However, as noted by Scholander et al. (1957), the root pressure values need to be considered in the context of the height of the plants. The measured values of *root pressure* near the base of the stems of Dilleniaceae (a maximum of 64 kPa in *Doliocarpus major*) was modest considering that those plants reached the canopy height of 18 m. Based on the above argument and these data, Ewers et al. (1997) concluded that the root pressure of 64 kPa would be adequate, given enough time, to refill embolized vessels of the roots and lower stems, at maximum height of just 7.1 m. *L. venustum* with 7.5-m height exhibited root pressure values up to 66 kPa, suggesting that xylem pressures might be quite adequate for refilling of the tracheids even in the upper parts of the leaves. On the other hand, *R. racemiflorum* climbed to only 4.5 m but it had xylem pressure values up to 120 kPa. Furthermore, Cochard et al. (1994) found positive root pressure values even at the most distal part of the stems in this species, suggesting that root pressures could serve to refill embolized vessels throughout the shoot. Further, tropical palm trees whose saps are used as beverages have been reported to have root pressures sufficient to pump water up to heights as large as 12.5 m (Davis 1961), and there is evidence that at least some tropical lianas have positive water pressure in their stem xylem at certain times (Scholander et al. 1957; Putz 1983). Since the wide vessels of tropical liana stems remain conductive for many years, it has been suggested that they might be refilled (reversal of air embolism) as a result of root pressure sufficient to dissolve emboli in vessels (Putz 1983; Ewers et al. 1991). It is, therefore, clear from the foregoing discussion that the root pressure phenomenon is of common occurrence and that its magnitude may vary in space and time and go up to generally 0.6–0.8 MPa depending upon the plant species and their habitat. Occasionally, root pressures amounting up to 6 MPa (White et al. 1958), sufficient to satisfy the need for refilling the freeze- and drought-induced embolism in plants (Tyree 2003a, b; Kundt and Gruber 2006; Holbrook and Zwieniecki 1999; Zwieniecki and Holbrook 2009), seem to establish and signify their existence as necessary mechanism for upward water transport in plants.

6 Morphology and Structural Anatomy of Root Per Se and Root Pressure

Undoubtedly, in recent years, our understanding of water uptake and transport within plant roots has been substantially improved by new tools, which operate at the molecular, cell, tissue, organ, plant, and ecosystem levels. Techniques such as cell and root pressure probes, stopped flow, and the use of transgenic plants and of stable isotopes provided a fast progress in water transport research (Steudle 1993, 2001; Kramer and Boyer 1995; Maurel 1997; Steudle and Peterson 1998; Tyerman et al. 1999, 2002; Ehleringer et al. 2000). A better understanding of the mechanisms of water uptake by plant roots should be vital for improving water-use efficiency in agriculture, horticulture, and forestry. The morphological and anatomical features of the roots, such as their diameter or length, the cell layer from exodermis to endodermis, and the degree of their suberization and the radial and axial water transport pathway have a great influence both on the hydraulic conductivity and root pressure. The recent works highlight the progress for those who want to update their understanding of basic mechanisms of root hydraulics and plant water relations (Zhao et al. 2004; Aroca et al. 2012; Lobet et al. 2013).

6.1 Distribution of Root Systems in Soil and Collection of Water and Solutes

Root morphology and root pressure are intimately linked to each other. The spatial distribution of roots in soil determines the ability of plants to take up soil water and nutrients in order to sustain plant growth and development (Fig. 2). Water uptake by plant roots is controlled or is even regulated by different physical and physiological processes. Water supplied to plants by their roots has a major influence on root pressure, hence the shoot water status, influencing plant growth and development (Meister et al. 2014). A large number of studies confirm that deeper root systems enable plants to access and collect water not available to shallow rooted plants and to pump into the main stem or trunk. Coarse as well as fine regulation may coexist for water uptake. Coarse regulation is physical in nature and strongly depends on root structure. These regulations mainly involve minimizing water loss and maximizing water uptake. Water uptake, on the other hand, is maximized by adjusting the allocation pattern, namely, increasing investment in the roots, enhancing root depth, or extending root system distribution (Fig. 2). Since root hairs are unicellular functional units of root system architecture, modulating root hair number and length is an alternative to improving root function depending on the soil type. The development of *Arabidopsis* root hairs (type 3 striped pattern) differs from rice root hair development due to the asymmetric cell division (type 2). The known root hair loci in rice share homology to *Arabidopsis* genetic counterparts. These include the auxin-regulated *OsWOX3A*, which negatively regulates root hair number and

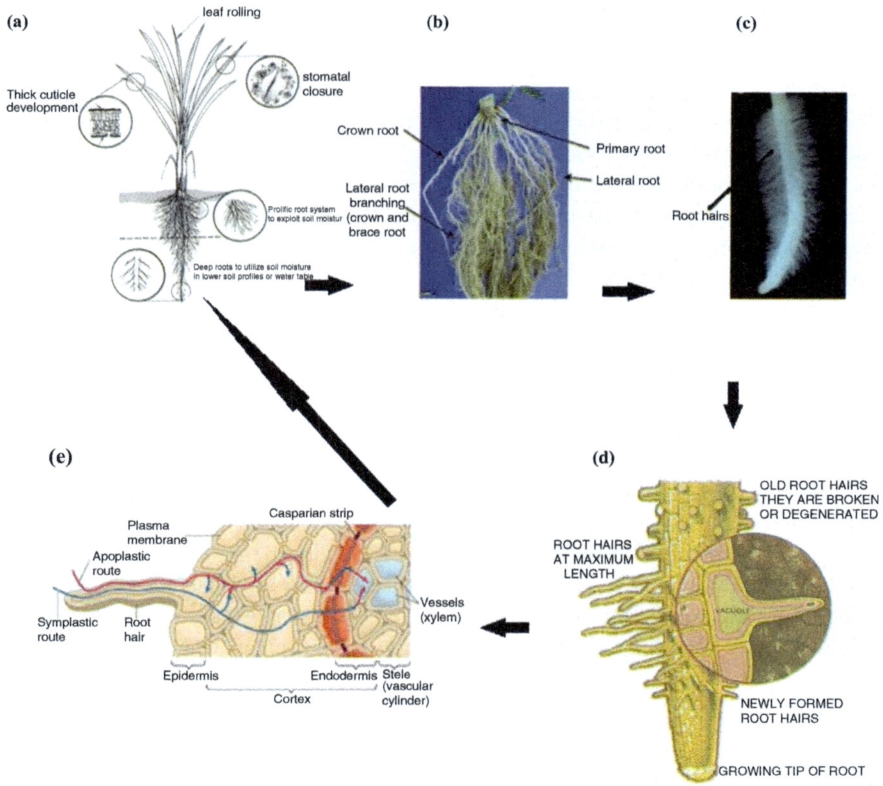

Fig. 2 Morphology and anatomy of root systems and water and solute collection and transport. (**a**) Whole plant with root and shoot [*Source*: O'Toole and Chang (1978)]. (**b**) Root system (*Source*: http://www.greenmanspage.com/guides/logistics.html). (**c**) Prolific root hairs (*Source*: https://www.pinterest.com/pin/506232814336189676/). (**d**) Magnified view of root hairs (*Source*: http://www.daviddarling.info/encyclopedia/T/trichome.html). (**e**) Radial flow of water and solutes in roots (*Source*: http://www.pleasanton.k12.ca.us/avhsweb/thiel/apbio/labs/plant_transport.html)

length but is a positive regulator of lateral root number, and a putative mannosyl-oligosaccharide glucosidase (*OsMOGS*), controlling both initiation and elongation of root hairs (Zhao et al. 2004; Baluska et al. 2000; Baluska and Mancuso 2013). Root hairs are an important component of root system architecture since they increase the surface area for uptake of water and nutrients and are one of the sites for plant–microbe interactions as well. Since the ontogeny of root hairs varies between species, know-how regarding how cell fate is specified will shed light on the ways to achieve epidermal cell type independent of root hair differentiation and whether it benefits productivity (Baluska et al. 2000; Baluska and Mancuso 2013). Fine regulation of water flow is achieved by aquaporins by phosphorylation and dephosphorylation or indirectly by regulating protein kinases and phosphatases or gating (Maurel et al. 2008).

6.2 Conduction of Water and Nutrients

Xylem is specialized for the conduction of water and mineral substances in the plant body. It is a heterogenous tissue made up of four different types of cellular elements such as xylem tracheids, xylem tracheae, xylem fibers, and xylem parenchyma. Of these, the tracheids and the tracheae are described as essential elements since they are directly involved in the translocation of water and mineral substances. Xylem fibers and xylem parenchyma are described as associated elements, since they are only supporting structures. The tracheids, the trachea, and the xylem fibers are nonliving components, while xylem parenchyma represents the only living component of the tissue. Xylem is commonly described as a dead, complex permanent tissue. Secondary xylem (wood) performs many functions, but chief among them is long-distance transport and mechanical support of the canopy. The studies on the comparison of the inter-conduit pits between angiosperm vessels and gymnosperm tracheids showed that the different torus–margo structure of conifer pit membrane was 60 times more efficient at water conduction than the homogenous pit membranes between angiosperm vessels and equally as safe against air seeding. The more efficient pits of conifers compensate for the shorter length of the single-celled tracheid. Consequently, a single-celled tracheid has approximately the same conducting capacity as a multicellular vessel of the same diameter (Tyree and Zimmermann 2002). In addition, perhaps the circulation of water and sugars via the phloem also seems to contribute to building up root pressure (Wegner 2014).

6.2.1 Physiological Anatomy of Root Pressure

Normally, it is not only water in the soil that is of significance to plant life, but nutrients are of vital importance, too. Soil consists of particles that contain masses of elements in solid form; these are not available to plants unless they are dissolved in the soil solutions (Fig. 2e). Chemical and physical erosion and fertilizers release elements from the soil particles, so making them available to plant. Also, soil particles, which normally have a negative electric charge, bind positively charged soil ions to their surface. Plant roots have developed an ingenious strategy to release these absorbed ions. Roots extrude protons through specific proton-ATPase complexes on the plasma membrane. These released protons are exchanged to the nutrient ions bound on the soil particles, since protons have a higher affinity to the soil particles than other ions. It is often thought that plants take up nutrients from the soil with the flow of water from the soil to the roots. This is not so. Charged nutrient ions cannot pass the plant plasma membrane; they have to be taken in through protein pores or channels in the membrane along an electrochemical gradient (Palmgren 2001). If they are taken in against a concentration gradient, the transport consumes energy, which is available in the form of the proton gradient formed by the proton-ATPase pump or by direct involvement of ATP in the transport process (Pedersen et al. 2007).

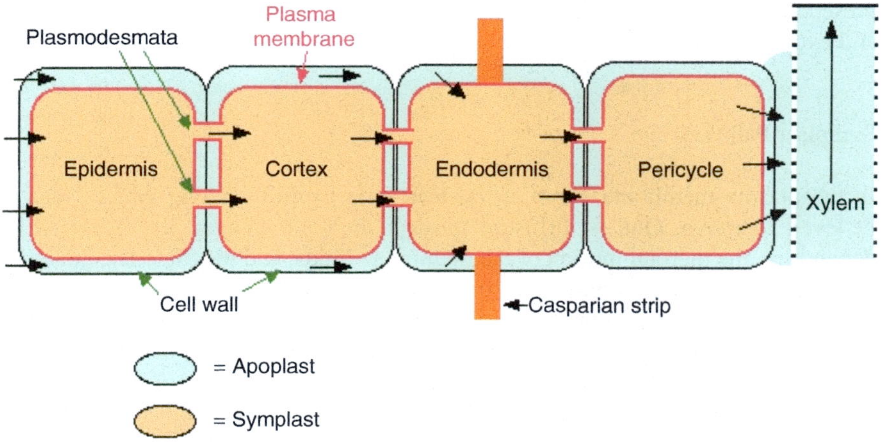

Fig. 3 Apoplastic, symplastic, and transcellular movement of water and solutes finally reaching the xylem of roots (*Source*: http://bankofbiology.blogspot.com/2014/07/comparison-between-active-absorption.html)

The water and solutes intake in the roots can follow two ways, i.e., apoplastic and symplastic with plasmodesmata (transcellular) linking the protoplasts from cell to cell. Movements in relation to the route of the epidermis to the endodermis of the root are called radial water transport (Fig. 3). The relative importance of these pathways is still a cause of much discussion, but there is some evidence for the suggestion that plants displaying low transpiration activity predominantly witness symplastic transport, while those displaying high transpiration activity witness a greater proportion of apoplastic transport (Boyer 1985; Steudle 2001). Another important detail in relation to these different pathways is relevant only in the outer layers of the root tissue, because in the endodermis, the water apoplastic flow is limited due to the Casparian strip. In this hydrophobic barrier, the radial and transverse endodermal cell walls are impregnated with lignin, suberin, structural wall proteins, and wax. In many plants such a barrier also occurs in the epidermal cells, forming a double-layered hydrophobic barrier in the roots (Enstone et al. 2003). It is important to note that the Casparian strip does not always establish a barrier that is totally impermeable to water and solutes coming from the soil. This can be observed, for example, by the development of young roots where pericycle growth can break parts of the endodermis and allow free access to water until the reconstitution of the tissue.

Apoplast Pathway

Here water passes from root hairs to xylem through the walls of intervening cells and middle lamella without crossing any membrane or cytoplasm. The pathway provides the least resistance to movement of water. However, it is interrupted by the presence of impermeable lignosuberin Casparian strips in the walls of endodermal and exodermal cells.

Symplast Pathway

In the plasma membrane, there are at least two pathways along which water is believed to move. One is diffusion through the lipid bilayer, a process which depends on the thermal motion of the membrane lipids, and the other is conduction through water channels which are established by polypeptides of the aquaporin type. Flow through these two pathways may be independently regulated (Henzler and Steudle 1995; Steudle and Henzler 1995). When considering water flow across a structure as complex as a root, the principal resistance in the pathway is often envisaged as being due to plasma membranes of cells in the pathway. A common view is that the plasma membrane of the endodermis having Casparian band may contribute the major part of this resistance, but there are different opinions about this matter (Kramer and Boyer 1995; Steudle and Peterson 1998). Water passes from cell to cell through their protoplasm with water and solutes also entering vacuoles en route. As mentioned earlier, the cytoplasms of the adjacent cells are connected through bridges called plasmodesmata. For entering into symplast, water has to pass through plasmalemma (cell membrane) at least at one place. It is also called transmembrane pathway. Symplastic movement is aided by cytoplasmic streaming of individual cells (Baluska et al. 2000; Baluska and Mancuso 2013). It is, however, slower than apoplastic movements. Both the pathways are involved in the movement across the root. Water flows via apoplast in the cortex, finally getting into the pericycle from where it enters the xylem. Mineral nutrients also have the same pathways as that of water. However, their absorption and passage into symplast mostly occurs through active absorption (Palmgren 2001; Pedersen et al. 2007). There exist numerous reports in the literature that show that radial uptake of water and that of solutes are uncoupled from each other (Munns 1985) and which suggests that ions (Na^+ and Cl^-) are transported across roots along the symplast (Lauchli et al. 2008). Once inside the xylem, the movement is purely along the pressure gradient, but there is different opinion about it too (Enns et al. 2000).

Pickard (2003a, b) presented a model of water flow through roots which incorporates hydrostatic pressure-driven flow through plasmodesmata. If turgor pressure in endodermal cells is in the range of pressures in other plant cells (0.4–1 MPa), a gradient in hydrostatic pressure should exist between the xylem/stele and the endodermis and enable water flow into the xylem even during endosmotic hydrostatic pressure relaxations. We cannot rule out the possibility that this water flow, being part of the cell-to-cell path and not mediated through aquaporins, contributes to slightly shorter halftimes in hydrostatic compared with osmotic experiments. However, the observations of Knipfer and Fricke (2010) emphasize membranes (and aquaporins) as control points for radial water transport in roots. The results also question the generally accepted idea that a special apoplastic, low-resistance pathway of water movement driven by hydrostatic gradients is required in roots to meet the transpirational water demand of the shoot. Since the present review is concerned with the "push-driven water ascent" in plants, the issue of "pull-driven water ascent" including the composite model of water transport (Steudle 2001) which has fallen under criticism requiring its revision (Zimmermann et al. 2004; Kundt and Gruber 2006; Knipfer and Fricke 2010) will not be discussed further.

6.2.2 Ultrastructural Anatomy of a Root Cell Per Se, Root-to-Shoot and Vice Versa Signaling, and Root Pressure

This section relates to the regulation of transport of water, nutrients, food, organic substances including hormones, and signaling between roots and shoots. After the upward transport, water is again required for taking the photosynthesis products from the leaves through the phloem to all sites of growth, in the buds, branches, stems, roots, blooms, and fruits. Water must therefore permanently circulate through every plant parts, as sap with dissolved materials of varying concentrations. Emphasizing the intricate relation between root and shoot, Kundt and Gruber (2006) opined that the "overlapping double saw-cut" experiment proves that a tree does not die when all its xylem vessels are severed (Preston 1952; Zimmermann et al. 2004), giving the theory of xylem functioning for water transport in isolation a deadly blow. The same is true for phloem's functioning when a young maple tree, whose stem was cut all the way around its periphery some 10 years ago, has continued growing ever since (Kundt and Gruber 2006). This proofs the strong interconnectivity between xylem and phloem. Although root pressures are generally the mechanism for embolism reversal in most of the studied plants (Tyree and Sperry 1989; Ewers et al. 1997; Fisher et al. 1997; Tyree and Zimmermann 2002; Tyree 2003a, b; Voicu et al. 2008), however, these may not be the only mechanism by which embolisms can be reversed in the xylem. Because xylem cavitations are largely a product of winter freezing (Cobb et al. 2007) and drought stress (Stiller et al. 2003; Singh et al. 2009a), the seasonal occurrence and species variation in root pressure are believed to function as a mechanism to repair cavitations in most of forest trees and crop plants. A recent study indicates that in the evergreen shrub *Laurus nobilis* xylem embolism can be reversed in the absence of positive root pressures, even at xylem water potentials of 30 kPa (Salleo et al. 1996). For that species, phloem transport through inter-trafficking between xylem and phloem appears to be essential to the refilling process, although the exact mechanism of refilling is unknown.

At the cellular level, the results of fluorescence resonance energy transfer (FRET) imaging in living maize protoplasts co-expressing plasma membrane intrinsic proteins 1 (PIP1s) and 2 (PIP2s) further support a model in which aquaporins of the two classes directly interact, very likely by heterotetramerization, to facilitate PIP1 trafficking. Whereas interaction-dependent trafficking of PIP1s and PIP2s offers a broad range of combinatorial regulations, a future challenge is to determine to what extent this process can dominate the expression of PIP1 or PIP2 homotetramers. Similar to other membrane proteins, PIP2 aquaporins are subjected to constitutive cycling. Their endocytosis is clathrin-dependent (Dhonukshe et al. 2007) and reduced by auxin (Siefritz et al. 2002). Export of PIP2 aquaporins from the endoplasmic reticulum is also critically controlled, and the role of a diacidic motif contained in the N-terminal tail of PIP2s was recently uncovered in maize and *Arabidopsis* (Fig. 4) (Maurel et al. 2008). The cellular mechanisms that determine aquaporin trafficking and their subcellular relocalization in response

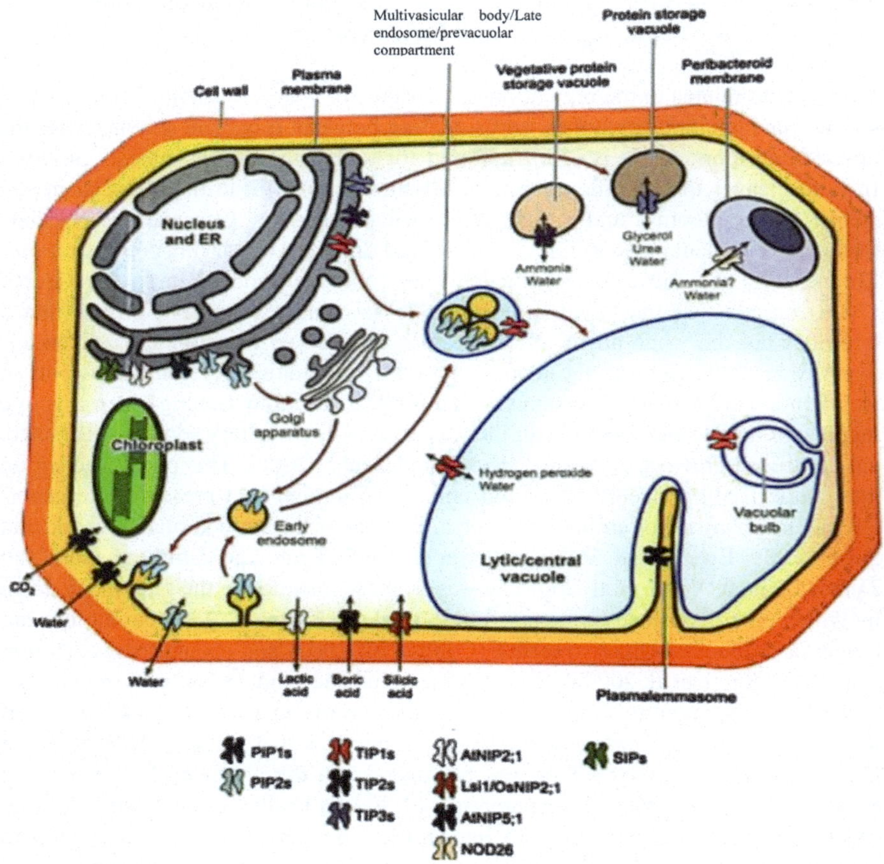

Fig. 4 The multiple cellular functions of plant aquaporins. The figure illustrates the variety of transport functions achieved by aquaporins in various subcellular compartments. The different aquaporins subclasses or isoforms are identified below the illustration in distinct colors. Isoforms of the plasma membrane intrinsic protein 1 (PIP1) and PIP2 subfamilies are thought to follow the secretory pathway, which carries cargo from the endoplasmic reticulum (ER) toward the plasma membrane through the Golgi apparatus. PIPs also undergo repeated cycles of endocytosis and recycling through endosomal compartments before being eventually targeted to the lytic vacuole through the multivesicular body [*Source*: Maurel et al. (2008), for further explanation, see Maurel et al. (2008)]

to stimuli will surely fuel intense investigations in the coming years. Additionally, the interplay of phytohormones produced in the roots and shoots in long-distance signaling evidenced by variations in xylem sap cytokinin concentrations, shoot auxin level, auxin transport, and auxin response seems to be operative for induction of root pressure. For instance, chemical signals, altered under and originating from roots, play an important role in the root-to-shoot communication in the movement of water from soil layers through roots and shoots to sustain plant growth. Therefore, the manner, not yet fully understood, for the initiation and control of root pressure is important in root-to-shoot signaling and vice versa. For example, expression of AtNCEDs, AtABA2, and AAO3 genes in phloem companion cells and xylem parenchyma cells of turgid plants is probably the main site of ABA biosynthesis in unstressed plants and ABA, and its precursors might be synthesized in vascular tissues and transported to target cells such as stomata, hydathodes, and sites of root pressure (Koiwai et al. 2004). ABA by way of its presence in vascular bundles, roots, and leaves might influence gating of aquaporins resulting in increased permeability to water hence its increased transport pressing the water to exude from stump and guttation.

7 Mechanism of Root Pressure

Till the last date of writing this review paper, there was no unanonymous agreement about the mechanism of root pressure in plants. Here, I am presenting a synthesis of osmotic and metabolic including biophysical and molecular aspects in an integrated manner stressing the need for both the water and solutes entering the plant roots to account for the mechanism of root pressure. The presence of solutes constitutes the osmotic aspect, but their uptake along with energy-driven uphill uptake of water and transport constitutes the metabolic including biophysical and molecular aspects (Dustmamatov et al. 2004; Zholkevich et al. 2007; Wegner 2014).

7.1 Osmotic Aspect of Root Pressure

The most generally accepted explanation of root pressure is based on active accumulation of solutes by root cells, their secretion into the xylem, and the subsequent osmotic movement of water along the water potential gradient thus established (Bai et al. 2007; Zhu et al. 2010). Thus, the "osmometer model" considers the tissue separating the xylem sap from the external medium as a semipermeable transport barrier or "membrane"; alternatively, the root symplast is accepted as a third, transitory compartment. At night, when there is almost no transpiration, root cells continue pumping mineral ions into the xylem of the stele. Meanwhile, the endodermis and exodermis help prevent the ions from leaking out. The resulting accumulation of minerals lowers the water potential within the stele.

Water flows in from the root cortex, generating root pressure, a push of xylem sap. In brief, the development of root pressure seems to follow or accompany an active transport of salt into the xylem conducting system, where the osmotic value (more negative osmotic potential) rises above that of the external solution. Thus, water so withdrawn from the external solution will depend on the difference between the osmotic pressure of the latter and that of the xylem vessels after intervening cells have reached full turgor. Since the parenchyma within the endodermis, being confined, is limited in extensibility on account of the structure of the endodermal cells, a strong hydrostatic pressure will therefore develop in this core of cells, sufficient to cause an excretion of water and solutes into the xylem vessels so long as the osmotic gradient persists. When the water and solutes enter the xylem, they are free to move upward in the vessels. Water may leak backward as far as the endodermis but no further on account of the suberized walls of the endodermis, which structure has been shown to prevent such a backward leakage. Priestley (1920) showed that the apical region of the root did not permit a backward leakage.

7.2 *Molecular Mechanism of Root Pressure*

There is a very strong group of Russian workers headed by Zholkevich who proposed the metabolic concept of root pressure apart from the osmotic concept discussed above which could not justify fully the uptake and upward movement of water and solutes implicated in root pressure (Dustmamatov et al. 2004; Zholkevich et al. 2007; Dustmamatov and Zholkevich 2008). Actually, it is the isotonic water flow or even radial water flow against an osmotic pressure gradient between the external medium and the guttation fluid or the exudate secreted by the root stump which paved the path for the involvement of metabolic process in root pressure (Oertli 1966; Zholkevich 1991; Schwenke and Wagner 1992; Enns et al. 2000; Pickard 2003a, b). It was Oertli (1966) who first proposed active water transport in plants taking place at the expense of metabolic energy defining the process characterized by the increased water potential, and this gain, he explained, must depend on the decrease in free energy in some metabolic process. This followed the work of Ginsburg (1971) on active water transport in a corn preparation who stated that water flow had two components, one osmotic and one non-osmotic. The non-osmotic flow was inhibited by cyanide. No correlation was found between water flow and solute flow suggesting that active water transport occurred in the root pressure. The Russian workers headed by Zholkevich at the K. A. Timiryazev Institute of Plant Physiology, Moscow, went steps further ahead accumulating a huge amount of data which suggested the involvement of metabolic process triggered by G-proteins during stimulatory action of neurotransmitters such as adrenalin and noradrenalin in root exudation, water transport, and creation of the root pressure (Mozhaeva and Pil'shchikova 1972; Zholkevich 1991; Dustmamatov et al. 2004; Zholkevich et al. 2007; Dustmamatov and Zholkevich 2008).

Now, very recently has come the work of Wegner (2014) on the horizon of root pressure research who has proposed an energetically driven "uphill water co-transport hypothesis" integrating osmotic and metabolic mechanisms with aquaporin-facilitated water transport within xylem parenchyma taking advantage of the free energy gradients of ions and sugars. In fact, the energy source for such pumping (ultimately ATP) was not exhausted for several days after supplies of photosynthate and reducing power from the shoot. The source of ATP must have been from substrate reserves and from tissue degradation. If the latter was significant, the author explained, it was also remarkable that the integrity of membranes, critical in maintaining the diffusion barrier between the xylem and the outside solution, was preserved for so long. This process could drive volume flow "energetically uphill" against the free energy gradient of water. According to the model of co-transport of water, solutes released by xylem parenchyma cells are subsequently retrieved from the sap at the expense of metabolic energy to maintain the concentration gradient that drives the water secretion (Fig. 5a, b). The salt release by co-transporters is an electroneutral process (Zeuthen and McAulay 2012) and would not interfere with K^+ reuptake by ion channels that require a membrane potential more negative than the Nernst potential of K^+, which is maintained by proton pump activity (Fig. 5a). Evidence for "simultaneous" uptake and release of K^+ has indeed been obtained for root tissue, using refined radioactive tracer techniques (Britto and Kronzucker 2006). Rapid, seemingly "futile cycling" of ions is apparently a common phenomenon at root membranes that was found for K^+, Na^+, and Cl^- and becomes more prominent at elevated concentrations of these ions. Futile cycling consumes metabolic energy, but its benefit for the plant seemed to be elusive and so far has remained an open question. Water secretion may be part of the answer. In his quantitative biophysical analysis of coupled ion and water transport by cation–chloride co-transporters type (CCC type), the framework of the thermodynamics of irreversible processes was conveniently applied to arrive at a quantitative expression for the hypothesis on root pressure developed by Wegner (2014). Overall volume flow across the plasma membrane of xylem parenchyma cells can be described as the sum of two separate components, namely, volume flow driven by the chemical potential of water and volume flow driven by the CCC transporter (Teakle and Tyerman 2010). Transporters of the CCC type known to mediate water secretion in mammalian cells have also been found in *Arabidopsis* and in rice. The mechanism proposed here for root pressure could also explain refilling of embolized vessels and contribute to long-distance water transport in trees when the cohesion–tension mechanism of water ascent fails. Possibly, several pathways for co-transport of water and ions (solutes) may coexist in the membrane to achieve the required rate of water secretion under various conditions, as previously also described for epithelia (Zeuthen 2010), though the reality is even more complex. However, the present hypothesis needs rigorous testing and verification in plants of different habitats and statures. Embolism repair requires vessels to be filled with xylem sap secreted by adjacent cells. It is generally believed that some overpressure has to be built up in a vessel during the refilling process in order to dissolve residual inclusions of gas completely; removal of cavitation nuclei appeared to be a prerequisite for a vessel to regain functionality (Holbrook and Zwieniecki 1999; Zwieniecki and Holbrook 2009; Nardini et al. 2011; Secchi and Zwieniecki 2011; Brodersen and McElrone 2013). Usually refilling occurs overnight when transpiration is low and little or no tension prevails in adjacent, functional vessels. This issue has been taken into account by this mechanism of root pressure.

8 Factors Affecting Root Pressure

Several factors such as drought, salinity, soil compaction, flooding, low temperatures, and light intensity; a number of mechanosignals such as pressure, wind, gravity, mechanical loading, etc.; and the nutritional status affect the hydraulic conductivity of the tissues, hence root pressure. The root pressure keeps on fluctuating generally within a range of 0.05–0.3 MPa and often goes up to 0.8 MPa or even 6 MPa (White et al. 1958; Kundt and Gruber 2006) varying with species, growing conditions, environmental and edaphic factors, etc. accompanied by seasonal and diurnal periodicity. Therefore, more than one cause of root pressure can

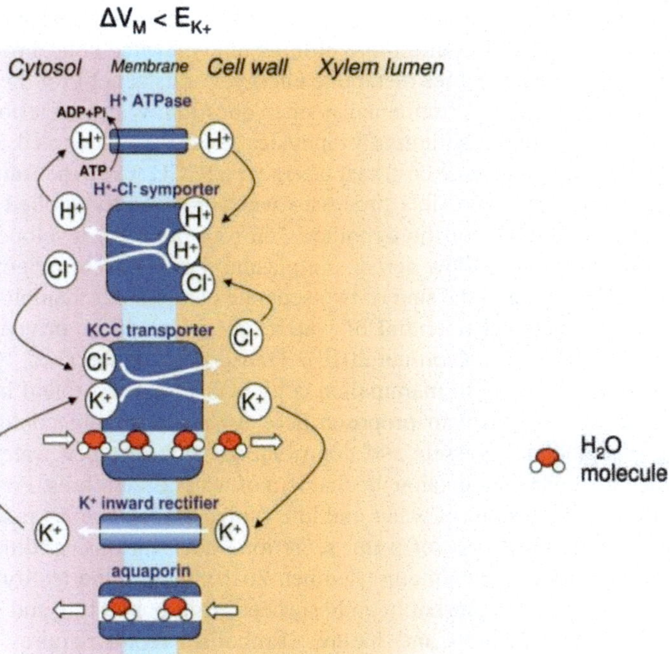

Fig. 5 (continued)

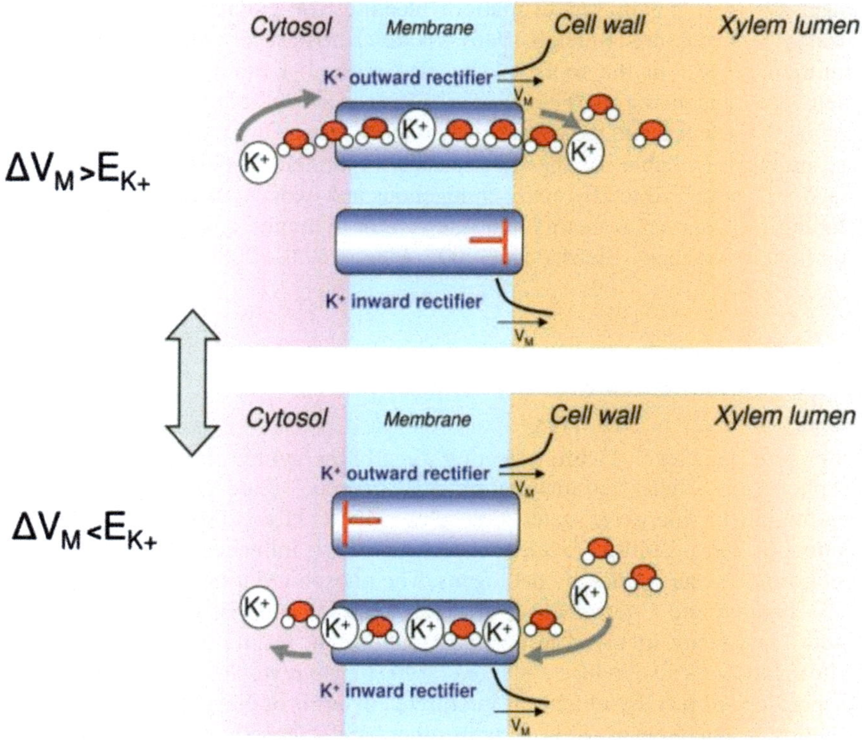

Fig. 5 (**a**) Hypothetical interplay of membrane transporters in the plasma membrane of xylem parenchyma cells for water secretion. Coupling between water and ion transport occurs in a potassium–chloride co-transporter type (KCC type) that translocates K^+ and Cl^- together with a fixed number of water molecules. Note that this transport is electrically silent. The ions are at least partly recycled via a K^+ inward-rectifying channel and a $Cl^- - 2H^+$ symporter, respectively. These processes are energized by the activity of a H^+ ATPase that maintains the H^+ gradient and hyperpolarizes the membrane to values more negative than E_{K+}. Aquaporins may to some extent short-circuit co-transport-driven water flow if their activity is not downregulated. Note that all transporters have been demonstrated to coexist in the plasma membrane of root stelar cells. ΔV_M = membrane potential of xylem parenchyma cell. E_{K+} = Nernst potential for K^+ [*Source*: Wegner (2014), for more details, see Wegner (2014)]. (**b**) Alternative model for water secretion that makes use of different water–ion coupling ratios in outward- and inward-rectifying K^+ channels. Arbitrarily, the K^+ outward rectifier is thought to carry three water molecules together with one K^+ ion, whereas the inward rectifier transports water and K^+ on a 1:1 basis. Both rectifiers operate alternately, coordinated by membrane potential oscillations. In this way, futile K^+ cycling is organized that drives a net water flow from the cytosol into the apoplast. Note that this K^+ cycling consumes metabolic energy when the membrane potential is hyperpolarized by proton pump activity. ΔV_M = membrane potential of xylem parenchyma cell. E_{K+} = Nernst potential for K [*Source*: Wegner (2014), for more details, see Wegner (2014)]

perhaps be effective in sap exudation, bleeding, or guttation due possibly to increased root pressure (Raleigh 1946). As described earlier, with regard to water absorption control in the roots, plants also present a family of water channel proteins, called aquaporins. These proteins have a critical role in water absorption, reducing the resistance to the water flow along the transcellular path. The number of these proteins available for the root surface is variable throughout the day. The aquaporins are controlled by many endogenous and exogenous factors of the roots including environmental factors that interfere in hydraulic conductance along the water flow by the plant (Siefritz et al. 2002; Maurel et al. 2008; Heinen et al. 2009).

8.1 Chemical Factors

At the molecular level, factors affecting directly the gating of aquaporins include phosphorylation, heterotetramerization, pH, cations, solute gradients, etc. apart from the kinetic energy of water molecules (Maurel et al. 2008; Benga 2009). In addition, the permeability of water channel proteins is influenced by nutrient stress, plant hormones, and attack by pathogens. The phosphorylation sites are located in the N-terminal and C-terminal segments and also in loop B. Calcium-dependent protein kinases are involved in phosphorylation that results in the pore opening. Hydroxyl radicals also induce a marked ($\geq$90 %) and reversible inhibition of water transport in *Chara* cells, which was interpreted in terms of direct oxidative gating of aquaporins (Henzler et al. 2004). On the other hand, cytosolic proteins decrease the water permeability of PIPs and tonoplast intrinsic proteins (TIPs). A coordinated inhibition of PIPs and, as a consequence, a general block of root water transport during anoxic stress (resulting from soil flooding) was attributed to closure of the channel after cell acidosis.

8.2 Nutrient Stress Factors

An unexplained interaction between cell and root hydraulic conductivity and the supply of certain mineral nutrients has been described in many species. At the cellular level, NO_3^- deprivation decreased hydraulic conductivity of cortical cells in cotton roots to a period of 4 days (Radin and Matthews 1989). At the whole-plant level, it decreased by a similar extent grown in P-deficient conditions (Radin and Eidenbock 1984) and by 20 % of control values over a 4-day period in SO_4^{2-} deprived barley roots (Karmoker et al. 1991). Short periods of oxygen starvation (Birner and Steudle 1993; Else et al. 1995) and variation in the supply of nutrients (N, P, S) and of NaCl (high salinity) are all characterized by marked changes in the hydraulic conductivities of roots or root cells. In the future, it will be an obvious challenge to find out how the diurnal variation in root hydraulic conductivity and aquaporin expression fit into the broader picture of variable root pressure which has been known for a long time.

8.3 Physical Factors

Physical pressure has been shown by "patch clamp" technique very clearly to cause prevention of ion transport across the membrane affecting turgor pressure of cells. Wind, gravity, touch, sound, snow loading, etc. have been found to affect aquaporins gating, hence water transport, affecting root pressure (Telewski 2006).

8.4 Genetic Factor

As described earlier, the phenomenon of root exudation occurs in a wide range of plant species which include herbaceous mesophytes, shrubs, and woody trees in angiosperms (Singh 2014b). Though there is lack of information on genotypic differences in root exudation among field and horticultural crops yet, the rate of root pressure differs among rice varieties (Lafitte and Courtois 2002). Similar varietal and species responses to root pressure were observed in tomato, orange, and watermelon (Mitchell et al. 1991; Dorais et al. 2001). Fujii and Tanaka (1957) examined the difference in the guttation and bleeding of seedlings of various varieties of rice. The increased guttation of the late-maturing varieties as compared to early-maturing ones was possible due to increased root pressure on account of efficient root metabolism. Recently, Singh et al. (2008, 2009b) also found large genotypic variability in guttation rate among modern rice cultivars which was correlated with their panicle sink potentials. This could also be due to perhaps increased root pressure though this trait was not quantified. Obviously, the genetic basis of variation in root pressure is not known and so is the case with ion channel types. Therefore, our knowledge is hampered by the lack of such information.

8.5 Aquaporin Factor

Globally, there appear to be eight major research centers, three in Germany, one each headed by Steudle (sadly, he died a few years ago), Schaffner, and Kaldenhoff, respectively, Maurel in France, Tyree in the USA (though he assumes multinational locations), Chaumont in Belgium, Tyerman in Australia, and Maeshima in Japan engaged in groundbreaking research in discovering, characterization, localization, structural chemistry, mechanism of action, and physiological role of aquaporins in fine-tuning of water uptake and transport in plants influenced by a number of chemico–mechanosensors for their survival and productivity in changing environment (Maurel et al. 2008; Maeshima and Ishikawa 2008; Kaldenhoff et al. 2008, 2014; Vandeleur et al. 2009; Heinen et al. 2009). Evidently, Steudle's laboratory has been the hub of attraction for aquaporin research in plants where scientists from all corners of the world tended to converge at one time or another, for example, Lafitte from IRRI, Tyerman from Australia, Chaumont from Belgium, Bohnert from the USA, Maurel from France, and Smith and Clarkson from England can be seen in a number of publications under joint authorships. By this narration, however, under no circumstances I mean to undermine the significance of excellent work currently being done in other laboratories as well (Benga 2009) which have been duly credited and described as and when necessary in this chapter.

The first water channel protein (WCP), called today aquaporin 1 (AQP1), was discovered in the red blood cell (RBC) membrane by Benga's group in 1985 in Romania (Benga et al. 1986a, b) followed by the first evidence regarding the existence of a WCP in plant membranes (Wayne and Tazawa 1990). In 1993, a protein from the vacuolar membrane (tonoplast) of *Arabidopsis thaliana* was identified as a WCP by Maurel and coworkers in France. Aquaporin activity is regulated at both the transcriptional and the posttranslational levels. Aquaporins are encoded by genes which display a remarkable degree of conservation across taxa and kingdoms, the most obvious homology stems from two loops that both harbor the signature amino acid motif Asn–Pro–Ala (NPA) (Maurel et al. 2008; Heinen et al. 2009). Based on structural studies with mammalian aquaporins, the NPA loops are presumed to form the pore which confers permeability to water driven by osmotic or hydraulic gradients across the membrane (Jung et al. 1994).

It is noted that responses to a change in environmental conditions can also be realized by other mechanisms, including aquaporin gating influenced by kinetic energy of water molecules, translocation of aquaporins into the membrane, and interactions of membrane proteins (Hedfalk et al. 2006; Zelazny et al. 2007; Maurel et al. 2008; Aroca et al. 2012). Finally, aquaporin functions need to be further integrated in the whole-plant physiology. This will first require a better understanding of how the various transport activities of aquaporins are coupled with those of other transport proteins. The chains of events that lead to control of aquaporin functions by hormones, local or long-distance signals, in response to mechanosensory stimuli, will also have to be elucidated. Finally, although the field of aquaporin research has already enlarged considerably, we may not be at the end of our surprises because of novel primary functions as diverse as cell proliferation. Although much has been learned about the possible physiological roles of aquaporins in plants, many questions remain unanswered (Baiges et al. 2002; Aroca et al. 2012).

8.6 *Environmental Factors*

The environmental conditions in general affect root pressure, but because of genetic differences related to internal cellular sensing networks, responses vary between plants in terms of the effect on root pressure. Which is why, conditions that discourage root pressure such as cold, dry aerated soil, etc. also reduce guttation which itself is the expression of root pressure. The change in humidity brings about variation in aquaporin activities. Thus, the change in humidity resulted in an accumulation of water channel proteins, and those proteins were still present 24 h later. This conclusion is supported by immunolabeling experiments, which revealed that PIP1 protein remained highly abundant in root cross-sections 28 h after the transfer to lower humidity. How exactly changes in the aboveground environment are transmitted to and sensed by roots remains unknown. The most parsimonious hypothesis is that root cells sense xylem pressure pulses (McElrone et al. 2007) or changes in water potential (Levin et al. 2009), and/or cell turgor (Hill et al. 2004), which all would correspond to changes in hydraulic conductivity affecting root pressure. Future work is required at unraveling the nature of this signaling process as how the signal is perceived by root aquaporins.

8.7 *Soil Factors*

8.7.1 Soil and Root Temperature

Root pressure being an energy-dependent process as discussed earlier, root temperature plays a dominant role in its regulation. Pedersen (1993, 1994) showed by measuring the rate of guttation, which is the manifestation of root pressure, that in submerged aquatic plants *Sparganium emersum* and *Lobelia dortmanna*, the acropetal water transport is clearly an active process confined to the roots. The flow is stopped by cooling the root compartment to 4 °C, and lowering the temperature from 15 to 10 °C reduced the guttation rate fivefold. This indicates that the water transport is dependent on root metabolism and that the driving force is restricted to the roots. Although periodicity in bleeding appears to be automatic in origin, a sharp increase in temperature will determine the time of occurrence of the maxima and minima (Fujii and Tanaka 1957). Under spring conditions, soil temperatures may remain several degrees higher than air temperatures at night. Here, intense radiation may warm the soil during the day, and a rapid cooling of the air at night produces optimal conditions for root pressure to occur (Frey-Wyssling 1941). Tropical areas with humid night air and warm soil also are particularly favorable for rapid root exudation. Thus, the phenomenon of root pressure does appear to be affected significantly by prevalent soil and root temperature suggesting the involvement of metabolic control of minerals uptake via membrane ATPases on the one hand and functioning of aquaporins for water influx into the roots, on the other.

8.7.2 Soil Moisture

Root exudation is very common during warm humid nights in plants growing in high soil moisture. These conditions favor low transpiration and high root pressure. Even at night after periods of water stress, absorption may not completely replace the water deficit in the plant, and then actual pressures would not be developed in the xylem (Stocking 1956). Thus, no root exudation was observed under conditions of soil moisture stress (Kramer and Boyer 1995). The root pressure mechanism, as measured by exudation, ceased in *Coleus*, sunflower, and tomato plants growing in sandy soil when about 45 % of the moisture available to intact plants still remained in the soil. It appears that root pressure probably is not developed in plants growing in soil containing less than about 45 % of the moisture in the range from moisture equivalent to permanent wilting percentage (Zaitseva et al. 1998). If in these instances water is added to the soil, root exudation soon follows. More recently, Singh et al. (2009b) have provided quantitative data on the relationship between soil moisture stress and guttation caused by root pressure in rice. The volumes of guttation fluid, an indirect measure of root pressure, were 19 μL, 56 μL, and 93 μL at leaf water potentials of −1.0, −0.5, and −0.2 MPa (watered), respectively. Lowered water potentials of roots seem to affect the gating as well as distribution of various isoforms of aquaporins (Katsuhara et al. 2008; Heinen et al. 2009) inhibiting the entry of water becoming not enough to cause hydrostatic pressure in the roots on the one hand and cause cavitation and embolism in plants on the other (Holbrook et al. 2001; Brodribb and Holbrook 2006).

9 Consequences and Implications of Root Pressure

The ascent of water in terrestrial transpiring plants, both tall and short, has been extensively studied and explained to occur in response to increasing water potential gradient between top and bottom of the plants, a concept popularly known as "cohesion–tension" mechanism (Dixon and Joly 1894; Scholander et al. 1965; Tyree 2003a). However, the upward movement of water in the absence or reduced transpiration, for example, in seedlings emerging through soil crust (Misra et al. 1986), upward movement of water and nutrients in submerged aquatic plants wherein no transpiration takes place (Pedersen 1993, 1994), nocturnal ascent of sap in tall bamboos (32–40-m high) (Zachary 2009; Cao et al. 2012), climbing forest vines (Ewers et al. 1997; Fisher et al. 1997), rice plants (Singh and Singh 1989; Singh et al. 2009a), and cracking of pavements and walls by grasses and tree roots (Grabosky et al. 2011) are known to take place due to positive hydrostatic pressure in roots. Water rises in plants not only during sunny daytime but also at night, during dry epochs, and in winter, before the leaves have unfolded. It also rises at the bottom of rain forests—manifested by guttation where transpiration is prevented by 100 % air moisture and almost complete darkness, as well as in plants kept locked up in closed glass containers for a whole season. In all such thirsty intervals, plants depend uniquely on their root pumps (Kundt and Gruber 2006). The significance and implications of root pressure in various fields of studies are briefly described hereunder.

9.1 Agronomic Significance

The phenomenon of root pressure has a number of practical and agronomic significance which will be briefly described here. It is a crucial factor to control tipburn in head lettuce by optimizing its magnitude in polyhouses (De Swaef and Bleyaert 2012). The radial and axial pressure components of root pressure of emerging seedlings help in reducing and overcoming resistance of compacted and sodic soils (Misra et al. 1986; Bengough and Mullins 1990) leading to proper seedling growth and optimum plant density. The roots exert radial as well as axial stress which causes tensile failure in significant magnitude helping the emergence of seedlings through the hard and compacted soils. The axial and radial root growth pressures measured in this manner were, for example, of comparable magnitudes registering 497, 289 and 238 kPa for pea, cotton, and sunflower seedlings of similar size, respectively (Misra et al. 1986). Obviously, this excess pressure allows plants to penetrate into the ground, split rock, lift concrete plates, and push away obstacles. Shoots from coconuts pierce their tough enclosing shells on account of high magnitude of hydrostatic pressure developed in the shoot and root of emerging seedlings. Root pressure helps in root growth causing its turgor-induced elongation through soil matrix. It also helps in pushing soils apart facilitating proper soil aeration which is essential for nutrient uptake and root respiration leading to proper root growth and its metabolism. Further, by pushing the soils apart, root pressure also enlarges pore spaces in soils increasing their water storage capacity, hence increasing the soil moisture content. Root pressure has also been implicated in some plant diseases (Johnson 1936). Further, due to increased root pressure, bursting of cells allows pathogens such as *Botrytis* and *Mycosphaerella* to infect the damaged tissue and spread around in the plant. Failure of refilling embolized tissues can lead to crown dieback disease.

Root pressure also aids in inter-trafficking of xylem and phloem saps. Recently, xylem and phloem saps have been reported to carry a number of proteins including transport proteins, enzymes, amino acids, and hormones apart from inorganic compounds (Singh and Singh 2013). The distribution of these chemicals is facilitated through inter-trafficking by root pressure and by circulating in the entire plant system to be utilized in organs they are required most (Burkle et al. 2003). Out of these two processes, the former is undoubtedly regulated by root pressure, at least under the conditions of the absence or reduced transpiration facilitating inter-trafficking of organic and inorganic substances through these conducting tissues. Therefore, root pressure is an essential component of inter-trafficking of chemicals required for proper plant growth and yield.

9.2 Horticultural Significance

In the wake of global warming and climate change, the reduction of energy use has become an important issue in glasshouse cultivation. It is, therefore, imperative to develop a modified crop management and new energy-efficient climate strategies that may create conditions favoring the buildup of optimum root pressure. Because the occurrence of excessive root pressure may cause considerable damage to the harvestable product (Heuvelink et al. 2008), the technique developed by De Swaef et al. (2013) could be of direct significance to growers, as an indicator when root pressure-associated problems might occur. Because these new energy-efficient technologies allow better glasshouse climate control, potentially unfavorable conditions could be avoided using an intelligently managed compromise between energy saving and optimal growing conditions. Because some important physiological problems in horticulture such as watery fruits in tomato, watermelon, muskmelon, etc. (Mitchell et al. 1991; Johnson et al. 1992; De Swaef et al. 2012) and glassiness in lettuce (Maaswinkel and Welles 1986) are currently attributed to excess root pressure, a critical level at which root pressure becomes undesired should be determined.

To date, the main problem for investigating this critical level, as well as the environmental factors driving root pressure, is the lack of appropriate systems for determining root pressure. Therefore, a nondestructive technique could allow root pressure to be estimated continuously while imposing a wide range of growing conditions. Recently, Clearwater et al. (2007) unraveled a correlation between the rootstock affinity to build up root pressure and the scion vigor after grafting in kiwi. For tomato, it is known in practice that rootstocks have an important effect on scion vigor and on the occurrence of root pressure-related problems. The noninvasive technique could be used to select rootstocks quantitatively with a different affinity for root pressure. As a result, better combinations of rootstocks and productive grafts could be made. Root pressure is also expected to play a beneficial role in the distribution of calcium to the leaves of cabbage and lettuce (Palzkill and Tibbitts 1977).

9.3 Use of Root Pressure in Physiological Investigations

9.3.1 Use of Root Pressure in Measuring Hydraulic Properties

The phenomenon of root pressure can be used for a number of physiological research investigations. Accordingly, the root pressure can be used as a tool to measure the hydraulic properties of the whole-plant system (Liu et al. 2009) in deciding the nature of root water flow, i.e., apoplastic or symplastic, in a number of crop plants of economic importance (Knipfer and Fricke 2010).

9.3.2 Use of Root Pressure in the Determination of Reflection Coefficients

The reflection coefficients may be determined by using root pressure as a tool. Deviation of the reflection coefficient from 1.0 may be attributable to a nonperfect semipermeability of the water uptake pathway between the medium and the xylem or to bypass flow at the endodermis (Steudle et al. 1993; Steudle and Peterson 1998). However, one should be cautious as the possibility of induction of some artifact may change the interpretation.

9.3.3 Use of Root Pressure in the Determination of Nutrients and Hormones Movement in Various Plant Organs

The root pressure can be used for tracing the path, velocity, and quantum of different nutrients, injurious heavy metals, pesticides, and hormones in plant organs at various growth stages for deciding their fate (Singh and Singh 2013). A convective viscous mass transport system can move nutrients in a bulk flow of water from root to shoot. Such a mass transport of water could be driven by root pressure which has been demonstrated in a wide range of terrestrial (Sperry 1983; Singh and Singh 1989; Zholkevich 1991; Dustmamatov et al. 2004; Feild et al. 2005; Singh et al. 2009a) and submerged aquatic plants (Pedersen 1993, 1997; Pedersen and Sand-Jensen 1997). Under moist or saturated soil conditions with high relative humidity of the atmosphere, the root systems of plants like rice, tomato, potato, etc. absorb excess of water by active uptake giving rise to the buildup of root pressure. Palzkill and Tibbitts (1977) and Dieffenbach et al. (1980) used successfully the root exudation carrying plant nutrients as a way to track the flow of water and mineral ions from the roots. It thus appears that the water transport in the xylem, brought about by root pressure, is in itself sufficient for long-distance mineral supply and that transpiration is not required for this function (Tanner and Beevers 2001). Root pressure-directed water transport, distributing required inorganic nutrients and phytohormones, both derived from the roots, ensures optimal plant growth in the absence of a transpiration stream.

Further, the important natural plant hormones such as auxins, GAs, cytokinins, and ABA have been characterized in xylem saps, hence transported to the sites of their requirement for regulating growth, development, and fruiting by root pressure (Fletcher and Mader 2007). The abovementioned physiological effects attributed to root pressure indicate the need to investigate this intriguing phenomenon further in depth. However, currently a "bottleneck" in root pressure research is the lack of a system that allows continuous and nondestructive measurements, especially for herbaceous plants.

9.3.4 Use of Root Pressure in the Maintenance of Shoot Water Storage

Undoubtedly, root pressure is a mechanism to "push" xylem sap up the plant. The contribution of root pressure giving rise to guttation in the maintenance of leaf water potential has been realized (Barrs 1966; O'Leary 1966), because within the vascular bundles, pressure pushes water into the main space available in the vessels, repairing embolisms caused by water stress (Singh and Singh 1989; Holbrook et al. 2001; Brodribb and Holbrook 2006; Singh et al. 2009a). Observations of strawberry plants suggest that the presence or absence of root pressure-generated guttation is related to plant water status (Takeda and Glenn 1989). Leaves with guttation had a predawn leaf water potential (PLWP) higher than −0.1 MPa, and those without guttation had PLWP lower than −0.1 MPa. Also, root pressure supplies water to the shoots of rice (Singh et al. 2009a), bamboos (Zachary 2009), and other trees as well (Sperry et al. 1987) during spring when plants are devoid of leaves, left with no provision for pulling water up by transpiration. Hence, it saves trees and crops from collapse due to water deficits (Sperry et al. 1987; Clearwater et al. 2007).

9.3.5 Use of Root Pressure in Developing Drought-Resistant Cultivars

This trait has been found to be associated with the amount of sap exuded from single cut tillers of rice, and initial analyses indicated lower root pressure to be linked with drought sensitivity (Lafitte and Courtois 2002; Lian et al. 2004). Xylem vessel cavitation was observed, and nighttime root pressure was hypothesized to be important for refilling of cavitated xylem vessels (Stiller et al. 2003). Cavitation-resistant juniper continues gas exchange during drought, extracting more water from the soil. But in doing so, the juniper eventually nevertheless suffers considerable cavitation, making it less competitive for water after the next rain. In this context, understanding the extensive mortality of piñon pine across the Southeastern USA in recent years is imperative (McDowell et al. 2008).

9.3.6 Use of Root Pressure in Enhancing Production of Recombinant Proteins, Allelochemicals, and Chelatins

Guttation is the best-known phenomenon attributed to root pressure which has a number of agricultural and pharmaceutical implications (Singh and Singh 2013; Singh 2014a, b). Recently, plants have emerged as one of the most promising general production platforms for pharmaceuticals and are now gaining widespread acceptance for the large-scale production of recombinant proteins (Komarnytsky et al. 2000). These problems have been addressed by engineering two related tobacco plant production systems in Raskin's laboratory in the USA to continuously secrete recombinant proteins from their roots called "rhizosecretion" into a simple hydroponic medium (Borisjuk et al. 1999). There is also leaf secretion called "phyllosecretion" which is caused by increased root pressure (Komarnytsky et al. 2000). Undoubtedly, these processes of nondestructive secretions are governed by root pressure. Apart from these, secretions of organic acids and transport of heavy metal chelatins (Sears 2013) and allelochemicals (Xiao et al. 2006) from roots are also regulated by root pressure. Therefore, the optimization of root pressure can play a significant role in the noninvasive production of pharmaceuticals, allelochemicals, and heavy metal chelatins which are of immense economic importance.

9.4 Ecological Significance

Ecological adaptation has enabled desert plants to even achieve root pressures of 6 MPa. They require an excess pressure inside the plant body, not to be established by suction (White et al. 1958; Kundt and Gruber 2006). Further, lianas (woody vines) are much more common in tropical than temperate ecosystems, but the reasons for this are unknown (Gentry 1991). Lianas have long been known to have thin stems and a high ratio of leaf area to transverse stem area (Putz 1983; Ewers and Fisher 1991). Wide xylem vessels, which in temperate areas are quite prone to freezing-induced embolism (Cochard and Tyree 1990; Sperry and Sullivan 1992; Sperry et al. 1994; Tyree et al. 1994), are also one of the characteristic features of tropical lianas (Berger 1931; Ewers and Fisher 1991). It is not known whether the wide vessels of tropical vines usually avoid embolism throughout the life of the stems or if they become embolized and are periodically refilled. If a perennial plant such as a liana could not refill its embolized vessels, its distribution might be limited to environments where embolism induction would be minimal. For instance, some species might be limited to environments with a very consistent water supply and without freezing temperatures. However, more is required to be known about the vulnerability of lianas to embolism and about the possibility of embolism reversal by means other than root pressure as well (Lens et al. 2013). In terms of global distribution, there is a strong inverse correlation between latitude and liana abundance. Thus, root pressure appears to play an important role in the distribution of plant species suiting to prevalent environment. Also, nocturnal refilling in bamboos is another example of ecological significance. Positive root pressure is the primary mechanism for hydraulic movement in *Guadua angustifolia* (Zachary 2009). Nocturnally pumped sap is stored within internodal cavities for daily use. Differences among tissue sap flow profiles suggest that root pressure-induced sap flow is mediated by internodal tissue, while subsequent distribution to leaves and branches is carried out by nodal tissue. The evolution of sophisticated micro-valves in gymnosperms is crucial to the success of this lineage and helps conifers compete effectively for water with angiosperms (Pittermann et al. 2005). Hydraulic lift, also a consequence of root pressure which adds moisture to otherwise dry soils, dramatically alters the microenvironment and influences community composition (Jackson et al. 2000; Meinzer et al. 2001).

10 Concluding Remarks and Future Perspectives

It has been very exciting to see the progress made over the past about 15 years in elucidating the molecular mechanisms of water uptake, transport, and the root pressure. Modification of root function should continue to be a productive area for future applications to agriculture, horticulture, and forestry which may eventually enhance yield and quality of these crops under sustainable ecosystem. To ensure success in this area, we will need to more fully understand how to best regulate the expression of water co-transporters and their substrates to create more efficient plants for water use that do not have any soil and yield penalties. The combination of continuous measurements of water flow and stem diameter and the mechanistic water flow and storage model allowed elucidation of the effects on mechanisms other than the C–T theory, such as root pressure, which is as yet difficult to measure noninvasively in situ. Therefore, the techniques described in this chapter could have important implications both in practice and in future root pressure research.

The study of roots has advanced to the point where modifications can be made to both architecture and function using molecular tools. However, a challenge will be to find the winning combination of shoot and root traits that can be successfully combined to benefit the whole-plant growth and productivity by optimizing root pressure. The development and use of noninvasive quantification of root pressure and high-throughput techniques such as pressure probe, radioactive tracer isotopes, fluorescence resonance energy transport, magnetic resonance imaging, computed tomography, X-ray crystallography, knockout gene transfer technology, etc. have helped elucidate the mechanisms involved in root morphology and physiology. However, for stress-induced embolism in plants of economic importance, simple, cheap, and dependable techniques are required for use in laboratory and field. Although a link has been established between root hydraulics and expression and function of aquaporins, the knowledge of cell-specific expression, specified location, and function of root aquaporins is largely lacking, mainly due to the high diversity of aquaporin isoforms in plants. In the future, analysis of single knockout aquaporin mutants will hopefully provide evidence for the multiple functions of aquaporins in water transport, their distribution, and root pressure development affecting growth and development of plants and in their adaptive response to chemical, physical, environmental, edaphic, and genetic factors. Further investigations are necessary to determine the molecular structure of the water pores and the mechanisms of their selectivity and gating. There are, to date, no novel approaches to estimate the relative contribution of the two water transport pathways (e.g., apoplastic and symplastic) to the overall water and solute uptake and hydraulic conductivity of roots. With respect to having a clear picture, models of water uptake by root systems are still only semiquantitative and require completion. In order to set up good quantitative physical models, we must combine mathematical principles and computer technology with biological principles, including some biophysics and biochemistry. Studies of root physiology and of key genes that regulate water transport and root pressure should complement traditional studies of shoot physiology and stomatal control of water loss, fundamental for the understanding of plant water balance and overall productivity.

Many researchers have contributed a large amount of valid work in the mechanisms of water uptake by roots. Yet, there are still a lot of aspects which need perfection and improvement. Intense research is under way in different laboratories to clarify mechanisms of plant hydraulics. Therefore, the future promises to see a much clearer picture of the basic mechanisms for water uptake by plant roots and root pressure. Since roots play an essential role in the acquisition of water and minerals from soils, modifying root traits such as root pressure in crops for enhanced agricultural, horticultural, and forest produce is imperative. The future for the modification of crop plant roots for water and nutrients collection for optimizing root pressure looks promising, but there are important challenges too, though examples have emerged showing that modifications to roots result in higher yield, enhanced quality of roots, and increased water productivity. Using combined "patch clamp" and "biosensor" techniques to determine the uptake and movement of ions in roots from the soil solution as well as within the roots could provide novel tools. How dynamic aboveground changes are perceived by roots and how root aquaporins are subsequently regulated is not well understood. Although much has been learned about the possible physiological roles of aquaporins in plants, many questions remain unanswered. It is hoped that the combination of aquaporin genetics integrated with plant physiology and energetics of root pressure will provide critical insights into the hydraulic conductance architecture in response to chemico–mechanosensory systems for the induction of root pressure.

While closing the chapter, I believe that the materials presented in this review would contribute significantly to a future solution of root pressure and its implications in the wake of advancing biotechnological and nanotechnological innovations and arrest, at least temporarily for some time to come, the legacy of the use of such terms as "riddle" and "enigma" of root pressure.

Acknowledgments The author is highly delighted and most grateful to Prof. U. Luettge, Editor of *Progress in Botany* for having being invited by him to write this chapter. He also extends his most sincere thanks to him for the pains taking and critical reading of the manuscript and forwarding useful comments and constructive suggestions for the improvement of the text. At the local level, the author wishes to thank Prof. Fassil Kebede, Dean, College of Agriculture & Rural Transformation, and Mr. Tesfaye Wossen, Head, Department of Plant Sciences, for continued moral support during the write-up of this chapter.

References

Ahmed AM, Kroener E, Holz M, Zarebanadkouki M, Carminati A (2014) Mucilage exudation facilitates root water uptake in dry soils. Funct Plant Biol 41:1129–1137

Ameglio T, Ewers FW, Cochard H, Martignac M, Vandame M, Bodet C, Cruiziat P (2001) Winter stem xylem pressure in walnut trees: effects of carbohydrates, cooling and freezing. Tree Physiol 21:387–394

Aroca R, Porcel R, Ruiz-Lozano JM (2012) Regulation of root water uptake under abiotic stress conditions. J Exp Bot 63:43–57

Azaizeh H, Gunse B, Steudle E (1992) Effects of NaCl and $CaCl_2$ on water transport across root cells of maize (*Zea mays* L.) seedlings. Plant Physiol 99:886–894

Bai X-F, Zhu J-J, Zhang P, Wang Y-H, Yang L-Q, Zhang L (2007) Na^+ and water uptake in relation to radial reflection coefficient of root in arrowleaf saltbush under salt stress. J Integr Plant Biol 49:1334–1340

Baiges I, Schäffner AR, Affenzeller MJ, Mas A (2002) Plant aquaporins. Physiol Plant 115:175–182

Balling A, Zimmermann U (1990) Comparative measurements of the xylem pressure of *Nicotiana* plants by means of the pressure bomb and pressure probe. Planta 182:325–338

Baluska F (1995) Structure and function of roots. Springer, Berlin

Baluska F, Mancuso S (2013) Root apex transition zone as oscillatory zone. Front Plant Sci 4:354. doi:10.3389/fpls.2013.00354

Baluska F, Salaj J, Mathur J, Braun M, Jasper F, Samaj J, Chua N-H, Barlow PW, Volkmann D (2000) Root hair formation: F-actin-dependent tip growth is initiated by local assembly of profilin-supported F-actin meshworks accumulated within expansin-enriched bulges. Dev Biol 227:618–632

Barber SA, Bouldin DR (1984) Roots, nutrients and water influx, and plant growth. ASA Publication, American Society of Agronomy, Madison, WI

Barrs HD (1966) Root pressure and leaf water potential. Science 152:1266–1268

Benga G (2009) Water channel proteins (later called aquaporins) and relatives: past, present, and future. IUBMB Life 61:112–133

Benga G, Popescu O, Pop VI, Holmes RP (1986a) *p*-Chloromercuribenzene-sulfonate binding by membrane proteins and the inhibition of water transport in human erythrocytes. Biochemistry 25:1535–1538

Benga G, Popescu O, Borza V, Pop VI, Muresan A, Mocsy I, Brain A, Wrigglesworth J (1986b) Water permeability of human erythrocytes. Identification of membrane proteins involved in water transport. Eur J Cell Biol 41:252–262

Bengough AG, Mullins CE (1990) Mechanical impedance to root growth—a review of experimental techniques and root growth responses. J Soil Sci 41:341–358

Berger W (1931) Das Wasserleitungssystem von krautigen Pflanzen, Zwergstaäuchern und Lianen in quantitativer Betrachtung. Beih Bot Centralbl 48:363–390

Birner TP, Steudle E (1993) Effects of anaerobic conditions on water and solute relations, and on active transport in roots of maize (*Zea mays* L.). Planta 190:474–483

Borisjuk NV, Borisjuk LG, Logendra S, Petersen F, Gleba YY, Raskin I (1999) Production of recombinant proteins in plant root exudates. Nat Biotechnol 17:466–469

Bose JC (1923) The physiology of the ascent of sap. Longmans, Green and Co., London

Boyer JS (1985) Water transport. Annu Rev Plant Physiol 36:473–516

Britto DT, Kronzucker HJ (2006) Futile cycling at the plasma membrane: a hallmark of low-affinity nutrient transport. Trends Plant Sci 11:529–534

Brodersen CR, McElrone AJ (2013) Maintenance of xylem network transport capacity: a review of embolism repair in vascular plants. Front Plant Sci 4:198

Brodribb TJ, Holbrook NM (2006) Declining hydraulic efficiency as transpiring leaves desiccate: two types of response. Plant Cell Environ 29:2205–2215

Burkle L, Cedzich A, Dopke C, Stransky H, Okumoto S, Gillissen B, Kuhn K, Frommer WB (2003) Transport of cytokinins mediated by purine transporters of the PUP family expressed in phloem, hydathodes, and pollen of *Arabidopsis*. Plant J 34:13–26

Cao KF, Yang SJ, Zhang YJ, Brodribb TJ (2012) The maximum height of grasses is determined by roots. Ecol Lett 15:666–672

Chamberlin TC (1916) The origin of the Earth. The University of Chicago Press, Chicago, IL

Clearwater MJ, Blattmann P, Luo Z, Lowe RG (2007) Control of scion vigor by kiwifruit rootstocks is correlated with spring root pressure phenology. J Exp Bot 58:1741–1751

Cobb AR, Choat B, Holbrook NM (2007) Dynamics of freeze-thaw embolism in *Smilax rotundifolia* (Smilacaceae). Am J Bot 94:640–649

Cochard H, Tyree MT (1990) Xylem dysfunction in *Quercus*: vessel sizes, tyloses, cavitation and seasonal changes in embolism. Tree Physiol 6:393–407

Cochard H, Ewers FW, Tyree MT (1994) Water relations of a tropical vine-like bamboo (*Rhipidocladum racemiflorum*): root pressures, vulnerability to cavitation and seasonal changes in embolism. J Exp Bot 45:1085–1089

Darwin C (1859) On the origin of species by means of natural selection or the preservation of favoured races in the struggle for life. John Murray, London

Davis TA (1961) High root pressure in palms. Nature 192:227–228

De Swaef T, Bleyaert P (2012). Is root pressure the crucial factor to control tipburn in head lettuce? http://www.inagro.be/ophalen_popup.aspx?lijst=Research&ID=11. Accessed 28 Jan 2015

De Swaef T, Verbist K, Cornelis W, Steppe K (2012) Tomato sap flow, stem and fruit growth in relation to water availability in rockwool growing medium. Plant Soil 350:237–252

De Swaef T, Hanssens J, Cornelis A, Steppe K (2013) Non-destructive estimation of root pressure using sap flow, stem diameter measurements and mechanistic modeling. Ann Bot 111:271–282

Dhonukshe P, Aniento F, Hwang I, Robinson DG, Mravec J et al (2007) Clathrin-mediated constitutive endocytosis of PIN auxin efflux carriers in *Arabidopsis*. Curr Biol 17:520–527

Dieffenbach H, Kramer D, Luttege U (1980) Release of guttation fluid from passive hydathodes of intact barley plants. I. Structural and cytological aspects. Ann Bot 45:397–401

Dixon HH, Joly J (1894) On the ascent of sap. Philosophical transactions of the royal society. Biol Sci 186:563–576

Dorais M, Gosselin A, Papadopoulos AP (2001) Greenhouse tomato fruit quality. Hortic Rev 26:239–306

Draye X, Kim Y, Lobet G, Javaux M (2010) Model-assisted integration of physiological and environmental constraints affecting the dynamic and spatial patterns of root water uptake from soils. J Exp Bot 61:2145–2155

Dustmamatov AG, Zholkevich VN (2008) Effects of $HgCl_2$ on principal parameters of exudation from maize detached root systems. Russ J Plant Physiol 55:814–820

Dustmamatov AG, Zholkevish VN, Kuznetsov VV (2004) Water pumping activity of the root system in the process of cross-adaptation of sunflower plants to hyperthermia and water deficiency. Russ J Plant Physiol 51:822–826

Ehleringer JR, Roden J, Dawson TE (2000) Assessing ecosystem-level water relations through stable isotopes ratio analysis. In: Sala OE, Jackson RB, Mooney HA, Howarth RW (eds) Methods in ecosystem science. Springer, New York, NY, pp 181–198

Else MA, Davies WJ, Malone M, Jackson MB (1995) A negative hydraulic message from oxygen deficient roots of tomato plants? Influence of soil flooding on leaf water potential, leaf expansion, and synchrony between stomatal conductance and root hydraulic conductivity. Plant Physiol 109:1017–1024

Enns LC, Canny MJ, McCully ME (2000) An investigation of the role of solutes in the xylem sap and in the xylem parenchyma as a source of root pressure. Protoplasma 211:183–197

Enstone DE, Peterson CA, Ma F (2003) Root endodermis and exodermis: structure, function, and responses to the environment. J Plant Growth Regul 21:335–351

Eshel M, Beeckman T (2013) Plant roots: the hidden half, 4th edn. CRC Press, Boca Raton, FL

Ewers FW, Fisher JB (1991) Why vines have narrow stems: histological trends in *Bauhinia* (Fabaceae). Oecologia 8:233–237
Ewers FW, Fisher JB, Fichtner K (1991) Water flux and xylem structure in vines. In: Putz FE, Mooney HA (eds) The biology of vines. Cambridge University Press, Cambridge, pp 127–160
Ewers FW, Cochard H, Tyree MT (1997) A survey of root pressures in vines of a tropical lowland forest. Oecologia 110:191–196
Feild TS, Arens NC (2007) The ecophysiology of early angiosperms. Plant Cell Environ 30:291–309
Feild TS, Sage TL, Czerniak C, Iles WJD (2005) Hydathodal leaf teeth of *Chloranthus japonicus* (Chloranthaceae) prevent guttation-induced flooding of the mesophyll. Plant Cell Environ 28:1179–1190
Fisher JB, Angeles GA, Ewers FW, López-Portillo J (1997) A survey of root pressure in tropical vines and woody species. Int J Plant Sci 158:44–50
Fletcher AT, Mader JC (2007) Hormone profiling by LC-QToF-MS/MS in dormant *Macadamia integrifolia*: correlations with abnormal vertical growth. Plant Growth Regul 26:351–361
Frey-Wyssling A (1941) Die guttation als aligemeine erscheinung. Berichte Der Schweizerischen Botanischen Gesellschaft 51:321–325
Fujii Y, Tanaka N (1957) Intensity of guttation in rice seedlings in relation to earliness or lateness of the variety. Jpn J Crop Sci 25:131–132
Gentry AH (1991) The distribution and evolution of climbing plants. In: Putz FE, Mooney HA (eds) The biology of vines. Cambridge University Press, Cambridge, pp 3–50
Ginsburg H (1971) Model for iso-osmotic water flow in plant roots. J Theor Biol 32:147–158
Grabosky JC, Smiley ET, Dahle GA (2011) Observed symmetry and force of *Plantanus* × *acerifolia* (Ait.) Willd. Roots occurring between foam layers under pavement. Arboricult Urban For 37:35–40
Hedfalk K, Hosefield T, Nyblom S, Johanson U, Kjellbom D, Neutze R (2006) Aquaporin gating. Curr Opin Struct Biol 16:447–456
Heinen RB, Ye Q, Chaumont F (2009) Role of aquaporins in leaf physiology. J Exp Bot 60:2971–2985
Henry A (2013) IRRI's drought stress research in rice with emphasis on roots: accomplishments over the last 50 years. Plant Root 7:5–19
Henzler T, Steudle E (1995) Reversible closing of water channels in *Chara* internodes provides evidence for a composite transport model of the plasma membrane. J Exp Bot 46:199–209
Henzler T, Ye Q, Steudle E (2004) Oxidative gating of water channels (aquaporins) in *Chara* by hydroxyl radicals. Plant Cell Environ 27:1184–1195
Heuvelink E, Bakker MJ, Marcelis LFM, Raaphorst M (2008) Climate and yield in a closed greenhouse. Acta Hortic 801:1083–1092
Hill AE, Shachar-Hill B, Shachar-Hill Y (2004) What are aquaporins for? J Membr Biol 197:1–32
Holbrook NM, Zwieniecki MA (1999) Embolism repair and xylem tension: do we need a miracle? Plant Physiol 120:7–10
Holbrook NM, Ahrens ET, Burns MJ, Zwieniecki MA (2001) In vivo observation of cavitation and embolism repair using magnetic resonance imaging (MRI). Plant Physiol 126:27–31
Jackson RB, Sperry JS, Dawson TE (2000) Root water uptake and transport: using physiological processes in global predictions. Trends Plant Sci 5:482–488
Javot H, Lauvergeat V, Santoni V, Laurent M, Guclu J, Vinh J, Heyes J, Franck KI, Schaffner AR, Bouchez D, Maurel C (2003) Role of a single aquaporin isoform in root water uptake. Plant Cell 15:509–522
Johnson J (1936) Relation of root pressure to plant disease. Science 84:135–136
Johnson RW, Dixon MA, Lee DR (1992) Water relations of the tomato during fruit growth. Plant Cell Environ 15:947–953
Jung JS, Preston GM, Smith BL, Guggino WB, Agre P (1994) Molecular structure of the water channel through aquaporins CHIP. The hourglass model. J Biol Chem 269:14648–14654

Kaldenhoff R, Ribas-Carbo M, Sans JF, Lovisolo C, Heckwolf M, Uehlein N (2008) Aquaporins and plant water balance. Plant Cell Environ 31:658–666
Kaldenhoff R, Kai L, Uehlein N (2014) Aquaporins and membrane diffusion of CO_2 in living organisms. Biochim Biophys Acta 1840:1592–1595
Karmoker JL, Clarkson DT, Saker LR, Rooney JM, Purves JV (1991) Sulphate deprivation depresses the transport of nitrogen to the xylem and hydraulic conductivity of barley (*Hordeum vulgare* L.) roots. Planta 185:2269–2278
Katsuhara M, Hanba YT, Shiratake K, Maeshima M (2008) Expanding roles of plant aquaporins in plasma membranes and cell organelles. Funct Plant Biol 35:1–14
Kirk GJD (1994) Rice roots: nutrient and water use. International Rice Research Institute, Los Banos, Philippines
Klepper B, Kaufmann MR (1966) Removal of salt from xylem sap by leaves and stems of guttating plants. Plant Physiol 41:1743–1747
Knipfer T, Fricke W (2010) Root pressure and a solute reflection coefficient close to unity exclude a purely apoplastic pathway of radial water transport in barley (*Hordeum vulgare*). New Phytol 187:159–170
Knipfer T, Das D, Steudle E (2007) During measurements of root hydraulics with pressure probe, the contribution of unstirred layers is minimized in the pressure relaxation mode: comparison with pressure clamp and high pressure flow meter. Plant Cell Environ 30:845–860
Koiwai H, Nakaminami K, Seo M, Toyomasu T, Koshiba T (2004) Tissue-specific localization of an abscisic acid biosynthetic enzyme, AAO3, in *Arabidopsis*. Plant Physiol 134:1697–1707
Komarnytsky S, Borisjuk N, Borisjuk L, Alam M, Raskin I (2000) Production of recombinant proteins in tobacco guttation fluid. Plant Physiol 124:927–933
Kramer PJ (1932) The absorption of water by root systems of plants. Am J Bot 19:148–164
Kramer PJ (1945) Absorption of water by plants. Bot Rev 11:310–355
Kramer PJ, Boyer JS (1995) Water relations of plants and soils. Academic, San Diego, CA
Kramer PJ, Currier HB (1950) Water relations of plant cells and tissues. Annu Rev Plant Physiol 1:265–284
Kramer PJ, Kozlowski TT (1979) The physiology of woody plants. Academia, Orlando, FL
Kundt W, Gruber E (2006) The water circuit of the plants. Do plants have hearts? Quant Biol 0603019:1–19
Lafitte HR, Courtois B (2002) Interpreting cultivar environment interactions for yield in upland rice: assigning value to drought-adaptive traits. Crop Sci 42:1409–1420
Lauchli A, James RA, Munns R, Huang C, McCully M (2008) Cellspecific localization of Na^+ in roots of durum wheat, and possible control points for salt exclusion. Plant Cell Environ 31:1565–1574
Lee KM, Driever SM, Heuvelink E et al (2012) Evaluation of diel patterns of relative changes in cell turgor of tomato plants using leaf patch clamp pressure probes. Physiol Plant 146:439–447
Lens F, Tixier A, Cochard H, Sperry JS, Jansen S, Herbette S (2013) Embolism resistance as a key mechanism to understand adaptive plant strategies. Curr Opin Biotechnol 16:287–292
Levin M, Resnick N, Rosianskey Y, Kolotilin I, Wininger S, Lemcoff JH, Cohen S, Galili G, Koltai H, Kapulnik Y (2009) Transcriptional profiling of *Arabidopsis thaliana* plants' response to low relative humidity suggests a shoot–root communication. Plant Sci 177:450–459
Lian HL, Yu X, Ye Q, Ding XS, Kitagawa Y et al (2004) The role of aquaporin RWC3 in drought avoidance in rice. Plant Cell Physiol 45:481–489
Liu BB, Steudle E, Deng X-P, Zhang S-Q (2009) Root pressure probe can be used to measure the hydraulic properties of whole root systems of corn (*Zea mays* L.). Bot Stud 50:303–310
Lobet G, Hachez C, Chaumont F, Javaux M, Draye X (2013) Root water uptake and water flow in the soil-root domain. In: Eshel M, Beeckman T (eds) Plant roots: the hidden half, 4th edn. CRC Press, New York, NY, pp 18–24
Lu P, Woo KC, Liu ZT (2002) Estimation of whole-plant transpiration of bananas using sap flow measurements. J Exp Bot 53:1771–1779
Lundegardh H (1944) Bleeding and sap movement. Arkiv for Botanik 31:1–56

Maaswinkel RHM, Welles GWH (1986) Factors influencing glassiness in lettuce. Neth J Agr Sci 34:57–65

Macduff JH, Bakken AK (2003) Diurnal variation in uptake and xylem contents of inorganic and assimilated N under continuous and interrupted N supply to *Phleum pretense* and *Festuca pratensis*. J Exp Bot 54:431–444

Maeshima M, Ishikawa F (2008) ER membrane aquaporins in plants. Eur J Physiol 456:709–716

Maurel C (1997) Aquaporins and water permeability of plant membranes. Annu Rev Plant Physiol Plant Mol Biol 48:399–429

Maurel C, Verdoucq L, Luu DT, Santoni V (2008) Plant aquaporins: membrane channels with multiple integrated functions. Annu Rev Plant Biol 59:595–624

Maurel C, Simonneau T, Sutka M (2010) The significance of roots as hydraulic rheostats. J Exp Bot 61:3191–3198

McDowell N, Pockman W, Allen C, Breshears D, Cobb N, Kolb T, Sperry JS, West A, Williams D, Yepez E (2008) Mechanisms of plant survival and mortality during drought. Why do some plants survive while others succumb to drought? New Phytol 178:719–739

McElrone AJ, Bichler J, Pockman WT, Addington RN, Linder CR, Jackson RB (2007) Aquaporin-mediated changes in hydraulic conductivity of deep tree roots accessed via caves. Plant Cell Environ 30:1411–1421

Meinzer FC, Clearwater MJ, Goldstein G (2001) Water transport in trees: current perspectives, new insights and some controversies. Environ Exp Bot 45:239–262

Meister R, Rajani MS, Ruzicka D, Schachtman DP (2014) Challenges of modifying root traits in crops for agriculture. Trends Plant Sci 19:779–788

Milburn JA, McLaughlin ME (1974) Studies of cavitation in isolated vascular bundles and whole leaves of *Plantago major* L. New Phytol 73:861–871

Miller DM (1985) Studies of root function in *Zea mays*: III. Xylem sap composition at maximum root pressure provides evidence of active transport into the xylem and a measurement of the reflection coefficient of the root. Plant Physiol 77:162–167

Miller-Rushing AJ, Primack RB (2008) Effects of winter temperatures on two birch (Betula) species. Tree Physiol 28:659–664

Misra RK, Dexter AR, Alston AM (1986) Maximum axial and radial growth pressures of plant roots. Plant Soil 95:315–326

Mitchell JP, Shennan C, Grattan SR, May DM (1991) Tomato fruit yields and quality under water deficit and salinity. J Am Soc Hort Sci 116:215–221

Mozhaeva LV, Pil'shchikova NV (1972) Nature of pumping water process by plant roots. Izvestiya Timiryazevskol Sel'skokhozyaistven-noi Akademii 3:3–15

Munns R (1985) Na+, K+ and Cl in xylem sap flowing to shoots of NaCl-treated barley. J Exp Bot 36:1032–1042

Nardini AL, Gullo MA, Salleo S (2011) Refilling embolized xylem conduits. Is it a matter of xylem unloading? Plant Sci 180:604–611

Neufeld HS, Grantz DA, Meinzer FC, Goldstein G, Crisosto GM, Cristosto C (1992) Genotypic variability in vulnerability of leaf xylem to cavitation in water-stressed and well-irrigated sugarcane. Plant Physiol 100:1020–1028

O'Leary JW (1966) Root pressure exudation from apical root segments. Nature 212:96–97

O'Leary JW, Kramer PJ (1964) Root pressure in conifers. Science 145:284–285

O'Toole JC, Chang TT (1978) Drought and rice improvement in perspective. IRRI Res Pap Ser Los Baños, Philippines 14:27

Oertli JJ (1966) Active water transport in plants. Physiol Plant 19:809–817

Ogata S, Saneoka H, Matsumoto K (1985) Nutritional-physiological evaluation of drought resistance of warm season forage species: comparative studies on root development water and nutrient absorption of forage species at various soil moisture levels. J Jpn Grassl Sci 31:263–271

Overton JB (1921) The mechanism of root pressure and its relation to sap flow. Am J Bot 8:369–374

Palmgren MG (2001) H^+-ATPases: powerhouses for nutrient uptake. Annu Rev Plant Physiol Mol Biol 52:817–845

Palzkill DA, Tibbitts TW (1977) Evidence that root pressure flow is required for calcium transport to head leaves of cabbage. Plant Physiol 60:854–856
Pedersen O (1993) Long-distance water transport in aquatic plants. Plant Physiol 103:1369–1375
Pedersen O (1994) Acropetal water transport in submerged plants. Bot Acta 107:61–65
Pedersen O (1997) The nature of water transport in aquatic plants. In: Pedersen O (ed) Freshwater biology. Priorities and development in Danish research. Gad, København, pp 196–207
Pedersen O, Sand-Jensen K (1997) Transpiration does not control growth and nutrient supply in the amphibious plant, *Mentha aquatica*. Plant Cell Environ 20:117–123
Pedersen BP, Buch-Pedersen MJ, Morth JP, Palmgren MG, Nissen P (2007) Crystal structure of the plasma membrane proton pump. Nature 450:1111–1114
Pfeffer W (1881) Pflanzenphysiologie. Ein Handbuch des Stoffwechsels und Kraftwechsels in der Pflanze. Erster Band: Stoffwechsel. Verlag Wilhelm Engelmann, Leipzig
Pickard WF (2003a) The riddle of root pressure. I. Putting Maxwell's demon to rest. Funct Plant Biol 30:121–134
Pickard WF (2003b) The riddle of root pressure. II. Root exudation at extreme osmolalities. Funct Plant Biol 30:135–141
Pittermann J, Sperry JS, Hacke UG, Wheeler JK, Sikkema E (2005) Torus-margo pits help conifers compete with angiosperms. Science 310:1924
Preston RD (1952) Movement of water in higher plants. In: Frey-Wyssling A (ed) Deformation and flow in biological systems. North Holland Publishing, Amsterdam, pp 257–321
Preston R (2007) The wild trees: a story of passion and daring. Anchor Books, California
Priestley JH (1920) The mechanism of root pressure. New Phytol 19:153–212
Putz FE (1983) Liana biomass and leaf area of a "tierra firme" forest in the Rio Negro Basin, Venezuela. Biotropica 15:185–189
Radin JW, Eidenbock MP (1984) Hydraulic conductance as a factor limiting leaf expansion of phosphorus-deficient cotton. Plant Physiol 75:372–377
Radin JW, Matthews MA (1989) Water transport properties of cortical cells in roots of nitrogen- and phosphorus-deficient cotton seedlings. Plant Physiol 89:264–268
Raleigh GJ (1946) The effect of various ions on guttation of the tomato. Plant Physiol 21:194–200
Renner O (1915) Theoretisches und Experimentelles zur Kohasionstherie der [Theoretical and experimental contributions to the cohesion theory of water transport]. Jahrbuch Wissenschaftliche Botanik 56:617–667
Renner O (1925) Zum Nachweis negativer Drucke im Gefäßwasser bewurzelter Holzgewächse. Flora 119:402–408
Sachs J (1887) Vorlesunguber Pflanzen-Physiologie, 2nd edn. Verlag Wilhelm Engelmann, Leipzig
Salleo S, Logullo MA, Depaoli D, Zippo M (1996) Xylem recovery from cavitation-induced embolism in young plants of Laurus nobilis: a possible mechanism. New Phytol 132:47–56
Scholander PS, Ruud B, Leivestad H (1957) The rise of sap in a tropical liana. Plant Physiol 32:1–6
Scholander PS, Bradstreet E, Hemmingsen E, Hammel H (1965) Sap pressure in vascular plants: negative hydrostatic pressure can be measured in plants. Science 148:339–346
Schwenke H, Wagner E (1992) A new concept of root exudation. Plant Cell Environ 15:289–299
Sears ME (2013) Chelation: harnessing and enhancing heavy metal detoxification—a review. Sci World J 2013:219840
Secchi F, Zwieniecki MA (2011) Sensing embolism in xylem vessels: the role of sucrose as a trigger for refilling. Plant Cell Environ 34:514–524
Siefritz F, Tyree MT, Lovisolo C, Schubert A, Kaldenhoff R (2002) PIP1 plasma membrane aquaporins in tobacco: from cellular effects to function in plants. Plant Cell 14:869–876
Singh S (2013) Guttation: path, principles and functions. Aust J Bot 61:497–515
Singh S (2014a) Guttation: quantification, microbiology and implications for phytopathology. In: Luttge U et al (eds) Progress in Botany, vol 75. Springer, Berlin, pp 187–214
Singh S (2014b) Guttation: new insights into agricultural implications. Adv Agron 128:7–135

Singh G, Singh TN (1989) Root-mediated water transport to the shoot of rice. Curr Sci 58:1134–1138
Singh S, Singh TN (1999) Root growth of wheat in simulated vertical and lateral splits of layered salt profiles in soil. Indian J Plant Physiol 4:73–78
Singh S, Singh TN (2000) Rooting ability and water relations of rice plant. Indian J Plant Physiol 5:1–6
Singh S, Singh TN (2013) Guttation: chemistry, crop husbandry and molecular farming. Phytochem Rev 12:147–172
Singh G, Singh TN, Singh S (1999) Trinodal rooting in rice: a new parameter on drought resistance. Indian J Plant Physiol 4:232–235
Singh S, Chauhan JS, Singh TN (2008) Guttation: a potential yield enhancing trait in rice. Curr Sci 95:455–456
Singh S, Singh TN, Chauhan JS (2009a) Water transport in crop plants with special reference to rice: key to crop production under global water crisis. J Crop Improv 23:194–212
Singh S, Singh TN, Chauhan JS (2009b) Guttation in rice: occurrence, regulation and significance in varietal improvement. J Crop Improv 23:351–365
Sperry JS (1983) Observations on the structure and function of hydathodes in *Blechnum lehmannii*. Am Fern J 73:65–72
Sperry JS (1993) Winter xylem embolism and spring recovery in *Betula cordifolia*, *Fagus grandifolia, Abies balsamea* and *Picea rubens*. In: Borghetti M, Grace J, Raschi A (eds) Water transport in plants under climatic stress. Cambridge University Press, Cambridge, pp 87–98
Sperry JS, Sullivan JEM (1992) Xylem embolism in response to freeze-thaw cycles and water stress in ring-porous, diffuse porous, and conifer species. Plant Physiol 100:603–613
Sperry JS, Holbrook NM, Zimmermann MH, Tyree MT (1987) Spring filling of xylem vessels in wild grapevine. Plant Physiol 83:414–417
Sperry JS, Donnelly JR, Tyree MT (1988) Seasonal occurrence of xylem embolism in sugar maple (*Acer saccharum*). Am J Bot 75:1212–1218
Sperry JS, Nichols KL, Sullivan JEM, Eastlack SE (1994) Comparative studies of xylem embolism in ring-porous, diffuse-porous and coniferous trees of northern Utah and interior Alaska. Ecology 75:1736–1752
Steudle E (1993) Pressure probe techniques: basic principles and application to studies of water and to studies of water and solute relations at cell, tissue and organ. In: Smith JAC, Griffihs H (eds) Water deficits: plant responses from cell to community. Bios Scientific Publishers, Oxford, pp 5–36
Steudle E, Jeschke WD (1983) Water transport in barley roots. Planta 158:237–248
Steudle E, Henzler T (1995) Water channels in plants: do basic concepts of water transport change? J Exp Bot 46:1067–1076
Steudle E (2000) Water uptake by roots: effects of water deficit. J Exp Bot 51:1531–1542
Steudle E (2001) The cohesion-tension mechanism and the acquisition of water by plant roots. Annu Rev Plant Physiol Plant Mol Biol 52:847–875
Steudle E, Meshcheryakov AB (1996) Hydraulic and osmotic properties of oak roots. J Exp Bot 47:387–401
Steudle E, Peterson C (1998) How does water get through roots? J Exp Bot 49:775–788
Steudle E, Oren R, Schulze ED (1987) Water transport in maize root. Plant Physiol 84:1220–1232
Steudle E, Murrmann M, Peterson CA (1993) Transport of water and solutes across corn roots modified by puncturing the endodermis. Plant Physiol 103:335–349
Stiller V, Lafitte HR, Sperry JS (2003) Hydraulic properties of rice and the response of gas exchange to water stress. Plant Physiol 132:1698–1706
Stocking CR (1956) Root pressure. In: Ruhland W (ed) Handbuch der pflazenphysiologie. Springer, Berlin, pp 581–595
Takeda F, Glenn DM (1989) Hydathode anatomy and the relationship between guttation and plant water status in strawberry (*Fragaria* x *ananassa* duch.). Acta Hortic (ISHS) 265:387–392

Tanner W, Beevers H (1999) Does transpiration have an essential function in long-distance ion transport in plants? Plant Cell Environ 13:745–750
Tanner W, Beevers H (2001) Transpiration, a prerequisite for long-distance transport of minerals in plants? Proc Natl Acad Sci USA 98:9443–9447
Teakle NL, Tyerman SD (2010) Mechanisms of chloride transport contributing to salt tolerance. Plant Cell Environ 33:566–589
Telewski FW (2006) A unified hypothesis of mechanoperception in plants. Am J Bot 93:1466–1476
Tyerman SD, Bohnert H, Maurel C, Steudle E, Smith JAC (1999) Plant aquaporins: their molecular biology, biophysics and significance for plant water relations. J Exp Bot 50:1055–1071
Tyerman SD, Niemietz CM, Bramley H (2002) Plant aquaporins: multifunctional water and solute channels with expanding roles. Plant Cell Environ 25:173–194
Tyree MT (2003a) Plant hydraulics: the ascent of water. Nature 423(6943)
Tyree MT (2003b) Hydraulic properties of roots. In: Kroon DH, Visser EJW (eds) Root ecology. Springer, Berlin, pp 125–149
Tyree MT, Sperry JS (1989) Vulnerability of xylem to cavitation and embolism. Annu Rev Plant Biol 40:19–36
Tyree MT, Zimmermann MH (2002) Xylem structure and the ascent of sap. Springer, New York, NY
Tyree MT, Fiscus EL, Wullschleger SD, Dixon MA (1986) Detection of xylem cavitation in corn under field conditions. Plant Physiol 82:597–599
Tyree MT, Yang S, Cruiziat P, Sinclair B (1994) Novel methods of measuring hydraulic conductivity of tree root systems and interpretation using AMAIZED: a maize-root dynamic model for water and solute transport. Plant Physiol 104:189–199
van Bavel MG, van Bavel CHM (1990) Dynagage installation and operational manual. Dynamax, Houston, TX
Vandeleur RK, Mayo G, Shelden MC, Gilliham M, Kaiser BN, Tyerman SD (2009) The role of plasma membrane intrinsic protein aquaporins in water transport through roots: diurnal and drought stress responses reveal different strategies between isohydric and anisohydric cultivars of grapevine. Plant Physiol 149:445–460
Voicu MC, Zwiazek JJ, Tyree MT (2008) Light response of hydraulic conductance in bur oak (*Quercus macrocarpa*) leaves. Tree Physiol 28:1007–1015
Wayne R, Tazawa M (1990) Nature of the water channels in the internodal cells of Nitellopsis. J Membr Biol 116:31–39
Wegner LH (2014) Root pressure and beyond: energetically uphill water transport into xylem vessels? J Exp Bot 65:381–393
Wei C, Tyree MT, Steudle E (1999) Direct measurement of xylem pressure in leaves of intact maize plants. A test of the cohesion-tension theory taking hydraulic architecture into consideration. Plant Physiol 121:1191–1205
White PR (1938) "Root pressure"—an unappreciated force in sap movement. Am J Bot 25:223–227
White PR, Schuler E, Kern JR, Fuller FH (1958) "Root pressure" in gymnosperms. Science 128:308–309
Xiao H, Peng S, Zheng Y, Mo J, Luo W, Zeng X, He X (2006) Interactive effects between plant allelochemicals, plant allelopathic potential and soil nutrients. J Appl Ecol 17:1747–1750
Zachary M (2009) Sap flow dynamics of a tropical, woody bamboo: deductions of physiology and hydraulics within *Guadua angustifolia*. PhD thesis, Washington University in St. Louis, USA
Zaitseva RI, Minashina NG, Sudnitsyn II (1998) Influence of capillary-sorptive and osmotic moisture pressure in chernozem on the growth and guttation of barley. Eurasian Soil Sci 31:1075–1082

Zelazny E, Borst JW, Muylaert M, Batoko H, Hemminga MA, Chaumont F (2007) FRET imaging in living maize cells reveals that plasma membrane aquaporins interact to regulate their subcellular localization. Proc Natl Acad Sci USA 104:12359–12364

Zeuthen T (2010) Water-transporting proteins. J Membr Biol 234:57–73

Zeuthen T, McAulay N (2012) Cotransport of water by Na^+–K^+–$2Cl^-$ cotransporters expressed in Xenopus oocytes: NKCC1 versus NKCC2. J Physiol 590:1139–1154

Zhao C-X, Xi-Ping D, Sui-QI Z, Qing Y, Steudle E, Lun S (2004) Advances in the studies on water uptake by plant roots. Acta Bot Sin 46:505–514

Zholkevich VN (1991) Root Pressure. In: Waisel Y, Eshel A, Kafkafi U (eds) Plant roots, the hidden half. Marcel-Dekker, New York, NY, pp 589–603

Zholkevich VN, Popova MS, Zhukovskaya NV (2007) Stimulatory effects of adrenalin and noradrenalin on root water-pumping activity and the involvement of G-proteins. Russ J Plant Physiol 54:790–796

Zhu JJ, Zimmermann U, Thurmer F, Haase A (1995) Xylem pressure in maize roots subjected to osmotic stress: determination of radial reflection coefficients by using the xylem pressure probe. Plant Cell Environ 18:906–912

Zhu JJ, Bai XF, Bu QM, Jiang XM (2010) An analysis to the driving forces for water and salt absorption in roots of maize seedlings under salt stress. Agr Sci China 9:806–812

Zimmermann U, Zhu JJ, Benkert R, Schneider H, Thurmer F, Zimmermann G (1995) Xylem pressure measurements in intact laboratory plants and excised organs: a critical evaluation of methods in the literature and the xylem pressure probe. In: Terazawa M, McLeoad CA, Tamia Y (eds) Tree sap. Hokkaido University Press, Sapporo, pp 59–70

Zimmermann U, Schneider H, Wegner LH, Haase A (2004) Water ascent in tall trees: does evolution of land plants rely on a highly metastable state? New Phytol 162:575–615

Zimmermann D, Reuss R, Westhoff M et al (2008) A novel, non-invasive, online-monitoring, versatile and easy plant-based probe for measuring leaf water status. J Exp Bot 59:3157–3167

Zwieniecki MA, Holbrook NM (2009) Confronting Maxwell's demon: biophysics of xylem embolism repair. Trends Plant Sci 14:530–534

Light- and CO_2-Dependent Systemic Regulation of Photosynthesis

Ryo Matsuda and Keach Murakami

Contents

Abstract Plant leaves do not only sense and respond to their local environment but also the environment experienced by the other leaves within the same plant. This long-distance signaling is involved in the systemic regulation of various photosynthesis-related phenotypic features of leaves. Here, we summarize the recent research on light- and CO_2-dependent, leaf-to-leaf systemic regulation. In the short term, leaves can pre-acclimate to excess light at the transcriptional level, in response to systemic signals from other leaves. Several substances, including reactive oxygen species and phytohormones, have been suggested to play key roles in the signaling pathway. In the long term, the light and CO_2 environment around mature leaves systemically regulates stomatal development, anatomical structure, and photosynthetic characteristics of young leaves. Possible mechanisms underlying the systemic regulation and the potential importance of systemic regulation in horticultural crop production are discussed.

R. Matsuda (✉) • K. Murakami
Department of Biological and Environmental Engineering, Graduate School of Agricultural and Life Sciences, The University of Tokyo, 1-1-1 Yayoi, Bunkyo, Tokyo 113-8657, Japan
e-mail: amatsuda@mail.ecc.u-tokyo.ac.jp

U. Lüttge et al. (eds.), *Progress in Botany*, Progress in Botany 77,
DOI 10.1007/978-3-319-25688-7_4

1 Introduction

Light and CO_2 are the two driving forces of photosynthesis. Plants respond to light intensity (photosynthetic photon flux density, PPFD), light quality (spectral distribution of light), and CO_2 concentration, via changes in various physiological and developmental characteristics. Recent evidence has indicated that leaves do not only sense and respond to the local environment around the leaf itself but also the environment to which other leaves of the plant are exposed. Thus, such kind of response system must involve signals that travel between leaves. This long-distance signaling, i.e., systemic regulation, has attracted increasing attention as a novel type of acclimation to environmental change. Accumulating research has revealed that diverse phenotypic features of leaves, such as tolerance to excess light (EL), stomatal development, anatomical structure, and photosynthetic characteristics, are systemically regulated in response to specific environmental factors.

In this review, we mainly focus on light- and CO_2-dependent, leaf-to-leaf systemic regulation of plant responses that are closely related to photosynthesis. Recent findings on systemic regulation that occurs in both the short term (minutes to hours) and the long term (days to weeks) are summarized, including possible mechanisms involved in the regulatory pathways. In addition, we briefly discuss the potential significance of systemic regulation from a horticultural viewpoint. There have been many studies on wound- and pathogen-induced systemic acquired resistance (SAR), which is beyond the scope of this article. However, recent review articles on this topic are available (Dempsey and Klessig 2012; Spoel and Dong 2012; Shah and Zeier 2013).

2 Short-Term Systemic Regulation

2.1 Systemic Acquired Acclimation to Excess Light

Although light is the energy source for photosynthesis, EL can be harmful to photosynthetic organisms. Light energy in excess of that required for CO_2 assimilation leads to over-reduction of the photosynthetic electron transport chain and accelerates the generation of reactive oxygen species (ROS) (Bowler et al. 1992; Niyogi 1999). ROS include the superoxide anion radical (O_2^-), which can be further converted to hydrogen peroxide (H_2O_2) and the hydroxyl radical ($^{\bullet}OH$) (Asada 1999), and singlet state oxygen ($^1O_2$) (Asada 2006). Generation of ROS can bring about direct oxidative damage to photosystem II (PSII) (Vass et al. 1992; Miyao et al. 1995; Keren et al. 1997) and/or inhibit the repair of damaged PSII (Nishiyama et al. 2006). The resulting reduction of the photosynthetic rate represents photoinhibition. Leaves have several protective systems to avoid excess light absorption and cope with oxidative stress (Ort 2001; Takahashi and Badger 2011). For example, the ROS scavenging system involves multiple enzymes and antioxidants, including superoxide dismutase and ascorbate peroxidase (APX), which catalyze the disproportionation of O_2^- to H_2O_2 and the reduction of H_2O_2 to water, respectively (Asada 1999).

It was reported that EL triggers cytosolic defensive reactions against ROS generation. The expression levels of genes for cytosolic APXs, *APX1* and *APX2*, in *Arabidopsis thaliana*, were rapidly upregulated within 15 min of EL exposure at a PPFD of 2,000 μmol m^{-2} s^{-1} (Karpinski et al. 1997). Furthermore, such upregulation of defense reactions does not only occur in directly exposed leaves but also in distal unexposed leaves. Exposing some portion (one-third) of the foliage to EL (2,700 μmol m^{-2} s^{-1} PPFD) while maintaining the remaining portion under weak light (200 μmol m^{-2} s^{-1} PPFD) increased the expression of *APX2*, associated with an increase in H_2O_2 content, in both EL-exposed and EL-unexposed leaves (Karpinski et al. 1999). The PSII of unexposed leaves was relatively tolerant to subsequent exposure to EL, and only slight reductions occurred in the maximum quantum yield of PSII (F_v/F_m) and the coefficient of photochemical quenching (q_p). However, the leaves directly exposed to EL showed significant reductions in the chlorophyll fluorescence parameters after prolonged exposure (Karpinski et al. 1999). This acclimative EL tolerance in distal leaves was termed systemic acquired acclimation (SAA) (Karpinski et al. 1999), and similar responses have been observed in several subsequent studies using *Arabidopsis* (Fryer et al. 2003; Mateo et al. 2004; Rossel et al. 2007; Mühlenbock et al. 2008; Szechyńska-Hebda et al. 2010; Gordon et al. 2013). In addition to EL, high- and low-temperature exposure is also reported to induce similar systemic acclimation responses (Gorsuch et al. 2010; Suzuki et al. 2013).

The SAA to EL in distal unexposed leaves is associated with various transcriptional changes (Rossel et al. 2007; Mühlenbock et al. 2008). The gene for the zinc finger transcription factor, *ZAT10*, reportedly plays a pivotal role in modulating the expression of SAA-regulated genes (Rossel et al. 2007; Gordon et al. 2013). A PPFD as low as 250 μmol m^{-2} s^{-1} suffices to induce SAA in *Arabidopsis*, with the extent of SAA depending on the PPFD level (Gordon et al. 2013). Recent studies have shown that SAA induction by EL is affected by the wavelength spectrum of EL (Szechyńska-Hebda et al. 2010; Gordon et al. 2013; Karpiński et al. 2013). In addition, SAA induced by local EL exposure is interrelated to other environmental factors, including air temperature and relative humidity around the plant (Gordon et al. 2013). Under natural conditions, sunlight exposure varies among the leaves of a plant. Therefore, SAA is thought to play a role in the pre-acclimation of shaded leaves in a plant to subsequent full sunlight exposure (Karpinski et al. 1999; Rossel et al. 2007; Gordon et al. 2013). To improve our understanding of the importance of SAA in nature, further research is needed to quantify the contribution of SAA to plant performance under realistically fluctuating environmental conditions.

2.2 Signal Transduction in Systemic Acquired Acclimation

Although the whole signal transduction pathway in SAA has not yet fully been understood, involved mechanisms have been suggested, which include starting points, signaling routes, signaling substances, and signal propagation systems. A key starting point for the signal transduction pathway in SAA is suggested to be the redox state of the plastoquinone pool in the photosynthetic electron transport chain (Karpinski et al. 1997, 1999; Mullineaux and Karpinski 2002; Mühlenbock et al. 2008). EL-induced *APX2* expression and H_2O_2 accumulation were particularly observed in bundle sheath cells of the vascular tissue, which may indicate local and systemic signaling via the vasculature (Fryer et al. 2003). Several candidates have been suggested to be involved in signal transduction of SAA to EL, including ROS, such as H_2O_2 (Karpinski et al. 1999; Fryer et al. 2003; Mateo et al. 2004; Rossel et al. 2007; Ślesak et al. 2007; Mühlenbock et al. 2008; Miller et al. 2009; Mittler and Blumwald 2015) and phytohormones, such as abscisic acid (ABA) (Fryer et al. 2003; Rossel et al. 2007; Suzuki et al. 2013; Mittler and Blumwald 2015) and jasmonate (JA) (Rossel et al. 2007). The systemic signaling pathways likely interact with other known signaling pathways (Pogson et al. 2008; Kangasjärvi et al. 2009). Systemic signal propagation systems that have been proposed, to date, include photoelectrophysiological signaling, which induces a specific pattern of changes in the electrical potential of plasma membranes (Karpinski and Szechynska-Hebda 2010; Szechyńska-Hebda et al. 2010; Karpiński et al. 2013), and the ROS waves, i.e., cascade of cell-to-cell communication events triggered by the initial burst of ROS (Mittler et al. 2011; Suzuki et al. 2013; Baxter et al. 2014). The signaling pathway of SAA is also likely to be closely linked with that of SAR (Mateo et al. 2004; Mühlenbock et al. 2008; Karpiński et al. 2013).

The dependence of SAA induction on the light spectrum (Szechyńska-Hebda et al. 2010; Gordon et al. 2013; Karpiński et al. 2013) may suggest photoreceptors to be involved. Red light is more effective in inducing systemic *APX2* expression than blue light (Szechyńska-Hebda et al. 2010), while *ZAT10* expression rather was induced by blue light (Gordon et al. 2013). However, *Arabidopsis* double mutants deficient in photoreceptors, cryptochromes (*cry1*/*cry2*), phototropins (*phot1*/*phot2*), and phytochromes (*phyA*/*phyB*), performed SAA, suggesting indirect secondary regulation of photoreceptors in SAA (Szechyńska-Hebda et al. 2010). Alternatively, the different light absorption spectrum between PSII and photosystem I (Evans 1986, 1987; Chow et al. 1990; Hogewoning et al. 2012) might cause the spectrum-dependent induction of SAA, because imbalanced excitation of the two photosystems causes changes in the redox state of the plastoquinone pool (Wagner et al. 2008). Detailed analysis of spectrum-dependent SAA induction is required with focus on photoreceptors and photosystem excitation balance.

3 Long-Term Systemic Regulation

3.1 Stomatal Development

Both CO_2 uptake from the atmosphere to the intercellular airspaces of leaves and water loss from plants predominantly occur through stomatal pores. Leaf gas exchange is largely determined by the extent of stomatal opening and stomatal density (SD, number of stomata per unit leaf area). Stomatal opening and closure are fast (in the order of seconds to minutes) in response to external environmental factors, including PPFD, light spectrum, CO_2 concentration, and atmospheric water vapor pressure deficit (Zeiger 1983; Assmann 1993; Schroeder et al. 2001). ABA plays an important role in instantaneous stomatal closure under water stress (Kim et al. 2010). Stomatal development in the leaf epidermis primarily occurs during leaf expansion, also being affected by environmental impact (Casson and Hetherington 2010). It is well documented that SD and/or stomatal index (SI, percentage of the number of stomata to the total number of stomata plus epidermal cells; Salisbury 1928) change when whole plants are placed in physically different environments (Casson and Gray 2008). The molecular mechanisms of stomatal development during leaf growth have extensively been studied and summarized in other review articles (Bergmann and Sack 2007; Nadeau 2009; Peterson et al. 2010; Pillitteri and Torii 2012).

Schoch et al. (1980) reported that shading of the expanded mature leaves only in cowpea (*Vigna sinensis*) decreased the SI in developing young leaves, compared with that in young leaves of unshaded plants. Such a decrease in SI was not observed when only young leaves were shaded. Studies in other species, including *Arabidopsis* (Lake et al. 2001; Coupe et al. 2006), tobacco (Thomas et al. 2003), and sorghum (Jiang et al. 2011), on this light intensity-dependent systemic regulation of stomatal development found decreased SD and/or SI in young leaves in response to shading of mature leaves. Besides light intensity, CO_2 concentration also has a systemic effect on stomatal development. Using gas-tight leaf cuvette/chamber systems that allowed mature leaves to be kept at a CO_2 concentration different from that of young leaves, it was shown that an elevated CO_2 concentration around mature leaves decreased the SD and/or SI in young leaves, whereas CO_2 enrichment around young leaves did not in *Arabidopsis* (Lake et al. 2001, 2002; Coupe et al. 2006) and hybrid poplar (*Populus trichocarpa* × *P. deltoides*) (Miyazawa et al. 2006, 2011). Determining stomatal development in young leaves by monitoring the CO_2 concentration around mature leaves may be ecologically meaningful, because young leaves still enclosed in buds may not sense the CO_2 concentration around the plant reliably (Lake et al. 2002).

ROS, phytohormones, sugars, and signals from photoreceptors have been proposed as candidates that can mediate the light- and CO_2-dependent systemic regulation of stomatal development. *Arabidopsis* mutants, fatty-acid desaturase-deficient *fad4*, ABA-deficient *aba1*, ethylene-insensitive *ein2*, and ascorbate-deficient *vtc1*, all showed altered CO_2-dependent systemic stomatal development, suggesting the possible involvement of the JA, ABA, ethylene, and ROS signaling pathways in long-distance CO_2 signaling (Lake et al. 2002). Pathways of sugar and phytohormone signaling were also suggested to be involved in systemic regulation, based on transcriptomic analysis (Coupe et al. 2006). Note, however, that shading and elevated CO_2 have opposing influences on the net photosynthetic rate of mature leaves and thereby the amount of sugars exported to young leaves, whereas both treatments negatively affect stomatal development in young leaves. Indeed, SI in young leaves was not correlated with net photosynthetic rate in mature leaves of the hybrid poplar (Miyazawa et al. 2006). It is therefore unlikely that sugars simply act as positive or negative regulators of systemic stomatal development (Casson and Gray 2008). More recently, the involvement of phytochrome B in light-dependent systemic regulation of stomatal development has been demonstrated in *Arabidopsis* (Casson and Hetherington 2014). They showed that SI in young leaves of the *Arabidopsis phyB-9* mutant did not decrease in response to shading only of the mature leaves, whereas such a decrease was observed in the *Arabidopsis* wild type.

Lake and Woodward (2008) found that SD alterations in young leaves of *Arabidopsis* in response to CO_2 concentration and relative humidity were quantitatively explained by the change in whole-plant transpiration rate. The change in SD was also positively correlated with changes in the leaf endogenous ABA concentration. Such quantitative relationships were seen in some hormonal mutants, but not so in the ABA-deficient mutant *aba1* (Lake and Woodward 2008). Such results may indicate a connection between the ABA-mediated short-term response of stomatal aperture and the long-term systemic regulation of stomatal development. Based on the results, models have been proposed that describe the interactive effects of transpiration, environment (light, CO_2, and humidity), and ABA signaling on stomatal development in young leaves (Lake and Woodward 2008; Chater et al. 2014). Such models can integrate the complex effects of environmental drivers on systemic regulation of stomatal development, through altered transpiration and thus water supply and demand in a plant. However, in the hybrid poplar, SI in young leaves was better correlated with stomatal conductance than with transpiration rate in mature leaves (Miyazawa et al. 2006). Roles of transpiration rate and stomatal conductance in systemic regulation of stomatal development demand for clarification.

3.2 Leaf Anatomical Structure

During development, leaf anatomy changes in response to PPFD. In general, leaves grown under high PPFD (sun leaves) have intensely developed palisade tissues and are thus thicker relative to those grown under low PPFD (shade leaves) (Björkman 1981; Yano and Terashima 2004). The thick sun leaves are considered to be advantageous to photosynthesis under high PPFD, via a larger mesophyll surface area per unit leaf area and thus a higher mesophyll conductance to CO_2 per unit leaf area (Terashima et al. 2001, 2006). Sun leaves also have chloroplasts with fewer grana thylakoids than shade leaves (Boardman 1977; Björkman 1981). The light environment of expanded mature leaves was demonstrated to systemically regulate part of the sun- and shade-type anatomy of developing young leaves.

Yano and Terashima (2001) studied the effects that variable PPFD exposure of young and mature leaves had on the anatomical characteristics of young leaves in an annual herb, *Chenopodium album*. It was demonstrated that cell differentiation in young leaves was determined by the PPFD on mature leaves, rather than by that on the young leaves: When mature leaves were shaded, young leaves showed shade-type anatomy (less cell layers in the palisade tissues, resulting in thinner leaves), irrespective of the PPFD experienced by the young leaves. A similar PPFD-dependent systemic effect on leaf anatomy was also seen in a C4 monocotyledonous plant, sorghum (Jiang et al. 2011). In sorghum, shading of whole plants or only mature leaves caused shade-type leaf anatomy in young leaves, with thinner leaves and a smaller contact area between the bundle sheath and mesophyll cells, whereas shading of only young leaves or full exposure did not (Jiang et al. 2011). Murakami et al. (2014) also reported that shading of only mature leaves led to a decrease in leaf mass per area in young leaves of common bean (*Phaseolus vulgaris*). This decline implies changes in the anatomy of young leaves, such as decreases in leaf thickness and/or the volumetric density of mesophyll cells.

Chloroplast ultrastructure in young leaves is, in contrast to leaf anatomy, dependent on the local PPFD on the leaves themselves, rather than that on mature leaves. Chloroplasts in young leaves developing in the shade showed more developed grana thylakoids, regardless of the PPFD on mature leaves (Yano and Terashima 2001; Jiang et al. 2011). Thus, leaf-level and chloroplast-level morphological characteristics seem to be regulated differentially in response to changes in light environment.

CO_2 concentration also has systemic effects on leaf anatomy. Elevating the CO_2 concentration around mature leaves of the hybrid poplar to 720 μmol mol^{-1} decreased both the fraction of mesophyll cells to total leaf volume and the mesophyll cell density of young leaves (Miyazawa et al. 2011). Unlike PPFD, the local CO_2 concentration around mature leaves did not influence leaf thickness of young leaves.

Some studies have pointed out that, within a leaf, vertical differences exist in anatomical sensitivity to systemic light and CO_2 signals. In sorghum, adaxial mesophyll thickness in young leaves was more sensitive to the shading of mature leaves than was abaxial mesophyll thickness (Jiang et al. 2011). In contrast, in young leaves of the hybrid poplar, abaxial spongy tissues were more influenced by systemic signaling of CO_2 than were adaxial palisade tissues: Elevated CO_2 concentration around mature leaves increased the thickness and decreased the cell density of spongy tissues without significantly affecting palisade tissues (Miyazawa et al. 2011). Thus, notwithstanding different intra-leaf sensitivities to systemic signals, the anatomical response seems to depend on environmental stimuli (light versus CO_2) and/or species.

3.3 Photosynthetic Characteristics

Numerous studies have examined changes in leaf photosynthetic characteristics when the whole plant is exposed to different PPFD levels or CO_2 concentrations. In general, shade leaves exhibit lower photosynthetic capacity under saturating PPFD, lower dark respiration rate, and a lower leaf nitrogen (N) content than sun leaves (Boardman 1977; Björkman 1981). Biochemical changes in photosynthetic components were also observed: Shade leaves have more light acquisition components, i.e., chlorophylls and light-harvesting complexes, and less light-use components for electron transport, ATP synthesis, and the Calvin cycle, such as ribulose-1,5-bisphosphate carboxylase/oxygenase (Rubisco), than have sun leaves (Boardman 1977; Terashima and Evans 1988; Makino et al. 1997; Hikosaka 2005). On the other hand, CO_2 concentrations above 390–400 μmol mol^{-1}, the current level of ambient air, often lead to downregulation of photosynthesis, associated with decreasing N and increasing carbohydrate contents in leaves (Stitt 1991; Makino and Mae 1999; Long et al. 2004).

Some studies have examined CO_2- and light-dependent systemic regulation of photosynthetic characteristics. Sims et al. (1998) studied the systemic effects of CO_2 concentration on photosynthetic characteristics. CO_2 concentration around two out of three young leaflets of soybean was controlled independently from that around the rest of the plant. They found that the in vivo maximum carboxylation capacity of Rubisco (V_{cmax}; Farquhar et al. 1980; von Caemmerer and Farquhar 1981) and the amount of Rubisco in the two young leaflets were unaffected by the local CO_2 concentration, but lowered when the rest of the plant was exposed to high CO_2 concentration. Similar results were reported by Araya et al. (2008), who constructed a local environment control system in which CO_2 concentration and PPFD of mature primary leaves of common bean could be controlled independently from those of young trifoliate leaves. They showed that treating mature primary leaves with high (1,000 μmol mol^{-1}) and low (150 μmol mol^{-1}) CO_2 concentrations while maintaining young trifoliate leaves at 400 μmol mol^{-1} CO_2 caused the net photosynthetic rate of the young leaves to decrease and increase, respectively. The changes in the young leaves were mainly attributed to changes in the initial slope of the response curve of the net photosynthetic rate to the intercellular CO_2 concentration, which reflects V_{cmax}. These two studies indicate that elevated and reduced CO_2 concentrations around mature leaves can lower and enhance the photosynthetic capacity of young leaves, respectively, which is associated with the changes in V_{cmax}. Miyazawa et al. (2011) reported that in young leaves of the hybrid poplar, leaf N content per unit leaf area decreased following the exposure of mature leaves to elevated CO_2. This may be related also to elevated CO_2-dependent systemic downregulation of photosynthesis; as in general, N content positively correlates with the photosynthetic capacity of C3 leaves (Evans 1989). In contrast, Coupe et al. (2006) observed no changes in chlorophyll fluorescence parameters, namely, actual quantum yield of PSII (ΦPSII) and q_p, in young leaves of *Arabidopsis* when mature leaves were exposed to elevated CO_2.

The effect of light environment on systemic regulation of photosynthesis appears to vary across experiments. Coupe et al. (2006) tested the effect of shading of mature leaves in *Arabidopsis* and found no changes in chlorophyll fluorescence parameters. They suggested the absence of systemic photosynthetic response in young leaves to shading or CO_2 enrichment of mature leaves. However, shading of mature primary leaves revealed increase in the net photosynthetic rate of young trifoliate leaves in common bean (Araya et al. 2008). In contrast, two studies reported shading of mature leaves to decrease the net photosynthetic rate of young leaves in sorghum (Jiang et al. 2011) and common bean (Murakami et al. 2014). Thus, although some plant species show systemic regulation of photosynthesis in response to light environment, the resulting responses of young leaves cannot be generalized. It is possible that the developmental stage of young leaves at which mature leaves are treated is important. Leaves in their early developmental stages may be more responsive to signals from mature leaves than those of later stages, as suggested by Murakami et al. (2014).

Shading decreases and elevated CO_2 increases photosynthetic rates and thereby the amount of sugars exported from leaves. Such light- and/or CO_2-dependent alterations in the export of sugars can be related to systemic regulation. It was proposed that an individual leaf senses its "photosynthetic status" relative to other leaves within the plant by monitoring its sugar concentration, because sugar concentration in the leaf is the result of the balance in demand by other leaves versus the leaf-internal production (Ono et al. 2001). Sucrose, the major sugar transported between leaves, does not only play a role as a substrate of energy and carbon metabolisms but also as a signaling molecule involved in leaf development and gene expression of metabolic processes (Wind et al. 2010; Tognetti et al. 2013). Photosynthate exported from mature leaves is therefore suggested to be a systemic signal regulating anatomical structure (Yano and Terashima 2001) and photosynthetic properties (Araya et al. 2008) of young leaves. Additionally, the redox state in the photosynthetic electron transport chain is suggested to participate in the systemic regulation of photosynthesis (Araya et al. 2008), as in the case in SAA.

Given that photoreceptors with different spectral sensitivities might play a role in light-dependent systemic regulation, a study recently attempted to investigate the relative contributions of PPFD and light spectrum to systemic regulation of photosynthetic characteristics (Murakami et al. 2014). Recent technological advances in light-emitting diodes (LEDs) have enabled us to irradiate small areas with monochromatic light at relatively high PPFD. Using handmade small LED panels (Fig. 1), the effects of PPFD and light spectrum employed to mature leaves on the photosynthetic characteristics of young leaves in common bean were examined. The results indicated that overall, PPFD had a greater effect on the amounts of leaf N, chlorophyll, and Rubisco than had light spectrum (Murakami et al. 2014). However, blue-light irradiation on mature leaves seemed to have a specific effect, increasing the Rubisco content of young leaves, compared with other wavebands tested. In general, blue light-containing irradiation tends to increase photosynthetic capacity when irradiating the whole plant during growth, compared with blue light-deficient irradiation (Goins et al. 1997; Matsuda et al. 2004, 2007, 2008;

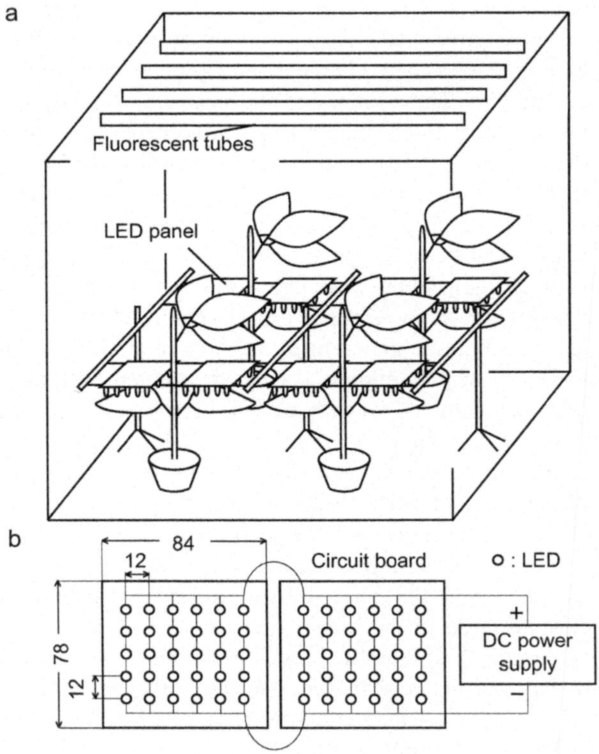

Fig. 1 Schematic diagram of the local light environment control system for mature leaves of *Phaseolus vulgaris*: (**a**) growth chamber and (**b**) the LED panel (in mm). Redrawn from Murakami et al. (2014) with slight modifications

Hogewoning et al. 2010). Further, blue light increases the amounts of biochemical components of photosynthesis, including Rubisco (Eskins et al. 1991, López-Juez and Hughes 1995; Matsuda et al. 2004). Although the involvement of some blue-light photoreceptors (CRY1, CRY2, and PHOT1) in such photosynthetic responses to blue light is unlikely (Weston et al. 2000), the action of blue light on local and systemic regulation of photosynthesis should be further explored.

4 Potential Effects of Systemic Regulation in Greenhouse Plant Production

Systemic regulation could potentially affect horticultural plant production in some specific cases. In fruit vegetable production in greenhouses, supplementary assimilation lighting using artificial light sources can be employed to promote dry matter productivity. Conventionally, supplemental lighting is applied at the top of the plant canopy using high-intensity discharge lamps, such as high-pressure sodium lamps. In closed canopies, the PPFD of supplemental lighting exponentially decreases with canopy depth, as does that of natural sunlight (Monsi and Saeki 2005). A considerable vertical gradient of PPFD is generated within the canopy, probably resulting in reduced efficiency of the supplemental lighting (Trouwborst et al. 2010). To generate a more uniform PPFD profile, intracanopy lighting has been developed for cucumber (Hovi et al. 2004; Hovi-Pekkanen and Tahvonen 2008; Pettersen et al. 2010; Trouwborst et al. 2010, 2011), tomato (Gunnlaugsson and Adalsteinsson 2006), and sweet pepper (Hovi-Pekkanen et al. 2006) production, where artificial light sources are placed within the canopy, so that the mature leaves are mainly irradiated. This intracanopy lighting may affect photosynthesis at the plant level, not only through improvement of the light environment within the canopy, but also through light-dependent systemic regulation of physiological and morphological characteristics of young leaves, as suggested by Trouwborst et al. (2010). LEDs are a promising light source for intracanopy lighting, because of their low operating temperature, low operating voltage, and physical robustness (Trouwborst et al. 2010). The spectrum of supplemental light created by different types of LEDs influences the photosynthesis of irradiated mature leaves (Murakami et al. 2013) and possibly young-leaf characteristics through systemic regulation. Further evaluation may be needed for optimizing PPFD and light spectrum of intracanopy lighting with LEDs, taking into account systemic regulation.

5 Conclusions and Future Perspectives

In the short term, leaves can pre-acclimate to EL that has been applied to other leaves within the same plant. In the long term, stomatal development, anatomical structure, and photosynthetic characteristics of young leaves are systemically regulated by environmental stimuli to mature leaves, including light and CO_2. Although part of the signaling mechanisms underlying this systemic regulation has been unveiled (Fig. 2), understanding still is limited. One question is about the extent to which signaling pathways of the short-term SAA and the long-term systemic regulation overlap. As described above, key players comprise ROS and ABA, probably acting in both pathways. ABA-mediated stomatal response is suggested to play an important role in both SAA (Mittler and Blumwald 2015) and systemic regulation of stomatal development (Lake and Woodward 2008; Chater et al. 2014). The light spectrum may have a regulatory role in EL-induced SAA (Szechyńska-Hebda et al. 2010; Gordon et al. 2013; Karpiński et al. 2013) and in light-dependent systemic regulation of stomatal development (Casson and Hetherington 2014) and photosynthetic characteristics (Murakami et al. 2014). Photoreceptors may partly be involved in the regulation. In addition to working with mutants and transformants of various signaling pathways, local environment control systems should be emphasized in analysis to broaden knowledge on systemic regulation. Systemic regulation in plant performance deserves attention both under natural conditions and in view of greenhouse horticultural production.

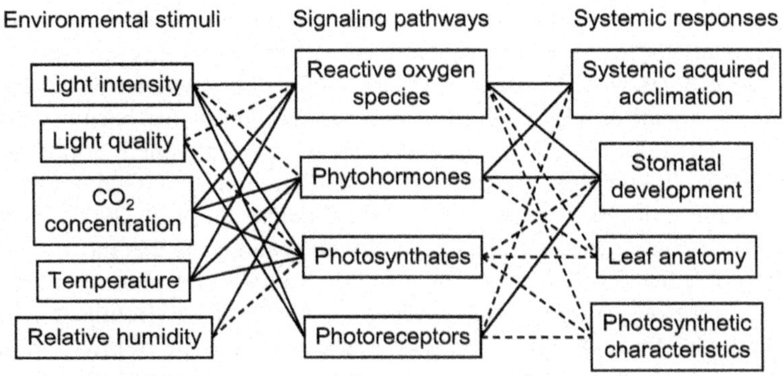

Fig. 2 Suggested relationships between environmental stimuli, signaling pathways, and systemic responses. *Solid* and *dashed lines* represent probable/direct and possible/indirect effects, respectively

Acknowledgments The authors are grateful to Prof. Kazuhiro Fujiwara (The University of Tokyo) for his invaluable suggestions.

References

Araya T, Noguchi K, Terashima I (2008) Manipulation of light and CO_2 environments of the primary leaves of bean (*Phaseolus vulgaris* L.) affects photosynthesis in both the primary and the first trifoliate leaves: involvement of systemic regulation. Plant Cell Environ 31:50–61

Asada K (1999) The water-water cycle in chloroplasts: scavenging of active oxygens and dissipation of excess photons. Annu Rev Plant Physiol Plant Mol Biol 50:601–639

Asada K (2006) Production and scavenging of reactive oxygen species in chloroplasts and their functions. Plant Physiol 141:391–396

Assmann SM (1993) Signal transduction in guard cells. Annu Rev Cell Biol 9:345–375

Baxter A, Mittler R, Suzuki N (2014) ROS as key players in plant stress signalling. J Exp Bot 65:1229–1240

Bergmann DC, Sack FD (2007) Stomatal development. Annu Rev Plant Biol 58:163–181

Björkman O (1981) Responses to different quantum flux densities. In: Lange OL, Nobel PS, Osmond CB, Ziegler H (eds) Physiological plant ecology I. Responses to the physical environment. Springer, Berlin, pp 57–107

Boardman NK (1977) Comparative photosynthesis of sun and shade plants. Annu Rev Plant Physiol 28:355–377

Bowler C, Van Montagu M, Inzé D (1992) Superoxide dismutase and stress tolerance. Annu Rev Plant Physiol Plant Mol Biol 43:83–116

Casson S, Gray JE (2008) Influence of environmental factors on stomatal development. New Phytol 178:9–23

Casson SA, Hetherington AM (2010) Environmental regulation of stomatal development. Curr Opin Plant Biol 13:90–95

Casson SA, Hetherington AM (2014) Phytochrome B is required for light-mediated systemic control of stomatal development. Curr Biol 24:1216–1221

Chater CCC, Oliver J, Casson S, Gray JE (2014) Putting the brakes on: abscisic acid as a central environmental regulator of stomatal development. New Phytol 202:376–391

Chow WS, Melis A, Anderson JM (1990) Adjustments of photosystem stoichiometry in chloroplasts improve the quantum efficiency of photosynthesis. Proc Natl Acad Sci USA 87:7502–7506
Coupe SA, Palmer BG, Lake JA, Overy SA, Oxborough K, Woodward FI, Gray JE, Quick WP (2006) Systemic signalling of environmental cues in *Arabidopsis* leaves. J Exp Bot 57:329–341
Dempsey DMA, Klessig DF (2012) SOS—too many signals for systemic acquired resistance? Trends Plant Sci 17:538–545
Eskins K, Jiang CZ, Shibles R (1991) Light-quality and irradiance effects on pigments, light-harvesting proteins and Rubisco activity in a chlorophyll- and light-harvesting-deficient soybean mutant. Physiol Plant 83:47–53
Evans JR (1986) A quantitative analysis of light distribution between the two photosystems, considering variation in both the relative amounts of the chlorophyll-protein complexes and the spectral quality of light. Photobiochem Photobiophys 10:135–147
Evans JR (1987) The dependence of quantum yield on wavelength and growth irradiance. Aust J Plant Physiol 14:69–79
Evans JR (1989) Photosynthesis and nitrogen relationships in leaves of C_3 plant. Oecologia 7:89–19
Farquhar GD, von Caemmerer S, Berry JA (1980) A biochemical model of photosynthetic CO_2 assimilation in leaves of C_3 species. Planta 149:78–90
Fryer MJ, Ball L, Oxborough K, Karpinski S, Mullineaux PM, Baker NR (2003) Control of *Ascorbate Peroxidase 2* expression by hydrogen peroxide and leaf water status during excess light stress reveals a functional organisation of *Arabidopsis* leaves. Plant J 33:691–705
Goins GD, Yorio NC, Sanwo MM, Brown CS (1997) Photomorphogenesis, photosynthesis, and seed yield of wheat plants grown under red light-emitting diodes (LEDs) with and without supplemental blue lighting. J Exp Bot 48:1407–1413
Gordon MJ, Carmody M, Albrecht V, Pogson B (2013) Systemic and local response to repeated HL stress-induced retrograde signaling in *Arabidopsis*. Front Plant Sci 3:303
Gorsuch PA, Sargeant AW, Penfield SD, Quick WP, Atkin OK (2010) Systemic low temperature signaling in Arabidopsis. Plant Cell Physiol 51:1488–1498
Gunnlaugsson B, Adalsteinsson S (2006) Interlight and plant density in year-round production of tomato at northern latitudes. Acta Hortic 711:71–75
Hikosaka K (2005) Nitrogen partitioning in the photosynthetic apparatus of *Plantago asiatica* leaves grown under different temperature and light conditions: similarities and differences between temperature and light acclimation. Plant Cell Physiol 46:1283–1290
Hogewoning SW, Trouwborst G, Maljaars H, Poorter H, van Ieperen W, Harbinson J (2010) Blue light dose-responses of leaf photosynthesis, morphology, and chemical composition of *Cucumis sativus* grown under different combinations of red and blue light. J Exp Bot 61:3107–3117
Hogewoning SW, Wientjes E, Douwstra P, Trouwborst G, van Ieperen W, Croce R, Harbinson J (2012) Photosynthetic quantum yield dynamics: from photosystems to leaves. Plant Cell 24:1921–1935
Hovi T, Näkkilä J, Tahvonen R (2004) Interlighting improves production of year-round cucumber. Sci Hortic 102:283–294
Hovi-Pekkanen T, Näkkilä J, Tahvonen R (2006) Increasing productivity of sweet pepper with interlighting. Acta Hortic 711:165–170
Hovi-Pekkanen T, Tahvonen R (2008) Effects of interlighting on yield and external fruit quality in year-round cultivated cucumber. Sci Hortic 116:152–161
Jiang CD, Wang X, Gao HY, Shi L, Chow WS (2011) Systemic regulation of leaf anatomical structure, photosynthetic performance, and high-light tolerance in sorghum. Plant Physiol 155:1416–1424
Kangasjärvi S, Nurmi M, Tikkanen M, Aro EM (2009) Cell-specific mechanisms and systemic signalling as emerging themes in light acclimation of C3 plants. Plant Cell Environ 32:1230–1240

Karpinski S, Escobar C, Karpinska B, Creissen G, Mullineaux PM (1997) Photosynthetic electron transport regulates the expression of cytosolic ascorbate peroxidase genes in *Arabidopsis* during excess light stress. Plant Cell 9:627–640

Karpinski S, Reynolds H, Karpinska B, Wingsle G, Creissen G, Mullineaux P (1999) Systemic signaling and acclimation in response to excess excitation energy in *Arabidopsis*. Science 284:654–657

Karpinski S, Szechynska-Hebda M (2010) Secret life of plants. From memory to intelligence. Plant Signal Behav 5:1391–1394

Karpiński S, Szechyńska-Hebda M, Wituszyńska W, Burdiak P (2013) Light acclimation, retrograde signalling, cell death and immune defences in plants. Plant Cell Environ 36:736–744

Keren N, Berg A, van Kan PJM, Levanon H, Ohad I (1997) Mechanism of photosystem II photoinactivation and D1 protein degradation at low light: the role of back electron flow. Proc Natl Acad Sci USA 94:1579–1584

Kim TH, Böhmer M, Hu H, Nishimura N, Schroeder JI (2010) Guard cell signal transduction network: advances in understanding abscisic acid, CO_2, and Ca^{2+} signaling. Annu Rev Plant Biol 61:561–591

Lake JA, Woodward FI (2008) Response of stomatal numbers to CO_2 and humidity: control by transpiration rate and abscisic acid. New Phytol 179:397–404

Lake JA, Woodward FI, Quick WP (2002) Long-distance CO_2 signalling in plants. J Exp Bot 53:183–193

Lake JA, Quick WP, Beerling DJ, Woodward FI (2001) Plant development: signals from mature to new leaves. Nature 411:154

Long SP, Ainsworth EA, Rogers A, Ort DR (2004) Rising atmospheric carbon dioxide: plants FACE the future. Annu Rev Plant Biol 55:591–628

López-Juez E, Hughes MJG (1995) Effect of blue light and red light on the control of chloroplast acclimation of light-grown pea leaves to increased fluence rates. Photochem Photobiol 61:106–111

Makino A, Mae T (1999) Photosynthesis and plant growth at elevated levels of CO_2. Plant Cell Physiol 40:999–1006

Makino A, Sato T, Nakano H, Mae T (1997) Leaf photosynthesis, plant growth and nitrogen allocation in rice under different irradiances. Planta 203:390–398

Mateo A, Mühlenbock P, Rustérucci C, Chang CCC, Miszalski Z, Karpinska B, Parker JE, Mullineaux PM, Karpinski S (2004) *LESION SIMULATING DISEASE 1* is required for acclimation to conditions that promote excess excitation energy. Plant Physiol 136:2818–2830

Matsuda R, Ohashi-Kaneko K, Fujiwara K, Goto E, Kurata K (2004) Photosynthetic characteristics of rice leaves grown under red light with or without supplemental blue light. Plant Cell Physiol 45:1870–1874

Matsuda R, Ohashi-Kaneko K, Fujiwara K, Kurata K (2007) Analysis of the relationship between blue-light photon flux density and the photosynthetic properties of spinach (*Spinacia oleracea* L.) leaves with regard to the acclimation of photosynthesis to growth irradiance. Soil Sci Plant Nutr 53:459–465

Matsuda R, Ohashi-Kaneko K, Fujiwara K, Kurata K (2008) Effects of blue light deficiency on acclimation of light energy partitioning in PSII and CO_2 assimilation capacity to high irradiance in spinach leaves. Plant Cell Physiol 49:664–670

Miller G, Schlauch K, Tam R, Cortes D, Torres MA, Shulaev V, Dangl JL, Mittler R (2009) The plant NADPH oxidase RBOHD mediates rapid systemic signaling in response to diverse stimuli. Sci Signal 2:ra45

Mittler R, Blumwald E (2015) The roles of ROS and ABA in systemic acquired acclimation. Plant Cell 27(1):64–70. doi:10.1105/tpc.114.133090

Mittler R, Vanderauwera S, Suzuki N, Miller G, Tognetti VB, Vandepoele K, Gollery M, Shulaev V, Breusegem FV (2011) ROS signaling: the new wave? Trends Plant Sci 16:300–309

Miyao M, Ikeuchi M, Yamamoto N, Ono T (1995) Specific degradation of the D1 protein of photosystem II by treatment with hydrogen peroxide in darkness: implications for the mechanism of degradation of the D1 protein under illumination. Biochemistry 34:10019–10026

Miyazawa SI, Livingston NJ, Turpin DH (2006) Stomatal development in new leaves is related to the stomatal conductance of mature leaves in poplar (*Populus trichocarpa* × *P. deltoides*). J Exp Bot 57:373–380

Miyazawa SI, Warren CR, Turpin D, Livingston NJ (2011) Determination of the site of CO_2 sensing in poplar: is the area-based N content and anatomy of new leaves determined by their immediate CO_2 environment or by the CO_2 environment of mature leaves? J Exp Bot 62:2787–2796

Monsi M, Saeki T (2005) On the factor light in plant communities and its importance for matter production. Ann Bot 95:549–567

Mühlenbock P, Szechyńska-Hebda M, Płaszczyca M, Baudo M, Mateo A, Mullineaux PM, Parker JE, Karpińska B, Karpiński S (2008) Chloroplast signaling and *LESION SIMULATING DISEASE1* regulate crosstalk between light acclimation and immunity in *Arabidopsis*. Plant Cell 20:2339–2356

Mullineaux P, Karpinski S (2002) Signal transduction in response to excess light: getting out of the chloroplast. Curr Opin Plant Biol 5:43–48

Murakami K, Matsuda R, Fujiwara K (2013) Effects of supplemental lighting to a lower leaf using light-emitting diodes with different spectra on the leaf photosynthetic rate in sweet pepper. J Agric Meteorol 69:55–63

Murakami K, Matsuda R, Fujiwara K (2014) Light-induced systemic regulation of photosynthesis in primary and trifoliate leaves of *Phaseolus vulgaris*: effects of photosynthetic photon flux density (PPFD) *versus* spectrum. Plant Biol 16:16–21

Nadeau JA (2009) Stomatal development: new signals and fate determinants. Curr Opin Plant Biol 12:29–35

Nishiyama Y, Allakhverdiev SI, Murata N (2006) A new paradigm for the action of reactive oxygen species in the photoinhibition of photosystem II. Biochim Biophys Acta 1757:742–749

Niyogi KK (1999) Photoprotection revisited: genetic and molecular approaches. Annu Rev Plant Physiol Plant Mol Biol 50:333–359

Ono K, Nishi Y, Watanabe A, Terashima I (2001) Possible mechanisms of adaptive leaf senescence. Plant Biol 3:234–243

Ort DR (2001) When there is too much light. Plant Physiol 125:29–32

Peterson KM, Rychel AL, Torii KU (2010) Out of the mouths of plants: the molecular basis of the evolution and diversity of stomatal development. Plant Cell 22:296–306

Pettersen RI, Torre S, Gislerød HR (2010) Effects of intracanopy lighting on photosynthetic characteristics in cucumber. Sci Hortic 125:77–81

Pillitteri LJ, Torii KU (2012) Mechanisms of stomatal development. Annu Rev Plant Biol 63:591–614

Pogson BJ, Woo NS, Förster B, Small ID (2008) Plastid signalling to the nucleus and beyond. Trends Plant Sci 13:602–609

Rossel JB, Wilson PB, Hussain D, Woo NS, Gordon MJ, Mewett OP, Howell KA, Whelan J, Kazan K, Pogson BJ (2007) Systemic and intracellular responses to photooxidative stress in *Arabidopsis*. Plant Cell 19:4091–4110

Salisbury EJ (1928) On the causes and ecological significance of stomatal frequency, with special reference to the woodland flora. Philos Trans R Soc B 216:1–65

Schoch PG, Zinsou C, Sibi M (1980) Dependence of the stomatal index on environmental factors during stomatal differentiation in leaves of *Vigna sinensis* L. 1. Effect of light intensity. J Exp Bot 31:1211–1216

Schroeder JI, Allen GJ, Hugouvieux V, Kwak JM, Waner D (2001) Guard cell signal transduction. Annu Rev Plant Physiol Plant Mol Biol 52:627–658

Shah J, Zeier J (2013) Long-distance communication and signal amplification in systemic acquired resistance. Front Plant Sci 4:30

Sims DA, Luo Y, Seemann JR (1998) Importance of leaf versus whole plant CO_2 environment for photosynthetic acclimation. Plant Cell Environ 21:1189–1196

Ślesak I, Libik M, Karpinska B, Karpinski S, Miszalski Z (2007) The role of hydrogen peroxide in regulation of plant metabolism and cellular signalling in response to environmental stresses. Acta Biochim Pol 54:39–50

Spoel SH, Dong X (2012) How do plants achieve immunity? Defence without specialized immune cells. Nat Rev Immunol 12:89–100
Stitt M (1991) Rising CO_2 levels and their potential significance for carbon flow in photosynthetic cells. Plant Cell Environ 14:741–762
Suzuki N, Miller G, Salazar C, Mondal HA, Shulaev E, Cortes DF, Shuman JL, Luo X, Shah J, Schlauch K, Shulaev V, Mittler R (2013) Temporal-spatial interaction between reactive oxygen species and abscisic acid regulates rapid systemic acclimation in plants. Plant Cell 25:3553–3569
Szechyńska-Hebda M, Kruk J, Górecka M, Karpińska B, Karpiński S (2010) Evidence for light wavelength-specific photoelectrophysiological signaling and memory of excess light episodes in *Arabidopsis*. Plant Cell 22:2201–2218
Takahashi S, Badger MR (2011) Photoprotection in plants: a new light on photosystem II damage. Trends Plant Sci 16:53–60
Terashima I, Evans JR (1988) Effects of light and nitrogen nutrition on the organization of the photosynthetic apparatus in spinach. Plant Cell Physiol 29:143–155
Terashima I, Hanba YT, Tazoe Y, Vyas P, Yano S (2006) Irradiance and phenotype: comparative eco-development of sun and shade leaves in relation to photosynthetic CO_2 diffusion. J Exp Bot 57:343–354
Terashima I, Miyazawa SI, Hanba YT (2001) Why are sun leaves thicker than shade leaves?—Consideration based on analyses of CO_2 diffusion in the leaf. J Plant Res 114:93–105
Thomas PW, Woodward FI, Quick WP (2003) Systemic irradiance signalling in tobacco. New Phytol 161:193–198
Tognetti JA, Pontis HG, Martínez-Noël GMA (2013) Sucrose signaling in plants: a world yet to be explored. Plant Signal Behav 8, e23316
Trouwborst G, Oosterkamp J, Hogewoning SW, Harbinson J, van Ieperen W (2010) The responses of light interception, photosynthesis and fruit yield of cucumber to LED-lighting within the canopy. Physiol Plant 138:289–300
Trouwborst G, Schapendonk AHCM, Rappoldt K, Pot S, Hogewoning SW, van Ieperen W (2011) The effect of intracanopy lighting on cucumber fruit yield—model analysis. Sci Hortic 129:273–278
Vass I, Styring S, Hundal T, Koivuniemi A, Aro EM, Andersson B (1992) Reversible and irreversible intermediates during photoinhibition of photosystem II: stable reduced Q_A species promote chlorophyll triplet formation. Proc Natl Acad Sci USA 89:1408–1412
von Caemmerer S, Farquhar GD (1981) Some relationships between the biochemistry of photosynthesis and the gas exchange of leaves. Planta 153:376–387
Wagner R, Dietzel L, Bräutigam K, Fischer W, Pfannschmidt T (2008) The long-term response to fluctuating light quality is an important and distinct light acclimation mechanism that supports survival of *Arabidopsis thaliana* under low light conditions. Planta 228:573–587
Weston E, Thorogood K, Vinti G, López-Juez E (2000) Light quantity controls leaf-cell and chloroplast development in *Arabidopsis thaliana* wild type and blue-light-perception mutants. Planta 211:807–815
Wind J, Smeekens S, Hanson J (2010) Sucrose: metabolite and signaling molecule. Phytochemistry 71:1610–1614
Yano S, Terashima I (2001) Separate localization of light signal perception for sun or shade type chloroplast and palisade tissue differentiation in *Chenopodium album*. Plant Cell Physiol 42:1303–1310
Yano S, Terashima I (2004) Developmental process of sun and shade leaves in *Chenopodium album* L. Plant Cell Environ 27:781–793
Zeiger E (1983) The biology of stomatal guard cells. Annu Rev Plant Physiol 34:441–475

Hierarchy and Information in a System Approach to Plant Biology: Explaining the Irreducibility in Plant Ecophysiology

Gustavo M. Souza, Suzana C. Bertolli, and Ulrich Lüttge

Contents

Abstract Hierarchy is considered as a central and essential aspect of systems biology. In a theoretical section on hierarchy, considerations include discussion of the simultaneous spatiotemporal operation in both top-down and bottom-up modes. Emergent biological systems are not reducible to their parts or modules. Reactions to signals may be different and even opposite at different integrated hierarchical levels. Therefore, the top-down and bottom-up flow of information through hierarchically organized systems is an essential feature supporting

G.M. Souza (✉)
Laboratory of Plant Intelligence and Ecophysiology "Ulrich Lüttge," University of Western São Paulo (UNOESTE), Presidente Prudente, São Paulo, Brazil
e-mail: gumaia.gms@gmail.br

S.C. Bertolli
Programa de Pós-graduação em Biologia Vegetal, Instituto de Biociências, Univ Estadual Paulista "Júlio de Mesquita Filho" (UNESP), Rio Claro, Brazil

U. Lüttge
Department of Biology, Technical University of Darmstadt, TUD, Schnittspahnstr. 3-5, 64287 Darmstadt, Germany

U. Lüttge et al. (eds.), *Progress in Botany*, Progress in Botany 77,
DOI 10.1007/978-3-319-25688-7_5

robustness and stability of the hierarchical systems. In an empirical section, examples are collected spanning scalar levels from electrons and whole plant physiology up the ecosystems. Analysis of hierarchy is appropriate for establishing systemic understanding models of the complex interactions between plants and their changing environment. However, the search for unique indicators that can be used to determine or predict a global plant behavior in response to environmental cues may be, actually, a search for a "holy grail."

1 Hierarchy: Theoretical Aspects

Seeing different worlds in the same world: epistemological considerations.

Because of the enormous complexity of nature, science attempts to build simplified models that allow some understanding of natural phenomena that is compatible with the technological level of data acquisition. Construction of heuristic models enables the simulation and prediction of certain aspects of nature. The basis of this scientific method was forged by René Descartes (1596–1650) in his "Discourse on the Method" in which he proposes the analysis or the decomposition of a problem into smaller parts. The parts should be more understandable separately, and once a consistent knowledge of the parts is obtained, integration of them accomplishes the synthesis reaching the solution of the problem initially proposed. All this is naturally immersed in a mechanical model of the universe, where all of the parts are linked in a linear way, and the whole is the sum of these parts (Mitchell 2009). This general theoretical model of thought, and its corresponding scientific method, allowed the development of a successful reductionist science, founded on the belief that the full understanding of complex phenomena would be embedded in the understanding of its constituent parts. However, accomplishing this is somewhat unlikely.

Biological systems are irreducible systems, showing emergent properties that arise from interactions among their components, which are strongly affected by the surrounding environment (Mazzocchi 2008; Sheth and Thaker 2014; Lüttge 2013; Souza and Lüttge 2014). This was already presented in Aristotle's philosophy by the aphorism "The whole is more than the sum of its parts." Therefore, it would be nearly impossible to reduce the observations on a particular scale of a system to a lower scale, since each particular scale of a system has its own properties. Now, a new class of models and epistemology has shown that many fundamental properties of complex systems, especially the biological ones, are emerging properties. Such properties would be, in general, those properties that are observed on a larger scale of the system (high hierarchical level) and that cannot be observed or inferred from the smaller scales of observation of the same system (low hierarchical level) (Mitchell 2009; Lüttge 2012). Actually, hierarchy seems to pervade all natural organizations (Corominas-Murtra et al. 2013).

Nevertheless, what exactly does "more than the sum of its parts" mean? What kind of effect would cause phenomena attributed to the emergence of nonreducible systemic properties? Supposedly, these properties come from the interactions between the different components of the system that alter the simple product of the parts as a linear sum of the particular characteristics of each component. Based on this principle, it must be assumed that the relationships between these components are not linear, so that interactions can generate complex self-organized behavior not directly inferred from specific properties of each component. Therefore, the key aspect in emerging phenomena lies in the interactions between the components of the system.

In general, in allowing the proposition of a generalizable model, we can take any type of interaction (chemical, physical, or biological) as an exchange of information between elements occupying a given space-time. In this broad context, one can assume that the interactions between the different scales of the system are also provided by exchanges of information. This review will discuss how information flows from one element to another and between scales, building systems based on complex hierarchical networks that, as a whole, are the essence of the system itself.

In the following, the basics of the theory of hierarchy will be presented. In sequence, the possibility to measure hierarchical properties on the basis of informational aspects is assessed (Corominas-Murtra et al. 2010), as it may allow a better understanding of emergent phenomena between different ecophysiological scales.

1.1 General Theory of Hierarchy: Bottom-Up Versus Top-Down Hierarchy

The Oxford Encyclopedic English Dictionary takes "hierarchy" as "a system in which grades of classes of status or authority are ranked one above the other." "Authority" primarily alludes to hierarchies in sociology. In a more formalistic vein, however, we may consider hierarchies quite generally in a terminology of systems leading to various classifications. With ranking of "one above the other," we have the choice of looking at hierarchy bottom up or top down. Whereas the hierarchical view of nestedness, like Chinese boxes or matryoshka dolls, allows understanding topological aspects of the system's structure, by representing them as graphs (Fig. 1), the "authority" view allows understanding the flow of information throughout a cascade of signaling affecting the functionality of the system (Corominas-Murtra et al. 2010), like in molecular signaling networks (Dietz et al. 2010) or ecological networks (Blonder et al. 2012).

Hierarchical systems are either or both topological (meaning, in fact, physical, chemical, or biological) or functional (meaning all phenomena of behavior) (Pattee 1970; Chauvet 1993). Although structure and function appear non-dissociable from each other because a biological function cannot be conceived without a structure to

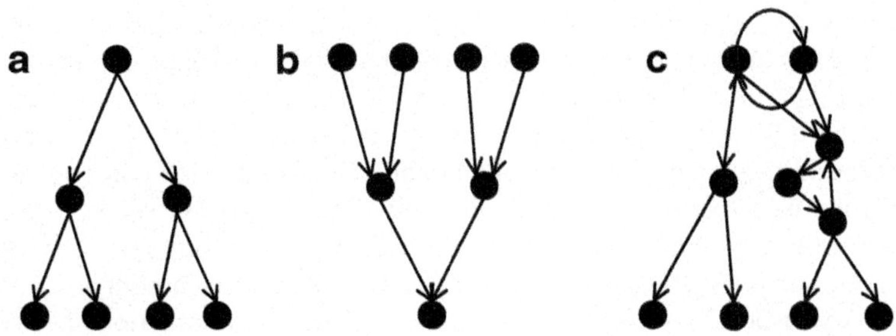

Fig. 1 Different hierarchical network organizations. (**a**) Ideal hierarchy seen as a treelike feedforward graph; (**b**) Antihierarchical structure as an inverted pyramid, such as drainage networks in river basins; (**c**) Hierarchical network with cycles in the feedforward graph

support it, the generation of functional organization involves hierarchical systems that do not necessarily coincide with corresponding structural systems (Chauvet 1993). The stability of a system, e.g., when it is subjected to a perturbation, results from the stability of both its topological and functional features.

Thus, hierarchies are also a matter of scaling and of structure–function relationships. Hierarchical ranking is not static but subject to spatiotemporal dynamics. In spatial dimensions, ranking may involve going through different scalar levels. In temporal dimensions, bottom up and top down often are not separated and can operate in accord. In Sect. 2 of this review, we shall collect a number of examples from plant biology at various scalar levels, i.e., at the atom level, at the molecular level, at the anatomical/morphological level, at the functional/physiological level, and at the ecological level.

1.2 Hierarchy and Information: Explaining the Irreducibility in Biological Systems

According to Simon (1962), a complex system can be defined as nested hierarchical networks of components organized as interconnected modules. Thus, in natural self-organized systems, hierarchy seems to be a pervasive and fundamental feature. Throughout the continued history of matter, new levels are superimposed on the individual units by the organization and integration of these units into a single hierarchical system (Novikoff 1945; Matyssek and Lüttge 2013). From the cell to the biosphere, the biological integration or "nestedness" culminates when we consider the entire biosphere of Earth as one supraorganism (Lüttge 2012; Ahl and Allen 1996) or, in accordance with the Lovelock (1979) concept, as "complex" Gaia. This means that Gaia represents the self-organizing and self-sustaining emergence (Lüttge 2012).

Hierarchy also seems to pervade a pattern of organization that allows decreasing the costs related to reliable information transmission, supporting, for instance, the efficiency of genetic and metabolic control in cellular networks (Corominas-Murtra et al. 2013; Ravasz et al. 2002).

Because hierarchy is a polysemous word, Corominas-Murtra et al. (2010, 2013) have proposed a formalism in order to both define and measure hierarchy. Thus, firstly, hierarchy is defined as a pattern of relations in a morphospace where, ideally, there is no ambiguity in who controls whom with a pyramidal structure in which the few control the many. A directed graph properly represents such treelike hierarchical structure (Fig. 1). Secondly, the morphospace where the graphs are embedded is a metric space defined from three coordinates: treeness (T), feedforwardness (F), and orderability (O). Treeness weighs how pyramidal is the structure and how unambiguous is the chain of command (T ranging from $-1 \leq T \leq 1$, where $T = 1$ is an ideal hierarchical pyramidal structure (Fig. 1a) and $T = -1$ is an antihierarchical structure (Fig. 1b)). The T values correspond to the measure of the entropy associated with the minimum information required to follow a path starting from some node in a top-down (forward entropy, Hf) or bottom-up (backward entropy, Hb) direction; when $Hf > Hb$, the system is considered hierarchical (Corominas-Murtra et al. 2013). Feedforwardness (F ranging from $0 \leq F \leq 1$) is a measure that weighs the impact of cyclic modules on the feedforward structure (Fig. 1c). The closer the cyclic modular is of the top of the network, the lesser is the hierarchical order of the system. Finally, orderability accounts for how orderable is the graph under study. O (ranging from $0 \leq O \leq 1$) is defined as the fraction of the nodes of the graph that do not belong to any cycle, taking part of the network that can be actually ordered, i.e., a hierarchy where the components of the system have causal relations between each other (ordered pairs) in a feedforward structure (Corominas-Murtra et al. 2013).

Corominas-Murtra et al. (2010) have shown how information theory provides a suitable framework to evaluate hierarchy in causal structures. Taking into account a pyramidal ordered tree structure as an ideal hierarchy (Fig. 1a), the authors derived a measure evaluating how far is a structure from this ideal model. This framework assumes that we have a hierarchy if there is no ambiguity in the chain of "command" from the top to the bottom in a network (the same would be valid for nested structures). However, natural networks do not match perfectly with such ideal hierarchy, often showing reversible feedback relationships. Roughly, this measure captures the trade-off between the complexity of the causal structure with the uncertainty related to the reversion of causal paths. In other words, hierarchy is a measure of complexity against the reversibility of the structure. Specifically, using two different measurements of entropy (mutual information and structural entropy), according to Corominas-Murtra et al. (2010), "*it is considered the balance between the richness of causal paths following the flow of causality (a top-down view)* versus *the uncertainty generated by the multiplicity of pathways connecting one element with the set of top nodes when trying to reverse a given causal flow (a bottom-up view)*".

This framework provides a rationale to understand why reductionism is limited when trying to explain emergent properties (high scale) from the components of the system (low scale). Assuming that biological systems are essentially complex hierarchical systems, properly represented by causal structures (but not ideal, since biological networks often present feedback loops—see Figs. 4 and 6), the access to the high properties of the system from its basic components is a very "entropic" strategy. In other words, the irreducibility of the biological emergent properties is, at least partially, explained by the uncertainty associated to the pathways in a bottom-up direction (scaling up the system).

1.3 The Role of Hierarchical Organization in System Stability

As stressed by Souza and Lüttge (2014), stability is an emergent phenomenon from interactions among plasticity, complexity, and diversity throughout plant ecophysiological scales. Therefore, the hierarchical organization of the system performs a fundamental role on the whole system stability. The stability of the system that is subjected to a perturbation results from the stability of the topological and functional features of the system. Subsequently, the problem is to determine how the system stays stable, while it grows and reproduces by restructuring both the levels of organization and distribution of functional interactions.

The more complex the system becomes, the more efficient and robust it is under the stable conditions, to which it is adapted. This is achieved via simultaneous fine-tuning of all the regulatory mechanisms to the parameters of the environment (Rojdestvenski et al. 1999). Once the system subunits are grouped spontaneously, this increases the stability of the subsystems at smaller scale, but the overall stability of the system is never fully achieved since new hierarchical levels may be added (or nested) indefinitely (Korn 1999; Ahl and Allen 1996). The upper levels of the system work as a framework for entities of other levels (Ahl and Allen 1996).

The subunits of a self-organizing biological system are susceptible to changes in the environment or "noise." However, the biological system is not too sensitive, because with that it could not have evolved to its present state (Perry 1995). Systems are self-organized when they are away from thermodynamic equilibrium that optimizes the flow of matter, energy, and information. The greater the environmental variability, the greater is the entropy of the system (Rojdestvenski et al. 1999). However, the hierarchical organization of the systems has the ability to optimize the use of energy by reducing losses and dissipating the entropy efficiently (Schneider and Kay 1994).

The effect of environmental heterogeneity (random fluctuations of physical factors of the environment), which interferes with plant responses to specific environmental factors, is making it difficult to forecast and understand the physiological processes of an organism as a whole (Bertolli and Souza 2013; Mittler and Blumwald 2010). Additionally, even if it were possible to place organisms in an environment perfectly constant, they would remain influenced by noise, because the noise is also originating inside the organisms themselves (Wagner 2005). The main cause of internal noise is thermal movement caused by heat, increasing the temperature inside the system. This internal noise affects, for example, the folding of macromolecules such as RNA and proteins. Thus, at temperatures beyond the physiological optimum, reactions of the body's metabolism may be severely constrained (Wagner 2005). As shown by Bertolli and Souza (2013), the level of environmental noise can significantly affect the physiological responses of plants to specific potential stress factors. Additionally, according to Atlan (1979), the observation of a disruptive noise at a lower hierarchical level can become an organizational factor at the global level or vice versa.

More peripheral subsystems, near the interface with surroundings, may act as a buffer against environmental disturbances and minimize the disruptive influence through the whole system (Ahl and Allen 1996). Moreover, to generate an adequate robustness, decoupling mechanisms of different regulatory processes must exist, such that the interactions between subsystems are substantially weaker than the interactions within the subsystem. Thus, changes in one subsystem cannot significantly affect the other subsystems (Rojdestvenski et al. 1999). The weak links between subsystems act as a buffer smoothing the noise propagation throughout the system hierarchy, improving overall stability (Csermely 2006).

2 Hierarchy: Empirical Examples

2.1 Electrons and Atoms

Hierarchical systems are built from modules, and they always are associated with emergence where modules are integrated and merge into new systems with novel properties and this can go through spatial scales from, e.g., sub-atom particles to the entire planet (Laughlin 2005; Lüttge 2012).

An example starting with electrons and atoms is a piece of metal iron. Atoms can be considered as stable modules or subunits that form the level below the whole piece of metal. The behavior of the lower-level atoms and their parts with their rate of vibration dissipates the condition of iron solidity. Solidity is an upper-level emergent property that is not itself a feature of the individual atoms at the lower level. Although at the upper level completely novel properties are expressed, the properties of the modules at the lower level influence the upper-level behavior (Trewavas 2006). Biological examples can be taken from photobiology where quanta of light, i.e., photons and electrons, are involved and on higher levels of integration complex processes are emergent, such as the phenomena of vision, photosynthesis, light-regulated morphogenesis, etc.

2.2 Hierarchical Principles in Molecular Structures

In his book "From the big bang to cyberspace", Köhler (2009: "Vom Urknall zum Cyberspace") explicitly talks about the hierarchical principle in molecular structures. He shows that the hierarchy from bottom to top starts with the order of atoms in molecules. Molecules become monomers or modules in macromolecules. Within macromolecules there are substructures of domains as we find it especially in proteins. Macromolecules can form aggregates as again it is particularly observed with proteins (Fig. 2). All of these hierarchical integrations are always associated with emergence of new properties and functions on the respective next higher scalar level (Lüttge 2012). The aggregation of subsystems as multienzyme complexes and calcium/calmodulin dependent enzymes creates the simple emergent properties of novel enzyme activities (Trewavas 2006). Tubulin or actin polymerization in the test tube raises the emergent behavior of isolated microtubules or filaments.

According to a dogma of biology, in complex systems emergence starts with deoxyribonucleic acid (DNA) where the required information is stored. DNA is

Fig. 2 Hierarchies of molecular structures

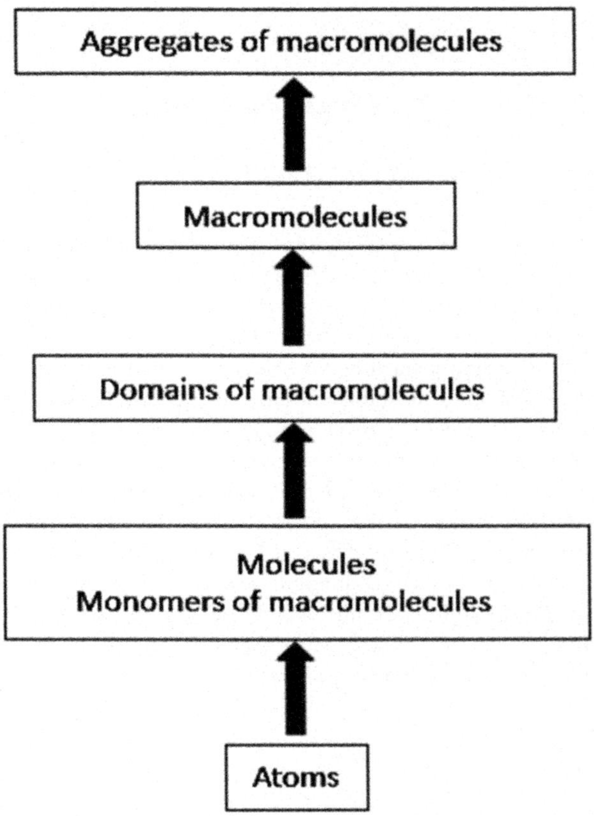

itself a starting point of hierarchies at increased scaling levels. It is forming the genes that can be clustered in sections of chromosomes embedded in entire chromosomes (Fig. 3). According to Köhler (2009), a major advantage of such bottom-up hierarchies is that there are only a small number of categories at the top. This should provide robustness. In molecular biology, these top categories are polynucleotides (ribonucleic and deoxyribonucleic acids, RNA and DNA), proteins, polysaccharides, and lipids, i.e., the major constituents required by cells at the next higher scalar level. We agree with Köhler that robustness given by the limitation of top categories is effective against moderate changes of the elements of the top categories maintaining top-down hierarchical order in the system. The analogy with social hierarchies is evident. However, this analogy implicitly teaches that the top-down hierarchy is also vulnerable. An ecological example, which we shall consider below, is the breakdown of the top "category" canopy with creation of gaps in forests and the dynamics of new hierarchies (Sect. 2.5.3).

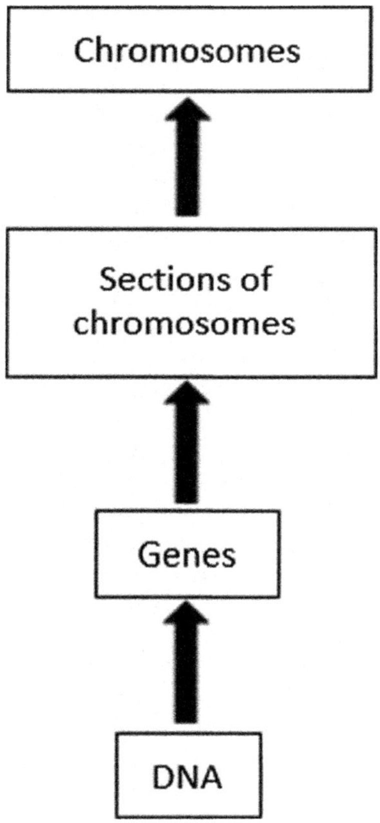

Fig. 3 Hierarchies of genomic structures

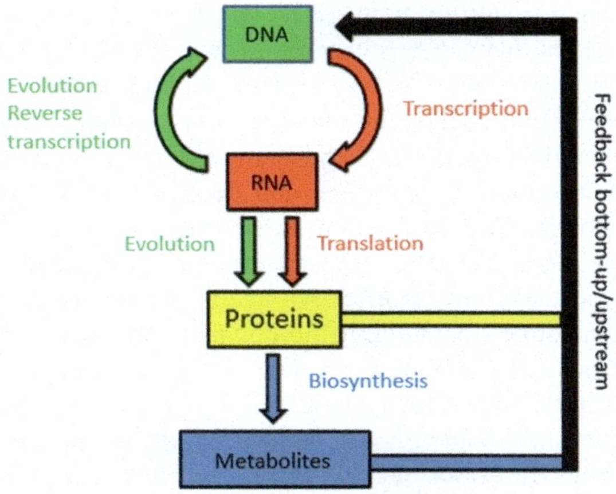

Fig. 4 Functional hierarchy within the level of macromolecules

Within the top-category scalar level of macromolecules, we have functional hierarchy (Fig. 4). Via transcription DNA hands down to RNA, which via translation hands down to proteins that in turn via biosynthesis generate organic molecules or metabolites. However, quite evidently, this is not a top-down—or as it is named in molecular biology "down-stream"—hierarchy. We know that by upstream feedback metabolites and proteins can affect the DNA level and regulate transcription exemplifying bottom-up hierarchy. Reverse transcription is upstream feedback from RNA to DNA. There is not a DNA dictatorship but a panel of governors at the hierarchical top. With its approaches of "omics", namely, genomics (DNA), transcriptomics (RNA), proteomics, and metabolomics, systems biology has opened a wide field of approaching the dissection of hierarchical organization and order within living organisms.

The development from "dictator" to "panel of governors" or the division of responsibilities is seen in the evolution of the top categories (Figs. 4 and 5). It is widely accepted that originally there was an RNA world where RNA fulfilled the three central functions (1) of storage of information, (2) of transfer and flexibility of using this information, and (3) of molecular tool or action molecule (catalyst or enzyme)-steering structures and functions. Reverse transcription and catalytic functions of RNA, which we know from the ribozymes, are extant reminiscence of that. Then there was a first division of labor where proteins took over the role of molecular catalyst tools. Finally the function of information storage was transferred to the DNA, which is a more stable molecule due to the loss of one OH group on the ribuloses in the macromolecule, while the role of information transfer and flexibility was retained by the RNA. Thus, the three top categories of the molecular hierarchy, i.e., RNA, DNA, and proteins, evolved as we know them now (Köhler 2009, Fig. 5).

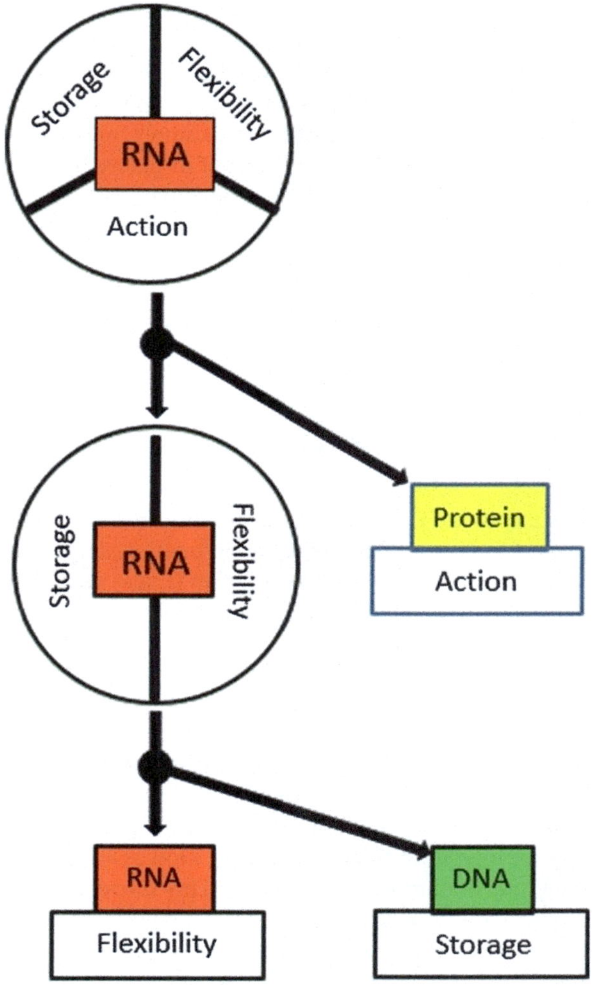

Fig. 5 Evolution of division of labor of the macromolecules RNA, DNA, and proteins

2.3 Cytology, Anatomy, and Morphology

There is a cooperative interaction of cellular organelles to the pattern of organization by which cells obtain their structure and function (Nederbragt 1997). Ravasz et al. (2002) showed that the metabolic network of compartments of cells is organized by integration of many small, highly connected topological modules that combine in a hierarchical manner into larger, less cohesive units, with their number and degree of clustering following a power law. This level of organization is much more complex. Understanding how its emergent properties are constructed is the important issue of systems biology (Trewavas 2006; Lüttge 2012).

Similar and analogous considerations can be made at the higher scalar levels above cells. The activity of cells is influenced by the activity of molecules produced by other cells of the body (Novikoff 1945). Cues, signals, and information are exchanged in building bottom-up and top-down hierarchies, thus influencing the entire organism. An illustrative example of top-down hierarchy is the apical dominance in the growth and gestalt of plants. The phytohormone auxin produced in apical organs (cells of buds) causes the elongation of stem cells and inhibits the outgrowth of lateral buds. As soon as after decapitation the hierarchical dominance of the apical bud is removed, a bottom-up effect gets in operation. Lateral buds are germinating and putting up branches in a race of which one is going to win the game and to establish a new top-down hierarchy (Fig. 34-11 in Lüttge et al. 2010).

2.4 Whole Plant Physiology

Despite of the molecular biology mainstream, recently an integrated view of the plants has been developed based on multi-scale network models, linking molecular networks to whole plant development (Lucas et al. 2011; Weston et al. 2012). Such perspective is particularly important when accessing plant responses to environmental cues.

In temporal dynamics of physiological performance, the observation of a disruptive noise at a lower hierarchical level can become an organizational factor at a higher level and vice versa. For instance, the responses of two plant species of different photosynthetic metabolisms (C_3 and C_4) were dependent of the scale level considered in relation to water deficit (Bertolli et al. 2014) or high temperature (Vítolo et al. 2012). On a top scale (i.e., parameters of growth), high temperature was limiting to the plants, while on a low scale (i.e., parameters of photosynthesis) high temperature was stimulating. This shows hierarchical level-dependent physiological adjustment (Vítolo et al. 2012). It is an illustrative example of the simultaneous temporal bottom-up and top-down flow of information in a hierarchical system. Thus, there is no simple and linear rule for the information fluxes among the different levels of organization of plants under a changing and constraining environment (Bertolli et al. 2014; Souza and Cardoso 2003).

Interestingly, however, there seems to be a regular relationship between metabolic energy and body mass (M) in all living beings (from unicellular to homeothermal organisms). According to Fujiwara (2003), the metabolic energy rate and the life span of the organisms is proportional to $M^{0.75}$ and to $M^{0.25}$, respectively. These simple scaling rules are achieved because, essentially, the same molecules, with similar internal energy, activation energy and free energy, compose all organisms. Thus, we suppose that the pervasive hierarchical patterns in biological systems are also underlying such a pervasive stable relation between energy and body mass of organisms, allowing organisms to stay stable under different environmental conditions.

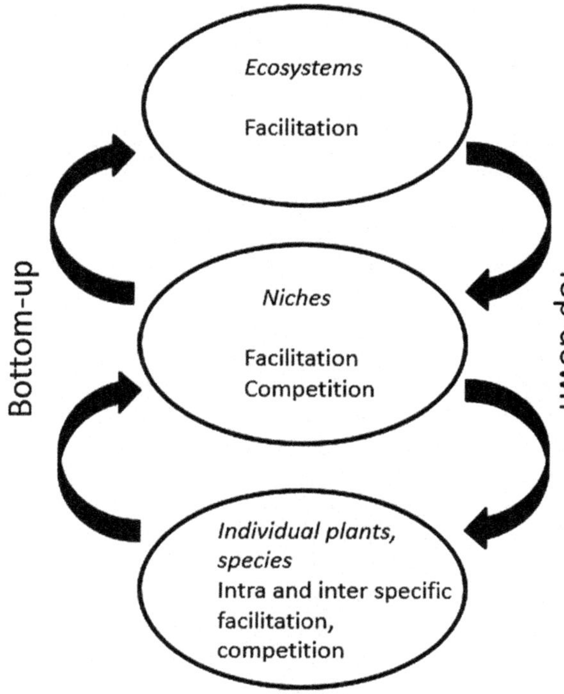

Fig. 6 Bottom-up and top-down hierarchies in ecological systems

2.5 Spatiotemporal Dynamics of Bottom-Up and Top-Down Hierarchies in Plant Ecology

Outstanding as well as apparently contrasting hierarchical functions in ecology are competition and facilitation. Hierarchical scalar levels are ecosystems, niches, and the species and individual plants (Fig. 6). The ecosystem level is usually characterized by heterogeneity, i.e., a large habitat complexity with a number of species and a certain species diversity. Complexity and diversity are somewhat lower in specific structural and functional niches. At the level of species and individual plants, we approach homogeneity.

2.5.1 The Bottom Level

The bottom level of species and individual plants is characterized by pair-wise interactions that may be both competition and facilitation. Competitions are direct interactions of species or of plants of the same species. An example of a hierarchy on a given scalar level and between individuals of a given species is the intraspecific hierarchy of dominant and suppressed trees in a forest stand. Based on the biological criteria of crown development and tree height, Kraft (1884) has distinguished five social classes of (1) predominant trees, (2) dominant trees, (3) low-dominant trees, (4) dominated trees, and (5) entirely overtopped trees (Pretzsch 2010). This is a hierarchy of social classes among the trees as the outcome of competition. The system of hierarchic classes is a qualitative estimate of the competitive status of individual trees within a stand. On the one hand, it is a sophisticated handout for application of the practice of thinning in silviculture. However, as Pretzsch (2010) shows in the detailed fascinating treatment in his book, on the other hand, it is also the basis for understanding the natural process of self-thinning in relation to competition and the availability and allocation of resources. Social hierarchy operates by regulating the relationship of death and survival. This follows the laws of exponential power rules with a declining number of trees per unit area versus an increase of average diameter, size, or weight (Pretzsch 2010).

For facilitation McIntire and Fajardo (2013) quote the definition of Bronstein (2009): "... an interaction in which the presence of one species alters the environment in a way that enhances growth, survival and reproduction of a second species". In this definition facilitation is a matter between individual plants or species, i.e., an event at the bottom level. We will have to extend this view below. However, there is a wealth of examples fitting the narrower form of the definition as given by Bronstein. There are individual plants that can function as pioneer plants on barren sites and facilitate the establishment of other plants under their protection. An example is *Clusia hilariana*. It stems from the Atlantic rain forest of Brazil and migrated into the coastal sandy plains, known as restingas, which were formed during the Quaternary Period (Scarano 2002, 2009). As a pioneer plant on the bare sand, it starts in vegetation islands and exerts facilitation generating denser vegetation (see review Lüttge et al. 2012). Plantations of *Eucalyptus* species are often sterile monocultures. However, with appropriate silvicultural management, which involves regular coppicing the *Eucalyptus* plants, the *Eucalyptus* trees can become facilitators under which a forest with high species diversity can develop (Feyera et al. 2002; Fetene and Beck 2004; Lüttge 2008; see review Lüttge et al. 2012). A prominent example of a pair-wise interaction of species is the germination of plants within the tanks formed by the leaves of bromeliads. Often one finds seedlings of the woody plant genus *Clusia* growing in those tanks (Scarano 2002). In these cases we can call the facilitators nurse plants (see review Lüttge et al. 2012). As the *Clusia* seedlings grow up to shrubs and trees, they outcompete and even kill the bromeliads, their facilitators and benefactors. The *Clusia*s then become facilitators themselves. The shrubs create vegetation islands (see above) from which also forests may develop. We can consider this as bottom-up hierarchy. However, it actually shows characteristic aspects of a succession. We may wish to distinguish both and to caution here for clarity of terminology. Successions are replacement of systems in time, one after the other, while hierarchies are interactions of coexisting systems in space.

Questions of successional patterns and causalities have been the central concerns of ecologists. Being aware of these mechanisms, Pickett et al. (1987) extended the concepts of models proposed by Connell and Slatyer (1977) using standards of hierarchy theory. Notwithstanding the quest for distinguishing succession and hierarchy, it remains interesting that Pickett et al. (1987) suggest a more complete enumeration of the succession causes and place them in a higher, i.e., a third hierarchical, level. The high levels within a hierarchy define the general and universal conditions under which the succession occurs: (1) availability of open space, (2) differential availability of species, and (3) differential performance of species at the site. The third hierarchical level is required to elucidate the mechanisms of succession at particular sites. Recognizing the appropriate level(s) in the hierarchy is critical for the successful explanation of successions, to design experiments and to elaborate detailed predictions, to construct models and to develop general theory.

Both competitors and facilitators can be considered as "condition modifiers" (McIntire and Fajardo 2013). They affect the distribution and allocation of all kinds of resources, such as nutrients, water, CO_2, light, and space. They modify biotic interactions, such as with herbivores, pathogens, pollinators, and diaspore dispersers. Facilitation and competition are often not strictly separated processes. They interchange and interact in time and space, i.e., there is a sophisticated balance between them (see review of Souza and Lüttge 2014; this series and references therein).

2.5.2 Bottom-Up Hierarchies

Bottom-up or scaling-up hierarchy is inherent in the approach of considering emergence (Lüttge 2012) by moving from microscopic symbiotic, i.e., mutualistic or parasitic, interactions, through ecosystems of different sizes up to zono-biomes and eventually the entire biosphere as increasingly integrated systems of holobionts (Matyssek and Lüttge 2013). The bottom-up hierarchy from the species and plant level upwards is effective in the creation of new niches by competition and facilitation (see review of Souza and Lüttge 2014) (Fig. 6). There is competition for niches. However, the creation of niches by facilitation is a predominant effect. In ecosystems facilitation has been widely underestimated as compared to competition. This is due to the long dominance of our thinking by the evolutionary selection theory of Charles Darwin and Alfred Russel Wallace.

2.5.3 Top-Down Hierarchies

Thus, at the top hierarchical ecosystem level in Fig. 6, facilitation appears to be the major interaction. The heterogeneity and biodiversity supports ecosystem functioning and stability (Souza and Lüttge 2014), where species can perform as "ecosystem engineers" eliciting hierarchical top-down effects from the ecosystem level to the niche and plant levels (Fig. 6).

Individual tree species can be at the bottom of facilitation hierarchies as we have seen in Sect. 2.5.1. Planting potential nurse trees can be considered anthropomorphic applied facilitation. Similarly silvicultural facilitation can be established at the top level with a diversity of planted trees as in the National Park of Mount Entoto at the rim of Addis Ababa at 2,600–3,100 m above sea level, where half a dozen of native tree species (*Acacia abyssinica*, *Hagenia abyssinica*, *Juniperus procera* (syn. *J. excelsa*), *Olea europaea*, *Podocarpus falcatus*, *Prunus africana*) are used (Ethiopian Heritage Trust, eht@ethionet.et). Bottom-up and top-down strategies can match for the same end in anthropogenic silviculture and in natural ecological progressions.

A hierarchical top-down dominance develops in forests. We can see it in the tropical rain forests with their high tree diversities (Lüttge 2008). Seedlings on the forest floor grow extremely slowly. They constitute a nursery of small plants whose growth is suppressed by the densely closed canopies of the mature trees. The hierarchical dominance is lifted when a gap is created. Then the seedlings of light-demanding pioneer species begin vigorous growth. There is fierce competition which is followed by a homeostatic phase. Then there is death of pioneer species, and another fierce competition sets in until late successional species take over in a new homeostatic phase (Jacobs 1988; Lüttge 2008). In the overall process, there are highly dynamic temporal changes of social hierarchies. The most important factor modulating these hierarchies is light. Light may act via temperature effects, especially diurnal temperature alternations. Canopy gaps lead to greater fluctuations of soil surface temperatures due to direct insolation. The major effect of closed canopies, however, is on light quality which regulates plant development via the red (R)-/far red (FR)-responsive phytochrome system of the plants. A higher proportion of R activates phytochrome and elicits phytochrome-regulated morphoses. A higher proportion of FR deactivates the phytochrome and the morphoses. The photosynthesis of closed canopies predominantly absorbs the red wavelengths of the penetrating light so that the R/FR ratio of the light is reduced to 0.5 as compared to the ratio of 1.2 of full sunlight (Vázquez-Yanes and Orozco-Segovia 1993).

A common approach to ecological hierarchy is the association of organisms in populations and of populations in communities (Guttman 1976; Nederbragt 1997). Communities or ecosystems may be either large or small. The physical processes that ecological systems must obey are strictly scaled in time and space (Allen and Hoekstra 1990). This leads to the recognition that even for the ecosystem level itself we can meet with hierarchical scaling up and scaling down. Scaling up is where communities and small ecosystems build larger systems and landscapes. An example of scaling down is where inside the larger ecosystem of a forest a dead organism, e.g., a fallen tree, forms an entire smaller ecosystem in the rotting log. Thus, scaling allows ecosystems occurring within ecosystems. Therefore, only with a very limited number of explanations, we can propose all the richness of ecological systems and their phenomena. Nevertheless, the communities and ecosystems have wave interference patterns between processes and organisms. Interfering with and accommodating each other, even though they occur at different scales in the landscape, they have different periodicities in their waved behavior of their bottom-up and top-down hierarchical dynamics.

In landscape ecology one often considers the ecological effects of spatial heterogeneity on the scale of large areas (Havelka 1997). By observing patterns in heterogeneous environments, ecologists can gain insight into how ecological processes are integrated over a range of scales bottom up or top down. Studies conducted at several scales may offer the best resolution of domains and patterns and their determinants (Wiens 1989). Patterns of fractal geometry are scale invariant (Liebovitch 1998). Analyses of landscape ecology over different fractal geometrical measurement scales have been used as a tool for understanding the vegetation structures (Johnson et al. 1995), the relations between species (Thistle et al. 2010) or between species or resources (Olff and Ritchie 2002). Thus, the fractal distributions may offer the quantification of the degrees of relation (positive or negative) between different entities that are distributed in a fractal manner.

3 Conclusions

Often models and the corresponding methods of study lie in understanding of components or modules, and the importance of hierarchical orders and emergent properties of systems are not implicated (Lüttge 2012). However, the responses of entire integrated systems to information (e.g., environmental perturbation) depend on the subsystem sensitivity, bearing in mind that subsystems may respond differentially to environmental perturbations (Souza and Cardoso 2003). The observation of different levels of organization or modules of an organism can reveal changes in the differential responses and leads to varying interpretations of the state of the organism as a whole (Souza and Cardoso 2003; Vítolo et al. 2012; Bertolli et al. 2014). This fact limits our understanding of the overall behavior of the systems based on limited target datasets. Moreover, as stressed in Sect. 1.2, because of the uncertainty associated to the pathways in a bottom-up direction of a non-ideal hierarchical system, as often the biological ones are, the full understanding of the high properties of the system from its basic components is unlikely.

Realizing these shortcomings, we may propose some issues to be addressed in the future. (1) Is there a specific or single organizational level that could represent the whole organism response to environmental cues? (2) What kind of dataset would allow us to capture the actual organism status? (3) If we consider that two hierarchical levels with at least one degree of separation (i.e., structural or functional) respond similarly or antagonistically to a disturbance, what is the biological meaning of this observation? (4) Are the conclusions of a multi-scale study, derived from partial datasets, "wrong" or "limited"?

Such questions are not easy to address. The reductionist approach of analysis of individual subsystems or modules as potentially providing varying interpretations about the effects of perturbations in cells, organisms, and ecosystems hampers an overarching diagnosis and deeper understanding of the interactions between systems and their context. Therefore, some studies have used a multivariate analysis through scales as a tool for the evaluation of the cross-scale relations in systems such as a whole integrated plant (Lüttge 2012; Vítolo et al. 2012). This analysis is an appropriate method for establishing models of systemic understanding of the complex interactions between plants and their changing environment. In short, the search for unique indicators that can be used to determine or predict a global plant behavior in response to environmental cues may be, actually, a search for a "holy grail" as argued by Vítolo et al. (2012).

References

Ahl V, Allen TFH (1996) Hierarchy theory. A vision, vocabulary, and epistemology. Columbia University Press, New York, NY

Allen TFH, Hoekstra TW (1990) The confusion between scale-defined levels and conventional levels of organization in ecology. J Veg Sci 1:5–12

Atlan H (1979) Entre le cristal et la fumée. Éditions du Seuil, Paris

Bertolli SC, Souza GM (2013) The level of environmental noise affects the physiological performance of Glycine max under water deficit. Theor Exp Plant Physiol 25:36–45

Bertolli SC, Mazzafera P, Souza GM (2014) Why is it so difficult to identify a single indicator of water stress in plants? A proposal for a multivariate analysis to access emergent properties. Plant Biol 16:578–585

Blonder B, Wey TW, Dornhaus A, James R, Sih A (2012) Temporal dynamics and network analysis. Methods Ecol Evol 3:958–972

Bronstein JL (2009) The evolution of facilitation and mutualism. J Ecol 97:1160–1170

Chauvet GA (1993) Hierarchical functional organization of formal biological systems: a dynamical approach. I. The increase of complexity by self-association increases the domain of stability of a biological system. Philos Trans R Soc Lond B Biol Sci 339(1290):425–444

Connell JH, Slatyer RO (1977) Mechanisms of succession in natural communities and their role in community stability and organization. Am Nat 111:1119–1144

Corominas-Murtra B, Goñi J, Rodríguez-Caso C, Solé R (2010) Hierarchy and information in feedforward networks. ArXiv eprints

Corominas-Murtra B, Goñi J, Solé RV, Rodríguez-Caso C (2013) On the origins of hierarchy in complex networks. Proc Natl Acad Sci USA 110:13316–13321

Csermely P (2006) Weak links: stabilizers of complex systems from proteins to social networks. Springer, Berlin

Dietz KJ, Jacquot JP, Harris G (2010) Hubs and bottlenecks in plant molecular signaling networks. New Phytol 188:919–938

Fetene M, Beck E (2004) Water relations of indigenous versus exotic tree species, growing at the same site in a tropical montane forest in southern Ethiopia. Trees 18:428–435

Feyera S, Beck E, Lüttge U (2002) Exotic trees as nurse-trees for the regeneration of natural tropical forests. Trees 16:245–249

Fujiwara N (2003) Origin of the scaling rule for fundamental living organisms based on thermodynamics. BioSystems 70:1–7

Guttman BS (1976) Is "Levels of Organization" a useful biological concept? BioScience 26:112–113

Havelka M (1997) A question of scale: the effects of environmental heterogeneity on populations. Coenoses 12:83–87

Jacobs M (1988) The tropical rain forest. Springer, Berlin

Johnson GD, Tempelman S, Patil GP (1995) Fractal based methods in ecology: a review for analysis at multiple spatial scales. Coenoses 10:123–131

Köhler M (2009) Vom Urknall zum Cyberspace. Wiley-VCH, Weinheim

Korn RW (1999) Biological organization—a new look at an old problem. BioScience 49:51–57
Kraft G (1884) Beiträge zur Lehre von den Durchforstungen, Schlagstellungen und Lichtungshieben. Klindworth's Verlag, Hannover
Laughlin RB (2005) A different universe—reinventing physics from the bottom down. Basic Books, New York, NY
Liebovitch LS (1998) Fractals and chaos. Simplified for the life sciences. Oxford University Press, Oxford
Lovelock J (1979) Gaia. A new look at life on earth. Oxford University Press, Oxford
Lucas M, Laplaze L, Bennett MJ (2011) Plant systems biology: network matters. Plant Cell Environ 34:535–553
Lüttge U (2008) Physiological ecology of tropical plants, 2nd edn. Springer, Heidelberg
Lüttge U (2012) Modularity and emergence: biology's challenge in understanding life. Plant Biol 14:865–871
Lüttge U (2013) Whole-plant physiology: synergistic emergence rather than modularity. Prog Bot 74:165–190
Lüttge U, Kluge M, Thiel G (2010) Botanik: Die umfassende Biologie der Pflanzen. Wiley-VCH, Weinheim
Lüttge U, Garbin ML, Scarano FR (2012) Evo-devo-eco and ecological stem species: potential repair systems in the planetary biosphere crisis. Prog Bot 74:191–212
Matyssek R, Lüttge U (2013) Gaia: The planet holobiont. Nova Acta Leopoldina 114(391): 325–344
Mazzocchi F (2008) Complexity in biology. EMBO Rep 9:10–14
McIntire EJB, Fajardo A (2013) Facilitation as an ubiquitous driver of biodiversity. New Phytol 201:403–416
Mitchell M (2009) Complexity—a guided tour. Oxford University Press, New York, NY
Mittler R, Blumwald E (2010) Genetic engineering for modern agriculture: challenges and perspectives. Annu Rev Plant Biol 61:443–462
Nederbragt H (1997) Hierarchical organization of biological systems and the structure of adaptation in evolution and tumorigenesis. J Theor Biol 184:149–156
Novikoff AB (1945) The concept of integrative levels and biology. Science 101:209–215
Olff H, Ritchie ME (2002) Fragmented nature: consequences for biodiversity. Landsc Urban Plan 58:83–92
Pattee HH (1970) The problem of biological hierarchy. In: Waddington CH (ed) Towards a theoretical biology, vol 3. Edinburgh University Press, Edinburgh, pp 117–136
Perry DA (1995) Self-organizing systems across scales. Tree 10:241–244
Pickett STA, Collins SL, Armesto JJ (1987) A hierarchical consideration of causes and mechanisms of succession. Vegetatio 69:109–114
Pretzsch H (2010) Forest dynamics, growth and yield. Springer, Heidelberg
Ravasz E, Somera AL, Mongru DA, Olivai ZN, Barabasi AL (2002) Hierarchical organisation of modularity in metabolic networks. Science 297:1551–1556
Rojdestvenski I, Cottam M, Youn-II P, Öquist G (1999) Robustness and time-scale hierarchy in biological systems. BioSystems 50:71–82
Scarano FR (2002) Structure, function and floristic relationships of plant communities in stressful habitats marginal to the Brazilian Atlantic rain forest. Ann Bot 90:517–524
Scarano FR (2009) Plant communities at the periphery of the Atlantic rain forest: rare-species bias and its risks for conservation. Biol Conserv 142:1201–1208
Schneider ED, Kay JJ (1994) Life as a manifestation of the second law of thermodynamics. Math Comput Model 19:25–48
Sheth BP, Thaker VS (2014) Plant systems biology: insights, advances and challenges. Planta 240:33–54
Simon HA (1962) The architecture of complexity. Proc Am Philos Soc 106:467–482
Souza GM, Cardoso JVM (2003) Toward a hierarchical concept of plant stress. Israel J Plant Sci 51:29–37

Souza GM, Lüttge U (2014) Stability as a phenomenon emergent from plasticity—complexity—diversity in eco-physiology. Prog Bot 76:211–239

Thistle ME, Schneider DC, Gregory RS, Wells NJ (2010) Fractal measures of habitat structure: maximum densities of juvenile cod occur at intermediate eelgrass complexity. Mar Ecol Prog Ser 405:39–56

Trewavas A (2006) A brief history of systems biology. Plant Cell 18:2420–2430

Vázquez-Yanes C, Orozco-Segovia A (1993) Patterns of seed longevity and germination in the tropical rain forest. Annu Rev Ecol Syst 24:69–87

Vítolo HF, Souza GM, Silveira J (2012) Cross-scale multivariate analysis of physiological responses to high temperature in two tropical crops with C3 and C4 metabolism. Environ Exp Bot 80:54–62

Wagner A (2005) Robustness and evolvability in living systems. Princeton University Press, Princeton, NJ

Weston DJ, Hanson PJ, Norby RJ, Tuskan GA, Wullschleger SD (2012) From system biology to photosynthesis and whole-plant physiology. Plant Signal Behav 7:260–262

Wiens JA (1989) Spatial scaling in ecology. Funct Ecol 3:385–397

Plants Shape the Terrestrial Environment on Earth: Challenges of Management for Sustainability

Ulrich Lüttge

Contents

Abstract Plants tend to occupy any suitable space available on Earth. They shape the inorganic terrestrial environment in a dynamic way through the geological ages. They affect the climate. They impact—with manifold kinds of biotic interactions—on the evolution of animals and microorganisms. They are the dominating primary producers of biomass on Earth and feed the other organisms. The Gaia concept of James Lovelock considers the entire biosphere as a supraorganism and postulates self-sustained stability. Plants play a major role in such self-management of nature. Natural self-management is juxtaposed with anthropogenic management, the former tending to sustain, the latter to exploit the biosphere. Anthropogenic management comprises agriculture and forestry. With relations to plants, the greatest challenge is intensified agriculture to feed 9.6 billion people by the year 2050.

U. Lüttge (✉)
Department of Biology, Technical University of Darmstadt, Schnittspahn-Straße 3-5, 64287 Darmstadt, Germany
e-mail: luettge@bio.tu-darmstadt.de

U. Lüttge et al. (eds.), *Progress in Botany*, Progress in Botany 77,
DOI 10.1007/978-3-319-25688-7_6

Faced with limited and declining resources, pollution, exploratory land use, socio-political ideologies, and by unavoidably contributing to some of these problems itself, agriculture is running into vicious cycles. Can agriculture and forestry learn from ecology? To which extent can ecological principles be introduced to them for securing sustained stability of productivity? Man assumes he "is the possessor of the planet, if not the owner," rather than the "tenant." Conversely, "the Gaia hypothesis implies that the stable state of our planet includes man as a part of, or partner in, a very democratic entity" (Lovelock, *Gaia: A new look at life on Earth*. Oxford University Press, Oxford, 1979). Can natural self-management and anthropogenic management be harmonized, given that mankind learns to conceive itself as part rather than owner of nature?

1 The Outset: Plants Modulate Physical Features of the Earth's Surface

Without the green photosynthesizing organisms, cryptogams and phanerogams, the terrestrial surface of the Earth would be barren land. Plants shape the physical features of the Earth's surface. Space is an essential resource for plants (Grams and Lüttge 2011), and plants tend to occupy any available surface. We can witness it in manifold ways when we observe plants colonizing apparently forbidding extreme sites such as cracks of rocks, the bare sand of dunes and deserts, coastal and inland salinas (Medina et al. 1989), and lava fields (Lüttge 2010). Even on wires of fences and electricity and telephone lines in the tropics, we may find populations of higher plants, so-called atmospheric bromeliads (Lüttge 2008; Grams and Lüttge 2011). As pioneers on such extreme sites, plants can create conditions around them which are suitable for other organisms, and new niches for the establishment of life are formed. For example, on sand plains such as the salinas at the Caribbean north coast of Venezuela (Medina et al. 1989) and the coastal restingas of Brazil (Lacerda et al. 1993; Duarte et al. 2005), vegetation islands can build up, inciting succession toward increasingly dense vegetation. We shall come back to this in this essay when we consider repair functions of plants with facilitation and successions (Sects. 2.5 and 2.6).

In global terms plants shape vegetation zones. A broader concept is that of the nine zonobiomes by Walter and Breckle (1984) which in addition to plants comprise all living beings. Let us run through them from the equator to the poles:

1. Equatorial zonobiome with evergreen wet tropical forests
2. Tropical zonobiome with summer rain and various types of tropical forests and savannas, with evergreen to deciduous vegetation
3. Subtropical arid zonobiome with desert climate
4. Mediterranean zonobiome with winter rain and evergreen laurel-like hard leaf and deciduous forests
5. Warm temperate zonobiome with temperate rainforests

6. Temperate nemoral zonobiome (from Latin *nemus* = grove) with summer-green deciduous broadleaf forests
7. Arid temperate continental zonobiome with steppes and prairies, i.e., grasslands subject to frost
8. Cold-temperate boreal zonobiome (from Latin *borealis* = northern) with the taiga of evergreen coniferous forests
9. Arctic and Antarctic zonobiome with the tundra dominated by evergreen and deciduous shrub species

We can see that wherever we may place ourselves on the globe, we can characterize the site by typical vegetation.

In the continents of Eurasia, Africa, and Northern and Southern America, the boundaries of these zonobiomes run approximately parallel to the latitudinal circles. Evidently the prime factor determining the zonobiomes is the climate. Conversely, plants also affect the climate. This is long known for local microclimate. However, recently also large-scale interactions of vegetation with climate have increasingly been zoomed into the focus of consideration. Such interactions include carbon budgets, the release of water vapor to the atmosphere due to transpiration, the cycling of nitrogen, and albedo functions of the vegetation. By photosynthesis, the vegetation functions as a sink for the greenhouse gas CO_2 that causes global warming (Pan et al. 2011), and especially forests constitute a large carbon store (Luyssaert et al. 2008). Transpiration supports cloudiness. The albedo effects determine reflection of radiation that otherwise would warm up the global climate. All of these effects are relevant to the problem of global change and are major aspects affecting sustainability.

With their photosynthesis, plants are the primary producers of biomass on Earth, and hence, they feed other organisms with organic substrates and energy. Table 1 exemplifies primary productivity on Earth and global energy relations. The immense extraterrestrial and quantitatively non-limiting energy source is the radiation of the sun. Energy incorporated by plants in gross primary productivity via photosynthesis is just a minute fraction of extraterrestrial availability. Primary productivity may be seen as "harvesting the sun."

For sustainability of global equilibria, it is essential that primary productivity by the plants remains stable. A particular aspect of this is agriculture needed to keep feeding a growing global population of 9.6 billion people as predicted for the year

Table 1 Global primary productivity and energy relations (rounded; after Lüttge et al. 2010)

	Oceans	Land surface	Total
Primary productivity ($\times 10^{15}$ g dry matter per year)	55	125	180
Input of solar energy to the Earth ($\times 10^{21}$ J per year)	1,600	700	2,300
Energy content of primary productivity[a] ($\times 10^{21}$ J per year)	1	2	3
Human energy consumption ($\times 10^{21}$ J per year)			0.3
Contribution of agriculture ($\times 10^{21}$ J per year)			0.015

[a]Assuming for the calculation that all dry matter produced is glucose and using the calorimetric energy content of glucose as 2.8×10^6 J/mole

2050 (Sect. 3.1). This is an existential challenge for mankind. Hence, in the following two major sections of this essay, I shall juxtapose the sustaining natural self-management by nature (Sect. 2) versus the exploitive anthropogenic management of nature (Sect. 3).

2 Natural Self-Management

2.1 *The Self-x Concept of Speck*

According to the self-x concept of Speck (Masselter et al. 2012; Speck et al. 2013a, b), living organisms possess a remarkable variety of self-supported self-x properties and self-x capacities. x can stand here for adaptation, organization, healing, and sealing of mechanical and functional injuries and for other repair functions. In bionics these self-x properties provide an amazing variety of bio-inspiration for the technical development of biomimetic materials and structures by engineers. Adaptation and self-repair as aspects of self-organization are also eminently important on large-scale complex, highly integrated systems like ecosystems. Here the x can be sustainability enabling persistence as an emergent feature. As we shall see, self-sustainability of ecosystems may be an important capacity of natural self-management.

2.2 *The Network: Plasticity–Diversity–Complexity–Stability*

Plasticity, diversity, complexity, and stability are interconnected with each other in a network which Souza and Lüttge (2014) called a quadruped. Stability enabling persistence in time is in the center of it. It is equivalent to sustainability. With stability we do not mean fixed steady states or thermodynamic equilibria because all organisms and any systems composed of organisms are open systems through which a continuous flow of material and energy is taking place. Hence, organisms and their systems constitute dynamic equilibria or pseudo-steady states. Maintaining such dynamic equilibria is the essence of stabilizing life in a way sustaining it (Souza and Lüttge 2014), i.e., warranting the persistence of life.

Plasticity is at the top of the quadruped. Plants cannot exert a choice on the conditions they are exposed to by escaping spatially; rather, they are bound to respond to the varying conditions at their site rooted. On such grounds, plasticity in stress response has been favored as a predominant feature of the evolution of plants. Flexibility of performance has allowed the occupation of the land by the plants (Sect. 1). Plastic responses can directly strengthen the persistence of individual plants and ecosystems in natural self-management. Plasticity can support stability via functional diversity and species diversity. Where plasticity supports niche width and niche occupation followed by segregation, it accelerates speciation and therefore diversity (West-Eberhard 1986, 1989, 2003; Solbrig 1994; Lüttge 1995a, b, 2000; Gehrig et al. 2001).

Considering the role of diversity, we may distinguish two hypotheses, the diversity–sustainability and the diversity–productivity hypothesis (Tilman et al. 1996). The latter is relevant for the assessment of the promises ecological principles may provide to agriculture (Sect. 4). There is much evidence of the role of diversity in stabilizing ecosystems (Tilman et al. 1996; Schläpfer and Schmid 1999). The contribution of diverse species (or genetic diversity in general terms) with different individual performances stabilizes ecosystems (Cottingham et al. 2001). Different requirements and performances are also accumulated with species constituting different functional groups (Hooper and Vitousek 1998). Compensation of losses of species and their functions in ecosystems under stress is more likely when there is large species diversity (Davies et al. 2012). Species without any obvious important functions for an ecosystem can be hidden in a large diversity and may take over and protect an ecosystem under new conditions given by environmental stress (Scherer-Lorenzen et al. 2005; Caprez et al. 2012; Körner 2012; Huck et al. 2013). Stabilizing effects of diversity are most prominent in grassland-type ecosystems (Schläpfer and Schmid 1999). Diversity sustains ecosystems in a multitude of ways. Species diversity can cause so-called portfolio effects named after the stock exchange when a broadly scattered portfolio in which different stocks oscillate non-synchronously prevents extreme losses or gains. Various facilitation effects are involved (Sect. 2.6), e.g., in resource allocation including mobilization of nutrients in the soil (Sect. 2.6), control of insect pests (Andow 1991; Altieri 1999), and pollination success (Fründ et al. 2010; Blüthgen and Klein 2011; Weiner et al. 2011). Hence, portfolio effects mitigate risk.

Diversity implies complexity. In systems reflecting functional biodiversity, the various species communicate in network pattern-type interactions, and the larger the richness of species, the higher is the complexity in communication and structure. Interactions in networks always comprise positive and negative feedbacks (Hütt and Lüttge 2005; Souza et al. 2009). There are alternatives via different pathways of connections of knots via edges. Thus, one-sidedness or dichotomy of options hardly exist, rather the multitude of possible reactions stabilizes network systems. This also involves redundancy in networks which buffers systems against external perturbations (Amzallag 2001; Edelman and Gally 2001).

2.3 The Biosphere/"Gaia" Concept of Lovelock: Self-Sustained Stability?

The scientific concept of Gaia was developed by the environmentalist James Lovelock in the 1960s to 1970s (Lovelock 1979). The ancient Greek goddess Gaia, the broad-breasted mother Earth, metaphorically stands for the entire biosphere comprising all life on Earth harbored by the lithosphere, hydrosphere, and atmosphere on its surface as seemingly caring for itself. In a grand optimistic vein, Lovelock (1979) considered Gaia "as a short-hand for the hypothesis [......] that the biosphere is a self-regulating entity with the capacity to keep our planet healthy by controlling the chemical and physical environment." Providing conditions supportive of life via feedback coupling mechanisms, the biosphere visualized as Gaia arrives at being one single living being, a self-regulating supraorganism.

One critique that was raised against the self-sustained persistence of Gaia, conceptually claiming life to stabilize itself on our planet, focused on the large structural and functional fluctuations among life forms which have evolved during the geological ages. In such terms history of life on Earth seems to prove biosphere to vary to an extent that does not allow talking of stabilization or equilibrium, as repeated waves of massive extinction of organisms had occurred throughout geological ages (Table 2). However, life is never static. Indeed, the extinction waves

Table 2 Geological ages with the five most massive extinction waves of species and the corresponding plant ages

Geological ages	Extinction waves	Plant ages
Ordovician 510		Proterophytic
	444	
Silurian 439		
Devonian 409		Palaeophytic
	364	Pteridophytes
Carbonic 364		
Permian 290		Mesophytic
	251 90 - 95 % of all existing species extinct, including the trilobites	Gymnosperms
Triassic 245		
	208 species extinction associated with appearance of the dinosaurs	
Jurassic 208		
Cretaceous 146		Neophytic
	65 75 % of all existing species extinct, including ammonites and dinosaurs, species extinction associated with tremendous proliferation of mammals	Angiosperms
Tertiary 65		

Numbers indicate million years in the past

elicited the appearance of new forms of life granting further evolution under such cataclysmic conditions (see "exaptation traits" below). Most prominent ones were the appearance of the dinosaurs and later the proliferation of the mammals after the disappearance of the dinosaurs. Hence, as argued in more detail by Matyssek and Lüttge (2013), we find that Gaia's self-sustainment in geological times applied to life as such and not to specific forms of life. All sustainment of systems of life arises from highly dynamic processes.

Some more specific examples of nature's natural self-management and self-sustainment will be given below in Sects. 2.5 and 2.6. Can such examples keep us optimistic? In Lovelock's biosphere/Gaia concept, we meet the ultimate extrapolation of the idea (1979). In the present essay, I juxtapose natural self-management to anthropogenic management. The latter interferes with the former. In acknowledging this interference, Lovelock has become much less optimistic 30 years after introducing the Gaia concept (Lovelock 2009).

2.4 Dynamic Self-Sustainment of Vegetation on Earth: The Plant Ages

Major extinction waves are mainly defined by the coming and going of animal species (Sect. 2.3, Table 2). It is often wrongly implied that because of their high plasticity, plants are not subject so much to extinctions. Plants began their conquest of space on terrestrial land with the psilophytes and the genus *Rhynia* 400 million years ago. The entire land mass of the Earth at that time was one conglomerate of the huge continent Pangaea before continental drift separated the current individual continents. The land was forbiddingly barren extreme desert. However, this was the beginning of plants shaping the geological terrestrial environment and making it suitable for their own and other life. Nevertheless, plants were also subject to extinction. After having started terrestrial life, the psilophytes already disappeared again in the Upper Devonian after the geologically relatively short period of 40 million years. Besides the geological ages, we encounter the plant ages of the Palaeophytic, the Mesophytic, and the Neophytic where the pteridophytes, gymnosperms, and angiosperms followed each other in dominance (Table 2). The green cover of the planet over the ages has proved to be very dynamic.

2.5 Repair Functions

The natural self-management of vegetation is obvious in many cases where we witness its acquisition of new land. Such is the case on coastal sand plains where some plants start small vegetation clumps or islands on the pure sand surface which in successions develop larger vegetation units and even forests. An example is

Clusia hilariana creating a new diverse community in the coastal restingas of Brazil (Dias and Scarano 2007; Scarano 2002, 2009; Lüttge et al. 2013). Other examples are maritime volcanic islands, such as the geologically older islands of Hawaii (70 million years old) or the Galapagos archipelago (5 million years old) or the younger famous Indonesian island Krakatau. The latter emerged in 1883 and 3 years later already accommodated 30 species of tropical vegetation. At the south coast of Iceland in November 1963, the island of Surtsey with currently 1.4 km^2 originated from submarine volcanic activities. In 1965 it already carried bryophytes and lichens, and in the first 20 years, 20 species of higher plants were growing on it (Magnússon et al. 2009).

Similarly degenerated land can be spontaneously recovered by vegetation, and in this case we observe the repair functions of plants. In the Brazilian Amazon region, bauxite washing tailings discharged into a lake created a new repelling habitat. Then spontaneous regeneration by native species, subsequently also supported by seedling-plantation management, reestablished vegetation (Scarano et al. 1998; Dias et al. 2012; Lüttge et al. 2013). In Germany in the Ruhr area, a large postindustrial landscape of 450 km^2 has been recovered for agriculture (40 %), forests (20 %), human settlements, and recreational purposes. The area has been called an "industry culture."[1 2] Spontaneous vegetation has integrated itself in the process. These are promising examples of the self-sustainment of the biosphere (sensu Gaia).

The participating plants can be named pioneer species, nurse plants, or ecosystem engineers. Neophytes or invaders may also be involved. The nurse plant syndrome comprises facilitation (Sect. 2.6) when plants shelter seedlings and young and adult individuals of other species throughout their ontogeny and enhance fitness, survival, and growth of associated species (Franco and Nobel 1989; Callaway et al. 2002; Bruno et al. 2003; Brooker et al. 2008). When the nurse plant effects reach beyond facilitation and modulate the physical space where other species live, and such direct effects last longer than the lifetime of the nurse plants themselves, we can call such nurse species ecosystem engineers (Hastings et al. 2006). Invaders are species that newly arrive in existing ecological systems (Elton 1958). They are not so welcome by ecologists, as they get established in their new host systems by outcompeting resident species, and thus, disturbing and modifying host systems.

These types of species are characterized by carrying what Gould (2002) calls exaptations. Exaptations are not traits explicitly selected in evolution for adaptation and for serving a special function under the conditions here and now. Rather, exaptation traits are "structures co-opted for utility from different sources of origin .... and not directly built as adaptations for their current function" (Gould 2002, p. 43). Under new environments with different challenges, the nonadaptive property of exaptations can bring with them exaptative surprises. Stem cells in the

[1] www.emscherlandschaftspark.de

[2] http://www.metropoleruhr.de/tr/freizeit-sport/natur-erleben/route-industrie

center of meristems in plants have various degrees of potency and are often totipotent (Weigel and Jürgens 2002). They have amazing regenerative power and repair functions. In analogy to the stem cells at the anatomical level in a previous essay in this series "Progress in Botany," we have suggested to name plants "stem species" (Lüttge et al. 2013). The term applies, when they have an exaptative pool of dormant traits which provide them with the potency to exert repair functions at the ecological level.

2.6 Competition and Facilitation

The repair functions discussed above often involve nurse effects and these are based on processes of facilitation. Competition and facilitation present a pair of strategies in biotic interactions which at a first glance are obviously counteracting each other. At closer inspection we shall see, however, that they are intimately related.

The mechanism of competition of plants is growth. There is a self-accelerating cycle between growth and competition. Competition for water and nutrients (Schenk 2006; Novoplansky 2009; Hodge 2009, 2010; Cahill 2013), under certain conditions for carbon, for light (Küppers 1989; Grams and Lüttge 2011; Cahill 2013), and even for space as a resource (Grams and Lüttge 2011; Grams et al. 2012; Grams 2013), supports growth while limiting neighboring plants. Vice versa growth is the dominant strategy of competition. This applies not only to interspecific but also to intraspecific competition. In a forest stand, individuals of the same tree species as an outcome of competition can form different social classes which were early defined by Kraft (1884) as follows: (1) predominant trees, (2) dominant trees, (3) low-dominant trees, (4) dominated trees, and (5) entirely overtopped trees (Pretzsch 2010). Originally this definition was meant for anthropogenic management as a handout for employing thinning in silviculture practice. However, much more than that, it is the basis for understanding an illustrative example of natural self-management of vegetation, namely, the process of self-thinning in relation to competition and the associated reallocation of resources available upon thinning. It regulates the relationship of death and survival, where the laws of exponential power rules relate a declining number of trees per unit area to the competition parameters of growth, such as increase of average diameter, size, or weight (Pretzsch 2010).

Facilitation is given when cohabiting species support each other as a result of an integrated outcome from more or less complex interactions reversing competition. This can reach as far as particular species providing nurse services to other species for getting established in the vicinity (DaSilva et al. 1995; Feyera et al. 2002; Grams and Lüttge 2011; Grams 2013). There is a plethora of examples for facilitation in a large variety of ecosystems as surveyed in a comprehensive review by Callaway (1995). Positive interactions among plants appear to be the rule rather than exceptions in most biomes. Facilitation enhances ecosystem functions and increases diversity and productivity.

Mechanisms of facilitation are manifold and can be complex multi-faceted modifications of the immediate local environment, having the capacity for creation of new ecological niches (Callaway 1995). A major aspect is resource modification, such as of light and temperature, of soil moisture, and of soil nutrients. Soil moisture is improved for shallow rooting species via "hydraulic lift" by more deeply rooting species which take up water from lower soil levels and lose it from their fine roots higher up (Richards and Caldwell 1987; Caldwell and Richards 1989; Williams et al. 1993). "Nutrient pumping" is associated when nutrients from deep soil levels taken up by deep rooting plants are released to the soil surface via litter-fall and through-fall (Richards and Caldwell 1987). Some species are more and others less capable of chemically mobilizing otherwise unavailable nutrients such as phosphorus, iron, zinc, and manganese. The former facilitate nutrient supply to the latter (Li et al. 2014). Such mobilization-based facilitative interactions enhance productivity and also can be an important benefit in agricultural strategies of intercropping (Sect. 4). Biological fixation of atmospheric nitrogen also belongs to the mechanisms of mineral-nutrient facilitation. In Ethiopia it was estimated that about 40 trees of N_2-fixing *Acacia*s per hectare provide sufficient N-fertilization to support pasture or crops in agroforestry (Lüttge 2008). "Plant defense guilds" (Atsatt and O'Dowd 1976) control herbivory with the facilitative effects of sheltering against or hiding from predators (Callaway 1995).

Competition and facilitation do not operate in isolation but are correlated through a continuum of feedback processes, as part of networks of comportment; they occur simultaneously within the same community (Callaway and Walker 1997; Callaway 1998, 2013; Lin et al. 2012; del Rio et al. 2014). With time facilitation can shift to competition when the beneficiaries grow up and outcompete their benefactors (Callaway 1995; Callaway and Walker 1997). Nurse plants may eventually be eradicated by their beneficiaries, e.g., when saplings of the tree genus *Clusia* germinating and growing in the tanks of bromeliads grow up to shrubs and trees (Dias and Scarano 2007). There may also be a sophisticated balance between competition and facilitation. The stress-gradient hypothesis (SGH) predicts that facilitation dominates in harsh and stressful environments of abiotic and/or biotic factors, whereas competition prevails in fertile resource-affluent environments (Bertness and Callaway 1994; Callaway and Walker 1997; Dangles et al. 2013). Often the balance is delicate, including temporal dynamics. In mixed-species forests of Central Europe, facilitation shifts to competition from growth-limiting to non-limiting years (Pretzsch 2013; del Río et al. 2014).

3 Anthropogenic Management

3.1 Agriculture Challenges: Feed 9.6 Billion People on Earth by the Year 2050

3.1.1 Anthropogenic Management and Human World Population

Anthropogenic management of vegetation originated 10–12 thousand years ago with the beginning of agriculture when the first groups of *Homo sapiens* changed from hunter-gatherers to growers of plants to reduce risk in the provision of food. At the same time, anthropogenic land use was initiated. Nowadays, the challenge of feeding people and the control of land use have become problems of existential dimensions in view of the excessive growth of human population on our globe (Table 3). For feeding a population of 9.6 billion people predicted by the year 2050, agricultural production of food would have to be increased by 2 % each year.

Population may be controlled in a humane way by family planning and birth control. If not successful, horrible human tragedies may add to it, e.g., natural catastrophes multiplied by local overpopulation, famine, epidemics such as AIDS and Ebola, and wars. In his more recent and more pessimistic view of the self-sustainment of Gaia (Sect. 2.3), *A Final Warning: The Vanishing Face of Gaia*, Lovelock (2009) writes "Mankind may survive but possibly with not more than a few hundred million people," which is the global population size of the year 1700.

Table 3 Human population on Earth: past, present, and predicted

Year	Number of people (billions)			Comments
1700	0.6			Historical
2015	7			Current; 1 billion suffering hunger, in sub-Saharan Africa 30 % of the population
2050	12.8			With current reproductive fertility
	9.6			Supposed that family planning will continue to succeed
2100	10.9	Global	100 %	Prediction made in 2014 by the United Nations
		Africa	39 %	
		India	14 %	
		China	10 %	
		Asia (rest)	19 %	
		Latin America	7 %	
		Europe	6 %	
		North America	4.4 %	
		Oceania	0.6 %	

3.1.2 Land Uses of Agriculture

Arable land on Earth currently covers about 14×10^6 km^2. 17 % are irrigated and carry 30–40 % of the global food production. Losses due to erosion, degradation, sealing, and urbanization are 10^4 km^2 per year. Losses are also due to agriculture itself under irrigation regimes. Due to evaporation leaving dissolved salt behind, even if the best freshwater is used for irrigation in arid regions, salinity is a great problem (Lüttge et al. 2010; von Braun 2011; Lüttge 2013), amounting to a land loss of 2×10^3 km^2 per year. This is one of the vicious cycles of intensified agriculture. The acquisition of new arable land is problematic; it is limited and destructive. Encroachment into natural landscape is in full progress, largely at the expense of savannas and forests especially in developing countries where the population growth and the demographic pressure are large (Table 3). Nevertheless, availability of arable land in absolute terms is stagnating at the global scale (Matyssek et al. 2013a) even though forested area continues to decline (Knoke and Hahn 2013). In view of the current problems of greenhouse gas emissions and global warming and the biodiversity and productivity crisis, the consequences of the change of land use by new acquisition of arable land for agriculture—if feasible at all—are not foreseeable (Lüttge 2013). Arable land is a most precious and limiting resource.

3.1.3 The Plant/Soil System: Water, Carbon, and Minerals

For plants on terrestrial land, water is the crucial resource for life. The water requirement of crops is high (Table 4). Thus, water use is an outstanding aspect of anthropogenic management. In agriculture water availability can become a problem (Sect. 3.1.2). Water use is intimately coupled with land use and plant growth not only in agriculture but also in establishment of recreational areas, sport fields, and other civilizing infrastructure. The water reserves of the oceans appear to be immense and almost infinite but are salty, whereas the available freshwater especially from rain and soils amounts to a fraction of less than 0.01 % of the global reservoir (Table 5). Techniques for gaining freshwater from seawater are very energy demanding. This is challenging energy politics in view of required increases in agricultural output, as resources vanish (also see minerals below) and pollution continues (Sect. 3.1.4).

Table 4 Requirements of water for some crops

Crop	Liter H_2O per kg C
Maize	700
Sugar beet	900
Wheat	1,050

Compiled from Haberl et al. (2012)

Table 5 Water reserves of the planet Earth

	Volume (million km^3)	% of total
Total global water	1,400	100
Freshwater	35	2.5
Glaciers and snow	24	1.7
Fresh groundwater	10.5	0.75
Ground ice and permafrost	0.3	0.02
Rain	0.1	0.007
Lakes	0.09	0.006
Soil	0.02	0.001
Atmosphere	0.01	0.0007
Wetlands	0.01	0.0007
Rivers	0.002	0.0001

Compiled after http://www.theglobaleducationproject.org/earth/fresh-water.php

Table 6 Global carbon pools

Pool	$\times 10^9$ t Carbon
Atmosphere	820
Terrestrial vegetation	800
Soil	1,600 (to 2,100)

Compiled from Haberl et al. (2012)

The quantities of carbon in the soil are higher than in the terrestrial vegetation and in the atmosphere (Table 6). Soil carbon is depleted by agricultural conversion and repeated soil disturbance related to harvests. Soil organic matter is essential for soil life and quality, so that soil carbon needs maintenance to sustain soil biota (Matson et al. 1997).

The debate of managing agricultural productivity for feeding mankind is often carbon centered. However, for the productivity of ecosystems in general, it is necessary to consider—aside from water resources—mineral nutrients in addition to photosynthetic primary productivity. In agricultural ecosystems nutrition determines both the quantitative aspects of yield and the quality of products. It is mandatory to consider the carbon cycle in intimate relation to the nutrient cycle (Körner 2013).

The dominating macronutrients are nitrogen and phosphorus. Natural resources of fertilizers are vanishing. With the Haber–Bosch technique, nitrogen fertilizer can be obtained from the N_2 reserves of the atmosphere. The planet's atmosphere contains about 80 % N_2. Its reduction to ammonia to produce fertilizer, however, is highly energy demanding, technologically mediated through temperatures of 500–600 °C and pressures of 250–350 bar, as well as by consuming the energy-carrier gas H_2 as reductant. Currently, the Haber–Bosch process uses 1.4 % of the world's energy consumption. The most economically exploitable reserves of phosphorus minerals are estimated to be exhausted in about a century (Cisse and Mrabet 2004; Cordell et al. 2009; Bünnemann et al. 2011; Vance and Chiou 2011). At present 95 % of P-rock mined is used for agriculture. Access to new reserves will consume high amounts of energy. Projects of recycling phosphorus excreted by livestock and humans back to agricultural land need to be developed (Oberson et al. 2011). The bottlenecks of mineral nutrition add to the expectation that intensified agriculture will increase still largely unforeseen but very likely challenges to energy politics.

3.1.4 Vicious Cycles: Agriculture and Pollution

The need to increase the output of agriculture for feeding a growing mankind leads into several vicious cycles where agriculture itself elicits negative feedback effects on its own sustainability. Increased erosion, reduced soil fertility, and lowered biodiversity are among such effects (Matson et al. 1997). Another important vicious cycle is between agriculture and pollution. The major atmospheric pollutants are listed in Table 7 (see also Matyssek et al. 2013b). Where high-technology agriculture requires energy, it contributes to production of the major greenhouse gas CO_2. Noteworthy is the close correlation between global growth of mankind and use of fossil energy sources, both skyrocketing since the early nineteenth century, which causes anxiety in view of the upcoming energetic bottleneck (Höök et al. 2011). By cultivation of rice in wet paddy rice fields and by cattle farming, agriculture contributes directly to the production of the very effective greenhouse gas CH_4. Intensive agriculture with nitrogen fertilizers leads to production of another effective greenhouse gas N_2O. Matson et al. (1997) estimate that only 40–60 % of the applied N is used by the crop plants and the rest remains in the soil or is leached out of the ecosystems (e.g., Durka et al. 1994). Some of the negative effects of these N compounds have feedback on agricultural productivity.

Table 7 Major atmospheric gaseous pollutants

Pollutant gas	Origin	Climate effectiveness	Anthropogenic climate effect (%)
Carbon dioxide CO_2	Fossil sources of energy, deforestation	1	50
Methane CH_4	Cattle farming, cultures of rice, new energy sources, marine methane clathrates	21	13
Nitrous oxide N_2O	Intensive agriculture, nitrogen fertilizers	206	5
Ozone O_3	Secondary pollutant originating from outputs of traffic, industry, and agriculture	2,000–16,000	5–12

All of them are greenhouse gases and cause global warming. In addition ozone is a strong oxidant, causes oxidative stress, and is an inhibitor of life functions especially in plants. The climate effectiveness of CO_2 is defined as 1 and the other gases are related to this. The anthropogenic climate effect is dependent on the concentration of the gases in the atmosphere and is given in per cent of total. After data of Matyssek et al. (2010)

Table 8 Examples of effects of O_3 on growth and yield of plants: reductions in %

	Leaf area	Stem productivity (DM)	Root biomass (DM)	Radial growth	Yield
Beech[a]		44			
Birch[b]	14	10	19	15	
Poplar[b]	19	17	24	14	
Wheat[c]					2–59
Rice[c]					7–55

DM dry matter
[a]Grünhage et al. (2013)
[b]Oksanen et al. (2013)
[c]Pandey et al. (2013): observations for India

Table 9 Vanishing advance of the development of crop yields: global gains (+) and losses (−) of yield in percent

Period	Wheat	Rice	Maize and soybean
1987–1997	+20	+17	
1997–2007	−1	+2	
Model predictions[a]			−30 to −46 −63 to −82

After Schlenker and Roberts (2009), Long and Ort (2010), Lüttge (2013)
[a]Depending on the global warming scenario applied

Ozone needs a special comment. The current public debate on environmental challenges mostly focuses on CO_2 problems and O_3 is largely overlooked. However, O_3 in the troposphere is a major and dominating problem; it is the most significant phytotoxic pollutant in our atmosphere (Ainsworth et al. 2008; Matyssek et al. 2013a; b; Weigel et al. 2014). It is a secondary pollutant gas, being produced in atmospheric chemistry from other pollutants, such as CO, CH_4, and volatile organic compounds (oxygenated hydrocarbons, isoprene, monoterpenes, sesquiterpenes) especially at strong solar irradiation and with nitrogen oxides as catalysts. Increased atmospheric CO_2 interacts with O_3 pollution (King et al. 2013) where warming, reduced cloudiness, and increased radiation accelerate O_3 formation. O_3 as a greenhouse gas feeds back on warming effects (Matyssek et al. 2013c). Inhibition of life functions of plants by O_3 amounts to several tens of percent (Table 8). Impacts of O_3 on vegetation can have long-term adverse effects on ecosystems as reviewed in last year's volume of this series (vol. 76: Weigel et al. 2014). In general after increments in the second half of the twentieth century, the yield of crops is now declining and model predictions are alarming (Table 9). This is due to various factors such as vanishing resources, warming, and pollution. The comparison of Tables 8 and 9 shows that with an effective control and reduction of O_3 pollution alone, we could make a huge step forward to being capable of feeding the growing mankind.

3.1.5 Sociopolitical Ideologies and Wishful Thinking

In view of the threatening apocalypse of not being able to feed humankind in the future, several sociopolitical ideologies need to be checked. The question is not if they might not be wishful, but if we can afford them. Among them are issues such as a sexual moral against birth control, preferences of energy sources, and increasing meat consumption. A more direct pertinence to the green plant cover of the planet is given by "bio-agriculture," "energy plants," and the hostility against "genetically modified organisms."

"Bio"-agriculture

We can distinguish three different levels: (1) biodynamic agriculture, (2) organic farming, and (3) agroecology. (1) This is anthroposophic occultism. (2) This is highly controversial. It appears sympathetic to increasing populations. It has growing markets for its products. It occupies 1 % of the world's total farmland and 8 % in Europe (Maeder et al. 2002). However, it often has low yield (Stanhill 1990) while demanding increased land use. Without application of fertilizers it may exhaust nutrients of the soil. It can become destructive for soils and agroecosystems and may not be sustainable (Lüttge 2013). (3) This is a scientifically and socioeconomically serious interdisciplinary approach applying ecological principles to agriculture and promising sustained functioning of agroecosystems. It generates rewarding questions and urgent needs for intensified agronomic research (see Sect. 4).

"Energy Plants"

The slogan "food or fuel" (Cassman and Liska 2007) is addressing the dilemma that the growth of crops for production of bioenergy interferes with the challenge of being able to feed humankind. Even if not directly food crops are used for bioenergy, as it is actually the case currently, any growth of energy plants of whichever species will always expand anthropogenic management of land use. It will compete with the use of arable land as one of the most scarce resources (Sect. 3.1.2) or it will lead to further destruction of cultured land and remaining natural ecosystems. Tropical rain forest is presently destroyed at a large scale especially in Indonesia to grow oil palms for satisfying the growing demand of the market of so-called "renewable" energy from plants. However, there is nothing renewable about that; it is an environmental disaster, as the destroyed rain forests are not renewable.

An example from temperate zones is the cultivation of maize for energy. In Germany maize covers 26×10^3 km^2 of which 9×10^3 km^2 are for production of biogas. The visual aspect of extended landscapes has been changed by the cultivation of maize. The output is not convincing. The overall costs of bioenergy

Table 10 Factors of energy return on investments (EROI) for energy products from maize and some other sources

Energy source	EROI factor
Maize	
Bioethanol	1–1.6
Bio-butanol	<1
Biogas	1.4–4.8
Firewood	10
Photovoltaics	7
Wind turbines	18
Hydropower	≥100

Compiled after Haberl et al. (2012). A factor of 1 corresponds to equal costs and gains of energy

comprise the amount of energy invested (energy input) during biomass production and processing to technically usable forms of energy. The relations can be quantified by the factor of EROI, meaning energy return on investments. The EROI factors of energy from maize are compared with some other energy sources in Table 10. Among the plant sources, solid firewood is by far superior. Solar cells of photovoltaics are very often changing land use, as one can see it in the landscapes. Covering large areas with solar cells may also have effects on the climate, e.g., by affecting the albedo. However, their EROI is far better than that of energy plants. As compared to energy from biomass, the same amount of energy can be gained from solar cells with a hundredth of consumption of area (Schulze and Körner 2012). Maize cultures have negative effects on the environment and will not be sustainable at the long run. The monocultures dramatically reduce biodiversity. Maize is shallow rooting and has dramatic demands on the soil. It is susceptible to drought and needs a lot of water and fertilizers with all the disadvantages especially related to the anthropogenic pollution associated with the nitrogen cycle (e.g., N_2O; Sect. 3.1.4).

Given the very important limitations of bioenergy, its chances must be seriously questioned (Leopoldina 2012).

Genetically Modified Organisms

Conventional breeding remains essential for improving the performance and productivity of crop plants. However, it is limited by the amount of genetic diversity in the germplasm of crop species (Century et al. 2008). Conversely, the rapid progress in the identification of genes offers a broad array of opportunities for improving crops by genetic engineering (Table 11; Century et al. 2008). Plants genetically modified by recombinant DNA technology (genetically modified organisms = GMOs) provide a great potential for handling the problem of feeding the growing mankind (Lüttge 2013). In some countries, such as many countries of Europe (Germany, France, and others) and in Japan, there is a pronounced hostility against GMOs. In other countries, such as the USA, Canada, India, and China, the mood is favorable. Hostility is mainly emotional rather than rational. The benefits

Table 11 Potential benefits from genetically modified organisms

• **Increase yield** in various ways
• **Develop new crops resistant to stresses**, such as of temperature, drought, nutrient limitation, salinity, high irradiance, toxic metals, and oxidative stress in general, and therefore suitable for growth on less favorable land, thus helping to acquire new arable land and to economize fertilizers
• **Reduce losses to pathogens and herbivores**, thus helping to reduce the use of pesticides, where an important already realized modification provides transgenic crop plants expressing the gene for the toxin of *Bacillus thuringiensis* (Bt) against the larvae of insects
• **Increase the resistance to viruses**, e.g., the realized development of virus-resistant *Papaya*
• **Minimize** often very considerable **postharvest losses**, e.g., the already achieved development of tomato with disturbed ethylene biosynthesis and better ripening control
• **Increase food quality** by increasing contents of essential minerals (such as iron and selenium) and vitamins (such as β-carotene for vitamin A by the introduction of "golden" rice) or by decreasing contents of unhealthy saturated fatty acids (as in genetically engineered soybean (*Glycine max*), or colza (*Brassica napus*))

of GMOs (Table 11) include improved resistance to biotic and abiotic stresses, control of diseases, more efficient resource use, increase of yield, and better quality and nutritional value of crops. The latter is essential because food security is not only a quantitative but also very much a qualitative problem.

Very ambitious projects are making C_3 photosynthesis more effective and introducing C_4 photosynthesis into C_3 crops. A target for manipulating C_3 plants is ribulose-bisphosphate carboxylase/oxygenase (RUBISCO) with the aim to suppress the oxygenase activity in relation to the carboxylase affinity and, thus, reduce losses due to photorespiration (see Lüttge 2013). The most ambitious aim is to introduce features of C_4 photosynthesis with its two times higher rates of CO_2 fixation than that of C_3 photosynthesis into C_3 crop plants such as rice and wheat. It is vigorously pursued by several working groups (Surridge 2002; Mitchell and Sheehy 2006; Hibberd et al. 2008; Reynolds et al. 2009; von Caemmerer and Evans 2010; Westhoff and Gowik 2010; Sage and Zhu 2011). Some are optimistic that this may succeed, while others are skeptic.

Techniques of genetic modifications need up to 5 years until new crop plants are available if one gene is the target, e.g., with many examples in Table 11 and RUBISCO. Where the complexity of multigene cascades is involved, the difficulties are tremendous, and unfavorable side effects in the engineered plants may arise. This is the major reason of skepticism in view of the potential success of introducing the C_4 syndrome into C_3 crop species. Nevertheless, plants with multiple transgenic traits, so-called stacked gene hybrids, are already cultivated, e.g., maize hybrids with multiple genes for insect control (Que et al. 2010; Bruns 2014). The progress cannot be halted; cultivation of GMOs is occupying increasing areas of arable land. Currently, the major GMO crops are maize (*Zea mays*), sweet maize (*Zea mays* var. *rugosa*), canola (*Brassica napus*), cotton (*Gossypium hirsutum*), soybeans (*Glycine max*), papaya (*Carica papaya*), sugar beet (*Beta vulgaris*), and squash (*Cucurbita pepo*). They cover 1 million km^2. Some benefits

Table 12 Environmental benefits of biotechnology-engineered crops

Reduced pesticide application	224×10^6 kg
Reduced environmental pesticide impact	14 %
Reduced liberation of CO_2[a]	960×10^6 kg

Data of Brookes and Barfoot (2006), quoted from Ainsworth et al. (2008)

[a]Ca. 0.01 % of the current annual anthropogenic CO_2 production of 8×10^{12} kg/year

of GMOs are listed in Table 12. The time needed for multigene engineering may last from 10 to 20 years. Thus, the time runs short for being prepared for food supply of larger populations by the year 2050 (Long and Ort 2010). GMO research needs to be accelerated now for the rescue from future catastrophes of hunger. We cannot afford hostility against GMOs.

3.2 *Forestry Challenges*

Globally forests occupy 40×10^6 km^2, and they cover about 30 % of the land surface of our globe. Natural or seminatural forests amount to 93 % of this, and 7 % are tree plantations. Forests contribute ca. 50 % of the terrestrial net primary production (tropical forests 33 %) and contain ca. 45 % of the entire terrestrial carbon (tropical forests 25 %, temperate forests 10 %) (Bonan 2008). The annual destruction of forests is 130×10^3 km^2, and this contributes 12–15 % of the annual carbon emissions, with the greenhouse gas CO_2 causing global warming (Knoke and Hahn 2013). Hence, an essential issue for climate change mitigation and sustained management of the global heritage is the reduction of carbon emissions from deforestation and degradation, especially in the tropics. Increasing again the cover of our globe by forests is another challenge of forestry. We distinguish reforestation, which is the reestablishment of forests on land that has been covered by forests before, and afforestation, which is planting new forests on degraded land.

Reestablishment of forests appears essential for global sustainability. However, the interactions of forests and climate are highly complex and simple linear conclusions are not justified. Counteractions of the benefits are increased dangers of disturbances by fires and insect calamities (Canadell and Raupach 2008). In tropical forests strong evapotranspiration contributes to relative cooling and the increase of cloudiness. Vegetation and especially forests have a low surface albedo, i.e., low reflection of solar irradiation, which causes warming. In tropical forests this effect is offset by the effects of evapotranspiration. However, in boreal forests the trees transpire less but much reduce the high albedo of snow cover and may therefore contribute to global warming (Bonan 2008). Socioeconomic implications add to the complexity (Canadell and Raupach 2008). The challenges on forestry for sustainable management are formidable and require intensified research and organizational/administrative skills.

Table 13 Comparison of advantages and disadvantages of exotic tree plantations (Lüttge 2008)

	Disadvantages	Advantages
Management	Harmful effects on physical, chemical, and biological soil properties, increasing the danger of forest fires	Experience with propagation and silviculture
Productivity and diversity	Displacement of local native vegetation	Initial fast growth and wood production
Community relations	Susceptibility to epidemic diseases and pests	Facilitation effects • Microclimate • Reduction of erosion • Enhancement of litter and humus production

Table 14 Naturally regenerating native woody species (number of stems/ha) in a plantation of *Eucalyptus saligna* in relation to plantation age and in an adjacent native forest in Ethiopia with a light penetration of 12–51 % in the plantation and 1–77 % in the native forest (Feyera et al. 2002; Lemenih and Teketay 2004; Lüttge 2008)

Plantation	
11 years	3,575
22 years	10,100
27 years	18,650
Native forest	9,658

Besides these quantitative aspects, our view on the shaping of our environment by plants must also consider the qualitative aspects of afforestation and reforestation. Biodiversity is a key property of forests in this respect (Sect. 2.2). Plantations on degraded pasture and range lands especially in the tropics frequently use exotic tree species such as *Eucalyptus*, *Acacia*, *Casuarina*, *Cupressus*, and *Pinus* (Feyera et al. 2002). In monocultures of these trees, biodiversity is suppressed including the undergrowth of the artificial forests. These monocultures do not really affect rehabilitation of the degraded land as they have pronounced disadvantages for the land, e.g., the enormous water demand of *Eucalyptus*. Some other disadvantages are listed in Table 13. However, this need not necessarily be so, because the use of the exotic trees also brings advantages with it, which appropriate management can benefit from. Some of these properties are also listed in Table 13. For management the major advantage is the fact that forestry has accumulated experience with propagation and silviculture of exotic tree species so that the facilitation effects exerted by these trees can be used to establish increasing biodiversity. A most important factor for the regeneration of undergrowth is the penetration of light (Geldenhuys 1997). Thus, mixing with tree species of contrasting canopy architectures, phenologies and belowground demands as well as careful thinning are essential management options supporting natural processes of succession (van Wyk et al. 1995). High biodiversity of up to more than 175 species can be established in various plantations similar to native forests (Table 14) (Parrotta 1993, 1995; Da Silva et al. 1995; Geldenhuys 1997; Keenan et al. 1997; Feyera et al. 2002; Grams and Lüttge 2011).

A remarkable case study is the regeneration of a forest of the native tree *Podocarpus falcatus* out of a plantation of *Eucalyptus saligna* in the Munessa-Shashemene Forest at Degaga in the eastern escarpment of the great rift valley of Ethiopia. The forestry management applies regular coppicing of the *Eucalyptus* in about 7-year rhythms. This supplies the local population of farmers with timber and with logs to build their huts and thus also takes care of acceptance and essential socioeconomic aspects. It ascertains light penetration (Feyera et al. 2002; Grams and Lüttge 2011). Seeds of *P. falcatus* are imported by birds. Ecophysiological measurements show that the photosynthetic capacity of *P. falcatus* is not much inferior to that of the *Eucalyptus* but the *Eucalyptus* uses much more water. Its evapotranspiration is up to six times higher than that of *Podocarpus* (Lüttge et al. 2003; Fetene and Beck 2004). Thus, as the coppicing gives the *Eucalyptus* a certain handicap versus the *Podocarpus*, a natural forest can regenerate.

Eucalyptus monocultures can perturb water relations of entire landscapes with adverse effects on agriculture and challenging the water supply of cities. In the National Park of Mount Entoto at the rim of Addis Ababa in Ethiopia at 2,600–3,100 m a.s.l., foresters therefore have begun an endeavor of reforestation replacing *Eucalyptus* with a diversity of more than half a dozen of native tree species, such as *Acacia abyssinica*, *Hagenia abyssinica*, *Juniperus procera*, *Olea europaea*, *Podocarpus falcatus*, and *Prunus africana* (Ethiopia Heritage Trust, eht@ethionet.et; Lüttge et al. 2013).

When primary forest is destroyed, we can never restore it again in its previous originality and diversity. Above- and belowground complexity of the diversity of participating organisms is much too large to reconstruct it after it is lost. However, with the approaches alluded to above reforestation can establish secondary forest of high biological value. This is actually not so new. In the German terminology sustainment or sustainability is "Nachhaltigkeit" and the term in fact originated from forestry. It was coined 300 years ago in 1713 by the Saxon forester Hans Carl von Carlowitz in a crisis of supplies of energy and materials when the European forests were recklessly exploited and when he wrote: "Therefore the greatest art, science, diligence and industriousness in this country will be in establishing a sustained conservation and plantation of the wood so that it gives continuous, maintained and sustained utility, because it is an indispensible matter without which the country in its essence may not continue to exist"[3] (von Carlowitz 1713). Currently Germany carries much more forest than at the times of von Carlowitz.

[3] Wird derhalben die größte Kunst, Wissenschaft, Fleiß und Einrichtung hiesiger Lande darinnen beruhen, wie eine sothane Conservation und Anbau des Holtzes anzustellen, daß es eine continuirliche beständige und nachhaltende Nutzung gebe, weiln es eine unentbehrliche Sache ist, ohne welche das Land in seinem Esse nicht bleiben mag.

4 Ecological Principles in Agriculture and Forestry: Can Natural Self-Management and Anthropogenic Management of Nature Be Harmonized?

In ecology the distinction is made between autecology, considering the ecological performance of individual plants or species, and synecology, referring to the community level of biocenoses. Aiming at sustainability the great challenge is that we might become able to apply the self-management principles of natural ecosystems to agriculture and forestry. Autecological evaluations in fact are always applied, when agriculture considers relations of individual crop species or their cultivars to resources, such as water, mineral nutrients, CO_2, light, or space, and to predators and pests. Genotype improvements by traditional breeding and molecular engineering aim at enhancing the ecophysiological performance of cultivated plants (Sect. 3.1). Another question is if we can extend management of ecological agriculture and forestry by employing synecological principles to them. Physiological ecology in general so far has mainly considered the autecological performance of plants. However, an extension to physiological synecology has been suggested (Lüttge and Scarano 2004, 2007; Lüttge 2005).

A key aspect of synecology is biodiversity. The role of diversity for sustainment has been addressed as an important issue (Sect. 2.2). The diversity–productivity relation (Tilman et al. 1996) will be of essential interest in ecological agriculture and forestry. This includes soil biota and their highly diverse communities of microorganisms and animals (Matson et al. 1997). Authors have repeatedly stressed that "biodiversity … effects that are beneficial to humans were found in a wide range of ecosystem contexts" (Schläpfer and Schmid 1999) and "loss of biodiversity from local communities may be detrimental to the ecosystem goods and services on which humans ultimately depend" (Hector et al. 1999). A meta-analysis without doubt showed biodiversity although not under all circumstances but in principle to have manifold beneficial effects (Schläpfer and Schmid 1999). Another major source of interference with sustainability is the currently dominating land-use policy, e.g., by the monocultures of maize for biofuels (section "Energy Plants") and forest burning for land conversion. Managed cultures of crop plants or forestry oriented at natural biodiversity can essentially foster risk diversification and mitigation (Knoke and Hahn 2013).

Intercropping or cocultivation, e.g., of cereal and legume species (Bedoussac and Justes 2010) or mixed stands of forest trees (Scherrer et al. 2011; Pretzsch 2013; Pretzsch et al. 2013; Del Río et al. 2014) often only have the very low "diversity" of two species (Richards et al. 2010). Nevertheless, over-yielding can be observed even with only two tree species, e.g., in mixed growth of spruce and beech as compared to monocultures (Grams et al. 2012; Pretzsch et al. 2014). However, agricultural ecosystems with higher crop diversity are of utmost interest (Altieri 1999) for increasing sustainability of ecosystem services and productivity (Balvanera et al. 2006; Tilman et al. 2006). Hooper and Vitousek (1998) assessed the role of species and functional-group diversity in the use of the soil resources of mineral nutrients and water. They studied four functional groups in a grassland which were distinguished by their seasonal performance in using soil resources at maximum intensity. A significant positive relationship resulted between diversity and relative resource use, because of the complementarity of the seasonal performance of the different groups during the year.

The diversity–productivity hypothesis has been confirmed in studies of species-rich grasslands (Tilman et al. 1996, 2001; Hector et al. 1999) with facilitation (Sect. 2.6) playing a foremost role. Over-yielding as compared to monocultures has been found to be related to species number per se and not only to the presence of functional groups, such as legumes fixing atmospheric nitrogen (Tilman et al. 2001).

Is ecological agriculture a promising approach to the great challenge of avoiding the threatening apocalypse of not being able to feed increasing humankind on Earth (Sect. 3.1)? In many cases—although not generally—ecological agriculture might bring about some immediate or short-term reduction of productivity. However, a trade-off must not be overlooked which offers profitable long-term return by minimizing degradation and sustaining of productivity over prolonged periods of time.

5 Conclusions

Plants cover the Earth with a green mantle. They modify the global heritage of geological processes. Without the plants there would be no other advanced life. The terrestrial land on Earth would have remained a barren desert. In geological history, plants shaped the terrestrial environment dynamically, as waves of geological plant ages, the Palaeophytic, the Mesophytic, and the Neophytic, followed each other. Vegetation zones characterizing latitudinal zonobiomes now determine the image of our globe much more than the inorganic geological background. Vegetation depends on global climate and exerts feedbacks that influence the climate. The ecological services of plants and vegetation to life on Earth are fundamental as a consequence of the self-sustained internal organization and regulation of the natural plant-dominated ecosystems.

Natural self-management is driven by plasticity, diversity, and complexity, which is extrapolated in the grand Gaia hypothesis of James Lovelock to the entire biosphere as a self-organizing, self-sustaining supraorganism. Self-management is very dynamic and complex on various scalar levels of bio-systemic organization. It involves competition and facilitation among plant species, functional groups, and guilds. Sustainability is supported by comportment of plants carrying repair functions.

Humankind is depending on the natural plant-mediated ecosystem sustainability, but exploiting this. Anthropogenic management is juxtaposed to the natural self-management. Large-scale interference by agriculture and forestry as well as abiotic land use is unavoidable given the huge and expanding human population of the globe. Ever-increasing demands on intensified agriculture initiate vicious cycles of gains and losses challenging sustainability.

Natural self-management and anthropogenic management of nature are opposing each other. The question if they can be reconciled is existential for humankind. Man should not counteract natural self-management. Man should not assume that he "is the possessor of the planet, if not the owner," rather than the "tenant" (Lovelock 1979). By contrast humankind should accept to be part of nature as Lovelock writes that "the Gaia hypothesis implies that the stable state of our planet includes man as a part of, or partner in, a very democratic entity." Thus, the question is if ecological principles of natural self-management can be merged with anthropogenic management of nature. It appears that this is not only possible but mandatory. Tremendous efforts of scientific research and sociopolitical evolution are urgently required.

Acknowledgment I thank Rainer Matyssek for critically reading the manuscript and for many valuable comments.

References

Ainsworth EA, Rogers A, Leakey DB (2008) Targets for crop biotechnology in a future high-CO_2 and high-O_3 world. Plant Physiol 147:13–19

Altieri M (1999) The ecological role of biodiversity in agroecosystems. Agric Ecosyst Environ 74:19–31

Amzallag GN (2001) Data analysis in plant physiology: are we missing the reality? Plant Cell Environ 24:881–890

Andow DA (1991) Vegetational diversity and arthropod population response. Annu Rev Entomol 36:561–586

Atsatt PR, O'Dowd DJ (1976) Plant defense guilds. Science 193:24–29

Balvanera P, Pfisterer AB, Buchmann N, He J-S, Nakashizuka T, Raffaelli D, Schmid B (2006) Quantifying the evidence for biodiversity effects on ecosystem functioning and services. Ecol Lett 9:1146–1156

Bedoussac L, Justes E (2010) The efficiency of durum wheat-winter pea intercrop to improve yield and wheat grain protein concentration depends on N availability during early growth. Plant Soil 330:19–35

Bertness MD, Callaway RM (1994) Positive interactions in communities. Trend Ecol Evol 9:191–193

Blüthgen N, Klein A-M (2011) Functional complementarity and specialization: the role of biodiversity in plant pollinator interactions. Basic Appl Ecol 12:282–291

Bonan GB (2008) Forests and climate change: forcings, feedbacks, and the climate benefits of forests. Science 320:1444–1449

Brooker RW, Maestre MT, Callaway RM, Lortie CL, Cavieres LA, Kunstler G, Liancourt P, Tielbörger K, Travis JMJ, Anthelme F, Armas C, Coll L, Corcket E, Delzon S, Forey E, Kikvidze Z, Olofsson J, Pugnaire F, Quiroz CL, Saccone P, Schiffers K, Seifan M, Touzard B, Michalet R (2008) Facilitation in plant communities: the past, the present and the future. J Ecol 96:18–34

Brookes G, Barfoot P (2006) Global impact of biotech crops: socio-economic and environmental effects in the first years of commercial use. Ag Bio Forum 9:139–151

Bruno JF, Stachowicz JJ, Bertness MD (2003) Inclusion of facilitation into ecological theory. Trends Ecol Evol 18:119–125

Bruns HA (2014) Stacked gene hybrids were not found to be superior to glyphosate-resistant or non-GMO corn hybrids. Crop Manag 13. doi:10.2134/CM-2013-0012-RS

Bünnemann EK, Oberson A, Frossard E (eds) (2011) Phosphorus in action: biological processes in soil phosphorus cycling, vol 26, Soil biology. Springer, Heidelberg

Cahill JF (2013) Plant competition: can understanding trait-behavior linkages offer a new perspective on very old questions? In: Matyssek R, Lüttge U, Rennenberg H (eds) The alternatives growth and defense: resource allocation at multiple scales in plants. Nova Acta Leopoldina 114 (391): 115–125

Caldwell MM, Richards JH (1989) Hydraulic lift: water efflux from upper roots improves effectiveness of water uptake by deep roots. Oecologia 79:1–5

Callaway RM (1995) Positive interactions among plants. Bot Rev 61:306–349

Callaway RM (1998) Competition and facilitation on elevation gradients in subalpine forests of the northern Rocky Mountains, USA. Oikos 82:561–573

Callaway RM (2013) Facilitation, competition and the organization of plant communities. In: Matyssek R, Lüttge U, Rennenberg H (eds) The alternatives growth and defense: resource allocation at multiple scales in plants. Nova Acta Leopoldina 114(391): 147–157

Callaway RM, Brooker RW, Choler P, Kikvidze Z, Lortie CJ, Michalet R, Paolini L, Pugnaire FI, Newingham B, Aschehoug ET, Armas C, Kikodze D, Cook BJ (2002) Positive interactions among alpine plants increase with stress. Nature 417:844–848

Callaway RM, Walker LR (1997) Competition and facilitation: a synthetic approach to interactions in plant communities. Ecology 78:1958–1965

Canadell JG, Raupach MR (2008) Managing forests for climate change mitigation. Science 320:1456–1457

Caprez R, Spehn E, Nakhutrishvili G, Körner C (2012) Drought at erosion edges selects for a "hidden" keystone species. Plant Ecol Divers. doi:10.1080/17550874.2011.600343

Cassman KG, Liska AJ (2007) Food and Fuel for all: realistic or foolish? Biofuel Bioprod Bior 1:18–23

Century K, Reuber TL, Ratcliffe OJ (2008) Regulating the regulators: the future prospects for transcription-factor-based agricultural biotechnology products. Plant Physiol 147:20–29

Cisse L, Mrabet T (2004) World phosphate production: overview and prospects. Phosphorus Res Bull 15:212–225

Cordell D, Drangert JO, White S (2009) The story of phosphorus: global food security and food for thought. Glob Environ Chang 9:292–305

Cottingham KL, Brown BL, Lennon JT (2001) Biodiversity may regulate the temporal variability of ecological systems. Ecol Lett 4:72–85

Dangles O, Herrera M, Anthelme F (2013) Experimental support of the stress-gradient hypothesis in herbivore-herbivore interactions. New Phytol 197:405–408

Da Silva MC Jr, Scarano FR, De Souza CF (1995) Regeneration of an Atlantic forest formation in the understory of an *Eucalyptus grandis* plantation in south-eastern. Brazil J Trop Ecol 11:147–152

Davies TW, Jenkins SR, Kingham R, Hawkins SJ, Hiddink JG (2012) Extirpation-resistant species do not always compensate for the decline in ecosystem processes associated with biodiversity loss. J Ecol 100:1475–1481

Del Río M, Schütze G, Pretzsch H (2014) Temporal variation of competition and facilitation in mixed species forests in Central Europe. Plant Biol 16:166–176

Dias ATC, Scarano FR (2007) *Clusia* as nurse plant. In: Lüttge U (ed) *Clusia*—a woody neotropical genus with remarkable plasticity and diversity. Springer, Heidelberg, pp 55–72

Dias ATC, Bozelli RL, Darigo RM, Esteves FA, Santos HF, Figueiredo-Barros MP, Nunes MFQS, Roland F, Zamith LR, Scarano FR (2012) Rehabilitation of a bauxite tailing substrate in Central Amazonia: the effect of litter and seed addition on flood-prone forest restoration. Restor Ecol 20:483–489

Duarte HM, Gessler A, Scarano FR, Franco AC, de Mattos EA, Nahm M, Rennenberg H, Rodrigues PJFP, Zaluar HLT, Lüttge U (2005) Ecophysiology of six selected shrub species in different plant communities at the periphery of the Atlantic Forest of SE—Brazil. Flora 200:456–476

Durka W, Schulze E-D, Gebauer G, Voerkeliust S (1994) Effects of forest decline on uptake and leaching of deposited nitrate determined from ^{15}N and ^{18}O measurements. Nature 372:765–767

Edelman GM, Gally JA (2001) Degeneracy and complexity in biological systems. Proc Natl Acad Sci USA 98:13763–13768

Elton CS (1958) The ecology of invasions by animals and plants. Methuen, London

Fetene M, Beck E (2004) Water relations of indigenous versus exotic tree species, growing at the same site in a tropical montane forest in southern Ethiopia. Trees 18:428–435

Feyera S, Beck E, Lüttge U (2002) Exotic trees as nurse-trees for the regeneration of natural tropical forests. Trees 16:245–249

Franco AC, Nobel PS (1989) Effect of nurse plants on the microhabitat and growth of cacti. J Ecol 77:870–886

Fründ J, Linsenmair KE, Blüthgen N (2010) Pollinator diversity and specialization in relation to flower diversity. Oikos 119:1581–1590

Gehrig H, Gaussmann O, Marx H, Schwarzott D, Kluge M (2001) Molecular phylogeny of the genus *Kalanchoë* (Crassulaceae) inferred from the nucleotide sequences of the IST-1 and IST-2 regions. Plant Sci 160:827–835

Geldenhuys CJ (1997) Native forest regeneration in pine and eucalypt plantations in Northern Province, South Africa. For Ecol Manag 99:101–115

Gould JG (2002) The structure of evolutionary theory. Harvard University Press, Cambridge, MA

Grams TEE (2013) A space-related perspective on plant-plant competition. In: Matyssek R, Lüttge U, Rennenberg H (eds) The alternatives growth and defense: resource allocation at multiple scales in plants. Nova Acta Leopoldina 114(391): 127–134

Grams TEE, Lüttge U (2011) Space as a resource. Prog Bot 72:349–370

Grams TEE, Daigo MJ, Winkler JB, Gayler S, Matyssek R (2012) Growth and space use in competitive interactions between juvenile trees. In: Matyssek R, Schnyder H, Oßwald W, Ernst D, Munch JC, Pretzsch H (eds) Growth and defence in plants. Resource allocation at multiple scales, vol 220, Ecological studies. Springer, Heidelberg, pp 273–286

Grünhage L, Matyssek R, Wieser G, Häberle K-H, Leuchner M, Menzel A, Dieler J, Pretzsch H, Grimmeisen W, Zimmermann L, Raspe S, Schröder M (2013) Flux based ozone risk assessment for adult beech and spruce forests. In: Matyssek R, Clarke N, Cudlin P, Mikkelsen TN, Tuovinen J-P, Wieser G, Paoletti E (eds) Climate change, air pollution and global challenges. Understanding and perspectives from forest research, vol 13, Developments in environmental sciences. Elsevier, Amsterdam, pp 251–266

Haberl H, Körner C, Lauk C, Schmid-Staiger U, Smetacek V, Schulze E-D, Thauer RK, Weiland P, Wilson K (2012) The availability and sustainability of biomass as an energy source. In: Bioenergy—chances and limits. German National Academy of Sciences Leopoldina, Halle (Saale), www.leopoldina.org, pp 9–42

Hastings A, Byers JE, Crooks JA, Cuddington K, Jones CG, Lambrinos JG, Talley TS, Wilson WG (2006) Ecosystem engineering in space and time. Ecol Lett 10:153–164

Hector A, Schmid B, Beierkuhnlein C, Caldeira MC, Diemer M, Dimitrakopoulos PG, Finn JA, Freitas H, Giller PS, Good J, Harris R, Högberg P, Huss-Danell K, Joshi J, Jumpponen A, Körner C, Leadley PW, Loreau M, Minns A, Mulder CPH, O'Donovan G, Otway SJ, Pereira JS, Prinz A, Read DJ, Scherer-Lorenzen M, Schulze E-D, Siamantziouras A-SD, Spehn EM, Terry AC, Troumbis AY, Woodward FI, Yachi S, Lawton JH (1999) Plant diversity and productivity experiments in European grasslands. Science 286:1123–1127

Hibberd JM, Sheehy JE, Langdale JA (2008) Using C_4 photosynthesis to increase the yield of rice—rationale and feasibility. Curr Opin Plant Biol 11:228–231

Hodge A (2009) Root decisions. Plant Cell Environ 32:628–640

Hodge A (2010) Roots: the acquisition of water and nutrients from the heterogeneous soil environment. Prog Bot 71:307–337
Höök M, Li J, Oba N, Snowden S (2011) Descriptive and predictive growth curves in energy system analysis. Nat Resour Res 20:103–116
Hooper DU, Vitousek PM (1998) Effects of plant composition and diversity on nutrient cycling. Ecol Monogr 68:121–149
Huck C, Körner C, Hitbrunner E (2013) Plant species dominance shifts across erosion edge-meadow grassland soils. Oecologia 171:693–703
Hütt M-T, Lüttge U (2005) Network dynamics in plant biology: current progress in historical perspective. Prog Bot 66:277–310
Keenan R, Lamb D, Woldring O, Irvine T, Jensen R (1997) Restoration of plant biodiversity beneath tropical tree plantations in Northern Australia. For Ecol Manag 99:117–131
King J, Liu L, Aspinwall M (2013) Tree and forest responses to interacting elevated atmospheric CO_2 and tropospheric O_3: a synthesis of experimental evidence. In: Matyssek R, Clarke N, Cudlin P, Mikkelsen TN, Tuovinen J-P, Wieser G, Paoletti E (eds) Climate change, air pollution and global challenges. Understanding and perspectives from forest research, vol 13, Developments in environmental sciences. Elsevier, Amsterdam, pp 179–208
Knoke T, Hahn A (2013) Global change and the role of forests in future land-use systems. In: Matyssek R, Clarke N, Cudlin P, Mikkelsen TN, Tuovinen J-P, Wieser G, Paoletti E (eds) Climate change, air pollution and global challenges. Elsevier, Amsterdam, pp 569–588
Körner C (2012) Biological diversity—the essence of life and ecosystem functioning. Nova Acta Leopoldina 116(394):147–159
Körner C (2013) Growth controls photosynthesis—mostly. Nova Acta Leopoldina 114 (391):273–283
Kraft G (1884) Beiträge zur Lehre von den Durchforstungen, Schlagstellungen und Lichtungshieben. Klindworth's Verlag, Hannover
Küppers M (1989) Ecological significance of above-ground architectural patterns in woody plants—a question of cost-benefit relationships. Trends Ecol Evol 4:375–379
Lacerda LD, Araujo DSD, Maciel NC (1993) Dry coastal ecosystems of the tropical Brazilian coast. In: van der Maarel E (ed) Dry coastal ecosystems: Africa; America, Asia and Oceania. Elsevier, Amsterdam, pp 477–493
Lemenih M, Teketay D (2004) Restoration of native forest flora in the degraded highlands of Ethiopia: constraints and opportunities. Sinet Ethiop J Sci 27:75–90
Leopoldina (2012) Bioenergy—chances and limits. German National Academy of Sciences Leopoldina, Halle (Saale), www.leopoldina.org
Li L, Tilman D, Lambers H, Zhang F-S (2014) Plant diversity and overyielding: insights from belowground facilitation of intercropping in agriculture. New Phytol 203:63–69
Lin Y, Berger U, Grimm V, Ji Q-R (2012) Differences between symmetric and asymmetric facilitation matter: exploring the interplay between modes of positive and negative plant interactions. J Ecol 100:1482–1491
Long S, Ort DR (2010) More than taking the heat: crops and global change. Curr Opin Plant Biol 13:241–248
Lovelock J (1979) Gaia. A new look at life on Earth. Oxford University Press, Oxford
Lovelock J (2009) The vanishing face of Gaia—a final warning. Basic Books, New York, NY
Lüttge U (1995a) Ecophysiological basis of the diversity of tropical plants: the example of the genus Clusia. In: Heinen HD, San José JJ, Caballero-Arias H (eds) Nature and human ecology in the neotropics. Scientia Guaianae 5:23–26
Lüttge U (1995b) *Clusia*: Ein Modellfall der ökologischen Plastizität in einer tropischen Gattung. In: Rundgespräche der Kommission für Ökologie BAdW. 10: 173–186. Bayerische Tropenforschung einst und jetzt. Dr. Pfeil München
Lüttge U (2000) Photosynthese-Physiotypen unter gleichen Morphotypen, Spezies und bei Klonen: Kann ökophysiologische Plastizität zur Entstehung von Diversität beitragen? Ber Reinhold Tüxen Ges 12:319–334

Lüttge U (2005) Genotypes—phenotypes—ecotypes: relations to crassulacean acid metabolism. Nova Acta Leopoldina 92(342):177–193
Lüttge U (2008) Physiological ecology of tropical plants, 2nd edn. Springer, Berlin
Lüttge U (2010) Struggle of plants with crassulacean acid metabolism (CAM) in topical environments under the action of dynamic networks of stressors. AoB PLANTS, 2010: 1–15. doi:10.1093/aobpla/plq005. http://aobplants.oxfordjournals.org/
Lüttge U (2013) The planet Earth: can it feed nine billion people? Nova Acta Leopoldina 114 (391):345–364
Lüttge U, Scarano FR (2004) Ecophysiology. Rev Bras Bot 27:1–10
Lüttge U, Scarano FR (2007) Synecological comparisons sustained by ecophysiological fingerprinting of intrinsic photosynthetic capacity of plants as assessed by measurements of light response curves. Braz J Bot 30:355–364
Lüttge U, Berg A, Fetene M, Nauke P, Peter D, Beck E (2003) Comparative characterization of photosynthetic performance and water relations of native trees and exotic plantation trees in an Ethiopian forest. Trees 17:40–50
Lüttge U, Garbin ML, Scarano FR (2013) Evo-devo-eco and ecological stem species: potential repair systems in the planetary biosphere crisis. Prog Bot 74:191–212
Lüttge U, Kluge M, Thiel G (2010) Botanik. Die umfassende Biologie der Pflanzen. Wiley-VCH, Weinheim
Luyssaert S, Schulze E-D, Börner A, Knohl A, Hessenmöller D, Law BE, Ciais P, Grace J (2008) Old-growth forests as global carbon sinks. Nature 455:213–215
Maeder P, Fliessbach A, Dubois D, Gunst L, Fried P, Niggli U (2002) Soil fertility and biodiversity in organic farming. Science 296:1694–1697
Magnússon B, Magnússon SH, Fridriksson S (2009) Developments in plant colonization and succession on Surtsey during 1999–2008. Surtsey Res 12:57–76
Masselter T, Bauer G, Gallenmüller F, Haushahn T, Poppinga S, Schmitt C, Seidel R, Speck O, Thielen M, Speck T (2012) Biomimetic products. In: Bar-Cohen Y (ed) Biomimetics. Nature-based innovation. CRC-Press, Boca Raton, FL, pp 377–429
Matson PA, Parton WJ, Power AG, Swift MJ (1997) Agricultural intensification and ecosystem properties. Science 277:504–509
Matyssek R, Lüttge U (2013) Gaia: the planet holobiont. Nova Acta Leopoldina 114 (391):325–344
Matyssek R, Fromm J, Rennenberg H, Roloff A (2010) Biologie der Bäume. Ulmer, Stuttgart
Matyssek R, Clarke N, Cudlin P, Mikkelsen TN, Tuovinen J-P, Wieser G, Paoletti E (eds) (2013a) Climate change, air pollution and global challenges. Understanding and perspectives from forest research, vol 13, Developments in environmental sciences. Elsevier, Amsterdam
Matyssek R, Clarke N, Cudlin P, Mikkelsen TN, Tuovinen J-P, Wieser G, Paoletti E (2013b) Climate change, air pollution and global challenges: understanding and perspectives from forest research. In: Matyssek R, Clarke N, Cudlin P, Mikkelsen TN, Tuovinen J-P, Wieser G, Paoletti E (eds) Climate change, air pollution and global challenges. Understanding and perspectives from forest research, vol 13, Developments in environmental sciences. Elsevier, Amsterdam, pp 3–16
Matyssek R, Wieser G, Fleischmann F, Grünhage L (2013c) Ozone research, *quo vadis*? Lessons from the free-air canopy fumigation experiment at Kranzberg Forest. In: Matyssek R, Clarke N, Cudlin P, Mikkelsen TN, Tuovinen J-P, Wieser G, Paoletti E (eds) Climate change, air pollution and global challenges. Understanding and perspectives from forest research, vol 13, Developments in environmental sciences. Elsevier, Amsterdam, pp 103–129
Medina E, Cram WJ, Lee HSJ, Lüttge U, Popp M, Smith JAC, Diaz M (1989) Ecophysiology of xerophytic and halophytic vegetation of a coastal alluvial plain in northern Venezuela. I. Site description and plant communities. New Phytol 111:233–243
Mitchell PL, Sheehy JE (2006) Supercharging rice photosynthesis to increase yield. New Phytol 171:688–693

Novoplansky A (2009) Picking battles wisely: plant behaviour under competition. Plant Cell Environ 32:726–741

Oberson A, Pypers P, Bünemann EK, Frossard E (2011) Management impacts on biological phosphorus cycling in cropped soils. In: Bünnemann EK, Oberson A, Frossard E (eds) Phosphorus in action: biological processes in soil phosphorus cycling, vol 26, Soil biology. Springer, Heidelberg, pp 431–458

Oksanen E, Keski-Saari S, Kontunen-Soppela S, Keinänen M (2013) Metabolomics and transcriptomics increase our understanding about defence responses and genotypic differences of northern deciduous trees to elevating ozone, CO_2 and climate warming. In: Matyssek R, Clarke N, Cudlin P, Mikkelsen TN, Tuovinen J-P, Wieser G, Paoletti E (eds) Climate change, air pollution and global challenges. Understanding and perspectives from forest research, vol 13, Developments in environmental sciences. Elsevier, Amsterdam, pp 309–329

Pan Y, Birdsey RA, Fang J, Houghton R, Kauppi PE, Kurz WA, Phillips OL, Shvidenko A, Lewis SL, Canadell JG, Ciais P, Jackson RB, Pacala SW, McGuire AD, Piao S, Rautiainen A, Sitch S, Hayes D (2011) A large and persistent carbon sink in the world's forests. Science 333:988–993

Pandey V, Oksanen E, Singh N, Sharma C (2013) Impacts of air pollution and climate change on plants: implications for India. In: Matyssek R, Clarke N, Cudlin P, Mikkelsen TN, Tuovinen J-P, Wieser G, Paoletti E (eds) Climate change, air pollution and global challenges. Understanding and perspectives from forest research, vol 13, Developments in environmental sciences. Elsevier, Amsterdam, pp 391–409

Parrotta JA (1993) Secondary forest regeneration on degraded tropical lands. The role of plantations as "foster ecosystems". In: Lieth H, Lohmann M (eds) Restoration of tropical forest ecosystems. Kluver, Dordrecht, pp 66–73

Parrotta JA (1995) Influence of overstory composition on understory colonization by native species in plantations on a degraded tropical site. J Veg Sci 6:627–636

Pretzsch H (2010) Forest dynamics, growth and yield. Springer, Heidelberg

Pretzsch H (2013) Facilitation and competition in mixed-species forests analyzed along an ecological gradient. Nova Acta Leopoldina 114(391):159–174

Pretzsch H, Roetzer T, Matyssek R, Grams TEE, Häberle K-H, Pritsch K (2014) Mixed Norway spruce (*Picea abies* [L.] Karst) and European beech (*Fagus sylvatica* [L.]) stands under drought: from reaction pattern to mechanisms. Trees 28:1305–1321

Pretzsch H, Schütze G, Uhl E (2013) Resistance of European tree species to drought stress in mixed versus pure forests: evidence for stress release by inter-specific facilitation. Plant Biol 15:483–495

Que Q, Chilton M-DM, de Fontes CM, He C, Nuccio M, Zhu T, Wu Y, Chen JS, Shi L (2010) Trait stacking in transgenic crops: challenges and opportunities. GM Crops 1:220–229

Reynolds M, Foulkes MJ, Slafer GA, Berry P, Parry MAJ, Snape JW, Angus WJ (2009) Raising yield potential in wheat. J Exp Bot 60:1899–1918

Richards AE, Forrester DI, Bauhus J, Scherer-Lorenzen M (2010) The influence of mixed tree plantations on the nutrition of individual species: a review. Tree Physiol 30:1192–1208

Richards JH, Caldwell MM (1987) Hydraulic lift: substantial nocturnal water transport between soil layers by *Artemisia tridentata* roots. Oecologia 73:486–489

Sage RF, Zhu X-G (2011) Exploiting the engine of C_4 photosynthesis. J Exp Bot 62:2989–3000

Scarano FR (2002) Structure, function and floristic relationships of plant communities in stressful habitats marginal to the Brazilian Atlantic rain forest. Ann Bot 90:517–524

Scarano FR (2009) Plant communities at the periphery of the Atlantic rain forest: rare-species bias and its risks for conservation. Biol Conserv 142:1201–1208

Scarano FR, Rios RI, Esteves FA (1998) Tree species richness, diversity and flooding regime: case studies of recuperation after anthropic impact in Brazilian flood-prone forests. Int J Ecol Environ Sci 24:223–225

Schenk HJ (2006) Root competition beyond resource depletion. J Ecol 94:725–739

Scherer-Lorenzen M, Körner C, Schulze E-D (eds) (2005) Forest diversity and function. Temperate and boreal systems, vol 176, Ecological studies. Springer, Berlin

Scherrer D, Bader MKF, Körner C (2011) Drought-sensitivity ranking of deciduous tree species based on thermal imaging of forest canopies. Agric For Meteorol 151:1632–1640
Schläpfer F, Schmid B (1999) Ecosystem effects of biodiversity: a classification of hypotheses and exploration of empirical results. Ecol Appl 9:893–912
Schlenker W, Roberts MJ (2009) Nonlinear temperature effects indicate severe damages to U.S. crop yields under climate change. Proc Natl Acad Sci USA 106:15594–15598
Schulze E-D, Körner C (2012) Nettoprimärproduktion und Bioenergie. In: Bioenergy—chances and limits. German National Academy of Sciences Leopoldina, Halle (Saale), www.leopoldina.org, pp 90–100
Solbrig OT (1994) Plant traits and adaptive strategies: their role in ecosystem function. In: Schulze E-D, Mooney HA (eds) Biodiversity and ecosystem function, vol 99, Ecological studies. Springer, Berlin, pp 97–116
Souza GM, Lüttge U (2014) Stability as a phenomenon emergent from plasticity—complexity—diversity in eco-physiology. Prog Bot 76:211–239
Souza GM, Ribeiro RV, Prado CHBS, Damineli DSC, Sato M, Oliveira MS (2009) Using network connectance and autonomy analyses to uncover patterns of photosynthetic responses in tropical woody species. Ecol Complex 6:15–26
Speck T, Bauer G, Flues F, Oelker K, Rampf K, Schüssele AC, von Tapavicza M, Bertling J, Luchgsinger R, Nellesen A, Schmidt AM, Mühlhaupt R, Speck O (2013a) Bio-inspired self-healing materials. In: Fratzl P, Dunlop JWC, Weinkamer R (eds) RSC Smart materials No. 4. Materials design inspired by nature: function through inner architecture. The Royal Society of Chemistry, pp 359–389
Speck T, Mühlhaupt R, Speck O (2013b) Self-healing in plants as bio-inspiration for self-repairing polymers. In: Binder WH (ed) Self-healing polymers. From principles to applications. Wiley-VCH, Weinheim, pp 61–89
Stanhill G (1990) The comparative productivity of organic agriculture. Agric Ecosyst Environ 30:1–26
Surridge C (2002) Agricultural biotech: the rice squad. Nature 416:576–578
Tilman D, Reich PB, Knops J, Wedin D, Mielke T, Lehman C (2001) Diversity and productivity in a long-term grassland experiment. Science 294:843–845
Tilman D, Reich PB, Knops JMH (2006) Biodiversity and ecosystem stability in a decade-long grassland experiment. Nature 441:629–632
Tilman D, Wedin D, Knops J (1996) Productivity and sustainability influenced by biodiversity in grassland ecosystems. Nature 379:718–720
Vance CP, Chiou T-J (eds) (2011) Focus issue on phosphorus plant physiology. Plant Physiol 156:987–1086
van Wyk GF, Everard DA, Geldenhuys CJ (1995) Forest ecotone development and succession: experimental results and guidelines for forest rehabilitation and protection. Report FOR DEA 876. Division of Forest Science and Technology, CSIR, Pretoria
von Braun J (2011) Das Welternährungsproblem heute und in der kommenden Generation. Akademie aktuell Bayer Akad Wiss 1:24–27
von Caemmerer S, Evans PR (2010) Enhancing C_3 photosynthesis. Plant Physiol 154:589–592
von Carlowitz HC (1713) Sylvicultura Oeconomica oder Haußwirthliche Nachricht und Naturmäßige Anweisung zur Wilden Baum-Zucht. J.F. Braun, Leipzig
Walter H, Breckle SW (1984) Ökologie der Erde. 2. Spezielle Ökologie der tropischen und subtropischen Zonen. Gustav Fischer, Stuttgart
Weigel D, Jürgens G (2002) Stem cells that make stems. Nature 415:751–754
Weigel HJ, Bergmann E, Bender J (2014) Plant-mediated ecosystem effects of tropospheric ozone. Prog Bot 76:395–438
Weiner CN, Wemer M, Linsenmair KE, Blüthgen N (2011) Land use intensity in grasslands: changes in biodiversity, species composition and specialization in flower visitor networks. Basic Appl Ecol 12:292–299

West-Eberhard MJ (1986) Alternative adaptations, speciation, and phylogeny (a review). Proc Natl Acad Sci USA 83:1388–1392

West-Eberhard MJ (1989) Phenotypic plasticity and origins of diversity. Annu Rev Ecol Syst 20:249–278

West-Eberhard MJ (2003) Developmental plasticity and evolution. Oxford University Press, Oxford

Westhoff P, Gowik U (2010) Evolution of C_4 photosynthesis—looking for the master switch. Plant Physiol 154:598–601

Williams K, Caldwell MM, Richards JH (1993) The influence of shade and clouds on water potential: the buffered behavior of hydraulic lift. Plant Soil 157:83–95

Shaping Theoretic Foundations of Holobiont-Like Systems

Wolfgang zu Castell, Frank Fleischmann, Tina Heger, and Rainer Matyssek

Contents

Abstract Acknowledging the fact that organisms never evolve in isolation, Zilber-Rosenberg and Rosenberg emphasized the concept of the holobiont, comprising a host organism together with all of its associated microorganisms. Considering the holobiont as being a unit of selection, the hologenome theory of evolution then leads to incorporate Lamarckian aspects into the cycle of adaptation and selection. Nevertheless, the concept of the holobiont carries an implicit temporal dependency. Similar contingencies can be identified for other ideas, e.g., the notion of a supraorganism. Building on ideas from computational thermodynamics and information theory leads to the concept of a holobiont-like system. This notion aims at

W. zu Castell (✉)
Scientific Computing Research Unit, Helmholtz Zentrum München, German Research Center for Environmental Health GmbH, Ingolstädter Landstraße 1, 85764 Neuherberg, Germany

Department of Mathematics, Technische Universität München, München, Germany
e-mail: castell@helmholtz-muenchen.de

F. Fleischmann
Section Pathology of Woody Plants, Center of Life and Food Sciences Weihenstephan, Technische Universität München, München, Germany

T. Heger
Restoration Ecology, Center of Life and Food Sciences Weihenstephan, Technische Universität München, München, Germany

R. Matyssek
Technische Universität München, Hans-Carl-von-Carlowitz-Platz, Wissenschaftszentrum Weihenstephan, 85354 Freising, Germany

U. Lüttge et al. (eds.), *Progress in Botany*, Progress in Botany 77,
DOI 10.1007/978-3-319-25688-7_7

capturing the essentials of a system of interacting biological agents, being driven by an evolutionary algorithm. The concept can be applied upon several scales, allowing to consider the holobiont sensu stricto as well as full ecosystems. It nicely frames within the metaphor of the adaptive cycle and, thus, leads to deeper insights into sustainability of biological systems.

1 The Holobiont Sensu Stricto

Natural organisms never evolve in isolation. Every individual is part of a rich network of interacting organisms determining the course of the individual's development. In particular, higher plants and higher animals are accompanied by an enormous variety of microorganisms which altogether form a coenobitic (i.e., symbiotic[1] sensu stricto) association. In the diversity of living organisms, we observe results from adaptation to a dynamically changing environment. The process realizing adaptation in natural systems is evolution taken in an algorithmic sense (see Dennett 1996). Evolution as a process is anchored in three crucial components: variation, heredity, and differential reproduction (Mayr 2002). Thus, evolution inherently operates on multiple scales ranging between the level of genes, being the carrier of heritable information, and that of populations of phenotypes unfolding the information. Although it is widely accepted that natural selection is intrinsic to evolution, there is quite some debate on what is the effective biological level of selection. Growing insights into epigenetic mechanisms reveal a deeper interrelationship between genes as one mode of coding information combined with environmental specifications (Goldberg et al. 2007).

Acknowledging the observation that selection extends beyond the level of the individual within coenobitic associations, Zilber-Rosenberg and Rosenberg (2008) formulated the hologenome theory of evolution. The central concept within this theory is the holobiont, originally defined as close association, temporal or permanent, of members of different species of organisms (Margulis 1993). Through the later use of the term by Rosenberg et al. (2007), a holobiont today is to be understood as the functional entity consisting of a "host organism" with all its associated symbiotic microorganisms (cf. Rosenberg et al. 2007). Hereby, Zilber-Rosenberg and Rosenberg deliberately advance beyond the conventional concept of coevolution. It is the genetic potential of the full association (see Fig. 1) which is subject to selection, and this potential changes throughout the life cycle of the host, i.e., via amplification or acquisition of microorganisms. Note that the classical definition of a holobiont focuses on host-microbe interactions, thereby neglecting

[1] We use the word "symbiotic" in the original sense of de Bary, i.e., in a neutral sense. Thus, symbiotic relations include mutualistic, commensalistic, and parasitic relationships.

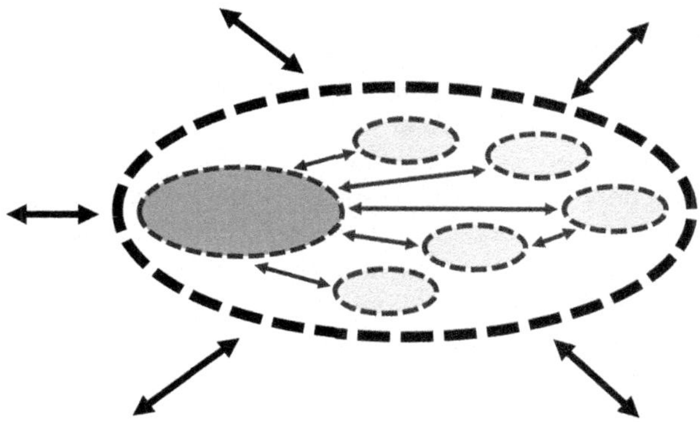

Fig. 1 The holobiont sensu stricto being defined as a host organism (*larger ellipse*) and all its associated microorganisms (*smaller ellipses*) together with all their interactions. Note that the boundary defining the system is contingent on the perspective of the observer

other associations, which also shape the host organism, particularly when considering plants.

According to Zilber-Rosenberg and Rosenberg (2008; Rosenberg et al. 2010), the hologenome theory is based on four general principles:

1. All animals and plants establish symbiotic (sensu coenobitic; see above) relationships with microorganisms.
2. Associated microorganisms are transmitted across developmental stages of the holobiont.
3. The association between host and microorganisms as an entity determines the fitness of the holobiont within its environment.
4. The hologenome is subject to variation through changes in the host and/or its microbial associates.

Interestingly, the last principle carries some Lamarckian aspects since variation in the holobiont can occur via additional mechanisms which operate beyond the level of the gene. These are amplification of existing microorganisms within the holobiont and acquisition of novel strains from the environment as well as abandonment of strains through horizontal drift (Rosenberg et al. 2009) as a consequence of information gained (i.e., "experience learnt") from environmental stress, just to name some examples.

Considering the three basic components of mutation, heredity, and selection, the four stated principles are well rooted in evolution theory, addressing the question of the representative unit of selection. Starting with neo-Darwinism, the gene-centered perspective on selection (Dobzhansky 1937; Mayr 1942; Dawkins 1976) aimed at explaining evolutionary adaptation solely on the level of the gene, since this is the level at which heritable information is transmitted to the next generation through the gametes. Somewhat more precise, adaptation is considered as a process changing the frequency of genetic alleles—more general chromosomal DNA—within a population. Opposing the gene-centric view, Wilson and Sober (1994) introduced a theory of multilevel selection, embracing the debate on group selection (Wynne-Edwards 1962, 1986; Wilson and Wilson 2008) versus kin selection (Fisher 1930;

Haldane 1955; Hamilton 1963, 1964). Both approaches aim at clarifying the puzzle of explaining cooperative behavior of social insects via Darwinian selection. Whereas kin selection is based on the concept of inclusive fitness, expanding the notion of fitness beyond the individual by taking fitness of related individuals into account, critics argue that the phenomenon could likewise be explained by accepting the point of view that natural selection reinforces traits that are favorable for a group of organisms in general. In some sense, the hologenome theory follows the gene-centric paradigm, considering the hologenome as the sum of inheritable information of the holobiont. Since associations among individuals, however, incorporate associations among individuals of the same kind, the concept of group selection is included to some extent.

Although being compelling, the hologenome concept raises questions. One of them results from the conceptual challenge regarding heredity. For example, the life cycles of eukaryotic and prokaryotic associates typically vary, proceeding at different timescales. Hence, what are the generations being addressed in the second principle stated above? The obvious answer, i.e., generations being defined by the host sensu Zilber-Rosenberg and Rosenberg, reinforces a bias already immanent in the definition of the holobiont. The so-called host, i.e., the dominant eukaryont, would be singled out from the association. This is an arbitrary choice made by the observer studying a particular type of complex system. Considering plants and associated mycorrhizal fungi, there is no logical necessity to rule out one over the other. Changing perspective in host-mycorrhiza interaction supports our view. While originally mycorrhiza has been considered to mainly unlock soil resources for the plant, there is growing evidence that the plant plays a major role in shaping the mycorrhizal community (Rennenberg et al. 2009; Pena et al. 2010).

Second, there is another inherent dependency on timescales. What exactly is considered as being an association—and what is "one generation" of an association? Do we enclose short-term events, e.g., infections through bacterial pathogens? We probably should, because if exemplifying the human immune system, being strongly path dependent, i.e., contingent on a unidirectional cause-effect chain, its development results from all contacts with microbial components (Eberl 2010).

Thus, we argue that the host-centric view within the common definition of the holobiont (sensu Zilber-Rosenberg and Rosenberg) is a choice made by the observer, resulting from the paradigm of observer dependence. The latter is an inherent property of any complex system (see Kay 2008) and, thus, conceptually unavoidable. Second, widening the concept of heredity beyond transmission of information across discrete generations (typically defined by meiotic cycles of individuals, maybe synchronized within populations), in particular through the incorporation of Lamarckian aspects (Rosenberg et al. 2009; Gilbert 2011; see above), there is no need to restrict the concept of the holobiont to the level of organisms. Such a claim gains further support by the argument that the notion of an organism is an intrinsically human concept, in particular when considering microorganisms (Ruse 1989; Rosselló-Mora and Amann 2001; Pepper and Herron 2008).

Similarly, other concepts of multi-organisms also carry intrinsic scale dependencies. A supraorganism[2] is defined as "a collection of single creatures that together possess the functional organization implicit in the formal definition of organism" (Wilson and Sober 1989, p. 339). The definition implicitly refers back to the scale of the organ level. Although there might be no consistent definition of an organ (see Ruse 1989), the commonly accepted concept of an organ implies some kind of local boundary. Thus, individuals comprising an organ-like component of a supraorganism are intrinsically assumed to be within spatial vicinity.

Both objections, i.e., the host-centric view and the gene-centric concept of heredity, thus are particular instances of scale dependency, scale being taken here in a spatiotemporal sense. Contrasting with both approaches, i.e., holobiont and supraorganism, we aim for a concept which is scale invariant, both in the temporal and spatial sense. It is through implicit scale dependency that difficulties arise which lead to controversy and conflicting debate. In order to derive such a scale-invariant concept, we employ ideas from information theory and computational thermodynamics. Neither aspect is new in ecology (see, e.g., Jørgensen and Svirezhev 2004). Through employing abstraction, we are able to extract essential characteristics of biological processes. Abstraction creates the grounds to us for recognizing relevant characteristics at various scales and in different scenarios, i.e., irrespective of scale-related process specificities.

After presenting a formal definition of our concept, we will introduce the major biological definitions needed from the perspective of computational thermodynamics in the following section. We will then proceed with arguing in favor of the suitability of the proposed concept. Finally, we will provide some hypotheses and show how they can be deduced from the theoretical foundations laid out in the preceding sections.

2 Defining a "Holobiont-Like" System

We propose the concept of a **holobiont-like system**, an evolving system of interacting agents. Agents are understood to be biological units (e.g., organs, individual organisms, or populations) that interact with each other and their (local) environments (e.g., soil structure, chemical neighborhood, other organisms within close vicinity). It is assumed that no single agent or group has a controlling function or "full view" of the system. It is through the interactions that the association of agents becomes observable as a system (see Fig. 2). Although, from the point of view of the observer it might seem like one associate or group

[2] Note that etymologically, the Latin word "supra" means "higher" in the sense of ordination, whereas "super" implies a spatial order. Thus, in contrast to the mainly used notion of "superorganism," we prefer to stay with the notion of a "supraorganism."

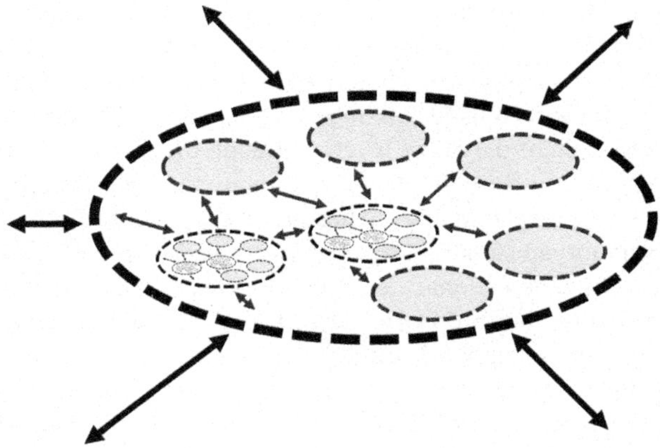

Fig. 2 The holobiont-like system as being defined as a system of interacting biological agents (*ellipses*) adapting through generic evolution. In contrast to the holobiont sensu stricto, no agent is singled out. The concept naturally allows being applied recursively, leading to a nested set of systems

is controlling the system, we consider systems lacking the prescribed role of a central conductor.

The second characteristic property of a holobiont-like system is its ability to adapt within a dynamic environment. We will consider evolution in a rather abstract, algorithmic sense. Hereby, adaptation is understood to be the generic property of a system to internally capture information (i.e., experience) and adopt actions based on the derived state of information. Therefore, adaptation is to be understood as the generic property of the system to react upon changes in its environment.[3] Generally speaking, adaptation of a system "may be at the individual level through learning, or it may be at the population level through differential survival and reproduction of the more successful individuals" (Axelrod 1997, p. 4). The kind of adaptation we are interested in is adaptation through evolution in a generic sense. In order to be able to consider evolution in a generic way on multiple scales, we employ the concept of information, being a measure for the internal entropy of the system. Information can be coded in various ways, e.g., within genes, structures, community composition, more generally any kind of pool, or in form of epigenetic modifications, and can be passed on through varying time resolution. **Generic evolution** then is the process of adaptation resulting from the interaction of three subprocesses:

1. **Generic mutation**: a process leading to transmissible random changes in the informational representation of a system, e.g., mutation or changes in the abundances of agents.
2. **Generic heredity**: the process of transmission of information over time, e.g., in genomes of agents through reproduction, in the abundances of agents, or structure.

[3] Therefore, "natural adaptation" in the genetic context of the theory of evolution is one example of adaptation of a system.

3. **Generic selection**: an autonomous process that uses the outcomes of local interaction to enhance a subset of components.

Clearly, this definition needs some explanation. The process of evolution of a holobiont-like system can be built on a reformulation of Lewontin's conception of the principles underlying evolution (Lewontin 1970):

1. Interacting agents in a population have different morphologies, physiologies, and behaviors, thus comprising phenotypic variation.
2. Different phenotypes have different rates of survival and different mechanisms to persist in variable and dynamic environments, leading to differential fitness.
3. The correlation between the predecessors and successors of individual agents in the contribution of each to future generations of the system, i.e., fitness (expressed through the proportion of the number of offspring within a population), is heritable.

Note that information can be transmitted through various channels, e.g., spatiotemporal fluxes of matter or energy between pools, signaling, reproduction, and organismic diversity. Within a dissipative system, any type of pattern, i.e., information, has a certain likelihood to emerge from pure noise (i.e., transmission without net outcome of informational structure). Hence, some patterns persist longer than what can be expected by randomness. Such persistence above random average can be observed. Heredity describes the process which leads to persistence of certain patterns. In the interesting case of genetic heredity, information stored in the gene code of a genome can persist beyond the life cycle of the individual and thus pass on to future generations. Thus, on the level of genes, information persists, both within the gene code as well as in epigenetic modifications, while on the level of the individual information, i.e., structure, resolves into entropy.

In a similar way, we are forced to carefully rethink our concept of fitness. A posteriori, we are able to observe persistence of a certain clustering of genes longer than what would be expected from random fluctuations. In biological terms, we are talking about the persistence of genetic traits over generations. Fitness is then often measured in terms of success of reproduction, i.e., success in preserving certain clusters of genes and hereby overcoming the spatiotemporal constraints of the individual. Similarly, fitness in asexual reproduction might be quantified via the amount of propagules being produced. Fitness thus turns into a characteristic of a group of genes. Dealing with groups of agents, with every agent carrying its own genetic information, we do not have such an obvious gene versus individual hierarchy anymore. Therefore, we cannot exploit the time discretization given through the life span of an individual in order to define fitness, e.g., in terms of frequency of a certain trait within generations. Fitness thus becomes the integration of system properties that allow the consortium to maintain its internal structure beyond the random fluctuation of entropy.

The core of Darwin's idea of evolution through natural selection is thus conserved in the proposed conceptual framework. Variation among the agents constituting a holobiont-like system gives rise to variation in the degree of interaction among the individuals. At the same time, variation of agents allows for the initiation of differing phenotypes with differential reproductive success. Hence, also another fundamental component of Darwin's idea is fulfilled. Both components give rise to diversity in network interaction, both internally and externally with the environment (Holland 1995; Levin 1998). As such, the potential for internal variability gives rise to the plasticity of the system in its ability to adapt to dynamically changing environmental conditions (Levin 1998).

Conversely, external variations—via selection—shape the ability of the associate network to persist longer than possible by pure chance. From the point of view of information, the system builds up regularities based on its "experience," i.e., preceding history in the flow of information. The selectively derived regularities then impact the future performance of the system, which feeds back into the system as the environment continues to pass through dynamic changes (see, e.g., Gell-Mann 1994). It is the persistence of the system over a certain period of time which gives the system its specific "character." Note that this again involves the observer. Recognizing a system as functional entity relies on our modes of perception which are highly scale dependent and intrinsic to human perception capacities, both in spatial and temporal terms (Kay 2008). Survival of a system then means that the functional entity we are observing manages to persist both spatially and temporally above the level of noise (with also the latter being defined by the perception of the observer).

Finally, adaptation within a persistent functional entity needs mechanisms to capture information within the system (Gell-Mann 1994). In principle, this can be achieved directly via some recognition process of factorial alterations, e.g., learning, or indirectly through reinforcement by repeated action. The classical concept of heredity refers to persistence of an organism as a type, not as an individual. This means, the captured information can be passed on, overcoming the temporal limits of the individual while guaranteeing system persistence on the level of its conceptual type, e.g., the species. It is the interplay of prevalent long-term patterns on various levels of biological organizations which the debate on the level of selection is all about. As such, there is further accordance with driving questions of ecological theory, although the latter are typically considered at short-term timescales.

3 Theoretical Foundation

Let us start with discussing some fundamental concepts of biological sciences within the framework of computational thermodynamics. Living systems clearly persist in a state of lower entropy compared to their environment. Thus, according to the second law of thermodynamics, such systems cannot exist without a flow of energy, which allows the system to establish and maintain its lowered internal entropy. The continuous flow of energy on Earth allows for the existence of structures, which consume energy, while augmenting the entropy of the universe. Schrödinger coined the notion of "feeding through negentropy" (Schrödinger

1944). Therefore, living systems locally alter the flow of energy within their environment and thus create an impact on other living systems in their vicinity. As such, any form of interaction influences the local energy landscape. Organisms persisting within this continuous energy flow are phenotypes expressing features which optimize some implicitly given function[4] of the local energy flow (Bar-Yam 1997). In this way, living organisms create what is called the *coupled fitness landscape*[5] (Kauffman and Johnsen 1991).

The potential of an organism to survive in a dynamically changing environment depends on its ability to utilize the local nonequilibrium conditions within its environment for maintaining its state of low entropy. Evolution on Earth has resulted in a variety of ways that make energy available to organisms, so that the informational content of the system may be sustained, e.g., upon fixation of energy from solar energy by means of photosynthesis. Following Bar-Yam (1997), living organisms reflect a local optimization of their usage of the variation in energy flow. It is not the absolute amount of free energy being used which determines the level of adaptation of the system, but the capability of the system to deal with dynamic variation of the local energy flow. Thus, fitness can be interpreted as the cost function resulting from the various optimizations performed by the organisms within a given proximity in the space-time continuum. Note that through this interpretation, fitness is a local concept. It cannot be decoupled from the spatio-temporal vicinity of the acting system.

Such a physical interpretation of fitness also decouples natural selection from the mechanism of organismic replication and, in particular, from any quantitative description of reproduction. Conserved information is one particular instance of order. Thus, conserving information means maintaining the state of low entropy. Therefore, fitness also depends on the ability of a system to adaptively persist through mechanisms of conserving information beyond the reproductive mechanisms, e.g., through ecologically successful niching. It is in this latter sense that heredity matters across all levels of biological organization.

With the given interpretation in mind, evolution can be understood as being a computational process (see Dennett 1996). Evolution is "a theory of information transfer, describing the process of transmitting messages containing biological information, with mutation a phenomenon of information change and a source of variation" (Zenil et al. 2012, p. 2174).

As mentioned above, persistence is ultimately linked to structural and temporal patterns constituting the entities we observe as being persistent. Patterns emerging from interaction of agents can have several sources. For example, nonlinear

[4] "Function" to be understood in the (abstract) mathematical/physical sense not as biological function. To be precise, the term "function of a variable" is used for a mapping of the variable into some space, without the need to further specify the concrete nature of the mapping.

[5] The landscape is "coupled" since it results from the superposition of the local fitness landscapes of each individual organism.

dynamics commonly leads to structures of self-similarity.[6] Similarly, patterns of coherence emerge from autonomous interaction in complex systems. However, as being local, patterns are contingent on a certain temporal and spatial scale. Thus, using the notion of a pattern implies observer dependency.

Following the ideas of generative science (Epstein 1999), we can take on a modeling perspective providing analytical tools to study nature. Therefore, in order to provide a formal definition of the systems we aim to analyze, we employ concepts from computational theory. A multi-agent system (MAS) is defined as the collection of a set of autonomous agents which interact through a set of rules, which is defined for each agent independently, thus implementing a local view for each agent (Wooldridge 2002). An agent itself is a persistent[7] object carrying a state. What exactly the state consists of results from the act of modeling. As such it depends on the observer or—more precisely—on whatever is considered worth of being represented.

MAS have been introduced to study decentralization in decision making and bottom-up solution strategies (Wooldridge 2002). Through their interactions, the agents change their states. The rules of the system as a whole result from the interaction of the agents. MAS have shown to produce complex interaction patterns (see, e.g., Hogeweg and Hesper 1983; Reynolds 1987; Palmer et al. 1994; Helbing et al. 2000). Within this context, complex can be defined in an etymological way as being hard to separate into parts (see Gershenson and Heylighen 2005). The essential characteristic of MAS is self-organization. Since each agent only has a local perspective, there is no governing system, determining the behavior of the system as a whole. Thus, the functionality and, as such, the character of the system as an entity emerges via self-organization through the interactions of its agents (see Heylighen 2013). "Self-organization establishes a relation between the behavior of the individual components and the structure and functionality of the system as a whole: simple interactions at the local level give rise to complex patterns at the global level. This phenomenon is called *emergence*" (Heylighen 2013, p. 121; highlighted in the original).

Note that in classical MAS, the rules of its agents are fixed (e.g., Conway's Game of Life, see Gardner 1970 for further details; swarm robotics, Şahin and Winfield 2008). Once we allow the rules of the agents to change through, e.g., some process of learning (i.e., in a sense of recognizing environmental change and storing recognition), we obtain a complex adaptive system (CAS). Holland (1992, 1995) defines a complex adaptive system[8] as a (typically large) collection of agents that interact and adapt. Through the possibility of the agents to adapt, the system shows self-similarity.[9] However, by doing so, the system inherently shows path

[6] Note in passing that self-similarity is one of the defining properties of fractals.

[7] In contrast to volatile.

[8] Holland uses the notion of *constrained generating procedures*.

[9] "Self-similarity" in the sense that the agents in both their states and their rules adapt as well as the system as a whole adapts in its composition of agents and their interactions.

dependency, i.e., the state of the system depends on the states the system has been going through in the past. Path dependency clearly can be conserved in the agents' memory, if they have anything like memory (i.e., in physical terms, not necessarily consciousness). For the particular example of plant memory, the reader is referred to Thellier and Lüttge (2013), Lüttge and Thellier (2016). Memory merely may be conserved in the system's structure. For example, the morphology of the skeleton of vertebrates results from their common evolutionary ancestor and, thus, conserves the lateral symmetries and essential characteristics resulting from the selection process the ancestor has been subject to. It is important to note that path dependency thus introduces a concept of memory which can be passed on during the further development of the system. We note in passing that path dependency comes along with intractability (Nikolic and Kasmire 2013). This means that the shortest model describing the system is the system itself. From intractability then follows unpredictability at least for all purposes of prediction and controllability.

Holland (1995) summarizes four basic properties which complex adaptive systems are showing:

1. Complex adaptive systems develop inhomogeneities in the way their basic elements are organized. Thus, aggregation can be observed, leading to patterns in space and time.
2. Complex adaptive systems evolve.[10] Through the process of adaptation, chance events are reinforced leading to nonlinearity.
3. The capability of responding to unforeseen changes in the environment is only possible if diversity is maintained.
4. Since interactions are commonly instantiated through energy flows, complex adaptive systems exhibit a homeostatic nature, in particular keeping flows of energy in balance.

As adaptation obviously is the key feature within the idea of a CAS, we need to take a closer look at this concept in the context of biological systems. Adaptation can be realized through several mechanisms, one of which is evolution. As mentioned above, the key components of evolution are mutation, heredity, and selection. Applying the abstract framework, we consider **(generic) evolution** in a broad sense, comprising three generic subprocesses. Evolution is based on variation which is provided by the diversity of agents comprising a system. Additional variation within the system is created through transmissible random changes in the informational representation of the system. This process is called **(generic) mutation**. While in classical neo-Darwinism chance has been claimed to be the major source of variation, recent contributions stress a dualism of chance and determinism (Buiatti and Buiatti 2008). Therefore, variation can be generated both through deterministic and random processes. For example, we have Mendel's Laws of Inheritance explaining frequencies of alleles in populations in a deterministic way. To generate variation within the genetic potential of a population, chance

[10] In the sense of adaptation through absorption of information (i.e., experience).

acts via mutation or sexual reproduction. The importance of chance results from the necessity to stabilize the system and counterbalance effects from energy dispersal (cf. Buiatti and Longo 2013).

(Generic) heredity describes a process of transmission of information over time. Thus, heredity leads to temporal persistence which can be observed. But heredity is more than just an occasional, temporal persistence. In order for an algorithm to optimize through variation and selection, information gained through the interaction of the system with its environment has to be captured within the system. Evolution on Earth resulted in highly optimized organisms which managed to adapt to specialized ecological niches. The successful "strategy" for searching through the virtual space of possible solutions requires a mechanism of maintaining information and thus introducing directedness in time. Without heredity, the simple trial-and-error strategy would be highly inefficient. However, through the transmission of information over time, "experience" gained through exploring the consequences of recent adaptations in the informational representation of the system will increase overall adaptiveness. The idea of generalizing the neo-Darwinian concept of heredity has gained growing interest with recent work in molecular biology, in particular in epigenetics (Bossdorf et al. 2008; Richards 2006; Lüttge and Thellier 2016). Jablonka and Lamb (2005) introduced the concept of evolution in four dimensions. Next to genetic evolution, comprising the classical neo-Darwinian concept, epigenetic inheritance, social learning, and symbolic communication are added as additional dimensions of transmission of information (cf. Jablonka and Lamb 2005)—with the last two dimensions representing outcome from the cultural evolution.

The third subprocess is selection. We can consider **(generic) selection** as an autonomous process that uses outcomes of local interaction to enhance a subset of components. Being an autonomous process, selection emerges from the interaction of systems with each other.

Note that the algorithmic nature of evolution does not lead to explanations why things are the way we see them. But the algorithm allows us to understand how things evolved over the course of time (Nikolic and Kasmire 2013). The gain in using an algorithmic approach lies in the ability to define essential biological concepts independent of temporal or spatial scales (see also Zenil et al. 2012), rather than giving an interpretation in a teleological sense. Thus, the interpretation of evolution as an algorithmic concept within the framework of adaptation in systems of interacting agents provides a conceptual definition conducive to analyzing holobiont-like principles on various scales.

4 Arguing for an Extension Beyond the Holobiont Sensu Stricto

Let us recall the definition of the holobiont as the functional entity of a "host organism" together with all its associated microorganisms (Sect. 2). Clearly, such an entity is a system of interacting agents, each of which acts without external control. Interactions can be manifold. There is no a priori assessment concerning the nature of the coenobitic relationships. Which of the interactions we consider as being part of the holobiont is up to the choice of the observer. The subjective perspective extends to the nature of the relationship between host and microbes. Whether a certain relationship can be termed mutualistic or parasitic, it "requires a clear appreciation of the spatial, temporal and taxonomic context in which these systems operate" (Herre et al. 1999, p. 49). Altogether, it is the observer assigning some purpose to the system (see also Gershenson and Heylighen 2003).

The holobiont operates within a certain environment which the entity of agents is exposed to. As such, the system is subject to adaptation. To give an example, consider vertebrates. There are many examples of species-specific bacteria which are essential for the maturation of the immune system (e.g., Mazmanian et al. 2005; Weiss et al. 2011; Buffie and Pamer 2013). Even parasites can contribute to increasing fitness (e.g., Herre et al. 1999). Thus, coenobitic relations affect the fitness of the system, i.e., the "holo-immunome" (cf. Dheilly 2014). Analog mechanisms also exist for plants, e.g., priming (Pozo and Azcón-Aguilar 2007) and induced resistance (Jung et al. 2012; Zamioudis and Pieterse 2012). Therefore, variation in abundances of microbiota through amplification and acquisition/abandonment and variation in the modes of interaction (altogether promoting ecological niching) have an effect on the ability of the holobiont to keep its overall level of internal entropy, i.e., its organization. "The theoretical framework provided by considering not only the host but also the parasite as a holobiont revealed that some interactions have been underestimated and others have not yet been explored" (Dheilly 2014, p. 1). For the example of the holobiont sensu stricto, classical selection leads to the enhancement of holobiontic associations. Concerning heredity, we should consider the broader setting of generic heredity. As mentioned in the introduction, the holobiont shows, next to genetic inheritance, additional modes of transmission of information, e.g., the community assembly as such. Information can be passed on vertically to the next generation, e.g., via gametes, as well as horizontally to accompanying individuals through interaction. For bacteria this could be quorum sensing, the capability of bacteria to sense their local cell density (Bassler and Losick 2006), just to give an example. Both vertical and horizontal transmissions thus change the informational representation of the holobiont. This can also be traced on an evolutionary timescale. For example, genome reduction may result from long-term symbiosis of bacteria with their symbiotic host (McCutcheon and Moran 2012) or the development of a highly specialized metabolic repertoire as in the case of the human gut commensal *Bacteroides thetaiotaomicron* (Benjdia et al. 2011). Summarizing, the holobiont can be seen as the prototype of a complex system adapting through an evolutionary process.

The general observations in the previous section clarified holobiont characteristics to be twofold: (1) On the ontological side, there is an association of biological agents together with their biotic–abiotic interactions. The association is taken to be a functional entity within a certain environment. (2) On the epistemological side, the evolutionary process drives the development of the system within its environment. The essential principles underlying this process can be directly deduced from the hologenome theory. Furthermore, interaction with microbes can affect epigenetically induced plasticity of plants on an evolutionary scale, although not being the only factor shaping epigenetic plasticity. Intrieri and Buiatti (2001) claim that introgression of genes from *Agrobacterium rhizogenes* had a major impact on the development of the genus *Nicotiana*. Thus, compared to the holobiont sensu stricto, the perspective of a holobiont-like system opens up a broader context. Buiatti concludes that the "data from plants suggest that selection operates at several levels of the hierarchical organization of life, and that fitness is determined by the effects of both epigenetic and genetic factors. I suggest that, since at every level (cell, organism, population, species, ecosystem) there are internal and external factors that affect fitness, there is a need for organisms to coordinate the networks of interactions that occur at the different levels" (Buiatti 2011, p. 257). Again, we stress that neither genetic nor epigenetic mechanisms provide the only modes of transmission. We deliberately aim to transcend beyond both of these dominantly gene-oriented modes.

Let us consider a second example. An ecosystem is the prototype of a complex adaptive system. The system is defined through a network of organisms as interacting with their biotic and abiotic environment at a denoted spatial location (cf. Tansley 1935). Current extensions include further aspects which are characteristic for a holobiont-like system. For example, Ellenberg et al. (1986) additionally concede to the system a limited capacity of self-regulation. Furthermore, some authors claim that emergent properties have to be viewed as characteristic for an ecosystem (see, e.g., Jørgensen and Müller 2000; Matyssek and Lüttge 2013; Lüttge 2016).

Coming back to the thermodynamic approach to ecosystem theory (see, e.g., Jørgensen and Svirezhev 2004), Aoki (1995) interprets ecological succession of a lake as evolution from oligotrophy to eutrophy and approaches succession via studying an increase of entropy. The entropy principle in living systems he proposes claims that entropy production in biological systems passes through at least two phases. An early phase in which entropy production increases over time is followed by a later period, where production decreases (cf. Aoki 1995). The concept has been picked up later, adding approaches to estimate entropy production in ecological systems such as food webs (Meysman and Bruers 2007).

Through the utilization of natural resources, organisms within an ecosystem interact, leaving an impact on the space of possible actions of neighboring species. Independently of being limiting or upon excess, changing resource availability implies interaction. Food webs constitute one example of structure emerging from the interaction of species within an ecosystem.

Species composition itself is capturing information about the ecosystem. Evolving from selection, speciation, dispersal, and drift (Vellend 2010), the community conserves information about the evolution of the system. “Experience” is preserved in existing ecological niches created through the actions/interactions of the individuals comprising the system. Adaptive fitness of the ecosystem as an entity depends, among others, on the species diversity the system attained to maintain as a result of successful ecological niching. Also other environmental aspects, e.g., spatial fragmentation, are captured within species composition. The latter aspect has been described by metapopulation theory (Hanski 2004; Alexander et al. 2012), although not without debate (Baguette 2004). It is the ability of an ecosystem to make natural resources available for the species constituting the system, which creates variation within the set of existing ecosystems. The Lamarckian aspects of transmission (Rosenberg et al. 2009) inherent in the hologenome theory will apply at the general ecosystem level. Microbial systems provide a particularly instructive example, since horizontal gene transfer and hereditary symbioses in eukaryonts provide mechanisms reaching beyond classical neo-Darwinian heredity (cf. Sapp 2011).

In contrast to standard definitions of an ecosystem, the view of the entity as a holobiont-like system includes the aspect of adaptation and/or evolution of the system as a whole. Thus, we emphasize the characteristics of the system to enhance its adaptive abilities and thus, contrasting more traditional, static perspectives. Dennett’s view of evolution as an algorithm has deep consequences. Through generic heredity, i.e., transmission of information, the system evolves over time, allowing to build on gained “experience.” As such, evolution is a realization of the principle of *competence without comprehension* (Dennett 2009). It defeats the traditional concept that any higher competence cannot be achieved without an a priori understanding. Evolution as an algorithm has been powerful enough to create all complex life on Earth through the combination of simple subprocesses. As such, the evolution of ecosystems has a direction, too. Ecosystem development moves on in cycles which proceed toward higher modes of “order” and complexity. Succession of pioneer species toward climax species reflects an increasing level of organization from rapid capturing toward retaining of resources, as reflected by refined ecological niching and increasing specialization (see Burkhard et al. 2011).

Let us briefly come back to the scale independency inherent in the definition of a holobiont-like system. There is system nestedness in multiple ways. Patterns emerging from the interaction of agents lead to structure on the level of the system and thus create the suite of actions for the behavior of the system as an entity. While providing stability, structure also limits the behavior of the agents on lower levels (*downward causation*) (see Gershenson and Heylighen 2003). Additionally, the choice of the observer to define the system boundaries creates dependencies in a horizontal sense. Boundaries of open systems need to be delimiting and penetrable at the same time. Therefore, there is always some fuzziness, since precise, (eco-) physiological boundaries can hardly ever be defined. Such kind of fuzziness is a prominent ecosystem feature. This feature is innate to the fact that systems need to be open in their energy exchanges with their environment, which forms the prerequisite for confining entropy.

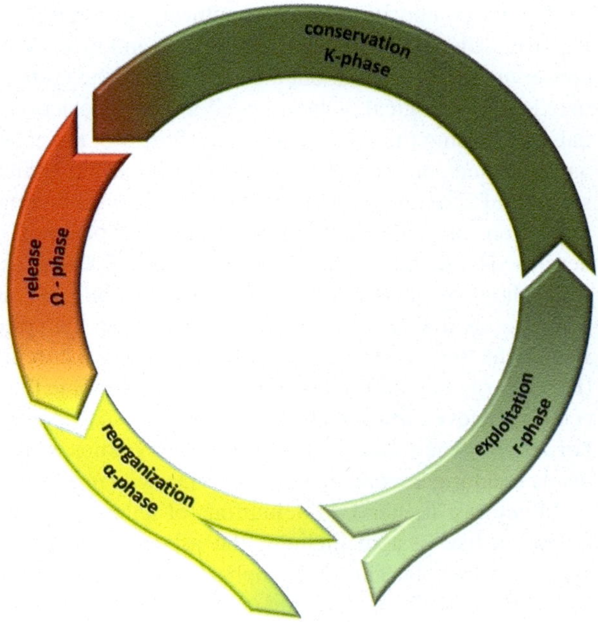

Fig. 3 The adaptive cycle (according to Gunderson and Holling 2002) consisting of a phase of exploitation (r-phase), followed by conservation (K-phase). Once the effort to sustain the state of high-level order raises too much, the system will switch into a phase of release (Ω-phase) followed by a period of reorganization (α-phase)

Systems provide context for other systems. Thus, systems form nested sets with vertical as well as horizontal relationships. Classically, ecosystem theory knows two phases of system development. First, there is an r-phase of exploitation of resources (where r stands for the rate in the standard model of population dynamics) followed by a K-phase of specialization and conservation of the attained structure and order (K refers to the capacity constant in the population dynamical model). The idea of an r-/K-phase generalizes the concept of r-/K-strategies, as introduced by MacArthur and Wilson (1967) in a natural way. Gunderson and Holling (2002) added two additional, typically shorter phases leading to an *adaptive cycle* of exploitation (see Fig. 3), conservation, release, and reorganization. Cells divide and die, individuals grow and vanish, species appear and disappear, and ecosystems evolve, persist, and collapse. Holling (2001) coined the notion of *panarchy*[11] as a metaphor to describe such interrelated sets of adaptive cycles (Fig. 4; see also Gunderson and Holling 2002). Through their development over time, systems pass through a sequence of *adaptive cycles* of exploitation. Evidently, cycles on hierarchically low levels are running on faster timescales compared to cycles on high levels of system organization. This conceptual background clearly applies to the

[11] The authors argue for the invention of a new term: "Since the word hierarchy is so burdened by the rigid, top-down nature of its common meaning, we prefer to invent another term that captures the adaptive and evolutionary nature of adaptive cycles that are nested one within the other across space and time scales. We call them panarchies, drawing on the image of the Greek god Pan—the universal god of nature" (Gunderson and Holling 2002, p. 74).

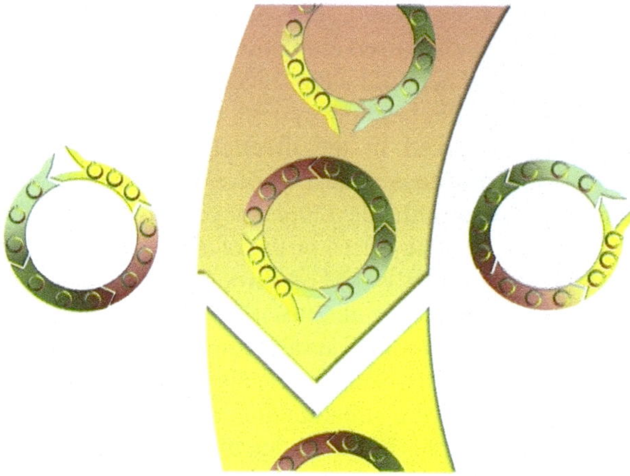

Fig. 4 The panarchy metaphor (Gunderson and Holling 2002) builds on the idea of the adaptive cycle. Systems passing through the cycle interact with other systems (horizontally) which might be within a different phase of the cycle. At the same time, systems are parts of other systems and consist of systems on a lower scale (vertical interaction). Thus, a nested set of horizontally and vertically interacting systems arises, forming a panarchy

concept of holobiont-like systems. Emerging through the interaction of agents, adaptive cycles of the holobiont-like system are determined by the adaptive cycles of the agents, which themselves constitute holobiont-like systems, too. It is interesting to note that through the association between eukaryotes and prokaryotes, these cycles become interlinked, reflecting a continuum in time and space for holobiont-like systems to exist (see Holling 2001).

It is thus reasonable to analyze a given holobiont-like system as being part in a panarchy of other systems. Disregarding these dependencies leads to skewed argumentation and debates hard to settle. Considering ecosystems as holobiont-like systems emphasizes the dynamic evolution of the system. The need to go through the full adaptive cycle follows from the algorithmic principle of evolution. In order to evolve, i.e., search through the (hypothetical) space of conceivably stable ecosystems, the evolutionary algorithm needs to instantiate examples and expose them to the process of selection. The collapse of the ecosystem ends this exposure. The information being transmitted to the "next generation," i.e., along the dimension of time, through generic heredity guarantees the evolutionary "experience" (see above) to be captured. Thus, a new status or even variant of ecosystem can develop through building up on the functions and performances being captured in previous rounds of the adaptive cycle (see also Burkhard et al. 2011).

Going downward on the biological scale, we can consider holobiont-like systems on the organ level. A prototype would be the human immune system. It is indeed questionable to consider the immune system as being an organ, since it is hard to localize the immune system within an organism. On the other side, the immune system has organ-like properties. It is built up from various cell types of different speciation, e.g., phagocytes, natural killer cells, lymphocytes, or T cells.

Through those cells, tasks are shared, e.g., building the innate and the adaptive immune system. Information is passed along through the system by direct cell–cell communication, the development of pattern recognition receptors and antigens, all reflecting the interaction of the immune system with its environment. The immunological memory is conserved in the diversity of B and T cells (Janeway et al. 2001). Persistent differences in adaptability of the system provide the foundation for current hypotheses in the development of autoimmune diseases, e.g., the *Hygiene Hypothesis* (Okada et al. 2010) or the *Old Friends Hypothesis* (Rook and Brunet 2005). Mutation is directly evoked through mechanisms such as *somatic hypermutation* (see, e.g., Janeway et al. 2001). Furthermore, the complex interplay of the innate with the adaptive immune system provides further sources of variation via proliferation of immune cells. Selection is also immediately present due to the interplay of inflammatory and anti-inflammatory mechanisms.

Another informative example can be a branch of a tree. The agents are given by different tissue types and/or microorganisms being present. As for generic mutation, we have classical genetic mutation within cells, as well as drift and dispersal of microorganisms. Furthermore, phenotypic modifications, e.g., plant gall, introduce further variation. Apart from the obvious modes of transmission, generic heredity at the organ level in this example includes spatial persistence through morphological structure. In a similar way, morphology adds to selection through, e.g., shading or restriction of water supply.

Although all classical modes of evolution on the genetic scale are still applicable within the context of holobiont-like systems, the generic counterparts of the evolutionary subprocesses typically comprise further modes of biological interaction. Thus, the concept of the holobiont-like system does not render the traditional neo-Darwinian concepts dispensable, but reaches beyond the gene-centric perspective, embracing a wider range of nongenetic mechanisms.

5 Hypotheses for Experimental Analysis

After all, do we actually need another concept? Focusing on both, system properties and scale independency opens a new view onto biological systems. In particular, the characteristic property of adaptation inherent in the definition of the holobiont-like system incorporates a perspective of evolutionary theory into ecological analysis. Conversely, the idea of nested adaptive systems developing and evolving over space and time allows explicitly addressing the different scales involved in common evolutionary theory and ecology.

An overarching hypothesis thus states that the view of plants as parts of holobiont-like systems improves the possibilities for explanation and prediction of ecological and evolutionary patterns and processes. Following the scale-independent view, we expect this hypothesis to hold across spatial and temporal scales. A gene-centered, evolutionary approach falls short in addressing the complexity of processes and mechanisms, determining the effectiveness of the phenotype when being confronted with the competitive situation within an ecosystem of interacting organisms. Improved knowledge in molecular biology has elucidated a variety of interacting functions, enfolding the potential lying in the gene. "No longer can the gene be thought of as inherently stable, discrete stretch of DNA that encodes information for producing a protein, and is copied faithfully before being passed on. [. . .] The stretch of DNA that is a 'gene' has meaning only within the system as a whole. And because the effect of a gene depends on its context, very often a change in a single gene does not have a consistent effect on the trait that it influences" (Jablonka and Lamb 2005, p. 7).

Theory building in plant sciences has traditionally been rooted in an autecological perspective. Shortcomings of, e.g., the *Growth-Differentiation-Balance Theory* (Herms and Mattson 1992; Matyssek et al. 2012a), address plant-internal resource allocation as a trade-off between the demands of growth versus stress defense. More precise, the trade-off between investments in primary versus secondary metabolism is associated with ecological costs. The latter might differ with changing environmental conditions. In contrast, experiments demonstrate the plant to possess enormous plasticity in regulating its resource allocation under dynamically changing ecological scenarios (Matyssek et al. 2012b). The concept of a holobiont-like system naturally frames the conceived plant system within a hierarchy of interactions, thus incorporating a systems biology perspective. Hereby, there is no limit on the considered level of biological organization. The panarchy metaphor locates adaptive cycles of proteins within cells, life cycles of cells within tissue, tissue within organisms, and the life span of the organisms itself within an ecosystem. The mechanisms shaping these adaptive cycles, enabling to preserve released free energy within the supra-system, will be different on each level. Nevertheless, generic factors driving the mechanisms as well as underlying principles are anticipated to be universal. Preservation of free energy may even appear as one thermodynamic reason in evolution toward advancing complex adaptive systems in biology, eventually linked within horizontally and vertically nested hierarchies.

Pattern organization within a hierarchy is a natural consequence of self-organization of systems of autonomously interacting agents: "Aggregation and hierarchical assembly are not imposed on complex adaptive systems, but emerge from local interactions through endogenous pattern formation" (Levin 1998, p. 432). Essential ingredients to enable self-organization within dissipative systems are a certain level of diversity, dynamically adapting interactions among agents, as well as effectiveness of an autonomous process counterbalancing the creation of diversity through selection (cf. Levin 1998). Thus, the subprocesses defining generic evolution within the holobiont-like system provide the basis for emerging patterns of self-organization. The perceived directedness of system development, passing through the adaptive cycle from the simple to the more complex, i.e., from an r-phase of exploitation toward a K-phase of specialization and conservation (cf. Gunderson and Holling 2002; Burkhard et al. 2011), can be explained through the action of an underlying evolutionary algorithm. Via hereditary processes, information is kept within the system. It is accumulated during various runs through the adaptive cycles on lower levels. Being embedded within a hierarchically upper adaptive cycle, information is persistently transmitted within the system. Various indicators and *ecological orientors* (Fath et al. 2004) can be employed to identify the state of the system at the various scales (see, e.g., Burkhard et al. 2011).

The idea of transmission of information both horizontally among biological units and vertically within the system is deeply rooted in biological theory. In particular, "change through use and disuse" (cf. Jablonka and Lamb 2005) is not foreign to Darwin's theory of evolution via natural selection. Although the historical development has defeated Lamarckism on the basis of lack of grounds for an hereditary mechanism transmitting acquired information, recent insights in, e.g., epigenetic functions, horizontal gene transfer, transmission of microorganisms, etc., have led to reconsider inheritance beyond the single gene (see, e.g., Jablonka and Lamb 2005; Zilber-Rosenberg and Rosenberg 2008).

Likewise, ecological theory has adopted a more dynamical perspective. Considering an ecosystem to be an adaptive system shifts the focus from individual organisms to organismic interactions. Thermodynamic ecosystem theory (Prigogine et al. 1972; Jørgensen and Svirezhev 2004) has been proposed as well as approaches based on network analyses (Fath and Patten 1999), just to give some examples. The aspect of adaptation across scales has been presented, among others, by Aoki (1995), who uses the term "evolution" to denote the development of ecosystems. Holling (1986) and Gunderson and Holling (2002) considered directedness in the evolution of ecosystems across scales while introducing their panarchy concept.

Acknowledging directedness in ecosystem development will also provide new perspectives. For example, the consideration of ecosystem maintenance has to be modified (see also Lüttge et al. 2016). Considering an ecosystem as proceeding through the adaptive cycle from less organized to higher complexity, the system cannot be maintained in a highly evolved state forever. The longer the system remains in a low-entropy state, vulnerability will rise under small, unforeseen perturbations, shifting the system into the following phase of release (cf. Burkhard et al. 2011). Thus, maintenance of ecosystem functions needs to warrant the constant flow of entropy at any level of self-organization.

Through the systems perspective, ecosystem functions emerge from the interaction of the organisms forming the system. One level further down the scale, these interactions are subject to dynamic change, too. The *Stress-Gradient Hypothesis* (Bertness and Callaway 1994; Brooker et al. 2008) postulates that beneficial organismic interaction increases along gradients of progressively limiting abiotic stress. On a broader perspective, interactions among certain organisms take place within the concerted action of many organismic associations. Maintenance of system functionality on the larger scale, however, imposes selective pressure on single interactions. Further on, selective pressure enforces a dynamic process on the set of interactions which might change the character of a particular single interaction, e.g., from mutualistic to parasitic. Thus, qualifiers such as "mutualistic" or "parasitic" might change during system development just because of the dynamic change of the state of the system.

From the overarching hypothesis stated at the beginning of the section, several sub-hypotheses can be deduced. First of all, we can study the holobiont-like system as a conceptual model and analyze its suitability in fostering ecological or evolutionary understanding. An important aspect of the model is the claim that emergent properties can be observed on the system level. Generic properties of complex adaptive systems can thus be studied in a scale-independent manner. We anticipate properties such as resilience, redundancy, or robustness to be universal on all scales. The holobiont-like system provides straight forward hypotheses for these functions to be realized. For example, resilience is postulated to be the consequence of the diversity of agents, buffering the system against various types of stress. In addition, agents might be partly exchangeable, allowing system functionality to remain unchanged although single agents might be lost. Accordingly, robustness is expected to result from the network of interactions, increasing beyond the robustness of the single agent.

Studying holobiont-like systems as evolutionary systems, multilevel selection theory is intrinsically integrated. Interactions are expected to be found within one level as well as among levels. Processes of selection acting upon those interactions comprise a conceptual part of the definition. Appealing to the panarchy of system nestedness, it might be speculated that positive feedback of selection at a lower level relates to adaptation on the higher level. General principles like the one just stated are claimed to be scale independent.

We have already highlighted ecological hypotheses to be addressed through the concept of holobiont-like systems. Abstracting from the *Stress-Gradient Hypothesis* (SGH), we claim that dynamic rather than static environments favor holobiont-like systems. Whereas the SGH postulates a change in the characteristics of interactions along a stress gradient, we can generally ask for determinants of the set of interactions of a holobiont-like system. Generally, it is the dynamic interaction of the agents as well as with their environment which forces each agent to constantly adjust its rules of involvement and adapt upon gained experience. Without dynamics in the environmental processes, stable trajectories will be found, leading the system to rest in its current state. In contrast, within dynamically changing environments, the autonomous process of selection will enfold its selective potential on the set of interactions which currently define the system. Counterbalancing the selective pressure, variation created within the system will lead to new potential, providing the basis for evolution to occur.

6 Conclusion

Timely developments both in ecological theory and in the theory of evolution have reached a state of maturation which begins to overcome the initially separating, apparent exclusiveness of each theory. In parallel, advancing insight into biological mechanisms and functioning has never ceased to unveil an ever growing richness and diversity. Historical claims that life on Earth may eventually be fully understood and ready for control and maintenance by man have turned out repeatedly to be falsified soon after being postulated (see, e.g., the 1988 fires in Yellowstone National Park, Turner et al. 2003, or the Exxon Valdez oil spill, Harwell and Gentile 2006). As the concept of the organism still has its scientific value in systemizing and classifying the richness of life on Earth, it encompasses limits of understanding as we leave the autecological perspective, which indeed is needed when striving for comprehension of the complex ecological cause-effect relationships as the mechanistic drivers of evolutionary processes. Evidently, ecology and evolution are intrinsically woven. In fact, such a move opens the conceptual dimension for grasping intellectually the ways in which life is actually self-perpetuating.

Various theoretical tools have in the meantime been introduced to ecological theory, enhancing insights into ecosystem functioning. Nevertheless, we are still far from understanding critical states in ecosystem functioning to an extent that may enable us to sustain such systems as well as their ecological services to mankind with its various perspectives and demands. We must learn that ecology and evolution represent dynamic and, hence, "progressive" rather than "conservative" phenomena. Such a perception imperatively requests the observer—in the shape of a researcher—to conceive crucial functionality of any kind of biotic systems in terms of multiply nested interaction networks.

We propose merging two conceptual frameworks so far mostly associated either with ecological or evolutionary research by introducing the integrative concept of holobiont-like systems. Reaching beyond the scale-limiting definition of the holobiont sensu stricto, we still adopt its natural focus on the interaction of autonomous agents as well as its implicit concept of hereditary transmission. As advancement, we gain scale independency, which allows incorporating evolutionary development as driven by ecological mechanisms into all levels of biological organization. The holobiont-like system thus becomes the conceptual core for exploring scale independency of biotic interaction networks. As a result, revelation of the multiple nestedness of informational control cycles is approached as envisaged by the panarchy metaphor. The latter allows comprehension of the operational grounds by which systems ensure persistence while allowing for advancement. On such grounds, hypotheses can be posed, as pointed out in this account, the assessment of which through empirical evidence and informational analysis will help to functionally explain the integrated "eco-evo" foundations of biotic interaction networks.

References

Alexander HM, Foster BL, Ballantyne F IV, Collins CD, Antonovics J, Holt RD (2012) Metapopulations and metacommunities: combining spatial and temporal perspectives in plant ecology. J Ecol 100:88–103

Aoki I (1995) Entropy production in living systems: from organisms to ecosystems. Thermochim Acta 250:359–370
Axelrod R (1997) The complexity of cooperation. Princeton University Press, Princeton, NJ
Baguette M (2004) The classical metapopulation theory and the real, natural world: a critical appraisal. Basic Appl Ecol 5:213–224
Bar-Yam Y (1997) Dynamics of complex systems. Addison-Wesley, Reading, MA
Bassler BL, Losick R (2006) Bacterially speaking. Cell 125:237–246
Benjdia A, Martens EC, Gordon JI, Berteau O (2011) Sulfatases and a radical S-adenosyl-L-methionine (AdoMet) enzyme are key for mucosal foraging and fitness of the prominent human gut symbiont, *Bacteroides thetaiotaomicron*. J Biol Chem 286:25973–25982
Bertness MD, Callaway R (1994) Positive interactions in communities. Trend Ecol Evol 9:191–193
Bossdorf O, Richards CL, Pigliucci M (2008) Epigenetics for ecologists. Ecol Lett 11:106–115
Brooker RW, Maestre FT, Callaway RM, Lortie CL, Cavieres LA, Kunstler G, Liancourt P, Tielbörger K, Travis JM, Anthelme F, Armas C, Coll L, Corcket E, Delzon S, Porey E, Kikvidze Z, Olofsson J, Pugnaire F, Quiroz CL, Saccone P, Schiffers K, Seifan M, Touzard B, Michalet R (2008) Facilitation in plant communities: the past, the present, and the future. J Ecol 96:18–34
Buiatti M, Buiatti M (2008) Chance vs. necessity in living systems, a false antinomy. Biol Forum 101:29–66
Buiatti M (2011) Plants: individuals or epigenetic cell populations? In: Gissis SB, Jablonka E (eds) Transformations of Lamarckism. From subtle fluids to molecular biology. MIT Press, Cambridge, MA, pp 251–260
Buiatti M, Longo G (2013) Randomness and multilevel interactions in biology. Theory Biosci 132:139–158
Buffie CG, Pamer EG (2013) Microbiota-mediated colonization resistance against intestinal pathogens. Nat Rev Immunol 13:790–801
Burkhard B, Fath BD, Müller F (2011) Adapting the adaptive cycle: Hypotheses on the development of ecosystem properties and services. Ecol Model 222:2878–2890
Dawkins R (1976) The selfish gene. Oxford University Press, Oxford
Dennett D (1996) Darwin's dangerous idea: evolution and the meanings of life. Simon & Schuster, London
Dennett D (2009) Darwin's "strange inversion of reasoning". Proc Natl Acad Sci USA 106:10061–10065
Dheilly NM (2014) Holobiont-holobiont interactions: redefining host-parasite interactions. PLoS Pathog 10, e1004093
Dobzhansky T (1937) Genetics and the origin of species. Columbia Press, New York, NY
Eberl G (2010) A new vision of immunity: homeostasis of the superorganism. Mucosal Immunol 3 (5):450–460
Ellenberg H, Mayer R, Schauermann JH (eds) (1986) Ökosystemforschung—Ergebnisse des Sollingprojekts 1966–1986. Ulmer Verlag, Stuttgart
Epstein JM (1999) Agent-based computational models and generative social science. Complexity 4(5):41–60
Fath BD, Jørgensen SE, Patten BC, Straškraba M (2004) Ecosystem growth and development. Biosystems 77:213–228
Fath BD, Patten BC (1999) Review of the foundations of network environ analysis. Ecosystems 2 (2):167–179
Fisher RA (1930) The genetical theory of natural selection. Clarendon Press, Oxford
Gardner M (1970) Mathematical games—the fantastic combinations of John Conway's new solitaire game "life". Sci Am 223:120–123
Gell-Mann M (1994) Complex adaptive systems. In: Cowan G, Pines D, Mellzer D (eds) Complexity: metaphors, models, and reality. SFI Studies in the sciences of complexity. Proc Vol XIX. Addison-Wesley, Boston, MA, pp 17–45

Gershenson C, Heylighen F (2003) When can we call a system self-organizing? In: Banzhaf W, Ziegler J, Christeller T, Dittrich P, Kim JT (eds) Advances in artificial life. Springer, Heidelberg, pp 606–614
Gershenson C, Heylighen F (2005) How can we think the complex? In: Richardson K (ed) Managing organizational complexity: philosophy, theory and application. Information Age Publishing, Charlotte, NC, pp 47–61
Gilbert SF (2011) Symbionts as an epigenetic source of heritable variation. In: Gissis SB, Jablonka E (eds) Transformations of Lamarckism. From subtle fluids to molecular biology. MIT Press, Cambridge, MA, pp 283–293
Goldberg AD, Allis CD, Bernstein E (2007) Epigenetics: a landscape takes shape. Cell 128:635–638
Gunderson L, Holling CS (2002) Panarchy: understanding transformations in human and natural systems. Island Press, Washington, DC
Haldane JBS (1955) Population genetics. New Biol 18:34–51
Hamilton WD (1963) The evolution of altruistic behaviour. Am Nat 97:354–356
Hamilton WD (1964) The genetical theory of social behaviour I, II. J Theor Biol 7:1–52
Hanski I (2004) Metapopulation theory, its use and misuse. Basic Appl Ecol 5:225–229
Helbing D, Farkas I, Vicsek T (2000) Simulating dynamical features of escape panic. Nature 407:487–490
Harwell MA, Gentile JH (2006) Ecological significance of residual exposures and effects from the Exxon Valdez oil spill. Integr Environ Assess Manag 2(3):204–246
Herms DA, Mattson WJ (1992) The dilemma of plants: to grow or defend. Q Rev Biol 67:283–335
Herre EA, Knowlton N, Mueller UG, Rehner SA (1999) The evolution of mutualisms: exploring the paths between conflict and cooperation. Trends Ecol Evol 14:49–53
Heylighen F (2013) Self-organization in communicating groups: the emergence of coordination, shared references and collective intelligence. In: Massip-Bonet À, Bastardas-Boada A (eds) Complexity perspectives on language. Communication and society. Springer, New York, NY, pp 117–149
Hogeweg P, Hesper B (1983) The ontogeny of the interaction structure in bumble bee colonies. Behav Ecol Sociobiol 12:271–283
Holland JH (1992) Adaptation in natural and artificial systems. MIT Press, Cambridge, MA
Holland JH (1995) Hidden order. How adaptation builds complexity. Addison-Wesley, Reading, MA
Holling CS (1986) The resilience of terrestrial ecosystems: local surprise and global change. In: Clark WC, Munn RE (eds) Sustainable development of the biosphere. Cambridge University Press, Cambridge, MA, pp 292–320
Holling CS (2001) Understanding the complexity of economic, ecological, and social systems. Ecosystems 4:390–405
Intrieri MC, Buiatti M (2001) The horizontal transfer of *Agrobacterium rhizogenes* genes and the evolution of the genus *Nicotiana*. Mol Phylogenet Evol 20:100–110
Jablonka E, Lamb MJ (2005) Evolution in four dimensions. MIT Press, Cambridge, MA
Janeway CA, Travers P, Walport M, Shlomchik MJ (2001) Immunobiology: the immune system in health and disease, 5th edn. Garland Science, New York, NY
Jørgensen SE, Müller F (eds) (2000) Handbook of ecosystem theories and management. CRC Publishers, New York, NY
Jørgensen SE, Svirezhev YM (2004) Towards a thermodynamic theory for ecological systems. Elsevier, Amsterdam
Jung SC, Martinez-Medina A, Lopez-Raez JA, Pozo MJ (2012) Mycorrhiza-induced resistance and priming of plant defenses. J Chem Ecol 38(6):651–664
Kauffman SA, Johnsen S (1991) Coevolution to the edge of chaos: coupled fitness landscapes, poised states, and coevolutionary avalanches. J Theor Biol 149:467–505

Kay JJ (2008) Framing the situation: developing a system description. In: Waltner-Toews D, Kay JJ, Lister NME (eds) The ecosystem approach. Complexity, uncertainty, and managing for sustainability. Columbia University Press, Chichester, NY, pp 15–35
Levin AS (1998) Ecosystems and the biosphere as complex adaptive systems. Ecosystems 1:431–436
Lewontin RC (1970) The units of selection. Annu Rev Ecol Syst 1:1–18
Lüttge U (2016) Plants shape the terrestrial environment on Earth—challenges of management for sustainability. In: Cánovas FM, Lüttge U, Matyssek R (eds) Progress in botany, vol 77. Springer, Heidelberg
Lüttge U, Souza GM, Bertolli SC (2016) Hierarchy and information in a system approach to plant biology. In: Cánovas FM, Lüttge U, Matyssek R (eds) Progress in botany, vol 77. Springer, Heidelberg
Lüttge U, Thellier M (2016) Roles of memory and the circadian clock in the ecophysiological performance of plants. In: Cánovas FM, Lüttge U, Matyssek R (eds) Progress in botany, vol 77. Springer, Heidelberg
MacArthur RH, Wilson EO (1967) The theory of island biogeography. Princeton University Press, Princeton, NJ
Margulis L (1993) Symbiosis in cell evolution: microbial communities in the Archean and Proterozoic Eons. W.H. Freeman, New York, NY
Matyssek R, Gayler S, zu Castell W, Oßwald W, Ernst D, Pretzsch H, Schnyder H, Munch JC (2012a) Predictability of plant resource allocation: new theory needed? In: Matyssek R, Schnyder H, Osswald W, Ernst D, Munch JC, Pretzsch H (eds) Growth and defence in plants—resource allocation at multiple scales, vol 220, Ecological studies. Springer, Berlin, pp 433–449
Matyssek R, Koricheva J, Schnyder H, Ernst D, Munch JC, Osswald W, Pretzsch H (2012b) The balance between resource sequestration and retention: a challenge in plant science. In: Matyssek R, Schnyder H, Osswald W, Ernst D, Munch JC, Pretzsch H (eds) Growth and defence in plants—resource allocation at multiple scales, vol 220, Ecological studies. Springer, Berlin, pp 3–24
Matyssek R, Lüttge U (2013) Gaia: the planet holobiont. Nova Acta Leopoldina NF 114 (391):325–344
Mayr E (1942) Systematics and the origin of species from a viewpoint of a zoologist. Harvard University Press, Cambridge, MA
Mayr E (2002) What evolution is. Orion Books, London
Mazmanian SK, Liu CH, Tzianabos AO, Kasper DL (2005) An immunomodulatory molecule of symbiotic bacteria directs maturation of the host immune system. Cell 122:107–118
McCutcheon JP, Moran NA (2012) Extreme genome reduction in symbiotic bacteria. Nat Rev Microbiol 10:13–26
Meysman FJR, Bruers S (2007) A thermodynamic perspective on food webs: quantifying entropy production within detrital-based ecosystems. J Theor Biol 249:124–139
Nikolic I, Kasmire J (2013) Theory. In: Dam KH, Nikolic I, Lukszo Z (eds) Agent-based modelling of socio-technical systems. Springer, Dordrecht, pp 11–71
Okada H, Kuhn C, Feillet H, Bach J-F (2010) The 'hygiene hypothesis' for autoimmune and allergic diseases: an update. Clin Exper Immun 160:1–9
Palmer RG, Arthur WB, Holland JH, LeBaron B, Tayler P (1994) Artificial economic life: a simple model of a stockmarket. Physica D 75:264–274
Pena R, Offermann C, Simon J, Maumann PS, Geßler A, Holst J, Dannenmann M, Mayer H, Kögel-Knabner I, Rennenberg H, Polle A (2010) Girdling affects ectomycorrhizal fungal (EMF) diversity and reveals functional differences in EMF community composition in a beech forest. Appl Environ Microbiol 76:1831–1841
Pepper JW, Herron MD (2008) Does biology need an organism concept? Biol Rev 83:621–627
Pozo MJ, Azcón-Aguilar C (2007) Unraveling mycorrhiza-induced resistance. Curr Opin Plant Biol 10:393–398

Prigogine I, Nicolis G, Babloyants A (1972) Thermodynamics of evolution. Phys Today. Part I 15:23–28, Part II 25:38–44
Rennenberg H, Dannenmann M, Gessler A, Kreuzwieser J, Simon J, Papen H (2009) Nitrogen balance in forest soils: nutritional limitation of plants under climate change stresses. Plant Biol 11(Suppl 1):4–23
Reynolds CW (1987) Flocks, herds, and schools: a distributed behavioral model. ACM SIGGRAPH Comp. Graphics 21(4):25–34
Richards EJ (2006) Inherited epigenetic variation—revisiting soft inheritance. Nat Rev Genet 7:395–401
Rook GAW, Brunet LR (2005) Microbes, immunoregulation, and the gut. Gut 54:317–320
Rosenberg E, Koren O, Reshef L, Efrony R, Zilber-Rosenberg I (2007) The role of microorganisms in coral health, disease and evolution. Nat Rev Microbiol 5:355–362
Rosenberg E, Sharon G, Zilber-Rosenberg I (2009) The hologenome theory of evolution contains Lamarckian aspects within a Darwinian framework. Environ Microbiol 11(12):959–2962
Rosenberg E, Sharon G, Atad I, Zilber-Rosenberg I (2010) The evolution of animals and plants via symbiosis with microorganisms. Environ Microbiol Rep 2(4):500–506
Rosselló-Mora R, Amann R (2001) The species concepts for prokaryotes. FEMS Microbiol Rev 25 (1):39–67
Ruse M (1989) Do organisms exist? Am Zool 29(3):1061–1066
Şahin E, Winfield A (2008) Special issue on swarm robotics. Swarm Intell 2:69–72
Sapp J (2011) Lamarckian leaps in the microbial world. In: Gissis SB, Jablonka E (eds) Transformations of Lamarckism. From subtle fluids to molecular biology. MIT Press, Cambridge, MA, pp 271–282
Schrödinger E (1944) What is life? Cambridge University Press, Cambridge
Tansley AG (1935) The use and abuse of vegetational concepts and terms. Ecology 16:284–307
Thellier M, Lüttge U (2013) Plant memory: a tentative model. Plant Biol 15:1–12
Turner MG, Rommer WH, Tinker DB (2003) Surprises and lessons from the 1988 Yellowstone fires. Front Ecol Environ 1(7):351–358
Vellend M (2010) Conceptual synthesis in community ecology. Q Rev Biol 85(2):183–206
Weiss BL, Wang J, Aksoy S (2011) Tsetse immune system maturation requires the presence of obligate symbionts in larvae. PLoS Biol 9, e1000619
Wilson DS, Sober E (1989) Reviving the superorganism. J Theor Biol 136:337–356
Wilson DS, Sober E (1994) Reintroducing group selection to the human behavioral sciences. Behav Brain Sci 17(4):585–654
Wilson DS, Wilson EO (2008) Evolution 'for the good of the group'. Am Sci 96(5):380–389
Wooldridge M (2002) Introduction to multiagent systems. Wiley, Chichester
Wynne-Edwards VC (1962) Animal dispersion in relation to social behaviors. Hafner Publishing, New York, NY
Wynne-Edwards VC (1986) Evolution through group selection. Blackwell, New York, NY
Zamioudis C, Pieterse CMJ (2012) Modulation of host immunity by beneficial microbes. Mol Plant Microbe Interact 25:139–150
Zenil H, Gershenson C, Marshall JAR, Rosenblueth DA (2012) Life as thermodynamic evidence of algorithmic structure in natural sciences. Entropy 14:2173–2191
Zilber-Rosenberg I, Rosenberg E (2008) Role of microorgnisms in the evolution of animals and plants: the hologenome theory of evolution. FEMS Microbiol 32:723–735

Advances in Genetic Diversity Analysis in Fruit Tree Crops

Nerea Larrañaga and José Ignacio Hormaza

Contents

Abstract Critical advances in the application of molecular tools to analyze genetic diversity in annual and perennial crops have taken place in the last two decades. Although in most cases the new technologies are first developed in annual crops, some particularities of most fruit crops, such as their perennial nature, long generation time, large individual size, or vegetative propagation, make the advantages of using these new approaches even more relevant in these species. In this work, the information available on the different strategies used to analyze genetic diversity in temperate and tropical fruit tree species using molecular approaches is reviewed. Special attention is given to the potential of next-generation sequencing and the combination of genomic tools with geographic information systems.

N. Larrañaga • J.I. Hormaza (✉)
Instituto de Hortofruticultura Subtropical y Mediterránea La Mayora (IHSM-UMA-CSIC), Estación Experimental La Mayora, Algarrobo-Costa, Málaga 29750, Spain
e-mail: ihormaza@eelm.csic.es

U. Lüttge et al. (eds.), *Progress in Botany*, Progress in Botany 77,
DOI 10.1007/978-3-319-25688-7_8

1 Introduction

Conservation and sustainable management of crop genetic resource diversity is the key for guaranteeing food security for future generations since this will determine the ability of populations and crops as a whole to adapt to environmental changes. This is becoming increasingly important in the current scenario of global climatic change where conserving genomic pools that will allow the development of resilient cultivars to a wide range of biotic and abiotic stresses is an increasing necessity. In this sense, it is important to keep in mind that conservation of plant genetic resources worldwide has been concentrated on a small group of species. Thus, from about 7,000 plant species that have been indeed used by humans (Hammer 2003), currently about 30 species contribute to more than 90 % of human nutrition globally, and just three cereals (wheat, rice, and maize) account for about two-thirds of human dietary needs (Cassman 1999). Moreover, in order to be effective in the long term, this conservation effort should also include crop wild relatives that can be source of interesting genes (Ford-Lloyd et al. 2011).

Of the 30 most important crops in terms of production worldwide, only five can be considered as fruit crops (bananas, oranges, grapes, apples, and mangoes) (Faostat 2014) although perennial crops account for about one-eighth of the world cultivated area (McClure et al. 2014). However, there are hundreds of neglected and underutilized fruit crops that are important in specific local markets and which cultivation can be extended in the future. As an example, it is considered that less than 5 % of the edible tropical fruit trees are cultivated commercially (Wijeratnam 2000) and just five species [banana (several species of the genus *Musa*), mango (*Mangifera indica*), pineapple (*Ananas comosus*), avocado (*Persea americana*), and papaya (*Carica papaya*)] comprise more than 90 % of the export of tropical fruits of those countries. In many cases, especially for underutilized fruit tree crops, most of the extant diversity is present in developing countries, and, consequently, the long-term conservation of their genetic resources is at risk since the capacity for studying and managing germplasm collections, including qualified personnel, is generally directly related to the degree of economic development of the country where the collection is situated.

Common characteristics of most fruit crops compared to domesticated annual crops are their perennial nature, long generation cycles, and long juvenile periods that often limit breeding programs and hinder genetic conservation efforts. Additionally, most fruit crops are highly heterozygous and usually propagated vegetatively favoring the cultivation of a few clones of selected cultivars. As a result, often the genetic diversity found in most of the commercially cultivated fruit tree crops is low (McClure et al. 2014). However, while most annual crops have been domesticated over a long period of time, only very few woody perennial tree species can be considered as truly domesticated. Consequently, most are separated by few generations from their wild relatives which can become interesting sources of additional genes.

The significant loss in biodiversity during the last decades has resulted in an increasing interest worldwide in conservation of plant genetic resources as shown through different national/international agreements and treaties such as the Convention on Biological Diversity in 1992 (http://www.cbd.int/history/) or the International Treaty on Plant Genetic Resources for Food and Agriculture in 2001 (http://www.planttreaty.org/), among many others. Genetic diversity, which can be defined as the amount of genetic variability among individuals of a variety, population, or species (Brown 1983), can be preserved through the conservation of germplasm. This can be performed both in situ and ex situ, two complementary approaches that, especially for the in situ component, should be integrated into a sustainable development strategy in the regions where biodiversity is present. In situ conservation of agricultural biodiversity can be carried out through the preservation of protected wild areas and/or the preservation of traditional cultivated fields ("on farm" conservation). The latter also includes the accompanying traditional agricultural methods and knowledge of local farmers and preserves interaction that takes place in ecosystems and, thus, the coevolution between all the elements that generate new diversity. Ex situ conservation programs preserve isolated genotypes (e.g., vegetative field collections, seeds, or in vitro culture banks). Most current ex situ germplasm collections show important limitations in the number of accessions and in the genetic variability and quality of the samples conserved. This situation is aggravated for certain species with recalcitrant seeds, such as many fruit trees, where conservation is not performed through seed collections but in living field collections which are very expensive to maintain. At least two complementary approaches can be envisaged to reduce maintenance costs in perennial fruit tree species. One approach involves the development of core collections that represent most of the diversity conserved in a small number of genotypes (van Hintum et al. 2000; Escribano et al. 2008). Another approach is based on a shift toward on-farm management of agrobiodiversity as a form of in situ conservation for the use of plant genetic resources in a dynamic production system (Hammer 2003). In any case, the establishment of appropriate measures for the conservation of the extant diversity in any crop species requires the availability of appropriate information on the distribution and structure of its current diversity.

Characterization of conserved and existent but as yet still unexplored genetic diversity is necessary to preserve the genetic resources of crop species for a sustainable use by future generations. The optimization of the management of genetic resources requires not only a precise identification and evaluation of the accessions already conserved but also to establish where hot spots of diversity exist in order to provide guidelines for future conservation of as yet non-collected variability. Recent advances in molecular genetics and the application of geographic information systems could represent new opportunities both for the management of biodiversity in situ and for the conservation of biodiversity in ex situ germplasm collections.

In this chapter we will focus on the most popular approaches used in recent years to study genetic diversity of fruit trees, with special attention to molecular techniques and the application of spatial analysis tools.

2 Molecular Markers and Genetic Diversity in Fruit Tree Crops

Traditionally, genotype identification and diversity studies in cultivated plant species including fruit trees have been carried out through morphological markers and phenotypic trait characterization. However, this is a slow and expensive process that limits the number of genotypes that can be studied as well as the accuracy of the observations. Some of those observations, like those related to phenological characters, need to be analyzed late in development, and some can affect other morphological characters or agronomical traits due to pleiotropic gene action (Le Corre and Kremer 2003). Although these observations can still be very useful tools to match phenotype with genotype and a good phenotyping is ultimately needed to link molecular markers with traits of interest, genotype characterization and genetic diversity studies in plants require research strategies that complement phenotypic characterization and estimation of diversity such as molecular techniques (Wünsch and Hormaza 2002). During the last two decades, fast and important advances in the methods used to study nucleic acids in both animals and plants have taken place resulting in the continuous development of different types of genetic markers. These allow reliable estimations of allelic richness reflected in the number and distribution of alleles per locus, which is the main parameter to measure genetic diversity. This information can be used to analyze the population structure of in situ and ex situ germplasm collections and wild stands of cultivated species and crop wild relatives in order to develop appropriate strategies to optimize the conservation of genetic diversity.

2.1 Main Types of Molecular Markers Used in Genetic Diversity Analyses in Fruit Trees

Molecular markers can be biochemical or based on DNA sequences. Biochemical markers involve the separation of proteins (allozymes and isozymes) into specific banding patterns by electrophoresis. They are codominant, and their main disadvantage is that the number of available enzymes is limited and, thus, their power to analyze diversity is reduced (Mondini et al. 2009). They have been used since the 1970s to analyze genetic diversity in plants (Brown 1978), but they can still provide valuable information for diversity studies in fruit trees shown in some recent works in apple (Wagner et al. 2014), hazelnut (Leinemann et al. 2013), or *Arbutus unedo* (Takrouni et al. 2012).

DNA-based markers, commonly known as molecular markers, have been continuously developed since the 1980s and can be associated or not to a functional trait; they can, in general terms, be classified into dominant (such as RAPDs or AFLPs) or codominant (such as RFLPs, microsatellites, or SNPs) markers depending on whether heterozygous and homozygous genotypes can be distinguished. RFLPs, widely used since Alec Jeffreys discovered the polymorphism in the fragment size after treating the DNA with restriction enzymes (Jeffreys 1979), have been applied in numerous works to assess fruit tree diversity; although due to the availability of new molecular approaches their application for genetic diversity studies is in disuse, some recent examples can still be found in fruit tree crops (Garcia-Ruiz et al. 2013).

Since the development of PCR in the mid-1980s by the Noble Prize winner in Chemistry 1993, Kary Mullis (Mullis et al. 1986), different types of PCR-based molecular markers have been developed. Among those routinely used in ecological, evolutionary, taxonomic, phylogenetic, and genetic diversity studies are randomly amplified polymorphic DNAs (RAPDs), in which fragments of genomic DNA are amplified using a decamer primer of random sequence and where polymorphism depends upon the presence or absence of an amplification product. They started to be used in different organisms in the late 1980s (Williams et al. 1990) and have been used extensively in a wide range of fruit tree crops; recent examples include pomegranate (Mansour et al. 2015), olive (Brake et al. 2014), apricot (Yilmaz et al. 2012), or mango (Samal et al. 2012). Amplified fragment length polymorphisms (AFLPs) (Vos et al. 1995) combine RFLP and PCR techniques. They have also been used in studies of diversity of different plant species (Mba and Tohme 2005); recent examples in fruit trees include *Prunus mira* (Li et al. 2014a), date palm (Sabir et al. 2014), or kiwifruit (Li et al. 2014b). Simple sequence repeats (SSRs) or microsatellites (Litt and Luly 1989), based in the presence of repetitive DNA sequences in eukaryote genomes flanked by specific conserved regions that allow their amplification by PCR, became the markers of choice for fingerprinting and diversity analyses in recent years (Ellegren 2004; Kalia et al. 2011; Arias et al. 2012). Some recent examples of their use for diversity studies in fruit trees include grapevine (Basheer-Salimia et al. 2014), apricot (Raji et al. 2014; Martin et al. 2011), chestnut (Beghe et al. 2013), litchi (Madhou et al. 2013), or pear (Sehic et al. 2012). Additional PCR-based markers used to some extent are inter-simple sequence repeats (ISSRs), applied, for example, in fig (Amel et al. 2004, 2005; Ikegami et al. 2009) or clementine (Breto et al. 2001); cleaved amplified polymorphisms (CAPs) used, for example, in *Citrus* spp. (Amar et al. 2011); sequenced characterized amplified regions (SCARs) used, for example, in persimmon (Cho et al. 2013); or sequence-tagged sites and expressed sequence tags (STS and ESTs, respectively) used, for example, in *Musa acuminata* (Passos et al. 2012) or *Actinidia* spp. (Crowhurst et al. 2008). Other less used molecular markers include start codon targeted polymorphisms (SCoT) that uses single primers designed to anneal to the flanking regions of the ATG initiation codon on both DNA strands and has been used in *Mangifera indica* (Gajera et al. 2014; Luo et al. 2011) and SRAP and TRAP, which stand for sequence-related amplified polymorphism and targeted region amplified polymorphism, respectively, that have been used in several fruit tree species. In the case of SRAP, recent works in fruit trees include apricot (Li et al. 2014c), persimmon (Jing et al. 2013a), or almond (Jing et al. 2013b), and, in the case of TRAP, apple (Guo et al. 2009). Another group of markers is based on mobile elements; examples include inter-retrotransposon amplified polymorphism (IRAP) that has been used in *Musa* spp. (Nair et al. 2005) or *Diospyros* spp. (Du et al. 2009a, b), the retrotransposon-microsatellite amplified polymorphism (REMAP) that has been used in *Diospyros* spp. (Du et al. 2009a, b) or *Prunus mume* (Shen et al. 2011), retrotransposon-based insertion polymorphism (RBIP) in *Pyrus pyrifolia* (Kim et al. 2012), or the retrotransposon-based sequence-specific amplification polymorphism (SSAP) in *Anacardium occidentale* (Syed et al. 2005) or *Diospyros* spp. (Du et al. 2009a, b).

More recently, simple nucleotide polymorphisms (SNPs), which are variations of single nucleotides which do not change the overall length of the DNA sequence in the region and occur throughout the genome, are being increasingly used (FAO 1997; de Vicente et al. 2004) especially due to the increasingly more economic availability of next-generation sequencing methods (see next section).

Whereas the previous markers usually target nuclear DNA, chloroplast and mitochondrial DNA can also be used to some extent to estimate genetic diversity although they are more useful at the interspecific level due to the usually low level of intraspecific polymorphism. Chloroplast and mitochondrial DNA analysis are of common use in barcode and phylogenetic studies in plant and animals, respectively, and they can be useful tools to assign unknown sample to the correct species and avoid errors that can affect the results obtained in genetic diversity analyses (Larrañaga and Hormaza 2015).

2.2 *Evolution of the Use of Different DNA-Based Markers for Diversity Analyses in Fruit and Nut Crops*

Although genetic diversity studies in annual crops clearly outnumber those in woody perennial crops, a great number of studies have been published on the application of molecular markers to study genetic diversity in temperate and tropical fruit tree crops. Thus, a basic search in the ISI web of knowledge for the seven most common types of markers (AFLP, ISSR, RAPD, SSR, RFLP, SNP +EST, and NGS), with the keyword "diversity", and 32 important genera of temperate and tropical tree fruit crops (*Actinidia*, *Anacardium*, *Annona*, *Averrhoa*, *Carica*, *Carya*, *Castanea*, *Citrus*, *Cocos*, *Cydonia*, *Dimocarpus*, *Diospyros*, *Eryobotria*, *Ficus*, *Juglans*, *Litchi*, *Malus*, *Mangifera*, *Musa*, *Olea*, *Passiflora*, *Persea*, *Phoenix*, *Pistacia*, *Prunus*, *Psidium* and *Vitis*) provides the following results as of February 2015: 1,229 studies for SSRs, 776 for RAPDs, 418 for AFLPs, 236 for RFLPs, 193 for ISSRs, 159 for SNP+EST, and 25 for NGS. These results are represented in Fig. 1 by year to show the variation in the number of studies with time. It is interesting to note that, although new molecular marker technologies are being developed continuously, some of the ones used 20 years ago (such as RAPDs) are still being used.

Molecular markers have been mainly used in fruit tree species commercially important at a global level; those include, for example, *Prunus* spp., *Malus* spp.,

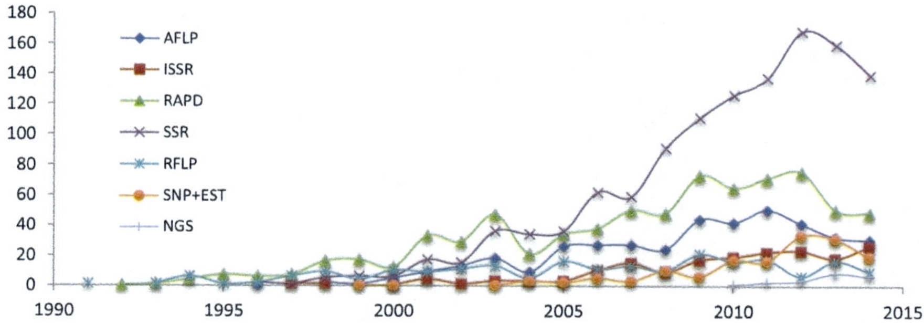

Fig. 1 Number of studies per year and per type of molecular marker shown by the Web of knowledge

Citrus spp., *Vitis vinifera, or Theobroma cacao*. However, some works have also been performed in crops that can be considered as "neglected" or "underutilized" species such as the Ethiopian banana, *Ensete ventricosum* (Birmeta et al. 2004); highland papayas, *Vasconcellea* sp. (van Droogenbroeck et al. 2004, 2006); cherimoya, *Annona cherimola* (Escribano et al. 2007; van Zonneveld et al. 2012); guava, *Psidium guajava* (Coser et al. 2012); or baobab, *Adansonia digitata* (Munthali et al. 2013). An additional set of studies have also focused on fruit tree crop wild relatives (CWR) such as *Rubus idaeus* (Graham et al. 2003); *Pistacia khinjuk* and *Pistacia atlantica* (Shanjani et al. 2009); *Olea europaea* subsp. *cuspidata* (Koehmstedt et al. 2011); *Malus sylvestris*, *M. orientalis*, and *M. sieversii* (Cornille et al. 2013); or *Citrus macroptera* (Malik et al. 2013). The characterization and conservation of genetic diversity in CWRs is becoming of increasing interest especially in the current context of climate change since they can be useful to incorporate into crops a range of useful genes to increase resiliency and productivity (Warschefsky et al. 2014). Clearly, additional resources and international cooperation networks should be devoted to address diversity studies and germplasm conservation programs in those two last groups, neglected and CWR species (Ford-Lloyd et al. 2011; Mba et al. 2012).

2.3 *Interpretation of the Results Obtained: Parameters and Software Tools*

The variances or polymorphisms shown by the different molecular markers can be analyzed with a range of different diversity parameters based (i) on the number of variants (number of percentage of polymorphic loci or allelic richness, i.e., mean number of alleles per locus) or (ii) on the frequencies of the alleles obtained (effective population size, expected and observed heterozygosities, or fixation index). Genetic diversity between different (sub)populations/collections is based on significant differences in allele frequencies among the different groups that can be estimated with F-statistics, Rst, Gst, or AMOVA, among others. In addition, graphic representation of genetic distances and cluster analyses are usually performed to visualize relationships among the different genotypes studied. Although this is starting to change due to the increasing availability of new generation sequencing technologies, usually just a small portion of the genome is explored in most genetic diversity studies, and, thus, the selection of a correct amount of individuals and loci to be analyzed is of great importance to make reliable extrapolation from the estimators to the natural populations or germplasm collections (de Vicente et al. 2004; Mondini et al. 2009; Porth and El-Kassaby 2014). Different software packages have been developed in the last two decades to obtain these genetic diversity measures; among them, some of the most commonly used are Genepop (Raymond and Rousset 1995), Popgene (Yeh et al. 1997), Arlequin (Excoffier et al. 2005), PowerMarker (Liu and Muse 2005), Ntsys-pc (Rohlf 2008), GenAlEx (Peakall and Smouse 2012), and Mega (Tamura et al. 2013). In addition, several software packages based in R (www.r-project.org) also allow the analysis of marker data for genetic diversity studies; among them, the most commonly used are Hierfstat (Goudet 2005), Adegenet (Jombart 2008), and DiveRsity (Keenan et al. 2013). The genetic structure of populations can be assessed using a method developed by Pritchard et al. (2000) implemented in the software program Structure. The program assigns probable individual haplotypes to multiple clusters relying on the allele frequencies at each locus identifying clusters of individuals on the basis of their genotypes at multiple loci using a Bayesian approach.

3 Perspectives of Next-Generation DNA Sequencing (NGS) for Studies of Genetic Diversity

3.1 Next-Generation Sequencing

During the 1970s, two different sequencing methods were developed by Frederick Sanger and Allan Maxam and Walter Gilbert, respectively (Sanger et al. 1977; Maxam and Gilbert 1977). The first became widely used for more than 25 years. In 2001, when the first draft of the human genome project was released, the scientific community started to look for cheaper and other approaches not based on cloning fragments (Mardis 2008). It was not until 2004 that next-generation DNA sequencing methods, sometimes classified in second and third generation automatic sequencers, depending whether prior amplification of the DNA template is needed (Glenn 2011), appeared in the market. These technologies provide several advantages and have revolutionized many areas in biology since they allow genome-wide characterization and profiling of mRNAs, small RNAs, transcription factor regions, structure of chromatin and DNA methylation patterns, microbiology, and metagenomics (Ansorge 2009). The main characteristic of these techniques is the possibility of performing massive sequencing in a parallel manner, which means that the number of sequences obtained in a single run is much higher than before. But probably the most important fact that is making possible the widespread use of these technologies is their continuously decreasing prices. Generally studies based on NGS have been applied to a single or to a small number of samples although multiplexing is allowing the application of NGS to larger data sets (Harrison and Kidner 2011).

Although the sequencing biochemistry differs among the different methods, the basic steps of most of the next-generation sequencers are similar, involving DNA fragmentation, in vitro adaptor ligation, generation of clonally clustered amplicons, and cyclic sequencing (Shendure and Ji 2008). Massively parallel scale of data production requires enormous computational analysis that includes image analysis, signal processing, background subtraction, base calling, and quality assessment to produce the final sequence reads for each run (Mardis 2008). Magi et al. (2010) reviewed several available software packages for the alignment of sequence reads to a reference, base-calling and/or polymorphism detection, de novo assembly from paired or unpaired reads, structural variant detection, and genome browsing.

3.2 Applications of NGS in Fruit and Nut Tree Crops

These NGS techniques are allowing to obtain sequences of entire genomes. Although the number of sequenced genomes is much higher in annual crops, there are some examples of published whole genome sequences in fruit tree crops such as grapevine (*Vitis vinifera*, Jaillon et al. 2007), papaya (*Carica papaya*, Ming et al. 2008), apple (*Malus domestica*, Velasco et al. 2010), date palm (*Phoenix dactylifera*, Al-Dous et al. 2011), cacao (*Theobroma cacao*, Argout et al. 2011; Motamayor et al. 2013), peach (*Prunus persica*, Arus et al. 2012; Verde et al. 2013), Japanese apricot (*Prunus mume*, Zhang et al. 2012), banana (*Musa acuminata*, D'Hont et al. 2012), Chinese white pear (*Pyrus bretschneideri*, Wu et al. 2013), orange (*Citrus sinensis*, Xu et al. 2013), or wild banana (*Musa balbisiana*, Davey et al. 2013), and the number of species will surely increase dramatically in the next years.

NGS technologies are permitting to reach every nucleotide on a genome and, thus, variants, common and rare, can be discovered with the appropriate sequencing read coverage and algorithm methods to identify them (Koboldt et al. 2013). In other words, these methodologies allow to detect genomic variation on a wide scale, from single nucleotide polymorphisms (SNPs) to copy number variations in large sequences blocks (Mardis 2008), being a technology with huge potential to allow the study of genetic/genomic diversity, even in non-model species (Garvin et al. 2010). In this sense, one of the most promising recent uses of new generation sequencing technologies is the genotyping-by-sequencing (GBS) approach that provides a high-throughput and cost-effective tool for a genome-wide analysis of genetic diversity (Spindel et al. 2013; Peterson et al. 2014). One of the main advantages of this approach is that it does not require a reference genome and allows the development of thousands of SNPs becoming an excellent tool especially for species in which genomic data are limited, such as most fruit tree crops. Deschamps et al. (2012) reviewed the applications of this technology in plant science.

Recently, new generation sequencing approaches have started to be used for diversity analyses in some fruit tree crops. Thus, in date palm (*Phoenix dactylifera*), nine different varieties were analyzed using NGS technologies, and 3.5 million polymorphic sites were found, including more than 10,000 genic copy number variations (Al-Dous et al. 2011). Kaya et al. (2013) analyzed the transcriptome and genome of five different olive genotypes (*Olea europaea*) obtaining 126,542,413 reads of cDNA. After assembling, SNPs were filtered and 2,987 high-quality putative SNPs were identified. From three European pear (*Pyrus communis*) cultivars, using NGS technologies, a subset of 1,096 SNPs were developed into informative markers by combining them with a set of 7,692 apple SNPs (Montanari et al. 2013).

Initiatives like the Genome Database for Rosaceae (GDR) (http://www.rosaceae.org/node/1) which is a curated and integrated web-based relational database that provides centralized access to Rosaceae genomics, genetics and breeding data and analysis tools to facilitate basic, translational, and applied Rosaceae research should be extended to other groups of fruit trees in the next future. In this sense, Russell et al. (2014) have recently developed an open-access, online database, tropiTree (http://bioinf.hutton.ac.uk/tropiTree). Here, simple sequence repeat (EST-SSR) markers for a range of 24 tropical tree species, encountered by bar-coded multiplexed paired-end Illumina NGS, are available to new genetic diversity studies. The sequences developed could also be used for SNP discovery. In fact, sequences already uploaded to databases could serve for developing this promising genome-wide genotyping application; for instance, Sun et al. (2013) downloaded sequences of two *Prunus mume* cultivars and developed SNPs, indels and SSRs from them.

4 Use of Geographic Tools and Landscape Genetics in the Study of Genetic Diversity

4.1 Geographic Information Systems

Spatial considerations have to be taken into account in order to understand evolution and, consequently, the development and maintenance of distinct genotypes, population differentiation, and, thus, diversity analysis and conservation (Jarvis et al. 2005). Understanding the spatial patterns of the genetic diversity and population structure of a species can contribute to improve knowledge of temporal and spatial dynamics (Thomas et al. 2012) and optimize diversity conservation and management projects by providing information on how diversity is distributed.

Guarino et al. (2002) stressed the importance of geographic information systems (GIS) as a tool to handle spatial data in graphic form like the location of the presence points and related different set of nonspatial data, such as species name and morphological, agronomical, or molecular data. GIS can then complement the information obtained from molecular markers to assess the influence of the environment on the presence of specific alleles providing new insights on genotype x environment interactions. GIS and ecogeographical analysis can then be applied for planning efficient germplasm collection, conservation, characterization, or management. The availability of increasing desktop processing power and the widespread use of GPS receivers allow a detailed study of the spatial components of diversity. There are several software packages available that can handle this type of information, and, besides allowing a spatial visualization of the data, they allow performing multiple statistical analyses. Among the free programs available online, some widely used are QGIS (http://www2.qgis.org/es/site/), DIVA-GIS (http://www.diva-gis.org/), or Grass (http://grass.osgeo.org/), whereas among those that require a license, the most popular is ArcGIS (https://www.arcgis.com/features). Several recent works analyzing genetic diversity of fruit trees have used a spatial approach for obtaining different diversity indexes both with morphological or molecular data: *Annona cherimola* (van Zonneveld et al. 2012) or *Theobroma cacao* (Thomas et al. 2012). The information obtained can be useful to propose different approaches for establishing priorities in conservation based on that diversity as has been shown in *Annona cherimola* (van Zonneveld et al. 2012) or *Prunus africana* (Vinceti et al. 2013).

4.2 Landscape Genetics

A step forward would be the application of landscape genetics, an approach that combines landscape ecology and population genetics to provide information about microevolutionary processes, such as gene flow and genetic drift and selection (Manel et al. 2003; Manel and Holderegger 2013). The main goal of this new approach is to combine the high resolution obtained from molecular information with spatial data and a variety of statistical methods to evaluate the role that landscape variables play in shaping genetic diversity and population structure (Storfer et al. 2007). A recent and thorough review (Storfer et al. 2010) has described the different topics that have been so far addressed by landscape genetics which include identifying specific barriers to dispersal, quantifying diversity, inferring the effects of landscape change, identifying migrants in relation to landscape conditions, estimating source-sink dynamics, predicting spread of disease or invasive species, and comparing observed genetic patterns between contemporary and historic landscapes. According to Holderegger et al. (2010), three different approaches are currently the most used to assess genetic flow, landscape distance/resistance, the overlay technique, and the assessment of contemporary gene flow. Storfer et al. (2010) reported that by 2010, landscape studies had been used mainly on vertebrates (62 %), invertebrates (18 %), and plants (14.5 %), followed by bacteria (3 %), viruses (3 %), lichens (1 %), and fungi (0.5 %). Recently, landscape genetics has started to be applied to studies of genetic diversity and gene structure in some fruit tree crops such as *Castanea sativa* (Martin et al. 2012) or *Juglans regia* (Pollegioni et al. 2014). A new toolbox for ArcGIS has been created by Vandergast et al. (2011) to map patterns of genetic divergence and diversity using a landscape genetic approach.

5 Conclusions

Conservation and sustainable use of genetic diversity in fruit tree crops is a global commitment for present and future generations. Locating neutral genetic diversity could be a starting point for selecting material for both conservation and agronomic assessment/breeding programs. As it has been shown in this review, studies of genetic diversity in most fruit trees still use a combination of traditional morphological and molecular markers, although NGS methods will likely replace them since they provide larger amount of information for a continuously decreasing price. In addition, bioinformatic systems to analyze the increasingly larger data sets are also continuously improving both in terms of costs and analyzing power. On the other hand, it is necessary to keep in mind that the huge amount of molecular data that are becoming available will still need phenotypic studies. A paradigm shift that is already taking place is that, in many cases, appropriate phenotyping rather than genotyping is becoming the main limiting factor. Association between nucleotide sequences and functional traits will be needed for agronomic, genetic mapping, functional diversity, breeding, or evolutionary studies. The combination of geographic information systems and vast molecular data sets that will be available in the following years as a result of NGS has a huge potential to both estimate the genetic diversity in a spatial context and to understand the structure and evolution of populations and domestication processes. This would be applicable to both wild and cultivated fruit trees.

Acknowledgments This work was supported by the Spanish Ministerio de Economía y Competitividad—European Regional Development Fund, European Union (AGL2013-43732-R), and INIA (RF2012-00010 and RFP2012-00016). NL was supported by an FPI fellowship from the Spanish Ministerio de Economía y Competitividad.

References

Al-Dous EK, George B, Al-Mahmoud ME, Al-Jaber MY, Wang H, Salameh YM, Al-Azwani EK, Chaluvadi S, Pontaroli AC, DeBarry J, Arondel V, Ohlrogge J, Saie IJ, Suliman-Elmeer KM, Bennetzen JL, Kruegger RR, Malek JA (2011) *De novo* genome sequencing and comparative genomics of date palm (*Phoenix dactylifera*). Nat Biotechnol 29:521–527

Amar MH, Biswas MK, Zhang Z, Guo WW (2011) Exploitation of SSR, SRAP and CAPS-SNP markers for genetic diversity of Citrus germplasm collection. Sci Hortic 128:220–227
Amel SH, Mokhtar T, Salwa Z, Jihene H, Messaoud M, Abdelmajid R, Mohamed M (2004) Inter-simple sequence repeat fingerprints to assess genetic diversity in Tunisian fig (*Ficus carica* L.) germplasm. Genet Resour Crop Evol 51:269–275
Amel SH, Khaled C, Messaoud M, Mohamed M, Mokhtar T (2005) Comparative analysis of genetic diversity in two Tunisian collections of fig cultivars based on random amplified polymorphic DNA and inter simple sequence repeats fingerprints. Genet Resour Crop Evol 52:563–573
Ansorge WJ (2009) Next-generation DNA sequencing techniques. New Biotechnol 25:195–203
Argout X, Salse J, Aury JM, Guiltinan MJ, Droc G, Gouzy J, Allegre M, Chaparro C, Legavre T, Maximova SN, Abrouk M, Murat F, Fouet O, Poulain J, Ruiz M, Roguet Y, Rodier-Goud M, Barbosa-Neto JF, Sabot F, Kudrna D, Ammiraju JSS, Schuster SC, Carlson JE, Sallet E, Schiex T, Dievart A, Kramer M, Gelley L, Shi Z, Bérard A, Viot C, Boccara M, Risterucci AM, Guignon V, Sabau X, Axtell MJ, Ma Z, Zhang Y, Brown S, Bourge M, Golser W, Song X, Clement D, Rivallan R, Tahi M, Akaza JM, Pitollat B, Gramacho K, D'Hont A, Brunel D, Infante D, Kebe I, Costet P, Wing R, McCombie WR, Guiderdoni E, Quetier F, Panaud O, Wincker P, Bocs S, Lanaud C (2011) The genome of *Theobroma cacao*. Nat Genet 43:101–108
Arias R, Borrone J, Tondo C, Kuhn D, Irish B, Schnell R (2012) Genomics of tropical fruit tree crops. In: Schnell RJ, Priyadarshan PM (eds) Genomics of tree crops. Springer, New York, pp 209–239
Arus P, Verde I, Sosinski B, Zhebentyayeva T, Abbott AG (2012) The peach genome. Tree Genet Genomes 8:531–547
Basheer-Salimia R, Lorenzi S, Batarseh F, Moreno-Sanz P, Emanuelli F, Grando MS (2014) Molecular identification and genetic relationships of Palestinian grapevine cultivars. Mol Biotechnol 56:546–556
Beghe D, Ganino T, Dall'Asta C, Silvanini A, Cirlini M, Fabbri A (2013) Identification and characterization of ancient Italian chestnut using nuclear microsatellite markers. Sci Hortic 164:50–57
Birmeta G, Nybom H, Bekele E (2004) Distinction between wild and cultivated enset (*Ensete ventricosum*) gene pools in Ethiopia using RAPD markers. Hereditas 140:139–148
Brake M, Migdadi H, Al-Gharaibeh M, Ayoub S, Haddad N, El Oqlah A (2014) Characterization of Jordanian olive cultivars (*Olea europaea* L.) using RAPD and ISSR molecular markers. Sci Hortic 176:282–289
Breto MP, Ruiz C, Pina JA, Asins MJ (2001) The diversification of *Citrus clementina* Hort. ex Tan., a vegetatively propagated crop species. Mol Phylogenet Evol 21:285–293
Brown AHD (1978) Isozymes, plant population genetic structure and genetic conservation. Theor Appl Genet 52:145–157
Brown W (1983) Genetic diversity and genetic vulnerability—an appraisal. Econ Bot 37:4–12
Cassman KG (1999) Ecological intensification of cereal production systems: yield potential, soil quality, and precision agriculture. Proc Natl Acad Sci USA 96:5952–5959
Cho KH, Cho KS, Han JH, Kim HR, Shin IS, Kim SH, Chun JA, Hwang HS (2013) Development of sequence characterized amplified region markers for cultivar identification in persimmon. Korean J Hortic Sci 31:798–806
Cornille A, Gladieux P, Giraud T (2013) Crop-to-wild gene flow and spatial genetic structure in the closest wild relatives of the cultivated apple. Evol Appl 6:737–748
Coser SM, da Silva Ferreira MF, Ferreira A, Mitre LK, Carvalho CR, Clarindo WR (2012) Assessment of genetic diversity in *Psidium guajava* L. using different approaches. Sci Hortic 148:223–229
Crowhurst RN, Gleave AP, MacRae EA, Ampomah-Dwamena C, Atkinson RG, Beuning LL, Bulley SM, Chagne D, Marsh KB, Matich AJ, Montefiori M, Newcomb RD, Schaffer RJ, Usadel B, Allan AC, Boldingh HL, Bowen JH, Davy MW, Eckloff R, Ferguson AR, Fraser LG, Gera E, Hellens RP, Janssen BJ, Klages K, Lo KR, MacDiarmid RM, Nain B, McNeilage MA,

Rassam M, Richardson AC, Rikkerink EHA, Ross GS, Schröder R, Snowden KC, Souleyre EJF, Templeton MD, Walton EF, Wang D, Wang MY, Wang YY, Wood M, Wu R, Yauk YK, Laing WA (2008) Analysis of expressed sequence tags from *Actinidia*: applications of a cross species EST database for gene discovery in the areas of flavor, health, color and ripening. BMC Genomics 9:351

D'Hont A, Denoeud F, Aury JM, Baurens FC, Carreel F, Garsmeur O, Noel B, Bocs S, Droc G, Rouard M, Da Silva C, Jabbari K, Cardi C, Poulain J, Souquet M, Labadie K, Jourda C, Lengellé J, Rodier-Goud M, Alberti A, Bernard M, Correa M, Ayyampalayam S, Mckain MR, Leebens-Mack J, Burgess D, Freeling M, Mbéguié-A-Mbéguié D, Chabannes M, Wicker T, Panaud O, Barbosa J, Hribova E, Heslop-Harrison P, Habas R, Rivallan R, Francois P, Poiron C, Kilian A, Burthia D, Jenny C, Bakry F, Brown S, Guignon V, Kema G, Dita M, Waalwijk C, Joseph S, Dievart A, Jaillon O, Leclercq J, Argout X, Lyons E, Almeida A, Jeridi M, Dolezel J, Roux N, Risterucci AG, Weissenbach J, Ruiz M, Glaszmann JC, Quétier F, Yahiaoui N, Wincker P (2012) The banana (*Musa acuminata*) genome and the evolution of monocotyledonous plants. Nature 488:213–217

Davey MW, Gudimella R, Harikrishna JA, Sin LW, Khalid N, Keulemans J (2013) A draft *Musa balbisiana* genome sequence for molecular genetics in polyploid, inter- and intra-specific *Musa* hybrids. BMC Genomics 14:683

De Vicente MC, López C, Fulton T (eds) (2004) Genetic diversity analysis with molecular marker data: learning module. International Plant Genetic Resources Institute (IPGRI), Rome

Deschamps S, Llaca V, May GD (2012) Genotyping-by-sequencing in plants. Biology 1:460–483

Du X, Zhang Q, Luo Z (2009a) Development of retrotransposon primers and their utilization for germplasm identification in *Diospyros* spp. (Ebenaceae). Tree Genet Genomes 5:235–245

Du XY, Zhang QL, Luo Z (2009b) Comparison of four molecular markers for genetic analysis in *Diospyros* L. (Ebenaceae). Plant Syst Evol 281:171–181

Ellegren H (2004) Microsatellites: simple sequences with complex evolution. Nat Rev Genet 5: 435–445

Escribano P, Viruel MA, Hormaza JI (2007) Molecular analysis of genetic diversity and geographic origin within an ex situ germplasm collection of cherimoya by using SSRs. J Am Soc Hortic Sci 132:357–367

Escribano P, Viruel MA, Hormaza JI (2008) Comparison of different methods to construct a core germplasm collection in woody perennial species with SSR markers. A case study in cherimoya (*Annona cherimola* Mill.), an underutilized subtropical fruit tree species. Ann Appl Biol 153:25–32

Excoffier L, Laval G, Schneider S (2005) Arlequin ver. 3.0: an integrated software package for population genetics data analysis. Evol Bioinform Online 1:47–50

FAO (1997) The state of the world's plant genetic resources for food and agriculture. Food and Agriculture Organization of the United Nations, Rome, p 501

Faostat (2014) FAO statistics database on the World Wide Web. http://faostat.fao.org/site/339/default.aspx. Accessed 12–2014

Ford-Lloyd BV, Schmidt M, Armstrong SJ, Barazani O, Engels J, Hadas R, Hammer K, Kell SP, Kang D, Khoshbakht K, Li Y, Long C, Lu BR, Ma K, Nguyen VT, Qiu L, Ge S, Wei W, Zhang Z, Maxted N (2011) Crop wild relatives-undervalued, underutilized and under threat? Bioscience 61:559–565

Gajera HP, Bambharolia RP, Domadiya RK, Patel SV, Golakiya BA (2014) Molecular characterization and genetic variability studies associated with fruit quality of indigenous mango (*Mangifera indica* L.) cultivars. Plant Syst Evol 300:1011–1020

Garcia-Ruiz MT, Mendoza-Castillo VM, Valadez-Moctezuma E, Muratalla-Lua A (2013) Initial assessment of natural diversity in Mexican fig landraces. Genet Mol Res 12:3931–3943

Garvin MR, Saitoh K, Gharrett AJ (2010) Application of single nucleotide polymorphisms to non-model species: a technical review. Mol Ecol Resour 10:915–934

Glenn TC (2011) Field guide to next-generation DNA sequencers. Mol Ecol Resour 11:759–769

Goudet J (2005) Hierfstat, a package for R to compute and test hierarchical F-statistics. Mol Ecol Notes 5:184–186

Graham J, Marshall B, Squire GR (2003) Genetic differentiation over a spatial environmental gradient in wild *Rubus ideaus* populations. New Phytol 157:667–675

Guarino L, Jarvis A, Hijmans RJ, Maxted N (2002) Geographic information systems (GIS) and the conservation and use of plant genetic resources. In: Engels JEA (ed) Managing plant genetic diversity. CAB International, Wallingford, pp 387–404

Guo L, Palumbo R, Zhang ZS, Wang GL, Tay D, Zhang DL, Shen X, Shu HR (2009) Target Region Amplification Polymorphism (TRAP) for evaluating genetic diversity in *Malus* Mill. genus. Hortscience 44:1117

Hammer K (2003) A paradigm shift in the discipline of plant genetic resources. Genet Resour Crop Evol 50:3–10

Harrison N, Kidner CA (2011) Next-generation sequencing and systematics: what can a billion base pairs of DNA sequence data do for you? Taxon 60:1552–1566

Holderegger R, Buehler D, Gugerli F, Manel S (2010) Landscape genetics of plants. Trends Plant Sci 15:675–683

Ikegami H, Nogata H, Hirashima K, Awamura M, Nakahara T (2009) Analysis of genetic diversity among European and Asian fig varieties (*Ficus carica* L.) using ISSR, RAPD, and SSR markers. Genet Resour Crop Evol 56:201–209

Jaillon O, Aury M, Noel B, Policriti A, Clepet C, Casagrande A, Choisne N, Aubourg S, Vitulo N, Jubin C, Vezzi A, Legeai F, Hugueney P, Dasilva C, Horner D, Mica E, Jublot D, Poulain J, Bruyère C, Billault A, Segurens B, Gouyvenoux M, Ugarte E, Cattonaro F, Anthouard V, Vico V, Del Fabbro C, Alaux M, Di Gaspero G, Dumas V, Felice N, Paillard S, Juman I, Moroldo M, Scalabrin S, Canaguier A, Le Clainche I, Malacrida D, Durand E, Pesole G, Laucou V, Chatelet P, Merdinoglu D, Delledonne M, Pezzotti M, Lecharny A, Scarpelli C, Artiguenave F, Pè ME, Valle G, Morgante M, Caboche M, Adam-Blondon AF, Weissenbach J, Quétier F, Wincke P (2007) The grapevine genome sequence suggests ancestral hexaploidization in major angiosperm phyla. Nature 449:463–467

Jarvis A, Yeaman S, Guarino L, Tohme J (2005) The role of geographic analysis in locating, understanding, and using plant genetic diversity. Methods Enzymol 395:279–298

Jeffreys AJ (1979) DNA sequence variants in the G gamma-, A gamma-, delta- and beta-globin genes of man. Cell 18:1–10

Jing ZB, Ruan X, Wang R, Yang Y (2013a) Genetic diversity and relationships between and within persimmon (*Diospyros* L.) wild species and cultivated varieties by SRAP markers. Plant Syst Evol 299:1485–1492

Jing ZB, Cheng J, Guo CH, Wang XP (2013b) Seed traits, nutrient elements and assessment of genetic diversity for almond (*Amygdalus* spp.) endangered to China as revealed using SRAP markers. Biochem Syst Ecol 49:51–57

Jombart T (2008) Adegenet: a R package for the multivariate analysis of genetic markers. Bioinformatics 24:1403–1405

Kalia RK, Rai MK, Kalia S, Singh R, Dhawan AK (2011) Microsatellite markers: an overview of the recent progress in plants. Euphytica 177:309–334

Kaya HB, Cetin O, Kaya H, Sahin M, Sefer F, Kahraman A, Tanyolac B (2013) SNP discovery by illumina-based transcriptome sequencing of the olive and the genetic characterization of Turkish olive genotypes revealed by AFLP, SSR and SNP markers. PLoS One 8(9):e73674

Keenan K, McGinnity P, Cross T, Crozier WW, Prodöhl PA (2013) DiveRsity: an R package for the estimation and exploration of population genetics parameters and their associated errors. Methods Ecol Evol 8:782–788

Kim H, Terakami S, Nishitani C, Kurita K, Kanamori H, Katayose Y, Sawamura Y, Saito T, Yamamoto T (2012) Development of cultivar-specific DNA markers based on retrotransposon-based insertional polymorphism in Japanese pear. Breed Sci 62:53–62

Koboldt DC, Steinberg KM, Larson DE, Wilson RK, Mardis ER (2013) The next-generation sequencing revolution and its impact on genomics. Cell 155:27–38

Koehmstedt AM, Aradhya MK, Soleri D, Smith JL, Polito VS (2011) Molecular characterization of genetic diversity, structure, and differentiation in the olive (*Olea europaea* L.) germplasm collection of the United States Department of Agriculture. Genet Resour Crop Evol 58: 519–531

Larrañaga N, Hormaza JI (2015) DNA barcoding of perennial fruit tree species of agronomic interest in the genus *Annona* (Annonaceae). Front Plant Sci 6:589

Le Corre V, Kremer A (2003) Genetic variability at neutral markers, quantitative trait loci and trait in a subdivided population under selection. Genetics 164:1205–1219

Leinemann L, Steiner W, Hosius B, Kuchma O, Arenhoevel W, Fussi B, Haase B, Kaetzel R, Rogge M, Finkeldey R (2013) Genetic variation of chloroplast and nuclear markers in natural populations of hazelnut (*Corylus avellana* L.) in Germany. Plant Syst Evol 299:369–378

Li TF, Liu JR, Xie YN, Wang QY, Meng FJ (2014a) Analysis of genetic diversity in *Prunus mira* Koehne ex Sargent populations using AFLP markers. Plant Syst Evol 300:475–482

Li D, Liu Y, Li X, Rao J, Yao X, Zhong C (2014b) Genetic diversity in kiwifruit polyploid complexes: insights into cultivar evaluation, conservation, and utilization. Tree Genet Genomes 10:1451–1463

Li M, Zhao Z, Miao X (2014c) Genetic diversity and relationships of apricot cultivars in North China revealed by ISSR and SRAP markers. Sci Hortic 173:20–28

Litt M, Luly JA (1989) A hypervariable microsatellite revealed by in vitro amplification of a dinucleotide repeat within the cardiac muscle actin gene. Am J Hum Genet 44:397–401

Liu K, Muse SV (2005) Power marker: an integrated analysis environment for genetic marker analysis. Bioinformatics 21:2128–2129

Luo C, He X, Chen H, Ou S, Gao M, Brown JS, Tondo CT, Schnell RJ (2011) Genetic diversity of mango cultivars estimated using SCoT and ISSR markers. Biochem Syst Ecol 39:676–684

Madhou M, Normand F, Bahorun T, Hormaza JI (2013) Fingerprinting and analysis of genetic diversity of litchi (*Litchi chinensis* Sonn.) accessions from different germplasm collections using microsatellite markers. Tree Genet Genomes 9:387–396

Magi A, Benelli M, Gozzini A, Girolami F, Torricelli F, Brandi ML (2010) Bioinformatics for next generation sequencing data. Genes 1:294–307

Malik SK, Uchoi A, Kumar S, Choudhary R, Pal D, Kole PR, Chaudhury R, Bhat KV (2013) Molecular characterization of *Citrus macroptera* Montr. (satkara): an endangered wild species from northeast India. Plant Biosyst 147:857–863

Manel S, Holderegger R (2013) Ten years of landscape genetics. Trends Ecol Evol 28:614–621

Manel S, Schwartz MK, Luikart G, Taberlet P (2003) Landscape genetics: combining landscape ecology and population genetics. Trends Ecol Evol 18:189–197

Mansour E, Ben Khaled A, Triki T, Abid M, Bachar K, Ferchichi A (2015) Evaluation of genetic diversity among South Tunisian pomegranate (*Punica granatum* L.) accessions using fruit traits and RAPD markers. J Agric Sci Biotechnol 17:109–119

Mardis ER (2008) Next-generation DNA sequencing methods. Annu Rev Genomics Hum Genet 9:387–402

Martin C, Herrero M, Hormaza JI (2011) Molecular characterization of apricot germplasm from an old stone collection. PLoS One 6(8):e23979

Martin M, Mattioni C, Molina JR, Alvarez JB, Cherubini M, Herrera MA, Villani F, Martin LM (2012) Landscape genetic structure of chestnut (*Castanea sativa* Mill.) in Spain. Tree Genet Genomes 8:127–136

Maxam AM, Gilbert W (1977) A new method for sequencing DNA. Proc Natl Acad Sci USA 74: 560–564

Mba C, Tohme J (2005) Use of AFLP markers in surveys of plant diversity. Methods Enzymol 395:177–201

Mba C, Guimaraes E, Ghosh K (2012) Re-orienting crop improvement for the changing climatic conditions of the 21st century. Agric Food Secur 1:1–17

McClure KA, Sawler J, Gardner KM, Money D, Myles S (2014) Genomics: a potential panacea for the perennial problem. Am J Bot 101:1780–1790

Ming R, Hou S, Feng Y, Yu Q, Dionne-Laporte A, Saw JH, Senin P, Wang W, Ly BV, Lewis KL, Salzberg SL, Feng L, Jones MR, Skelton RL, Murray JE, Chen C, Qian W, Shen J, Du P, Eustice M, Tong E, Tang H, Lyons E, Paull RE, Michael TP, Wall K, Rice DW, Albert H, Wang ML, Zhu YJ, Schatz M, Nagarajan N, Acob RA, Guan P, Blas A, Wai CM, Ackerman CM, Ren Y, Liu C, Wang J, Wang J, Na JK, Shakirov EV, Haas B, Thimmapuram J, Nelson D, Wang X, Bowers JE, Gschwend AR, Delcher AL, Singh R, Suzuki JY, Tripathi S, Neupane K, Wei H, Irikura B, Paidi M, Jiang N, Zhang W, Presting G, Windsor A, Navajas-Pérez R, Torres MJ, Feltus FA, Porter B, Li Y, Burroughs AM, Luo MC, Liu L, Christopher DA, Mount SM, Moore PH, Sugimura T, Jiang J, Schuler MA, Friedman V, Mitchell-Olds T, Shippen DE, dePamphilis CW, Palmer JD, Freeling M, Paterson AH, Gonsalves D, Wang L, Alam M (2008) The draft genome of the transgenic tropical fruit tree papaya (*Carica papaya* Linnaeus). Nature 452:991–996

Mondini L, Noorani A, Pagnotta MA (2009) Assessing plant genetic diversity by molecular tools. Diversity 1:19–35

Montanari S, Saeed M, Knaebel M, Kim Y, Troggio M, Malnoy M, Velasco R, Fontana P, Won K, Durel CE, Perchepied L, Schaffer R, Wiedow C, Bus V, Brewer L, Gardiner SE, Crowhurst RN, Chagné D (2013) Identification of *Pyrus* single nucleotide polymorphisms (SNPs) and evaluation for genetic mapping in European pear and interspecific *Pyrus* hybrids. PLoS One 8(10):e77022

Motamayor JC, Mockaitis K, Schmutz J, Haiminen N, Livingstone D III, Cornejo O, Findley SD, Zheng P, Utro F, Royaert S, Saski C, Jenkins J, Podicheti R, Zhao M, Scheffler BE, Stack JC, Feltus FA, Mustiga GM, Amores F, Phillips W, Marelli JP, May GD, Shapiro H, Ma J, Bustamante CD, Schnell RJ, Main D, Gilbert D, Parida L, Kuhn DN (2013) The genome sequence of the most widely cultivated cacao type and its use to identify candidate genes regulating pod color. Genome Biol 14:r53

Mullis K, Faloona F, Scharf S, Saiki R, Horn G, Erlich H (1986) Specific enzymatic amplification of DNA in vitro: the polymerase chain reaction. Cold Spring Harb Symp Quant Biol 51: 263–273

Munthali CRY, Chirwa PW, Changadeya WJ, Akinnifesi FK (2013) Genetic differentiation and diversity of *Adansonia digitata* L. (baobab) in Malawi using microsatellite markers. Agrofor Syst 87:117–130

Nair AS, Teo CH, Schwarzacher T, Harrison PH (2005) Genome classification of banana cultivars from South India using IRAP markers. Euphytica 144:285–290

Passos MAN, de Oliveira CV, Emediato FL, de Camargo TC, Souza MT, Matsumoto T, Renno Azevedo VC, Ferreira CF, Amorim EP, de Alencar Figueiredo LF, Martins NF, de Jesus Barbosa Cavalcante M, Baurens FC, da Silva OB, GJ P Jr, Pignolet L, Abadie C, Ciampi AY, Piffanelli P, Miller RN (2012) Development of expressed sequence tag and expressed sequence tag-simple sequence repeat marker resources for *Musa acuminata*. AoB Plants 2012:pls030. doi:10.1093/aobpla/pls030

Peakall R, Smouse P (2012) GenAlEx 6.5: genetic analysis in Excel. Population genetic software for teaching and research—an update. Bioinformatics 28:2537–2539

Peterson GW, Dong Y, Horbach C, Fu Y (2014) Genotyping-by-sequencing for plant genetic diversity analysis: a lab guide for SNP genotyping. Diversity 6:665–680

Pollegioni P, Woeste KE, Chiocchini F, Olimpieri I, Tortolano V, Clark J, Hemery GE, Mapelli S, Malvolti ME (2014) Landscape genetics of Persian walnut (*Juglans regia* L.) across its Asian range. Tree Genet Genomes 10:1027–1043

Porth I, El-Kassaby YA (2014) Assessment of the genetic diversity in forest tree populations using molecular markers. Diversity 6:283–295

Pritchard JK, Stephens M, Donnelly P (2000) Inference of population structure using multilocus genotype data. Genetics 155:945–959

Raji R, Jannatizadeh A, Fattahi R, Esfahlani MA (2014) Investigation of variability of apricot (*Prunus armeniaca* L.) using morphological traits and microsatellite markers. Sci Hortic 176: 225–231

Raymond M, Rousset F (1995) GENEPOP (version 1.2): population genetics software for exact tests and ecumenicism. J Hered 86:248–249

Rohlf FJ (2008) NTSYSpc: numerical taxonomy system, ver. 2.20. Exeter Publishing, Setauket, NY

Russell JR, Hedley PE, Cardle L, Dancey S, Morris J, Booth A, Odee D, Mwaura L, Omondi W, Angaine P, Machua J, Muchugi A, Milne I, Kindt R, Jamnadass R, Dawson IK (2014) TropiTree: an NGS-based EST-SSR resource for 24 tropical tree species. PLoS One 9:e102502

Sabir JSM, Abo-Aba S, Bafeel S, Zari TA, Edris S, Shokry AM, Atef A, Gadalla NO, Ramadan AM, Al-Kordy MA, El-Domyati FM, Jansen RK, Bahieldin A (2014) Characterization of ten date palm (*Phoenix dactylifera* L.) cultivars from Saudi Arabia using AFLP and ISSR markers. C R Biol 337:6–18

Samal KC, Jena RC, Swain SS, Das BK, Chand PK (2012) Evaluation of genetic diversity among commercial cultivars, hybrids and local mango (*Mangifera indica* L.) genotypes of India using cumulative RAPD and ISSR markers. Euphytica 185:195–213

Sanger F, Nicklen S, Coulson AR (1977) DNA sequencing with chain-terminating inhibitors. Proc Natl Acad Sci USA 74:5463–5467

Sehic J, Garkava-Gustavsson L, Fernandez-Fernandez F, Nybom H (2012) Genetic diversity in a collection of European pear (*Pyrus communis*) cultivars determined with SSR markers chosen by ECPGR. Sci Hortic 145:39–45

Shanjani PS, Mardi M, Pazouki L, Hagidimitriou M, Avanzato D, Pirseyedi SM, Ghaffari MR, Nekoui SMK (2009) Analysis of the molecular variation between and within cultivated and wild *Pistacia* species using AFLPs. Tree Genet Genomes 5:447–458

Shen Y, Ding X, Wang F, Cai B, Gao Z, Zhang Z (2011) Analysis of genetic diversity in Japanese apricot (*Prunus mume* Sieb. et Zucc.) based on REMAP and IRAP molecular markers. Sci Hortic 132:50–58

Shendure J, Ji H (2008) Next-generation DNA sequencing. Nat Biotechnol 26:1135–1145

Spindel J, Wright M, Chen C, Cobb J, Gage J, Harrington S, Lorieux M, Ahmadi N, McCouch S (2013) Bridging the genotyping gap: using genotyping by sequencing (GBS) to add high-density SNP markers and new value to traditional bi-parental mapping and breeding populations. Theor Appl Genet 126:2699–2716

Storfer A, Murphy MA, Evans JS, Goldberg CS, Robinson S, Spear SF, Dezzani R, Delmelle E, Vierling L, Waits LP (2007) Putting the 'landscape' in landscape genetics. Heredity 98: 128–142

Storfer A, Murphy MA, Spear SF, Holderegger R, Waits LP (2010) Landscape genetics: where are we now? Mol Ecol 19:3496–3514

Sun LD, Zhang QX, Xu ZD, Yang WR, Guo Y, Lu JX, Pan HT, Cheng TR, Cai M (2013) Genome-wide DNA polymorphisms in two cultivars of mei (*Prunus mume* Sieb. et Zucc.). BMC Genet 14:98

Syed NH, Sureshsundar S, Wilkinson MJ, Bhau BS, Cavalcanti JJV, Flavell AJ (2005) Ty1-copia retrotransposon-based SSAP marker development in cashew (*Anacardium occidentale* L.). Theor Appl Genet 110:1195–1202

Takrouni MM, Ali IB, Messaoued C, Boussaid M (2012) Genetic variability of Tunisian wild strawberry tree (*Arbutus unedo* L.) populations interfered from isozyme markers. Sci Hortic 146:92–98

Tamura K, Stecher G, Peterson D, Filipski A, Kumar S (2013) MEGA6: molecular evolutionary genetics analysis version 6.0. Mol Biol Evol 30:2725–2729

Thomas E, van Zonneveld M, Loo J, Hodgkin T, Galluzzi G, van Etten J (2012) Present spatial diversity patterns of *Theobroma cacao* L. in the neotropics reflect genetic differentiation in Pleistocene refugia followed by human-influenced dispersal. PLoS One 7:e47676

van Droogenbroeck B, Kyndt T, Maertens I, Romeijn-Peeters E, Scheldeman X, Romero-Motochi JP, Van Damme P, Goetghebeur P, Gheysen G (2004) Phylogenetic analysis of the highland papayas (*Vasconcellea*) and allied genera (Caricaceae) using PCR-RFLP. Theor Appl Genet 108:1473–1486 van Droogenbroeck B, Kyndt T, Romeijn-Peeters E, van Thuyne W, Goetghebeur P, Romero-Motochi JP, Gheysen G (2006) Evidence of natural hybridization and introgression between *Vasconcellea* species (Caricaceae) from southern Ecuador revealed by chloroplast, mitochondrial and nuclear DNA markers. Ann Bot 97:793–805

van Hintum T, Brown AHD, Spillane C, Hodgkin T (2000) Core collections of plant genetic resources. IPGRI Technical Bulletin No. 3. International Plant Genetic Resources Institute, Rome

van Zonneveld M, Scheldeman X, Escribano P, Viruel MA, Van Damme P, Garcia W, Tapia C, Romero J, Siguenas M, Hormaza JI (2012) Mapping genetic diversity of cherimoya (*Annona cherimola* Mill.): application of spatial analysis for conservation and use of plant genetic resources. PLoS One 7:e29845

Vandergast AG, Perry WM, Lugo RV, Hathaway SA (2011) Genetic landscapes GIS Toolbox: tools to map patterns of genetic divergence and diversity. Mol Ecol Resour 11:158–161

Velasco R, Zharkikh A, Affourtit J, Dhingra A, Cestaro A, Kalyanaraman A, Fontana P, Bhatnagar SK, Troggio M, Pruss D, Salvi S, Pindo M, Baldi P, Castelletti S, Cavaiuolo M, Coppola G, Costa F, Cova V, Dal Ri A, Goremykin V, Komjanc M, Longhi S, Magnago P, Malacarne G, Malnoy M, Micheletti D, Moretto M, Perazzolli M, Si-Ammour A, Vezzulli S (2010) The genome of the domesticated apple (*Malus* x *domestica* Borkh.). Nat Genet 42:833–839

Verde I, Abbott AG, Scalabrin S, Jung S, Shu S, Marroni F, Zhebentyayeva T, Dettori MT, Grimwood J, Cattonaro F, Zuccolo A, Rossini L, Jenkins J, Vendramin E, Meisel LA, Decroocq V, Sosinski B, Prochnik S, Mitros T, Policriti A, Cipriani G, Dondini L, Ficklin S, Goodstein DM, Xuan P, Del Fabbro C, Aramini V, Copetti D, Gonzalez S, Horner DS, Falchi R, Lucas S, Mica E, Maldonado J, Lazzari B, Bielenberg D, Pirona R, Miculan M, Barakat A, Testolin R, Stella A, Tartarini S, Tonutti P, Arús P, Orellana A, Wells C, Main D, Vizzotto G, Silva H, Salamini F, Schmutz J, Morgante M, Rokhsar DS (2013) The high-quality draft genome of peach (*Prunus persica*) identifies unique patterns of genetic diversity, domestication and genome evolution. Nat Genet 45:487–494

Vinceti B, Loo J, Gaisberger H, van Zonneveld M, Schueler S, Konrad H, Kadu CAC, Geburek T (2013) Conservation priorities for *Prunus africana* defined with the aid of spatial analysis of genetic data and climatic variables. PLoS One 8:e59987

Vos P, Hogers R, Bleeker M, Reijans M, van de Lee T, Hornes M, Fritjters A, Pot J, Peleman J, Kuiper M, Zabeau M (1995) AFLP: a new technique for DNA fingerprinting. Nucleic Acids Res 23:4407–4414

Wagner I, Maurer WD, Lemmen P, Schmitt HP, Wagner M, Binder M, Patzak P (2014) Hybridization and genetic diversity in wild apple (*Malus sylvestris* (L.) Mill) from various regions in Germany and from Luxembourg. Silvae Genet 63:81–94

Warschefsky E, Penmetsa RY, Cook DR, von Wettberg EJB (2014) Back to the wilds: tapping evolutionary adaptations for resilient crops through systematic hybridization with crop wild relatives. Am J Bot 101:1791–1800

Wijeratnam RSW (2000) Identification of problems in processing of underutilized fruits of the tropics and their solutions. Acta Hortic 518:237–240

Williams JGK, Kuberik AR, Livak KJ, Rafalski JA, Tingey SV (1990) DNA polymorphisms amplified by arbitrary primers are useful as genetic markers. Nucleic Acids Res 18:6531–6535

Wu J, Wang Z, Shi Z, Zhang S, Ming R, Zhu S, Khan MA, Tao S, Korban SS, Wang H, Chen NJ, Nishio T, Xu X, Cong L, Qi K, Huang X, Wang Y, Zhao X, Wu J, Deng C, Gou C, Zhou W, Yin H, Qin G, Sha Y, Tao Y, Chen H, Yang Y, Song Y, Zhan D, Wang J, Li L, Dai M, Gu C, Wang Y, Shi D, Wang X, Zhang H, Zeng L, Zheng D, Wang C, Chen M, Wang G, Xie L, Sovero V, Sha S, Huang W, Zhang S, Zhang M, Sun J, Xu L, Li Y, Liu X, Li Q, Shen J, Wang J, Paull RE, Bennetzen JL, Wang J, Zhang S (2013) The genome of the pear (*Pyrus bretschneideri* Rehd.). Genome Res 23:396–408

Wünsch A, Hormaza JI (2002) Cultivar identification and genetic fingerprinting of temperate fruit tree species using DNA markers. Euphytica 125:59–67

Xu Q, Chen LL, Ruan X, Chen D, Zhu A, Chen C, Bertrand D, Jiao WB, Hao BH, Lyon MP, Chen J, Gao S, Xing F, Lan H, Chang JW, Ge X, Lei Y, Hu Q, Miao Y, Wang L, Xiao S, Biswas MK, Zeng W, Guo F, Cao H, Yang X, Xu XW, Cheng YJ, Xu J, Liu JH, Luo OJ, Tang Z, Guo WW, Kuang H, Zhang HY, Roose ML, Nagarajan N, Deng XX, Ruan Y (2013) The draft genome of sweet orange (*Citrus sinensis*). Nat Genet 45:59–66

Yeh FC, Yang RC, Boyle TBJ, Ye ZH, Mao JX (1997) POPGENE, the user-friendly shareware for population genetic analysis. Molecular Biology and Biotechnology Centre, University of Alberta, Edmonton

Yilmaz KU, Paydas-Kargi S, Dogan Y, Kafkas S (2012) Genetic diversity analysis based on ISSR, RAPD and SSR among Turkish apricot germplasms in Iran Caucasian eco-geographical group. Sci Hortic 138:138–143

Zhang Q, Chen W, Sun L, Zhao F, Huang B, Yang W, Tao Y, Wang J, Yuan Z, Fan G, Xing Z, Han C, Pan H, Zhong X, Shi W, Liang X, Du D, Sun F, Xu Z, Hao R, Lv T, Lv Y, Zheng Z, Sun M, Luo L, Cai M, Gao Y, Wang J, Yin Y, Xu X, Cheng T, Wang J (2012) The genome of *Prunus mume*. Nat Commun 3:1318

Transposon Activation Tagging in Plants for Gene Function Discovery

Matthias Fladung

Contents

Abstract In the last 20 years, activation tagging or gain-of-function mutagenesis has become a very powerful tool to reveal gene functions in plant reverse genetics. The idea behind activation tagging is to overexpress endogenous genes by random insertion of a DNA sequence as a tag-carrying enhancer or promoter, but without changing transcript patterns. The approaches employed so far mainly comprise two different DNA molecules as tag, either T-DNA or transposable elements (transposons). T-DNA activation tagging is strongly based on classical transformation approaches and is only feasible for plant species with well-developed transformation protocols. The basis for transposon tagging is the breakthrough observation that transposable elements are active in heterologous plant species following transformation and are able to pass chromosome boundaries. Thus, only few transgenic lines are needed for transposon-based activation tagging. Examples for successful transposon activation tagging are provided for some plant species, with particular focus on the tree genus *Populus*.

M. Fladung (✉)
Thünen-Institute of Forest Genetics, Sieker Landstr. 2, 22927 Grosshansdorf, Germany
e-mail: matthias.fladung@ti.bund.de

U. Lüttge et al. (eds.), *Progress in Botany*, Progress in Botany 77,
DOI 10.1007/978-3-319-25688-7_9

1 Introduction

Transposable elements (TEs) or transposons were first identified in maize (*Zea mays*) by Barbara McClintock in 1948 (McClintock 1950). She investigated the unstable inheritance of mosaic color patterns of maize seeds and found two mobile genetic loci that could change their positions on the chromosome: dissociator (Ds) and *activator* (*Ac*). Later, many other transposons were identified in maize, but also in many other plant and animal species (Huang et al. 2012), e.g., different transposon classes as listed in the Arabidopsis Information Resource (TAIR) database (https://www.arabidopsis.org/index.jsp), or the *Tam* transposable elements in *Antirrhinum* (Döring and Starlinger 1986; Gierl and Saedler 1992), or the so-called P elements, a very famous family of transposons in the fruit fly (*Drosophila melanogaster*) (Spradling and Rubin 1982), but also in mammals (including humans) (Giordano et al. 2007), in yeasts (Boeke and Sandmeyer 2009), and in bacteria (http://www.ndsu.edu/pubweb/~mcclean/plsc431/transelem/trans5.htm).

Irrespective of their occurrence in any organism, transposons are separated into two classes, Class I (retrotransposons) and Class II (DNA transposons). Both differ in the cut-and-paste mechanism of transposition, i.e., retrotransposons involve an RNA intermediate and a reverse transcriptase, while transposition of Class II DNA transposons is catalyzed by a special transposase enzyme. Transposons from both classes transpose either "autonomously" or "nonautonomously." An autonomous transposition means that transposons are able to excise and reintegrate on their own (Fig. 1), while nonautonomous transposons are not themselves mobile and need other transposons to move. A prominent example for an autonomous transposon is

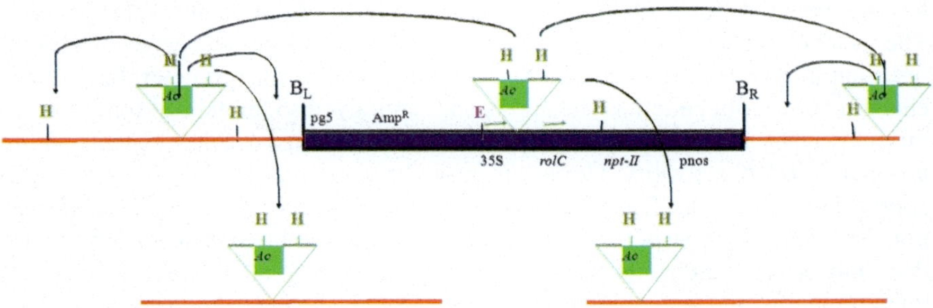

Fig. 1 Schematic summary of autonomous transposition of a transposable element. A DNA transposon excises out of an original position and reintegrates into new genomic position either close or distant (even on a unrelated chromosome) to the original position, with possibility of subsequent transpositions out of the new position. In case that reintegration occurs near a gene locus, the resulting variant is either dominant (and thus detectable) or (in most cases) recessive and not be expressed because of the heterozygous allelic situation. In the latter situation, self-crossing is needed for creating homozygous alleles and an expressible recessive mutation. The transposon sequence can then be used as a probe to screen a genomic library or to design specific primer for PCR-based screens

the already mentioned *activator* element (*Ac*) discovered by B. McClintock, while the Ds element lacks the transposase and is not able to transpose; thus, it is called nonautonomous.

It has been shown that transposons comprise a large percentage of the eukaryotic genome, e.g., about 12 % in *Drosophila* and 35 % in humans (Labrador and Corces 1997). The function of transposons is still under discussion (Fedoroff 2000). First they were considered to be "junk" DNA without any function or to be "selfish" DNA (Charlesworth et al. 1994). However, it has been shown that transposons are in particular present in heterochromatic regions of the genome (Dimitri and Junakovic 1999). Examples have been presented that excision and in particular reinsertion of transposons provide a considerable source of genomic regulatory variation, which might play a significant role during evolution (Fedoroff 2012; Lisch 2013). The ability of transposons to move and to reinsert somewhere in the genome has led to a number of successful gene isolations in these species, following a gene-tagging protocol (Walbot 1992). Here, I will review the scientific activities that to apply transposons as tags for gain-of-function, or activation tagging approaches in heterologous plant species to unravel the functions of genes. The approaches are compared with T-DNA-based activation tagging, and examples for successful T-DNA and transposon activation tagging are described and compared for the tree genus *Populus*.

2 Transposon Tagging in Host or Heterologous Plant Species

For maize, extensive transposon collections are available in different databases (e.g., http://www.maizegdb.org/ or http://maizetedb.org/~maize/). These provide researchers the opportunity to initiate large transposon tagging experiments either to unravel genome structure or to analyze gene functions (Walbot 1992; Brutnell 2002). The best investigated systems including *Ac*/*Ds*, *En*/*Spm*, and *Mu* have been widely used for transposon tagging and exist in multiple copies within the maize genome (Settles 2009). Several important genes have been isolated following transposon tagging in maize, e.g., *OPAQUE*-2 (Schmidt et al. 1987), *BRONZE2* (*Bz2*; McLaughlin and Walbot 1987), *TASSELSEED2* (*TS2*; DeLong et al. 1993), and *GLOSSY2* (Tacke et al. 1995). Transposons have been well investigated and applied for tagging approaches and gene discovery in only few other plant species, e.g., floral homeotic genes in snapdragon (*Antirrhinum majus*, Martin et al. 1985; Carpenter and Coen 1990) and the *ANTHOCYANIN2* gene in petunia (Quattrocchio et al. 1999).

More successful was the extension of the host range of the well-developed maize transposable elements to species containing no or only weakly characterized endogenous transposons. One of the first approaches was the genetic transfer of the maize transposon *Ac* to tobacco, with the breakthrough observation that this maize element is functional in transgenic tobacco (Baker et al. 1986, 1987; Spena et al. 1989). Functionally active means that both excision and reintegration of the foreign transposon have to be demonstrated in the transgenic plants. Also the inactive *Ac* derivative Ds could be successfully transferred to tobacco, and induced transposition of Ds has been demonstrated in crosses with *Ac* transgenic tobacco plants (Hehl and Baker 1989). Besides tobacco, the functionality of the heterologous transposons has been demonstrated for *Arabidopsis* (Van Sluys et al. 1987; Cardon et al. 1993), carrot (Van Sluys et al. 1987), potato (Knapp et al. 1988; Pereira et al. 1991), rice (Murray et al. 1991), and tomato (Jones et al. 1992; Yoder et al. 1988), as well as poplar (Fladung and Ahuja 1997). The observation that maize transposable elements remain capable of transposition when transformed into other plant species suggests that these elements can be used as gene tags in plants that do not necessarily have characterized endogenous elements (Ostergaard and Yanofsky 2004; Chuck et al. 1993; Belzile and Yoder 1992; Whitham et al. 1994; James et al. 1995; Lawrence et al. 1995; Martienssen 1998; Greco et al. 2001; Scholz et al. 2001).

Based on these results, protocols for insertional mutagenesis were established in various annual plant species (Parinov et al. 1999; Meissner et al. 2000; Greco et al. 2003; McKenzie and Dale 2004). Two different transposable element systems (*Ac*/*Ds* and *En*/*Spm*) have been most frequently employed to establish successful transposon tagging systems in the genome of heterologous (transgenic) plant species with the main aim of tagging genes (Rommens et al. 1991; Coomber and Feldmann 1993; Hehl 1994). Insertional mutants were produced and some of the tagged genes were successfully isolated, e.g., in *Arabidopsis* a male sterility gene (Aarts et al. 1993), the *DRL1* locus (developmental mutant; Bancroft et al. 1993), or the *FATTY ACID ELONGATION1* (*FAE1*) gene (James et al. 1995); in petunia a flower color gene (Chuck et al. 1993); in tomato the *Cf-9* gene for resistance to *Cladosporium fulvum* (Jones et al. 1994); and in tobacco the resistance gene N that mediates resistance to tobacco mosaic virus (TMV) (Dinesh-Kumar et al. 1995).

The genetic transfer of a transposable element into the genome of a tree species was first reported by Howe et al. (1991, 1994) for poplar. However, no evidence could be provided by these authors that the transferred transposon is mobile, i.e., excises out of its original position and reintegrates somewhere in the genome. By employing the *rolC* reporter gene system developed by Schmülling et al. (1988) for tobacco, Fladung (1990) for potato, and Nilsson et al. (1996) and Fladung et al. (1996) for poplar (Fig. 2), the genomic integration of the *Ac* transposon as well as its excision and reintegration (transposition) was indicated visually and confirmed by Northern blot experiments (Fladung and Ahuja 1997; Kumar and Fladung 2003a). Transposition of the *Ac* element could then be confirmed in sequencing of empty donor sites revealing in most cases precise excisions of *Ac* (Fladung 1999) and PCR and Southern blot experiments (Fladung 1999; Kumar and Fladung 2003a).

Fig. 2 Leaf of 35S-*Ac-rolC* transgenic poplar revealing *pale* and *dark green* sectors. *Large* and *small pale green* sectors result from *Ac* transposition early and late during leaf development, respectively

3 Applicability of the Transposable Elements for Transposon Tagging in Heterologous Plant Species

Even if indications exist that transposons are mobile in transgenic plant species and reintegrate in or near coding regions of genes, it would be useful to calculate the frequency of insertional mutagenesis. In *Arabidopsis*, Parinov et al. (1999) analyzed Ds flanking sequences from 931 independent transposed lines and found that from 511 lines (55 %), flanking sequences were identical or homologous to DNA or protein sequences stored in public databases. Disruptions within known or putative genes were indicated for 354 lines (38 %) (Parinov et al. 1999). Meissner et al. (2000) produced 2,932 tomato F_3 families with transposed and stabilized Ds elements and suggested a high rate of Ds insertions into genes. Out of 50 sequenced Ds insertion sites, 28 were similar to known genes or ESTs (56 %). In rice, new *Ac* transpositions were generated at a frequency of 15–50 % in different lines, with two-thirds of *Ac*-tagged sites showing homology to sequences in public databases that were categorized as probable genes (Greco et al. 2001). The overall frequency of genomic reinsertion of the *Ac* transposon in poplar varies from 46 % to 73 % and into coding sequences by nearly 60 % (Fladung 2011). The frequencies reported for the different plant species are sufficient to claim the suitability of a heterologous transposon system (in particular *Ac* and Ds elements) for tagging approaches.

Another question regarding the usability of the heterologous transposon system to tag genes is either the physical distance of the transposon jump or the ability of the transposon to cross chromosome barriers and thus to jump to other chromosomes. This question is of particular importance regarding the minimum number of independent primary transposon-carrying transgenic lines required to establish a reliable transposon tagging system. In *Arabidopsis*, high numbers of different *Ac* and Ds insertion sites were analyzed, and generally, the mapped insertions appeared to be evenly dispersed throughout the genome, even though insertion site clusters and significant preferences for reinsertion sites have been identified (Fig. 1; Parinov et al. 1999; Raina et al. 2002; Ito et al. 2005). Following mapping of Ds element inserts in *Brassica oleracea*, McKenzie and Dale (2004) could localize the original Ds insertion site as well as 17 Ds insertions, which were spread over six (out of nine) linkage groups. For poplar, Fladung (2011) analyzed the position sites of *Ac* reinsertions in three independent *Ac* transposition lines and confirmed the earlier findings. It was demonstrated that 52–82 % of all transposition events involve reintegration sites located on other chromosomes. Neither a preference to any chromosome nor "hot spots" of transposon reintegration sites could be found (Fladung 2011).

However, regardless of the system described, the highest numbers of insertion sites of the *Ac* or Ds elements were, as expected, in the chromosomes on which the original start loci were located. But despite this restriction, in consequence, only a few primary transposon-carrying transgenic lines are sufficient to initiate a large transposon tagging experiment to saturate a genome with transposon reintegration sites.

4 Transposon Versus T-DNA Tagging

Besides employing transposable elements for gene tagging in transgenic plants, an alternative tagging strategy is the use of the insertional event of the *Agrobacterium*-mediated T-DNA itself as a tag. This idea is nearly as old as the technique of gene technology in plants (Koncz et al. 1989), based on the observation that some of the regenerated transformed plants had a particular morphology which could have resulted from the inactivation of plant genes by T-DNA insertion (André et al. 1986). For a T-DNA-based tagging approach, the receptive plant species should be easily transformable (in most cases in tissue culture) to obtain a high number of tagged lines (Qu et al. 2008). Indeed, this has been shown to be true for only a limited number of plant species like *Arabidopsis*, tobacco, rice, poplar, and few others (Fladung 2014).

A second prerequisite for T-DNA tagging is that the T-DNA insertions are random and stable. Szabados et al. (2002) assessed the efficiency of T-DNA insertion mutagenesis and sequenced flanking regions of 1,000 T-DNA insertions, locating their positions in the *Arabidopsis* genome. The majority of T-DNA insertions were found in chromosomal regions of high gene density (Szabados et al. 2002). A similar result was obtained for rice, where more than 1,000 T-DNA tag-flanking sequences were mapped on 12 rice chromosomes (Chen et al. 2003). The authors could clearly show that the T-DNA tags were not randomly spread on rice chromosomes and were preferentially inserted in gene-rich regions. In both reports, obvious biases were found for the insertions in the 5′ and 3′ regulatory regions outside the coding regions (Szabados et al. 2002; Chen et al. 2003). The reported distribution patterns and biases for T-DNA integration in *Arabidopsis* and rice may be a result from the mechanism of T-DNA integration itself.

Nevertheless, T-DNA-based tagging has been shown to be successful in only few heterologous plant species. Initial attempts comprised Ti-plasmid vectors containing promoterless reporter genes suitable for trapping of plant genes. For tobacco, André et al. (1986) and Koncz et al. (1989) constructed a vector with a promoterless APH(3′)II reporter sequence as a T-DNA tag. Koncz et al. (1989) showed that on average, frequency of successful reporter gene expression was about 30 % in both *Arabidopsis* and tobacco. One year later, the same group reported the isolation of a gene encoding a novel chloroplast protein by T-DNA tagging in *Arabidopsis* (Koncz et al. 1990). Also in *Arabidopsis*, Feldmann (1991) produced more than 8,000 independent transformants and screened them for visible alterations in phenotype. More than 1,000 putative mutants could be detected. In rice, Jeon et al. (2000) generated about 25,700 taggings using a promoterless β-*GLUCURONIDASE* (GUS) reporter gene. Only about 2 % of tested organs revealed GUS expression; however, this large population of T-DNA-tagged lines was foreseen as a resource to discover new genes in rice. By also employing a promoterless GUS reporter gene, Fobert et al. (1994) generated a transgenic tobacco plant that expressed GUS activity only in developing seed coats. The authors concluded that T-DNA insertion occurred along with the cryptic promoter controlling a seed coat-specific gene.

Comparing the two strategies for gain-of-function mutagenesis, transposon, and T-DNA tagging, both methods have advantages and disadvantages for the generation of mutant phenotypes with the long-term goal of gene cloning (Walbot 1992; Fladung and Polak 2012). Usually, there is a compromise between the ease of mutant generation and the ease of cloning (Walbot 1992). For T-DNA tagging, numerous transformations have to be performed to obtain a high number of tagged lines (Qu et al. 2008). The situation is more convenient for transposon tagging since the transposon, once transferred, is able to jump to other chromosomes, and thus only a few primary transposon transgenic lines are required (Fladung 2011; Fladung and Polak 2012). However, most transposable elements behave autonomously, thus a system is needed to control transposon activity.

Insertion of numerous copies of both tags, T-DNA and transposon, increases the probability of producing more mutants; however, this strategy requires more effort to match a particular element to the mutant phenotype. On the other hand, the aim of cloning genes with available phenotypes justifies the higher work investment (Walbot 1992). The strategy for gene cloning for both T-DNA and transposon tagging is firstly to phenotype an existing tagged population and secondly to determine the new genomic insertion loci of the tag (Fladung and Polak 2012).

To saturate the genome with T-DNA tags is time consuming and only possible with plant species which are very easily transformable, e.g., *Arabidopsis*. This has, among other things, led to the commercial availability of huge collections of T-DNA-tagged transgenic *Arabidopsis* lines, e.g., the GABI-Kat system developed by Li et al. (2003) and Rosso et al. (2003). With a simple online search tool for gene code (line ID or GenBank accession number), annotation text or sequence-based search using BLAST, a gene hit search result page, appears containing all available information, e.g., on a possible responding GABI-Kat line ID, a link to the FST sequence, and a link to a visualization of the gene (http://www.gabi-kat.de/). In case a hit is obtained, seeds from the respective *Arabidopsis* mutant can be ordered. Unfortunately, GABI-Kat database and seed storage facility were closed by the end of December 2014 due to a grant shutdown.

5 Activation Tagging

There is, however, an important limitation for classical T-DNA and the transposon tagging strategy. Insertion of these elements as a tag in coding regions or promoters normally results mostly in recessive loss-of-function mutants by disrupting the gene. Loss-of-function mutagenesis is very successful in plants, where a self-fertilization system exists, e.g., tobacco and *Arabidopsis*. In plants with self-incompatibility barriers or dioecious sexuality (a situation present in most of the *Populus* species), or even in plants with in part very long vegetative cycles (e.g., trees), loss-of-function mutagenesis is less practicable (Fig. 1). Other limitations of loss-of-function mutagenesis are the low probability of tagging genes leading to early lethality or redundantly acting genes (Weigel et al. 2000). Genes absolutely required during multiple stages of the life cycle and thus essential for early survival, or whose function can be compensated for by alternative regulatory pathways, and so are not always easy to identify by loss-of-function mutagenesis (Bolle et al. 2011).

The concern of structural redundancy of genes has become in particular evident along with the increasing number of sequencing efforts of eukaryotic genomes. In many of these genomes, complete duplication events have been postulated through sequence similarity studies, i.e., many duplicated genes exist, potentially also functioning redundantly. Insertion of tagging elements in functionally redundant genes could hamper the detection of respective phenotypes, because one or more other family members can provide the same function and complement the mutant phenotype. By employing loss-of-function mutagenesis, genes that are not absolutely required for a certain pathway can still be identified if such genes are sufficient to activate that pathway.

A solution to circumvent the difficulties described is the so-called gain-of-function strategy, i.e., mutants are obtained based on increased levels of gene expression (Walden et al. 1994; Kakimoto 1996; Weigel et al. 2000; Nakazawa et al. 2003; Ayliffe and Pryor 2007). Gain-of-function phenotypes are dominant and are caused either by a spontaneous mutation/genomic rearrangement or by insertion of a tag (heterologous mutagenesis), all leading to constitutive overexpression of a gene and thus increased production of the respective protein. Homologous mutations that alter levels of gene expression were described for dominant *Antp* mutants in *Drosophila* (Schneuwly et al. 1987) and for dominant ethylene response mutants in *Arabidopsis* (Chang et al. 1993). Genomic rearrangements could also be responsible for bringing genes under the control of new promoters or enhancers that could result in an increase of expression of the respective genes.

Pioneering work in heterologous gain-of-function mutagenesis was initiated by J. Schell and colleagues from the Max Planck Institute for Plant Breeding in Cologne (Germany) who constructed a T-DNA vector with four copies of an enhancer element from the constitutively active cauliflower mosaic virus (CaMV) 35S promoter. This T-DNA vector was genetically transferred to tobacco to tag genes involved in the action of the plant growth substance auxin (Hayashi et al. 1992; Walden et al. 1994). The idea was that the enhancer elements integrate near coding regions and cause transcriptional activation of these genes (Suzuki et al. 2001). There is also no need to identify lines homozygous for the transgene because gain-of-function phenotypes are dominant. Bolle et al. (2011) mentioned that mutant phenotypes induced by loss-of-function and activation tagging approaches are often complementary to each other. This strategy can therefore be employed for studies of redundantly acting genes (Jeon and An 2001).

The presence of a vector (either T-DNA or transposable element) in any given genomic position, leading to activation of a nearby gene due its structural composition, was also the reason to term the gain-of-function strategy "activation tagging." Activation tagging has been shown to be a powerful tool to induce gain-of-function mutants in different plant species (Weigel et al. 2000). Important work has been performed with both T-DNA and transposable elements as tags in *Arabidopsis*, rice, and poplar, but also in other plant species.

5.1 Arabidopsis

Activation tagging was mainly developed for *Arabidopsis* by using T-DNA as tag. In most cases, the T-DNA carrying enhancers or promoters (e.g., four tandemly arranged copies of the CaMV 35S enhancer; Tani et al. 2004) were inserted randomly into the genome. Programs were initiated to establish large collections of activation-tagged populations. Schneider et al. (2005) reported the production of an activation tagging population in *Arabidopsis* (so-called TAMARA lines) and its possible application in the identification of dominant developmental and metabolic mutations. Another collection is the RIKEN *Arabidopsis* full-length (RAFL) cDNA resource that developed a large-scale database called RARGE, containing information from transcriptome to phenome (Sakurai et al. 2005). The *Arabidopsis* mutants contained full-length cDNAs in a correct orientation between the CaMV 35S promoter and the NOS terminator (Seki et al. 2002).

Examples for the isolation of single *Arabidopsis* genes by activation tagging have been published in numerous reports, e.g., the flowering-time gene *FLOWERING LOCUS T* (*FT*) (Kardailsky et al. 1999), the patatin-like *STURDY* gene (Huang et al. 2001), a conserved *MYB* regulator of phenylpropanoid biosynthesis (Borevitz et al. 2000), and the *LEAFY PETIOLE* gene affecting the leaf petiole development (van der Graaff et al. 2000). Masaki et al. (2005) isolated an

Arabidopsis activation-tagged line with increased expression of the *ASML2* gene coding for a protein belonging to a class of the "CONSTANS, CONSTANS-like, TOC1"- (CCT-) domain that activates subset of sugar-inducible genes. To study the regulation of tapetum formation and pollen development in *Arabidopsis*, an activation-tagged *SHI-RELATED SEQUENCE* gene mutant was identified revealing disrupted anther dehiscence and abnormal floral organ development (Kim et al. 2010). The isolation of specific *Arabidopsis* mutants has also been described with mutations in, e.g., flavin monooxygenases (Woodward et al. 2005) and genes involved in photosynthesis that are expressed in calli (Niwa et al. 2006), sexual reproduction (Perrella et al. 2006), root patterning (Nakajima et al. 2006), and salt tolerance (Ahmad et al. 2007). Kang et al. (2010) screened an *Arabidopsis* activation-tagged population to identify genes that suppress a weak brassinosteroid receptor mutant.

An alternative gain-of-function gene tagging approach has been described by Ichikawa et al. (2006) by introducing the "full-length cDNA overexpressing (FOX) gene hunting system". The authors produced about 10,000 independent transgenic *Arabidopsis* lines carrying full-length cDNAs under control of the CaMV 35S promoter by in planta transformation. Possible morphological mutants were detected in 1,487 of a total of 15,547 transformants.

Another technique for activation tagging has been introduced by Pogorelko et al. (2008) describing a new and easier way to analyze *Arabidopsis* gain-of-function mutants by transferring two different vectors. The first vector contained multiple copies of the transcriptional enhancer from the cauliflower mosaic virus 35S gene flanked by two loxP sites, and the second vector contained the *CRE* gene (Pogorelko et al. 2008). In about 10 % of the double transgenic lines, constitutive ectopic expression of genes adjacent to the T-DNA insertion were observed causing development of the mutant phenotype. Also, reversion of the mutants to the wild-type phenotype after removing the CaMV enhancer has been demonstrated (Pogorelko et al. 2008). Aboul-Soud et al. (2009) super-transformed an already transgenic *Arabidopsis* line carrying the PR1::*LUCIFERASE* gene with activation T-DNA tags to screen for constitutive LUC activity. LUC imaging was then used to identify activated disease resistance 2 (*adr2*) lines. Waki et al. (2013) developed a GAL4-based targeted activation tagging system in *Arabidopsis* to identify genes with regulatory functions that are not readily identified by conventional screening methods, e.g., seedling lethality.

A few approaches have employed transposon systems for activation tagging in *Arabidopsis*. Marsch-Martinez et al. (2002) could identify four dominant activation-tagged morphological mutants using the En-I maize transposon system. Rosin et al. (2008) produced over 15,000 Ds-tagged *Arabidopsis* lines, each harboring a CaMV 35S::*LUCIFERASE* gene and a *lac* operator repeat within the Ds-tagging cassette. The combination of both systems in one reporter cassette enables a visual recognition of tagged loci in a living plant and, at the same time, determination of position effects (Rosin et al. 2008).

5.2 Rice and Cereals

Rice has become a model species for monocot plants because of the accumulating genome sequence information for this species (Goff et al. 2002; Yu et al. 2002). Many approaches have been undertaken to establish different activation-tagged populations in rice, and T-DNA flanking sequence-tag databases have been established following the molecular analysis of the isolated mutants (Jeong et al. 2006; Charng et al. 2007). The construction of large-scale activation tagging lines has been described by Mori et al. (2003). From these lines, a lesion mimic variant designated as leaf-spotted mutant was characterized (Mori et al. 2007) or a short grain mutant was isolated (Mori et al. 2006).

Jeong et al. (2002) have developed a new T-DNA-based activation tagging vector that can be used for promoter trapping and activation tagging of rice genes. Wan et al. (2009) developed a powerful transformation approach in rice with the pER38 activation tagging vector containing tandemly arranged double CaMV 35S enhancers next to the right border of T-DNA. Out of more than 50,000 individual transgenic rice plants generated, about 400 dominant mutants were selected (Wan et al. 2009). Lee et al. (2011) reported the generation of rice lines with an increased content of nicotianamine, a key compound in metal transport pathway and cell homeostasis, by activation tagging of the *NICOTIANAMINE SYNTHASE 2* (*OsNAS2*) gene. As a consequence of *OsNAS2* overexpression, the tagged line contained up to 20-fold more nicotianamine and 2.7-fold more zinc (Lee et al. 2011).

Another application of T-DNA activation tagging has been described for rice to identify glutamate receptor-like genes that enhance drought tolerance in plants (Lu et al. 2014a). More than 200,000 activation-tagged rice lines were produced with an expected coverage of more than 90 % of the rice genes. Indeed, one line identified revealed improved drought tolerance, and molecular analyses confirmed integration of the activation tag in a region with two glutamate receptor-like genes (Lu et al. 2014a).

For barley, the use of an activation tagging system based upon the maize *Ac*/Ds transposable element system was published (Ayliffe et al. 2007). Here, the modified Ds element system carries two maize polyubiquitin promoters and transposes at frequencies ranging from 0 % to 52 % per family (Ayliffe et al. 2007). Based upon their analyses, the authors conclude that this system is applicable to all aspects of plant development and biogenesis. Ayliffe and Pryor (2009) described the establishment of an activation tagging system in barley (*Hordeum vulgare* L.) based upon a maize transposable element that carries two highly expressed cereal promoters. The authors have also tested this system in wheat and confirmed transposon mobility.

5.3 Other Plant Species

Activation tagging constructs have also been transferred to plant species other than the ones described previously and have been tested for their usability for mutant isolation. Zubko et al. (2002) identified a *Petunia* T-DNA activation-tagged line that showed cytokinin-specific effects including enhanced shooting, reduced apical dominance, and delayed senescence and flowering. A T-DNA-tagged tomato line could be isolated that revealed an intense purple pigmentation based on activation of the transcriptional regulator gene *ANT1* of the anthocyanin biosynthetic pathway (Mathews et al. 2003). Also for tomato, Carter et al. (2013) developed an *Ac*/Ds transposon system for activation tagging and produced a population of 25 T0 plants from a single transformed line regenerated in tissue culture. From this T0, a T1 population was generated consisting of 11,000 selfed and cv M82 backcrossed progeny (Carter et al. 2013). Insertion sites of the transposon tags were determined and transposed lines carrying only the Ds element spanning all 12 tomato chromosomes were selected.

Transposon-based activation tagging has successfully been established, and activation-tagged mutants identified in other plant species have been employed for genetic improvement, e.g., in diploid strawberry and monoploid potato (Lu et al. 2014b), *Phalaenopsis*, and *Doritaenopsis* (Tsay et al. 2012). In tobacco, Ahad et al. (2003) screened a T-DNA activation-tagged population and selected mutants with tolerance to aryl carbamates (a blocker of microtubule assembly) and chilling tolerance that could survive for several months at 3 °C. And in *Lotus japonica*, a model legume, a T-DNA-based activation tagging approach has been described using a multifunctional vector for gene and activation tagging (Imaizumi et al. 2005).

But even for exotic plant species, e.g., *Salvia miltiorrhiza*, an important herb in traditional Chinese medicine, a T-DNA activation tagging mutagenesis system was successfully established (Lee et al. 2008). In a subsequent paper, Ho et al. (2013) reported the isolation of *Salvia miltiorrhiza* transgenic lines revealing higher yields of tanshinones, a medical-active substance formed in the roots. In another exotic plant species, *Catharanthus roseus*, van der Fits et al. (2001) isolated regulators of genes that are involved in the biosynthesis of secondary metabolites of the terpenoid indole alkaloid (TIA) class by T-DNA activation tagging.

6 T-DNA and Transposon Activation Tagging in *Populus*

6.1 T-DNA

The hemizygous status of transferred tags makes the approach of predominantly acting recessive loss-of-function mutagenesis nearly impracticable in perennial, or even in long-lived, plant species like trees. The reason is that either the long vegetative cycles or self-incompatibility systems or dioecy impedes the creation of homozygous alleles. Therefore, the usefulness of a gain-of-function approach that is simply based on the dominant nature of this approach for trees was already discussed a decade ago (Fladung et al. 2004; Busov et al. 2005). Examples of successful T-DNA- and transposon-based activation tagging mutagenesis have only been reported thus far for poplar (Busov et al. 2011; Fladung and Polak 2012; Fladung 2014).

Busov et al. (2011) reviewed the feasibility of T-DNA-based activation tagging as a forward genetic tool for *Populus*. Mutant phenotypes discovered so far include a variety of morphological and physiological traits, including leaf size and morphology, crown architecture, stature, vegetative dormancy, and tropic responses (Busov et al. 2011). Molecular analyses confirmed that the integrated T-DNAs are distributed more or less evenly among the 19 chromosomes, and insertions occurred in a region of up to 13 Kbp surrounding the coding region of the genes. However, a high portion of mutants only became visible after the second year of field cultivation, but not in vitro and not during greenhouse cultivation (Busov et al. 2011).

Further, as discussed earlier, the problem of functional redundancy has also become particularly apparent for the genus *Populus*, as sequencing of the *Populus* genome has revealed a recent duplication event (8–13 million years; Tuskan et al. 2006). According to the most recent information of this sequencing project (Phytozome v10; JGI v3.0 gene annotation of assembly v3), the main genome assembly is approximately 422.9 Mb arranged in 1,446 scaffolds. 181 scaffolds are larger than 50 kb in size, representing approximately 97.3 % of the genome. The number of genomic loci containing protein-coding transcripts is 41,335 (http://www.phytozome.net/poplar.php, Tuskan et al. 2006), but the function of the majority of the transcripts is unknown (Tuskan, pers. communication).

Only few reports have been published in the past describing T-DNA- and transposon-based activation in poplar. One of the first publications describes a T-DNA activation tagging approach that revealed a dwarf transgenic hybrid poplar individual among a population of about 600 independent activation-tagged transgenic lines (Busov et al. 2003). Molecular analyses of this line showed that the dwarf phenotype was related to the overexpression of the poplar GA2ox gene (*PtaGA2ox1*). Groover et al. (2004) reported a gene and enhancer trap discovery system for poplar trees by employing a promoterless *β-GLUCURONIDASE* (*GUS*) reporter gene. The idea was that GUS expression in tissues is obtained when the promoterless *GUS* gene is activated following insertion into a transcribed region of a chromosomal gene. Other activation tagging poplar populations were generated by Harrison et al. (2007). The largest activation tagging poplar population contained about 1,800 independent transgenic lines, and the mutant frequency reported was about 2.4 %. Mutants obtained comprised developmental abnormalities as well as alterations in leaf and stem structure and overall stature (Harrison et al. 2007).

Plett et al. (2010) identified a mutant line with increased foliar trichome density in a T-DNA activation tagging poplar population. The mutant revealed a rate of photosynthesis twice as high as in wild-type poplars, a slight increase in growth rate, and a significantly increased resistance to feeding by insect larvae. The phenotype was attributed to an increased expression of the *PtaMYB186* gene, homologous to *Arabidopsis MYB106*, a known regulator of trichome initiation (Plett et al. 2010). Five other activation-tagged mutant lines were described by Trupiano et al. (2013), showing changes in their adventitious rooting. In one mutant line, upregulation of a gene encoding a transcription factor of the AP2/ERF family of unknown function (PtaERF003) was shown to be responsible for the observed rooting phenotype (Trupiano et al. 2013). Finally, Yordanov et al. (2014) also identified the gene *EARLY BUD-BREAK 1* (*EBB1*) regulating the reactivation of meristem activity after winter dormancy in poplar by T-DNA-based activation tagging.

6.2 Transposon

Ideas to establish a "general" transposon-based activation tagging protocol in poplar were proposed by Fladung et al. (2004), based on the observation that the maize transposable autonomous *Ac* element is functionally active in transgenic aspen-*Populus* (Fladung et al. 1997; Kumar and Fladung 2003b; Fladung 2011). By employing the nonautonomous element Ds, Suzuki et al. (2001) developed an "Activation Tagging Ds" (ATDs) system containing four tandem repeats of enhancer fragments from the CaMV 35S promoter and, at both terminal positions, outward directed 35S promoters, each flanked by Ds ends (Fig. 3a). This ATDs element was combined with the phenotypically selectable *rolC* reporter gene (Fladung et al. 1997; Fladung 1999; Kumar and Fladung 2001), which was cloned

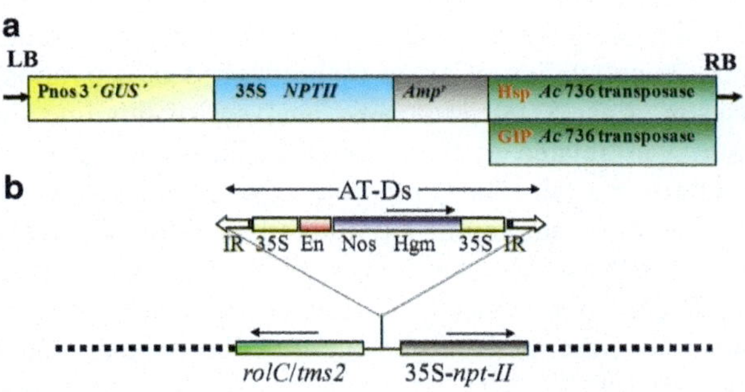

Fig. 3 Activation tagging approach developed for poplar. (**a**) The activation tagging construct harboring the nonautonomous Ds element (ATDs) contains two CaMV 35S promoters and four tandem repeats of enhancer fragments (En) of the 35S promoter, flanked by terminal inverted repeats (IR) (Suzuki et al. 2001). Either *rolC* or *tms2* gene is outside of ATDs, and the 35S promoter keeps these genes active under non-excised conditions. The marker gene 35S-*npt-II* allows selection of ATDs transgenic plants. (**b**) Controlling transposase expression by use of either a heat shock-inducible promoter (Hsp) (Balcells et al. 1994) or a glucocorticoid-inducible promoter (GIP, Ouwerkerk et al. 2001)

outside of the ATDs element so that it is under the control of one CaMV 35S promoter and thus active when the ATDs is not excised (Fladung and Polak 2012). A second construct harbored the *Ac* transposase controlled by the heat shock promoter (HSP) Gmhsp17.5-E from soybean (Czarnecka et al. 1985) (Fig. 3b).

First, HSP::*TRANSPOSASE* transgenic aspen lines were established. The RT-PCR approach was followed to check for transposase transcription in five transgenic HSP::*TRANSPOSASE* and one control line (N55-1). All six lines were heat shocked at 37 °C for 24 h culture under continuous light. RNA was isolated from leaves, and the quality and amount of the DNAse digested RNA was sufficient for RT-PCR (not shown). To check for DNA contamination, a reaction was performed without the AMV reverse transcriptase. An induction of transposase transcription could be observed in all five HSP::*TRANSPOSASE* transgenic lines. Lines N66-2 and N66-5 were chosen for super-transformation with the ATDs construct (not shown). To exclude the possibility that transposase transcription also occurs in non-heat-shocked-treated plants due to leakage of the HSP promoter, non-induced leaves were included in RT-PCR experiments. No transposase transcription could be detected in non-induced leaves (not shown).

In total, 23 double transgenic lines were obtained following super-transformation of N66-2 and N66-5 with the ATDs-*rolC* element (Fladung and Polak 2012). Based on these 23 lines, Fladung and Polak (2012) established an activation tagging population consisting of about 12,000 individuals and detected 29 different phenotypic variants, most of them remaining stable for at least 12 months in tissue culture (Fig. 4a) and/or in the greenhouse (Fig. 4b, c). Visible phenotypic mutant frequency (including leaf and stem phenotypic alterations) based on positive ATDs-transposed individuals was about 1 % (Fladung and Polak 2012). Silencing effects due to multiple copy integration could explain the overall relatively low frequency of ATDs transposition of only 30 % from total; however, no correlation could be found between copy number and mutant frequency in the double transgenic lines carrying 1–4 copies of ATDs (Fladung and Polak 2012). Putting together the phenotypic variants with the other lines without any phenotypic alteration, 57 genomic sequences were obtained flanking the ATDs at the new positions following ATDs transposition. Of these, 32 sequences (56 %) could be successfully annotated to putative *P. trichocarpa* transcripts. The results confirmed earlier findings (a) that the new ATDs insertion loci are unlinked to those harboring the original donor locus (Fladung 2011) and (b) that the *Ac* element frequently inserts in or near coding regions (Kumar and Fladung 2003a).

An alternative approach for transposon activation tagging is the replacement of the phenotypic *rolC* reporter gene with a selectable one. An option is the use of the negative selectable marker gene *INDOLE-3-ACETAMIDE HYDROLASE* (*TMS2*) from *Agrobacterium tumefaciens*. This *TMS2* gene catalyzes the conversion of biologically inactive auxin amides, such as naphthaleneacetamide [NAM], into active auxins, which are toxic to plants at elevated concentrations, but in the absence of the auxin amides, *tms2* transgenic plantlets grow normally (Karlin-Neumann et al. 1991). Similar to ATDs-*rolC*, the *tms2* gene is also under the control of one CaMV 35S promoter in non-ATDs-transposed plants, and thus

Fig. 4 (**a**) Two putative chlorophyll-defective variant calli (*arrows*) following transposon activation. (**b**) *Pale green* leaf (*right*) compared to normal *green* control leaf (*left*). (**c**) Variant plant revealing dwarf phenotype

transcribed non-ATDs-transposed plants are therefore able to metabolize NAM to active IAA that leads to callus formation instead of plant regeneration. In consequence, only tissues revealing ATDs excision will regenerate to plantlets. This strategy allows the efficient selection of only plants with ATDs reintegration somewhere in the genome.

7 Conclusions

Both T-DNA and transposon tags have been shown to be appropriate to harbor promoter and/or enhancer sequences for successful activation tagging. These elements are able to activate genes when they become positioned in their vicinity. T-DNA is highly recommended as a tag for plant species with highly efficient transformation protocols, whereas transposon-based activation tagging approaches are suitable for species with poorly developed transformation protocols. For the theoretical aim of tagging every gene in a tree genome, a transposon that is able to pass chromosome barriers is more suited as a tag than T-DNA, since with exception of *Arabidopsis*, plant transformation is time consuming and cost intensive. In contrast, only few independent transposon-carrying transgenic lines are required for the establishment of large activation tagging populations. Activation tagging or gain-of-function mutagenesis is the only way to tag genes in species with long vegetative periods such as trees, or those with special sexuality (self-incompatibility systems, dioecy). By employing selectable reporter genes, very efficient transposon-based activation tagging approaches can still be established.

Acknowledgments I would like to thank Dr. Trevor Fenning (Northern Research Station, Midlothian, Scotland, UK) and Dina Führmann (Thünen-Institute, Scientific Information Centre, Braunschweig, Germany) for critically reading the manuscript and English language editing. Thanks also to all scientists and technicians who contributed to this work on transposon-based activation tagging (S. Kumar, F. Deutsch, O. Polak) and funding agencies (Federal Ministry of Education and Research [BMBF], German Research Foundation [DFG]) for financial support.

References

Aarts MGM, Dirkse WG, Stiekema WJ, Pereira A (1993) Transposon tagging of a male sterility gene in Arabidopsis. Nature 363:715–717. doi:10.1038/363715a0

Aboul-Soud MAM, Chen XW, Kang JG, Yun BW, Raja MU, Malik SI, Loake GJ (2009) Activation tagging of ADR2 conveys a spreading lesion phenotype and resistance to biotrophic pathogens. New Phytol 183:1163–1175. doi:10.1111/j.1469-8137.2009.02902.x

Ahad A, Wolf J, Nick P (2003) Activation-tagged tobacco mutants that are tolerant to antimicrotubular herbicides are cross-resistant to chilling stress. Transgenic Res 12:615–629. doi:10.1023/A:1025814814823

Ahmad A, Niwa Y, Hatakeyama M, Goto S, Kobayashi K, Kobayashi H (2007) Saturated activation tagging for hunting salt-tolerant genes in Arabidopsis. Plant Cell Physiol 48:S140

André D, Colau D, Schell J, Van Montagu M, Hernalsteens JP (1986) Gene tagging in plants by a T-DNA insertion mutagen that generates APH(3′)II-plant gene fusions. Mol Gen Genet 204:512–518. doi:10.1007/BF00331033

Ayliffe MA, Pryor AJ (2007) Activation tagging in plants—generation of novel, gain-of-function mutations. Aust J Agric Res 58:490–497. doi:10.1071/Ar06154

Ayliffe MA, Pryor AJ (2009) Transposon-based activation tagging in cereals. Funct Plant Biol 36:915–921. doi:10.1071/Fp09130

Ayliffe MA, Pallotta M, Langridge P, Pryor AJ (2007) A barley activation tagging system. Plant Mol Biol 64:329–347. doi:10.1007/s11103-007-9157-8

Baker B, Schell J, Loerz H, Fedoroff N (1986) Transposition of the maize controlling element activator in tobacco. Proc Natl Acad Sci USA 83:4844–4848

Baker B, Coupland G, Fedoroff N, Starlinger P, Schell J (1987) Phenotypic assay for excision of the maize controlling element *Ac* in tobacco. EMBO J 6:1547–1554

Balcells L, Sundberg E, Coupland G (1994) A heat-shock promoter fusion to the *Ac* transposase gene drives inducible transposition of a Ds element during Arabidopsis embryo development. Plant J 5:755–764. doi:10.1111/j.1365-313X.1994.00755.x

Bancroft I, Jones JDG, Dean C (1993) Heterologous transposon tagging of the Drl1 locus in Arabidopsis. Plant Cell 5:631–638, http://dx.doi.org/10.1105/tpc.5.6.631

Belzile F, Yoder JI (1992) Pattern of somatic transposition in a high copy *Ac* tomato line. Plant J 2:173–179. doi:10.1046/j.1365-313X.1992.t01-40-00999.x

Boeke JD, Sandmeyer SB (2009) Yeast transposable elements. In: The molecular and cellular biology of the yeast saccharomyces: genome dynamics, protein synthesis, and energetics. Cold Spring Harbor monographs, vol 21A, pp 193–261. doi:10.1101/087969363.21A.193

Bolle C, Schneider A, Leister D (2011) Perspectives on systematic analyses of gene function in *Arabidopsis thaliana*: new tools, topics and trends. Curr Genomics 12:1–14. doi:10.2174/138920211794520187

Borevitz JO, Xia YJ, Blount J, Dixon RA, Lamb C (2000) Activation tagging identifies a conserved MYB regulator of phenylpropanoid biosynthesis. Plant Cell 12:2383–2393. doi:10.1105/tpc.12.12.2383

Brutnell TP (2002) Transposon tagging in maize. Funct Integr Genomics 2:4–12. doi:10.1007/s10142-001-0044-0

Busov VB, Meilan R, Pearce DW, Ma C, Rood SB, Strauss SH (2003) Activation tagging of a dominant gibberellin catabolism gene (*GA2-oxidase*) from poplar that regulates tree stature. Plant Physiol 132:1283–1291. doi:10.1104/pp. 103.020354

Busov V, Fladung M, Groover A, Strauss SH (2005) Insertional mutagenesis in *Populus*: relevance and feasibility. Tree Genet Genomes 1:135–142. doi:10.1007/s11295-005-0019-8

Busov V, Yordanov Y, Gou J, Meilan R, Ma C, Regan S, Strauss S (2011) Activation tagging is an effective gene tagging system in Populus. Tree Genet Genomes 7:91–101. doi:10.1007/s11295-010-0317-7

Cardon GH, Frey M, Saedler H, Gierl A (1993) Definition and characterization of an artificial *En*-based *Spm*-based transposon tagging system in transgenic tobacco. Plant Mol Biol 23:157–178. doi:10.1007/Bf00021428

Carpenter R, Coen ES (1990) Floral homeotic mutations produced by transposon-mutagenesis in *Antirrhinum majus*. Genes Dev 4:1483–1493. doi:10.1101/gad.4.9.1483

Carter JD, Pereira A, Dickerman AW, Veilleux RE (2013) An active *Ac*/Ds transposon system for activation tagging in tomato cultivar M82 using clonal propagation. Plant Physiol 162:145–156. doi:10.1104/pp. 113.213876

Chang C, Kwok SF, Bleecker AB, Meyerowitz EM (1993) *Arabidopsis* ethylene-response gene *ETR*1: similarity of product to two-component regulators. Science 262:539–544. doi:10.1126/science.8211181

Charlesworth B, Sniegowski P, Stephan W (1994) The evolutionary dynamics of repetitive DNA in eukaryotes. Nature 371:215–220. doi:10.1038/371215a0

Charng YC, Wu G, Hsieh CS, Chuang HN, Huang JY, Yeh LC, Shieh YH, Tu J (2007) The inducible transposon system for rice functional genomics. Bot Stud 48:1–11

Chen S, Jin W, Wang M, Zhang F, Zhou J, Jia Q, Wu Y, Liu F, Wu P (2003) Distribution and characterization of over 1000 T-DNA tags in rice genome. Plant J 36:105–113. doi:10.1046/j.1365-313X.2003.01860.x

Chuck G, Robbins T, Nijjar C, Ralston E, Courtney-Gutterson N, Dooner HK (1993) Tagging and cloning of a petunia flower colour gene with the maize transposable element *Activator*. Plant Cell 5:371–378, http://dx.doi.org/10.1105/tpc.5.4.371

Coomber SA, Feldmann KA (1993) Gene tagging in transgenic plants. In: Kung SD, Wu R (eds) Transgenic plants, vol 1. Academic, New York, NY, pp 225–240

Czarnecka E, Gurley WB, Nagao RT, Mosquera LA, Key JL (1985) DNA-sequence and transcript mapping of a soybean gene encoding a small heat-shock protein. Proc Natl Acad Sci USA 82:3726–3730. doi:10.1073/pnas.82.11.3726

DeLong A, Calderon-Urrea A, Dellaporta SL (1993) Sex determination gene *TASSELSEED2* of maize encodes a short-chain alcohol dehydrogenase required for stage-specific floral organ abortion. Cell 74:757–768. doi:10.1016/0092-8674(93)90522-R

Dimitri P, Junakovic N (1999) Revising the selfish DNA hypothesis: new evidence on accumulation of transposable elements in heterochromatin. Trends Genet 15:123–124. doi:10.1016/S0168-9525(99)01711-4

Dinesh-Kumar SP, Whitham S, Choi D, Hehl R, Corr C, Baker B (1995) Transposon tagging of tobacco mosaic-virus resistance gene-N—its possible role in the *Tmv*-N-mediated signal-transduction pathway. Proc Natl Acad Sci USA 92:4175–4180. doi:10.1073/pnas.92.10.4175

Döring HP, Starlinger P (1986) Molecular genetics of transposable elements in plants. Annu Rev Genet 20:175–200

Fedoroff NV (2000) Transposons and genome evolution in plants. Proc Natl Acad Sci USA 97:7002–7007. doi:10.1073/pnas.97.13.7002

Fedoroff NV (2012) Transposable elements, epigenetics, and genome evolution. Science 338:758–767. doi:10.1126/science.338.6108.758

Feldmann KA (1991) T-DNA insertion mutagenesis in Arabidopsis: mutational spectrum. Plant J 1:71–82. doi:10.1111/j.1365-313X.1991.00071.x

Fladung M (1990) Transformation of diploid and tetraploid potato clones with the *rolC* gene of *Agrobacterium rhizogenes* and characterization of transgenic plants. Plant Breed 104:295–304. doi:10.1111/j.1439-0523.1990.tb00439.x

Fladung M (1999) Gene stability in transgenic aspen-*Populus*. I. Flanking DNA sequences and T-DNA structure. Mol Gen Genet 260:574–581. doi:10.1007/s004380050931

Fladung M (2011) Analysis of re-integrated *Ac* element positions in the genome of *Populus* provides a basis for *Ac*/Ds-transposon activation tagging in trees. Trees 25:551–557. doi:10.1007/s00468-010-0511-0

Fladung M (2014) Prospects of using a modified *Ac*/Ds transposon system from maize for activation tagging in the tree species *Populus*. In: Ramawat KG, Mérillon JM, Ahuja MR (eds) Tree biotechnology. CRC Press, Boca Raton, FL, pp 469–482

Fladung M, Ahuja MR (1997) Excision of the maize transposable element *Ac* in periclinal leaves of *35S-Ac-rolC* transgenic aspen-*Populus*. Plant Mol Biol 33:1097–1103. doi:10.1023/a:1005788706864

Fladung M, Polak O (2012) *Ac*/Ds-transposon activation tagging in poplar: a powerful tool for gene discovery. BMC Genomics 13:61. doi:10.1186/1471-2164-13-61

Fladung M, Muhs HJ, Ahuja MR (1996) Morphological changes observed in transgenic *Populus* carrying the *rolC* gene from *Agrobacterium rhizogenes*. Silvae Genet 45:349–354

Fladung M, Großmann K, Ahuja MR (1997) Alterations in hormonal and developmental characteristics in transgenic *Populus* conditioned by the *rolC* gene from *Agrobacterium rhizogenes*. J Plant Physiol 150:420–427. doi:10.1016/S0176-1617(97)80092-2

Fladung M, Deutsch F, Hönicka H, Kumar S (2004) DNA and transposon tagging in aspen. Plant Biol 6:5–11. doi:10.1055/s-2003-44745

Fobert PR, Labbe H, Cosmopoulos J, Gottlob-McHugh S, Ouellet T, Hattori J, Sunohara G, Iyer VN, Miki BL (1994) T-DNA tagging of a seed coat-specific cryptic promoter in tobacco. Plant J 6:567–577. doi:10.1046/j.1365-313X.1994.6040567.x

Gierl A, Saedler H (1992) Plant transposable elements and gene tagging. Plant Mol Biol 19:39–49. doi:10.1007/BF00015605

Giordano J, Ge Y, Gelfand Y, Abrusán G, Benson G, Warburton PE (2007) Evolutionary history of mammalian transposons determined by genome-wide defragmentation. PLoS Comput Biol 3 (7), e137. doi:10.1371/journal.pcbi.0030137

Goff SA, Ricke D, Lan TH et al (2002) A draft sequence of the rice genome (*Oryza sativa* L. ssp. *japonica*). Science 296:92–100. doi:10.1126/science.1068275

Greco R, Ouwerkerk PBF, Taal AJC, Favalli C, Beguiristain T, Puigdomenech P, Colombo L, Hoge JHC, Pereira A (2001) Early and multiple *Ac* transpositions in rice suitable for efficient insertional mutagenesis. Plant Mol Biol 46:215–227. doi:10.1023/A:1010607318694

Greco R, Ouwerkerk PBF, De Kam RJ, Sallaud C, Favalli C, Colombo L, Guiderdoni E, Meijer AH, Hoge JHC, Pereira A (2003) Transpositional behaviour of an *Ac*/Ds system for reverse genetics in rice. Theor Appl Genet 108:10–24. doi:10.1007/s00122-003-1416-8

Groover A, Fontana JR, Dupper G, Ma C, Martienssen R, Strauss S, Meilan R (2004) Gene and enhancer trap tagging of vascular-expressed genes in poplar trees. Plant Physiol 134:1742–1751, http://dx.doi.org/10.1104/pp.103.034330

Harrison EJ, Bush M, Plett JM, McPhee DP, Vitez R, O'Malley B, Sharma V, Bosnich W, Seguin A, MacKay J, Regan S (2007) Diverse developmental mutants revealed in an activation tagged population of poplar. Can J Bot 85:1071–1087. doi:10.1139/B07-063

Hayashi H, Czaja I, Lubenow H, Schell J, Walden R (1992) Activation of a plant gene by T-DNA tagging: auxin-independent growth *in vitro*. Science 258:1350–1353. doi:10.1126/science.1455228

Hehl R (1994) Transposon tagging in heterologous host plants. Trends Genet 11:385–386. doi:10.1016/0168-9525(94)90041-8
Hehl R, Baker B (1989) Induced transposition of Ds by a stable *Ac* in crosses of transgenic tobacco plants. Mol Gen Genet 217:53–59. doi:10.1007/BF00330942
Ho HS, Vishwakarma RK, Chen ECF, Tsay HS (2013) Activation tagging in *Salvia miltiorrhiza* can cause increased leaf size and accumulation of tanshinone I and IIA in its roots. Bot Stud 54:37. doi:10.1186/1999-3110-54-37
Howe GT, Strauss SH, Goldfarb B (1991) Insertion of the maize transposable element *Ac* into poplar. In: Ahuja MR (ed) Woody plant biotechnology. Plenum, New York, NY, pp 283–294
Howe GT, Goldfarb B, Strauss SH (1994) *Agrobacterium*-mediated transformation of hybrid poplar suspension cultures and regeneration of transformed plants. Plant Cell Tiss Org Cult 36:59–71. doi:10.1007/BF00048316
Huang SS, Cerny RE, Bhat DS, Brown SM (2001) Cloning of an Arabidopsis patatin-like gene, *STURDY*, by activation T-DNA tagging. Plant Physiol 125:573–584. doi:10.1104/Pp.125.2.573
Huang CRL, Burns KH, Boeke JD (2012) Active transposition in genomes. Annu Rev Genet 46:651–675. doi:10.1146/annurev-genet-110711-155616
Ichikawa T, Nakazawa M, Kawashima M, Iizumi H, Kuroda H, Kondou Y, Tsuhara Y, Suzuki K, Ishikawa A, Seki M, Fujita M, Motohashi R, Nagata N, Takagi T, Shinozaki K, Matsui M (2006) The FOX hunting system: an alternative gain-of-function gene hunting technique. Plant J 48:974–985. doi:10.1111/j.1365-313X.2006.02924.x
Imaizumi R, Sato S, Kameya N, Nakamura I, Nakamura Y, Tabata S, Ayabe SI, Aoki T (2005) Activation tagging approach in a model legume, *Lotus japonicus*. J Plant Res 118:391–399. doi:10.1007/10265-005-0231-5
Ito T, Motohashi R, Kuromori T, Noutoshi Y, Seki M, Kamiya A, Mizukado S, Sakurai T, Shinozaki K (2005) A Resource of 5,814 dissociation transposon-tagged and sequence-indexed lines of Arabidopsis transposed from start loci on chromosome 5. Plant Cell Physiol 46:1149–1153. doi:10.1093/pcp/pci112
James DW Jr, Lim E, Keller J, Plooy I, Ralston E, Dooner HK (1995) Directed tagging of the Arabidopsis *FATTY ACID ELONGATION1* (*FAE1*) gene with the maize transposon activator. Plant Cell 7:309–319, http://dx.doi.org/10.1105/tpc.7.3.309
Jeon JS, An G (2001) Gene tagging in rice: a high throughput system for functional genomics. Plant Sci 161:211–219. doi:10.1016/S0168-9452(01)00414-9
Jeon JS, Lee S, Jung KH et al (2000) T-DNA insertional mutagenesis for functional genomics in rice. Plant J 22:561–570. doi:10.1046/j.1365-313x.2000.00767.x
Jeong DH, An SY, Kang HG, Moon S, Han JJ, Park S, Lee HS, An KS, An GH (2002) T-DNA insertional mutagenesis for activation tagging in rice. Plant Physiol 130:1636–1644. doi:10.1104/Pp.014357
Jeong DH, An SY, Park S, Kang HG, Park GG, Kim SR, Sim J, Kim YO, Kim MK, Kim SR, Kim J, Shin M, Jung M, An GH (2006) Generation of a flanking sequence-tag database for activation-tagging lines in japonica rice. Plant J 45:123–132. doi:10.1111/j.1365-313X.2005.02610.x
Jones JDG, Bishop G, Carroll B, Dickinson M, English J, Harrison K, Jones D, Scofield S, Thomas CM (1992) Prospects for establishing a tomato gene tagging system using the maize transposon *Activator* (*Ac*). Proc R Soc Edinb 99B:107–119, http://dx.doi.org/10.1017/S0269727000005534
Jones DA, Thomas CM, Hammond-Kosack KE, Balint-Kurti PJ, Jones JD (1994) Isolation of the tomato *cf-9* gene for resistance to *Cladosporium fulvum* by transposon tagging. Science 266:789–793. doi:10.1126/science.7973631
Kakimoto T (1996) *CKI1*, a histidine kinase homolog implicated in cytokinin signal transduction. Science 274:982–985. doi:10.1126/science.274.5289.982
Kang B, Wang H, Nam KH, Li JY, Li JM (2010) Activation-tagged suppressors of a weak brassinosteroid receptor mutant. Mol Plant 3:260–268. doi:10.1093/Mp/Ssp099

Kardailsky I, Shukla VK, Ahn JH, Dagenais N, Christensen SK, Nguyen JT, Chory J, Harrison MJ, Weigel D (1999) Activation tagging of the floral inducer *FT*. Science 286:1962–1965. doi:10.1126/science.286.5446.1962

Karlin-Neumann GA, Brussian JA, Tobin EM (1991) Phytochrome control of the *tms2* gene in transgenic Arabidopsis: a strategy for selecting mutants in the signal transduction pathway. Plant Cell 3:573–582

Kim SG, Lee S, Kim YS, Yun DJ, Woo JC, Park CM (2010) Activation tagging of an Arabidopsis *SHI-RELATED SEQUENCE* gene produces abnormal anther dehiscence and floral development. Plant Mol Biol 74:337–351. doi:10.1007/s11103-010-9677-5

Knapp S, Larondelle Y, Rossberg M, Furtek D, Theres K (1988) Transgenic tomato lines containing Ds elements at defined genomic positions as tools for targeted transposon tagging. Mol Gen Genet 243:666–673

Koncz C, Martini N, Mayerhofer R, Koncz-Kalman Z, Körber H, Redei GP, Schell J (1989) High-frequency T-DNA-mediated gene tagging in plants. Proc Natl Acad Sci USA 86:8467–8471. doi:10.1073/pnas.86.21.8467

Koncz C, Mayerhofer R, Koncz-Kalman Z, Nawrath C, Reiss B, Redei GP, Schell J (1990) Isolation of a gene encoding a novel chloroplast protein by T-DNA tagging in *Arabidopsis thaliana*. EMBO J 9:1337–1346

Kumar S, Fladung M (2001) Gene stability in transgenic aspen (Populus). II. Molecular characterization of variable expression of transgene in wild and hybrid aspen. Planta 213:731–740. doi:10.1007/s004250100535

Kumar S, Fladung M (2003a) Somatic mobility of the maize element *Ac* and its usability for gene tagging in aspen. Plant Mol Biol 51:643–650. doi:10.1023/a:1022505808929

Kumar S, Fladung M (2003b) Forest tree transgenesis and functional genomics: from fast forward to reverse genetics. Silvae Genet 52:229–232

Labrador M, Corces VG (1997) Transposable element-host interactions: regulation of insertion and excision. Annu Rev Genet 31:381–404. doi:10.1146/annurev.genet.31.1.381

Lawrence GJ, Finnegan EJ, Ayliff MA, Ellis JG (1995) The *L6* gene for flax rust resistance is related to the *Arabidopsis* bacterial resistance gene *RPS2* and the tobacco viral resistance gene *N*. Plant Cell 7:1195–1206. http://dx.doi.org/10.1105/tpc.7.8.1195

Lee CY, Agrawal DC, Wang CS, Yu SM, Chen JJW, Tsay HS (2008) T-DNA activation tagging as a tool to isolate *Salvia miltiorrhiza* transgenic lines for higher yields of tanshinones. Planta Med 74:780–786. doi:10.1055/s-2008-1074527

Lee S, Persson DP, Hansen TH, Husted S, Schjoerring JK, Kim YS, Jeon US, Kim YK, Kakei Y, Masuda H, Nishizawa NK, An G (2011) Bio-available zinc in rice seeds is increased by activation tagging of nicotianamine synthase. Plant Biotechnol J 9:865–873. doi:10.1111/j.1467-7652.2011.00606.x

Li Y, Rosso MG, Strizhov N, Viehoever P, Weisshaar B (2003) GABI-Kat SimpleSearch: a flanking sequence tag (FST) database for the identification of T-DNA insertion mutants in *Arabidopsis thaliana*. Bioinformatics 19:1441–1442. doi:10.1093/bioinformatics/btg170

Lisch D (2013) How important are transposons for plant evolution? Nat Rev Genet 14:49–61. doi:10.1038/nrg3374

Lu GH, Wang XP, Liu JH et al (2014a) Application of T-DNA activation tagging to identify glutamate receptor-like genes that enhance drought tolerance in plants. Plant Cell Rep 33:617–631. doi:10.1007/s00299-014-1586-7

Lu N, Carter JD, Medina TB, Holt SH, Manrique-Carpintero NC, Upham KT, Pereira A, Shulaev V, Veilleux RE (2014b) Transposon based activation tagging in diploid strawberry and monoploid derivatives of potato. Plant Cell Rep 33:1203–1216. doi:10.1007/s00299-014-1610-y

Marsch-Martinez N, Greco R, Van Arkel G, Herrera-Estrella L, Pereira A (2002) Activation tagging using the *En*-I maize transposon system in Arabidopsis. Plant Physiol 129:1544–1556. doi:10.1104/Pp.003327

Martienssen RA (1998) Functional genomics: probing plant gene function and expression with transposons. Proc Natl Acad Sci USA 95:2021–2026

Martin C, Carpenter R, Sommer H, Saedler H, Coen ES (1985) Molecular analysis of instability in flower pigmentation of *Antirrhinum majus*, following isolation of the *Pallida* locus by transposon tagging. EMBO J 4:1625–1630

Masaki T, Tsukagoshi H, Mitsui N, Nishii T, Hattori T, Morikami A, Nakamura K (2005) Activation tagging of a gene for a protein with novel class of CCT-domain activates expression of a subset of sugar-inducible genes in *Arabidopsis thaliana*. Plant J 43:142–152. doi:10.1111/j.1365-313X.2005.02439.x

Mathews H, Clendennen SK, Caldwell CG, Liu XL, Connors K, Matheis N, Schuster DK, Menasco DJ, Wagoner W, Lightner J, Wagner DR (2003) Activation tagging in tomato identifies a transcriptional regulator of anthocyanin biosynthesis, modification, and transport. Plant Cell 15:1689–1703. doi:10.1105/Tpc.012963

McClintock B (1950) The origin and behavior of mutable loci in maize. Proc Natl Acad Sci 36:344–355

McKenzie N, Dale PJ (2004) Mapping of transposable element Dissociation inserts in *Brassica oleracea* following plant regeneration from streptomycin selection of callus. Theor Appl Genet 109:333–341. doi:10.1007/s00122-004-1629-5

McLaughlin M, Walbot V (1987) Cloning of a mutable *bz2* allele of maize by transposon tagging and differential hybridization. Genetics 117:771–776

Meissner R, Chague V, Zhu Q, Emmanuel E, Elkind Y, Levy AA (2000) A high throughput system for transposon tagging and promoter trapping in tomato. Plant J 22:265–274. doi:10.1046/j.1365-313x.2000.00735.x

Mori M, Tomita C, Sugimoto K, Ooka H, Onodera H, Kajiwawra H, Tanaka H, Sekimoto H, Hirochika H, Kikuchi S (2003) Construction of the large scale activation tagging lines and characterization of a lesion mimic mutant. Plant Cell Physiol 44:S129

Mori M, Nakamura H, Ichikawa H (2006) Characterization of the short grain mutant (*Sg1*) isolated by rice activation tagging. Plant Cell Physiol 47:S177

Mori M, Tomita C, Sugimoto K, Hasegawa M, Hayashi N, Dubouzet JG, Ochiai H, Sekimoto H, Hirochika H, Kikuchi S (2007) Isolation and molecular characterization of a *Spotted leaf 18* mutant by modified activation-tagging in rice. Plant Mol Biol 63:847–860. doi:10.1007/s11103-006-9130-y

Murray EE, Rocheleau T, Eberle M, Stock C, Sekar V, Adang M (1991) Transposition of the maize *activator* element in transgenic rice plants. Plant Mol Biol 16:1035–1050

Nakajima K, Miyashima S, Waki T, Hashimoto T (2006) Identification of genes implicated in Arabidopsis root patterning by a GAL4/UAS activation tagging system. Plant Cell Physiol 47:S97

Nakazawa M, Ichikawa T, Ishikawa A, Suzuki K, Kobayashi H, Tsuhara Y, Kawashima M, Muto S, Matsui M (2003) Activation tagging, a novel tool to dissect the functions of a gene family. Plant Cell Physiol 44:S128

Nilsson O, Moritz T, Sundberg B, Sandberg G, Olsson O (1996) Expression of the Agrobacterium *rolC* gene in a deciduous forest tree alters growth and development and leads to stem fasciation. Plant Physiol 112:493–502, http://dx.doi.org/10.1104/pp.112.2.493

Niwa Y, Goto S, Nakano T, Sakaiya M, Hirano T, Tsukaya H, Komeda Y, Kobayashi H (2006) Arabidopsis mutants by activation tagging in which photosynthesis genes are expressed in dedifferentiated calli. Plant Cell Physiol 47:319–331. doi:10.1093/Pcp/Pci242

Ostergaard L, Yanofsky MF (2004) Establishment gene function by mutagenesis in *Arabidopsis thaliana*. Plant J 39:682–696. doi:10.1111/j.1365-313X.2004.02149.x

Ouwerkerk PBF, de Kam RJ, Hoge JHC, Meijer AH (2001) Glucocorticoid-inducible gene expression in rice. Planta 213:370–378. doi:10.1007/s004250100583

Parinov S, Sevugan M, Ye D, Yang WC, Kumaran M, Sundaresan V (1999) Analysis of flanking sequences from Dissociation insertion lines: a database for reverse genetics in *Arabidopsis*. Plant Cell 11:2263–2270, http://dx.doi.org/10.1105/tpc.11.12.2263

Pereira A, Aarts M, van Agtmaal S, Stiekema WJ, Jacobson E (1991) Waxy variegation in transgenic potato. Maydica 36:323–327

Perrella G, Cremona G, Consiglio F, Errico A, Bressan RA, Conicella C (2006) Screening for mutations affecting sexual reproduction after activation tagging in *Arabidopsis thaliana*. J Appl Genet 47:109–111. doi:10.1007/BF03194608

Plett JM, Wilkins O, Campbell MM, Ralph SG, Regan S (2010) Endogenous overexpression of *Populus* MYB186 increases trichome density, improves insect pest resistance, and impacts plant growth. Plant J 64:419–432. doi:10.1111/j.1365-313X.2010.04343.x

Pogorelko GV, Fursova OV, Ogarkova OA, Tarasov VA (2008) A new technique for activation tagging in Arabidopsis. Gene 414:67–75. doi:10.1016/j.gene.2008.02.008

Qu S, Desai A, Wing R, Sundaresan V (2008) A versatile transposon-based activation tag vector system for functional genomics in cereals and other monocot plants. Plant Physiol 146:189–199. doi:10.1104/pp. 107.111427

Quattrocchio F, Wing J, van der Woude K, Souer E, de Vetten N, Mol J, Koes R (1999) Molecular analysis of the *anthocyanin2* gene of *Petunia* and its role in the evolution of flower color. Plant Cell 11:1433–1444, http://dx.doi.org/10.1105/tpc.11.8.1433

Raina S, Mahalingam R, Chen FQ, Fedoroff N (2002) A collection of sequenced and mapped Ds transposon insertion sites in *Arabidopsis thaliana*. Plant Mol Biol 50:93–110. doi:10.1023/A:1016099215667

Rommens CMT, Kneppers TJA, Haring MA, Nijkamp HJJ, Hille J (1991) A transposon tagging strategy with *Ac* on plant-cell level in heterologous plant species. Plant Sci 74:99–106

Rosin FM, Watanabe N, Cacas JL, Kato N, Arroyo JM, Fang Y, May B, Vaughn M, Simorowski J, Ramu U, McCombie RW, Spector DL, Martienssen RA, Lam E (2008) Genome-wide transposon tagging reveals location-dependent effects on transcription and chromatin organization in *Arabidopsis*. Plant J 55:514–525. doi:10.1111/j.1365-313X.2008.03517.x

Rosso MG, Li Y, Strizhov N, Reiss B, Dekker K, Weisshaar B (2003) An *Arabidopsis thaliana* T-DNA mutagenized population (GABI-Kat) for flanking sequence tag-based reverse genetics. Plant Mol Biol 53:247–259. doi:10.1023/B:PLAN.0000009297.37235.4a

Sakurai T, Satou M, Akiyama K, Iida K, Seki M, Kuromori T, Ito T, Konagaya A, Toyoda T, Shinozaki K (2005) RARGE: a large-scale database of RIKEN Arabidopsis resources ranging from transcriptome to phenome. Nucleic Acids Res 33:D647–D650. doi:10.1093/nar/gki014

Schmidt RJ, Burr FA, Burr B (1987) Transposon tagging and molecular analysis of the maize regulatory locus *opaque-2*. Science 238:960–963. doi:10.1126/science.2823388

Schmülling T, Schell J, Spena A (1988) Single genes from *Agrobacterium rhizogenes* influence plant development. EMBO J 7:2621–2629

Schneider A, Kirch T, Gigolashvili T, Mock HP, Sonnewald U, Simon R, Flugge UI, Werr W (2005) A transposon-based activation-tagging population in *Arabidopsis thaliana* (TAMARA) and its application in the identification of dominant developmental and metabolic mutations. FEBS Lett 579:4622–4628. doi:10.1016/j.febslet.2005.07.030

Schneuwly S, Kuroiwa A, Gehring WJ (1987) Molecular analysis of the dominant homeotic *Antennapedia* phenotype. EMBO J 6:201–206

Scholz SC, Lörz H, Lütticke S (2001) Transposition of the maize transposable element *Ac* in barley (*Hordeum vulgare* L.). Mol Gen Genet 264:653–661. doi:10.1007/s004380000351

Seki M, Narusaka M, Kamiya A, Ishida J, Satou M, Sakurai T, Nakajima M, Enju A, Akiyama K, Oono Y, Muramatsu M, Hayashizaki Y, Kawai J, Carninci P, Itoh M, Ishii Y, Arakawa T, Shibata K, Shinagawa A, Shinozaki K (2002) Functional annotation of a full-length Arabidopsis cDNA collection. Science 296:141–145. doi:10.1126/science.1071006

Settles AM (2009) Transposon tagging and reverse genetics. In: Kriz AL, Larkins BA (eds) Molecular genetic approaches to maize improvement, biotechnology in agriculture and forestry, vol 63. Springer, Berlin, pp 143–159

Spena A, Aalen RB, Schulze SC (1989) Cell-autonomous behavior of the *rolC* gene of *Agrobacterium rhizogenes* during leaf development: a visual assay for transposon excision in transgenic plants. Plant Cell 1:1157–1164, http://dx.doi.org/10.1105/tpc.1.12.1157

Spradling AC, Rubin GM (1982) Transposition of cloned P elements into *Drosophila* germ line chromosomes. Science 218:341–347. doi:10.1126/science.6289435

Suzuki Y, Uemura S, Saito Y, Murofushi N, Schmitz G, Theres K, Yamaguchi I (2001) A novel transposon tagging element for obtaining gain-of-function mutants based on a self-stabilizing *Ac* derivative. Plant Mol Biol 45:123–131. doi:10.1023/A:1006455130098

Szabados L, Kovács I, Oberschall A, Ábrahám E, Kerekes I, Zsigmond L, Nagy R, Alvarado M, Krasovskaja I, Gál M, Berente A, Rédei GP, Haim AB, Koncz C (2002) Distribution of 1000 sequenced T-DNA tags in the Arabidopsis genome. Plant J 32:233–242. doi:10.1046/j.1365-313X.2002.01417.x

Tacke E, Korfhage C, Michel D, Maddaloni M, Motto M, Lanzini S, Salamini F, Döring HP (1995) Transposon tagging of the maize *Glossy2* locus with the transposable element *En/Spm*. Plant J 8:907–917. doi:10.1046/j.1365-313X.1995.8060907.x

Tani H, Chen X, Nurmberg P, Grant JJ, SantaMaria M, Chini A, Gilroy E, Birch PR, Loake GJ (2004) Activation tagging in plants: a tool for gene discovery. Funct Integr Genomics 4:258–266. doi:10.1007/s10142-004-0112-3

Trupiano D, Yordanov Y, Regan S, Meilan R, Tschaplinski T, Scippa GS, Busov V (2013) Identification, characterization of an AP2/ERF transcription factor that promotes adventitious, lateral root formation in *Populus*. Planta 238:271–282. doi:10.1007/s00425-013-1890-4

Tsay HS, Ho HM, Gupta SK, Wang CS, Chen PT, Chen ECF (2012) Development of pollen mediated activation tagging system for *Phalaenopsis* and *Doritaenopsis*. Electron J Biotechnol 15(4):9. doi:10.2225/vol15-issue4-fulltext-1

Tuskan GA, Difazio S, Jansson S et al (2006) The genome of black cottonwood, *Populus trichocarpa* (Torr. & Gray). Science 313:1596–1604. doi:10.1126/science.1128691

van der Fits L, Hilliou F, Memelink J (2001) T-DNA activation tagging as a tool to isolate regulators of a metabolic pathway from a genetically non-tractable plant species. Transgenic Res 10:513–521

van der Graaff E, Den Dulk-Ras A, Hooykaas PJJ, Keller B (2000) Activation tagging of the *LEAFY PETIOLE* gene affects leaf petiole development in *Arabidopsis thaliana*. Development 127:4971–4980

Van Sluys MA, Temp J, Fedoroff N (1987) Studies on the introduction and mobility of the maize *activator* element in *Arabidopsis thaliana* and *Daucus carota*. EMBO J 6:3881–3889

Waki T, Miyashima S, Nakanishi M, Ikeda Y, Hashimoto T, Nakajima K (2013) A *GAL4*-based targeted activation tagging system in *Arabidopsis thaliana*. Plant J 73:357–367. doi:10.1111/Tpj.12049

Walbot V (1992) Strategies for mutagenesis and gene cloning using transposon tagging and T-DNA insertional mutagenesis. Annu Rev Plant Physiol Plant Mol Biol 43:49–82

Walden R, Fritze K, Hayashi H, Miklashevichs E, Harling H, Schell J (1994) Activation tagging—a means of isolating genes implicated as playing a role in plant-growth and development. Plant Mol Biol 26:1521–1528. doi:10.1007/978-94-011-0239-1_16

Wan SY, Wu JX, Zhang ZG, Sun XH, Lv Y, Gao C, Ning YD, Ma J, Guo YP, Zhang Q, Zheng X, Zhang CY, Ma ZY, Lu TG (2009) Activation tagging, an efficient tool for functional analysis of the rice genome. Plant Mol Biol 69:69–80. doi:10.1007/s11103-008-9406-5

Weigel D, Ahn JH, Blazquez MA, Borevitz JO, Christensen SK, Frankhauser C, Ferrandiz C, Kardailsky I, Malancharuvil EJ, Neff MM, Nguyen JT, Sato S, Wang ZY, Xia Y, Dixon RA, Harrison MJ, Lamb CJ, Yanofsky MF, Chory J (2000) Activation tagging in *Arabidopsis*. Plant Pysiol 122:1003–1013, http://dx.doi.org/10.1104/pp.122.4.1003

Whitham S, Dinesh-Kumar SP, Choi D, Hehl R, Corr C, Baker B (1994) The product of the tobacco mosaic virus resistance gene *N*: similarity to toll and the interleukin-1 receptor. Cell 78:1101–1115. doi:10.1016/0092-8674(94)90283-6

Woodward C, Bemis SM, Hill EJ, Sawa S, Koshiba T, Torii KU (2005) Interaction of auxin and *ERECTA* in elaborating Arabidopsis inflorescence architecture revealed by the activation tagging of a new member of the *YUCCA* family putative flavin monooxygenases. Plant Physiol 139:192–203, http://dx.doi.org/10.1104/pp.105.063495

Yoder JI, Palys J, Alpert K, Lassner M (1988) *Ac* transposition in transgenic tomato plants. Mol Gen Genet 213:291–296

Yordanov YS, Ma C, Strauss SH, Busov VB (2014) *EARLY BUD-BREAK 1* (*EBB1*) is a regulator of release from seasonal dormancy in poplar trees. Proc Natl Acad Sci USA 111:10001–10006. doi:10.1073/pnas.1405621111

Yu J, Hu S, Wang J et al (2002) A draft sequence of the rice genome (*Oryza sativa* L. ssp. *indica*). Science 296:79–92. doi:10.1126/science.1068037

Zubko E, Adams CJ, Machaekova I, Malbeck J, Scollan C, Meyer P (2002) Activation tagging identifies a gene from *Petunia hybrida* responsible for the production of active cytokinins in plants. Plant J 29:797–808. doi:10.1046/j.1365-313X.2002.01256.x

Evolution of the Flowering Pathways

Eva Lucas-Reina, M Isabel Ortiz-Marchena, Francisco J. Romero-Campero, Myriam Calonje, José M. Romero, and Federico Valverde

Contents

Abstract Flowering plants are some of the most successful organisms on Earth, particularly those used in agriculture due to the widespread distribution produced by farming activities. The correct moment of the year to flower is a crucial decision as it strongly compromises the success of the progeny and is thus strictly controlled. Crops have been artificially selected to flower in those conditions better adapted for human production, and many genes related to flowering time are selected as targets for breeding programs. These characteristics reflect a complex regulatory pathway

---

Eva Lucas-Reina and M Isabel Ortiz-Marchena contributed equally to this work.

E. Lucas-Reina • M.I. Ortiz-Marchena • M. Calonje • J.M. Romero • F. Valverde (✉)
Institute for Plant Biochemistry and Photosynthesis, Plant Development Unit, CSIC and Universidad de Sevilla, 49th Americo Vespucio Avenue, 41092 Seville, Spain
e-mail: federico.valverde@ibvf.csic.es; http://www.ibvf.csic.es/en/molecular-basis-flowering-photoperiod-and-metabolism

F.J. Romero-Campero
Department of Computational Sciences and Artificial Intelligence, Research Group in Natural Computing, Universidad de Sevilla, Reina Mercedes Avenue, 41012 Seville, Spain

U. Lüttge et al. (eds.), *Progress in Botany*, Progress in Botany 77,
DOI 10.1007/978-3-319-25688-7_10

that has to respond both to predictable and unexpected changes in the environment. This plasticity confers the flowering plants with a genetic toolkit to adapt to varied habitats and changing environmental conditions. Recent advances in massive acquisition of data from many different species belonging to the green eukaryotic lineage allow us to make an evolutionary approach to the main mechanisms that influence the floral transition and how flowers are formed in modern plants. This work will review some of these aspects from the floral transition to the floral organogenesis.

1 Introduction

The flowering transition is one of the most important developmental decisions that a plant has to take during its life cycle. An incorrect decision to flower has a strong negative influence on the capacity of the plant to transmit its genes to the next generation, and thus it is strictly regulated (Casal et al. 2004). This decision is strongly influenced by external and internal cues among which light, temperature and nutrient signals are probably the most influential (Amasino 2010). In order to understand the complex signaling events that promote or inhibit flowering, different pathways have been proposed and excellent reviews have been recently published (Smeekens et al. 2010; Huijser and Schmid 2011; Andrés and Coupland 2012; Song et al. 2012a, b, c), but they can all be directly or indirectly grouped into three groups (Fig. 1). The light pathway integrates those signals derived from the light quality, day length, or the circadian clock. The internal signals comprise those provided by hormones, nutrients (sugar, nitrogen compounds, etc.) and age. The temperature signals include the so-called autonomous pathway, the ambient temperature signals and the vernalization signals. These pathways will be described in more detail below.

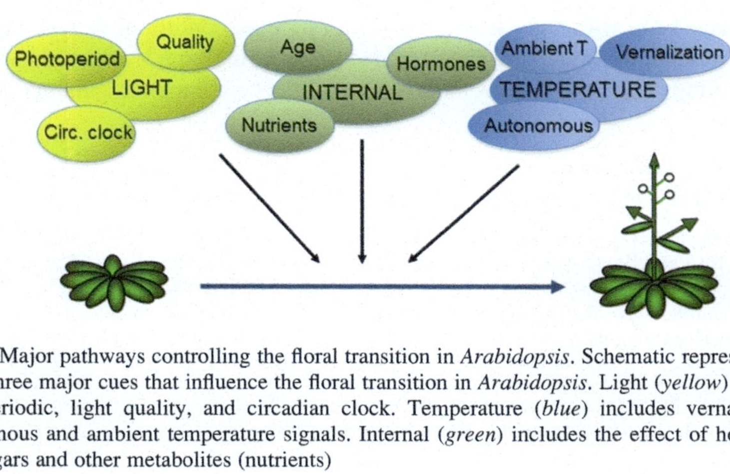

Fig. 1 Major pathways controlling the floral transition in *Arabidopsis*. Schematic representation of the three major cues that influence the floral transition in *Arabidopsis*. Light (*yellow*) includes photoperiodic, light quality, and circadian clock. Temperature (*blue*) includes vernalization, autonomous and ambient temperature signals. Internal (*green*) includes the effect of hormones, age, sugars and other metabolites (nutrients)

Most of the plants will flower when one, or a combination of these signals, reaches the threshold that triggers the floral transition. This is coordinated by a network of genes that is highly conserved throughout evolution (Romero-Campero et al. 2013). In this work we will review recent knowledge about the gene networks that control the flowering pathways as well as floral organogenesis and how can we trace back this gene toolkit into the evolutionary story of plants. It will allow us to understand the origin of the flowering pathways and why they have reached such complexity in angiosperms. Inevitably, *Arabidopsis thaliana* will be the model to follow, as most of the flowering work has been done in this small brassica. Nevertheless, we will try to extrapolate this information into other plants representing different phylogenetic relationships and evolutionary steps within the green eukaryote lineage.

We will also review the process of floral organogenesis because it is chronologically and locally connected to the last stages of the floral transition within the shoot apical meristem (SAM). In this way, many of the late genes involved in the floral transition, including the floral integrators, control the early stages of floral formation. This assures the continuity in the signaling process necessary to achieve the successful step-by-step hierarchy of floral organogenesis.

2 The Evolution of the Photoperiod Pathway

The amount of incident light at a particular point on most of the Earth's surface changes throughout the year resulting in the different seasons, particularly in the middle half of the hemispheres where most of the human population is concentrated. Animals and plants have developed throughout their evolution molecular tools consisting in genes and signaling networks that transduce day length information (or photoperiod) into the regulation of key developmental and metabolic processes. This capacity is known as photoperiod response (Bradshaw and Holzapfel 2007).

2.1 Photoperiod Pathway in Vascular Plants

One of the most conserved day length responses among plants is the photoperiodic flowering pathway (Romero-Campero et al. 2013). *CONSTANS* (*CO*) is the central gene in this pathway and promotes flowering by inducing the expression of the florigen *FLOWERING LOCUS T* (*FT*) gene (Valverde 2011). Recent advances in genomics of vascular plants have allowed researchers to identify several genes that control flowering in species such as potato (Martínez-García et al. 2002), tomato (Corrales et al. 2014), sorghum (Murphy et al. 2011), rice (Yano et al. 2000) and *Jatropha* (Yang et al. 2011). Nevertheless, the long-day (LD) facultative plant

Arabidopsis thaliana is the model organism where most studies have been performed (Amasino 2010).

In *Arabidopsis*, *CO* and *FT* expression are regulated by circadian and photoperiodic regulatory elements. In this sense, CYCLING DOF FACTOR (CDF) proteins are a group of four DOF transcription factors that bind to the *CO* and *FT* promoters negatively regulating their expression (Imaizumi et al. 2005; Fornara et al. 2009; Song et al. 2012c). At the end of a LD, the blue light-dependent GI-FKF1 complex induces CDF degradation (Rubio and Deng 2007), allowing FLOWERING BHLHs (FBHs) to enhance *CO* expression (Ito et al. 2012) and thereby *FT* induction. Moreover, GIGANTEA (GI) is involved in *FT* induction in a CO-independent way (Sawa and Kay 2011; Srikanth and Schmid 2011). *CO* expression is also regulated at the transcriptional level by the circadian clock whose core is constituted in *Arabidopsis* by the genes *CIRCADIAN CLOCK ASSOCIATED 1* (*CCA1*), *LATE ELONGATED HYPOCOTYL* (*LHY*), and *TIMING OF CAB EXPRESSION 1* (*TOC1*) (McClung 2014). Additionally, *CO* is posttranslationally regulated by the 26S proteasome due to the action of two E3 ubiquitin ligases with Ring Finger domains: CONSTITUTIVE MORPHOGENIC 1 (COP1) during the night and HIGH EXPRESSION OF OSMOTICALLY RESPONSIVE GENES 1 (HOS1) during the morning (Jang et al. 2008; Lazaro et al. 2012). Moreover, light has an important role in the regulation of *CO* expression. The photoreceptor PHYTOCHROME B (PHYB) promotes CO degradation by red light, whereas CRYPTOCHROMES 1 and 2 (CRY1, CRY2) and PHYTOCHROME A (PHYA) promote its stability through a blue light signal (Valverde et al. 2004) specifically during the daylight period. This complex regulatory network determines that *CO* mRNA coincides with a high stable protein level during the evening of a LD (external coincidence model) (Andrés and Coupland 2012) triggering the expression of the florigen *FT* gene. However, depending on its geographical location, plants have developed different regulatory mechanisms to anticipate photoperiod changes. For example, in short-day (SD) plants, such as rice (*Oryza sativa*), an *FT* homolog (*HEADING DATE 3a, HD3A*) is induced in SD by a CO homolog (HD1), whereas in LD HD1 behaves as a repressor of *HD3A* (Turck et al. 2008). Additionally, transgenic rice overexpressing *CDF* homologs (*OsDOF12*) induces *HD3A* expression only under LD conditions in a *HD1*-independent manner (Li et al. 2009). In this species, GI promotes *HD1* expression although it is yet unknown whether this regulation is direct or through a FKF1/CDF route similar to the one in *Arabidopsis* (Higgins et al. 2010).

The regulatory differences observed in vascular plants may reflect an evolutionary divergence produced by the needs to adapt to specific environmental conditions. This could explain the emergence of new regulatory genes involved in the same processes or the change in function of a specific gene. For example, in rice EARLY HEADING DATE 1 (EHD1), a B-type response regulator, induces *HD3A* transcription in SD conditions, independently of HD1 (Doi et al. 2004) and the GRAIN NUMBER, PLANT HEIGHT, and HEADING DATE 7 (GHD7) rice protein plays a key role in the photoperiod pathway (Xue et al. 2008). Nevertheless, no putative *Arabidopsis* orthologs of these genes have been identified so far. In potato, similar genes to those that control the floral transition also regulate other biological pathways such as tuberization. Both processes are finally controlled by two different *FT*-like paralogues, *StSP3D* that promotes flowering and *StSP6A* that regulates tuber formation (Navarro et al. 2011) in two separated transduction pathways. *StSP3D* and *StSP6A* respond to different photoperiod conditions involving the StGI-StFKF1 complex, StCDF, and StCO protein (Kloosterman et al. 2013). Interestingly, in neutral-day plants, where flowering time is not controlled by photoperiod, CDFs are involved in other biological processes not related to the photoperiod response. For example, in tomato, *SlCDFs* are induced in response to abiotic stress conditions. Nevertheless, the *SlCDF* heterologous expression in *Arabidopsis* delays flowering by reducing *CO* and *FT* transcript levels. This suggests that the ability of SlCDFs to control the photoperiod response is conserved although it is not involved in the floral transition in these plants (Corrales et al. 2014).

Strikingly, flowering gene regulatory networks from a wide range of photosynthetic organisms share a large set of orthologs. This suggests that the photoperiodic gene regulatory network evolved very early in the green evolutionary linage constituting an ancestral network. The current photosynthetic organisms have then inherited this gene network from these common ancestors.

2.2 *Compared Evolution of Photoperiodic Signaling in Green Algae and Land Plants*

The latest results from our group and others (Serrano et al. 2009; Romero-Campero et al. 2013) have demonstrated an exclusive origin of the photoperiod response in algae of the Chlorophyceae class, which would have then evolved into the complex pathway of modern plants. In this section we will try to dissect the evolutionary processes involved.

2.2.1 Homolog Genes in *Chlamydomonas*

CONSTANS Homolog

Chlamydomonas reinhardtii is considered to be a living representative of the common ancestor that gave rise to the green eukaryotic lineage. The first gene related to the photoperiod pathway identified in the *Chlamydomonas* genome was a single-copy *CO* homolog, called *CrCO* (Serrano et al. 2009; Valverde 2011). *CrCO* was shown to be involved, among other mechanisms, in processes controlled by the circadian clock, such as starch synthesis and cell growth. Surprisingly, transgenic plants overexpressing *CrCO* under a constitutive or phloem-specific promoter, flowered earlier than WT and in a similar way to plants overexpressing the original *CO* gene. *CrCO* can, thus, complement *co* mutation. In contrast, *CO like 1* (*COL1*) is unable to complement *co* mutation in spite of being evolutionarily more related to *CO* than *CrCO*. Possibly, CO and CrCO share key structural similarities that are not reflected in the alignment of their sequences, which shows very low general amino acid identity. This constitutes an example of the limitations of using solely sequence similarity when detecting potential orthologs (Romero-Campero et al. 2013). Recently, evidence of the high relevance of the *CrCO* gene in the algae transcriptome has been suggested by gene co-expression analysis. It has been shown that the *CrCO* gene constitute a hub gene in a gene co-expression network constructed based on RNA-seq data from a wide range of relevant physiological conditions (Romero-Campero et al. 2013). A single-copy *CrCO* gene has evolved into numerous *CONSTANS-LIKE* (*COLs*) gene families in *Physcomitrella* (*PpCOLs*) and *Arabidopsis* (*AtCOLs*), establishing complex and robust networks with greater numbers of hub genes in both species (Romero-Campero et al. 2013). This diversification of the *COL* family in *Physcomitrella* and vascular plants and the high overlapping between their functions indicate that the different biological processes in which CrCO, PpCOLs, and AtCOLs are involved are highly conserved across evolution (Romero-Campero et al. 2013). Additionally, COLs may have a wide repertoire of plant-specific light-dependent functions besides those already described (Valverde 2011) such as axillary ramification (Wang et al. 2013), bud dormancy (Böhlenius et al. 2006), and tuber growth (González-Schain et al. 2012).

CDF Homologs

The genome of *Chlamydomonas* contains another single-copy gene called *CrDOF* that seems to be part of the ancestral photoperiod pathway. *CrDOF* evolution has produced a numerous gene family, the DOF transcription factors (TFs), following a similar evolutionary history as *CrCO*. DOFs are specific TFs in vascular plants (Moreno-Risueño et al. 2007) and are not present in animal or fungi genomes. Specifically, *Arabidopsis* has 36 DOF proteins (Noguero et al. 2013) including the small family of four CDFs (Imaizumi et al. 2005; Fornara et al. 2009). In *Chlamydomonas*, CrDOF is regulated, in a similar way as CDFs in *Arabidopsis*, by the circadian clock and photoperiodic mechanisms. Additionally, like the CDFs, CrDOF controls *CrCO* transcription. Nevertheless, in contrast to the CDF function in *Arabidopsis*, CrDOF activates *CrCO* expression in *Chlamydomonas* by binding to its promoter. In addition, CrDOF controls important physiological processes in the algae exhibiting a surprisingly dual function, repressor or activator, depending on the day length. In this way CrDOF is able to induce cellular division by activating CrCO in SD, whereas in LD CrDOF represses the cell cycle progression to mitosis in a CrCO-independent manner. CrDOF phenocopies CDF function in *Arabidopsis* so that transgenic plants expressing *CrDOF* under different tissue-specific promoters delay flowering by suppressing *CO* and *FT* expression. Finally, RNA-seq data analysis revealed an apparent functional overlap between CrDOF and DOF proteins. These results reflect again how the functions of proteins involved in photoperiodic responses are extremely conserved across evolution.

The diversification and subsequent acquisition of new regulatory domains by CrDOF (which has only a DOF domain and nuclear localization signal) to vascular plant DOF factors could explain the new regulatory processes in which CDFs and other DOF proteins are involved (Lucas-Reina et al. 2015).

2.2.2 Putative Homologs

Several putative *Chlamydomonas* orthologs of *Arabidopsis* genes involved in the photoperiod response have been identified using non-curated bioinformatic methods such as the BBH (bidirectional best hit) method (Table 1). Their involvement in the photoperiod response in *Chlamydomonas* and their interactions with *CrCO* and *CrDOF* are yet to be validated experimentally. Here we analyzed the conservation of the co-expression patterns among these genes by comparing them to the co-expression patterns of homologs from *Arabidopsis* (Fig. 2).

Circadian Clock Genes

Approximately 30 putative genes have been identified in *Chlamydomonas* that are involved in the control of circadian processes. These genes are called *RHYTHM OF CHLOROPLAST* (*ROC*). Some of the codified proteins are specific from algae; others present conserved domains with plant circadian clock proteins (Matsuo and Ishiura 2011). Strikingly, other ROCs present domains similar to those found only in animal proteins involved in circadian rhythm control (Schulze et al. 2010). Specifically, ROC40 has a MYB domain similar to CCA1 and LHY proteins and ROC66, which presents B-box and CCT domains similar to CO, to COL1, involved in circadian clock (Ledger et al. 2001) and COL9 (Matsuo and Ishiura 2011). ROC66 CCT domain is also similar to the CCT domain from *Arabidopsis* TOC1 (Matsuo and Ishiura 2011). Besides the sequence similarity that *ROC40* and *ROC66* show with *CCA1*/*LHY* and *TOC1*, these two *Chlamydomonas* genes also exhibit similar co-expression patterns as their putative *Arabidopsis* orthologs. *CCA1*/*LHY* and *TOC1* present a negative co-expression pattern in *Arabidopsis*, which seems to be conserved in *Chlamydomonas*, as *CrLHY* and *CrTOC1* show a negative co-expression pattern (Fig. 2).

The conservation of the circadian clock core genes, *CCA1*/*LHY* and *TOC1*, has also been found in the green algae *Ostreococcus tauri*, although in this case, their expression patterns differ from those in the *Arabidopsis* genes (Bouget et al. 2014).

Photoreceptors

Light perception in plants is carried out by a set of different photoreceptors. One of them is the phototropin (PHOT) involved in physiological processes like phototropism and stomatal opening. On the other hand, cryptochromes (CRYs) and

Table 1 Genes involved in the photoperiod response in *Arabidopsis* and *Chlamydomonas*

Gene name	*Arabidopsis thaliana*	*Chlamydomonas reinhardtii*
CO	At5g15840	g6302
COL1	At5g15850	g6302
FT	At1g65480	Not identified
CDF1	At5g62430	Cre12.g521150
CDF3	At3g47500	Cre12.g521150
FBH1	At1g35460	Cre14.g620850
FBH4	At2g42280	Cre14.g620850
ZTL	At5g57360	Cre12.g518800
FKF1	At1g68050	Cre12.g518800
GI	At1g22770	Not identified
TOC1	At5g61380	g16738
LHY	At1g01060	Cre06.g275350
CCA1	At2g46830	Cre06.g275350
CRY1	At4g08920	Cre06.g295200
CRY2	At1g04400	Not identified
HOS1	At2g39810	g16152
COP1	At2g32950	Cre02.g098100

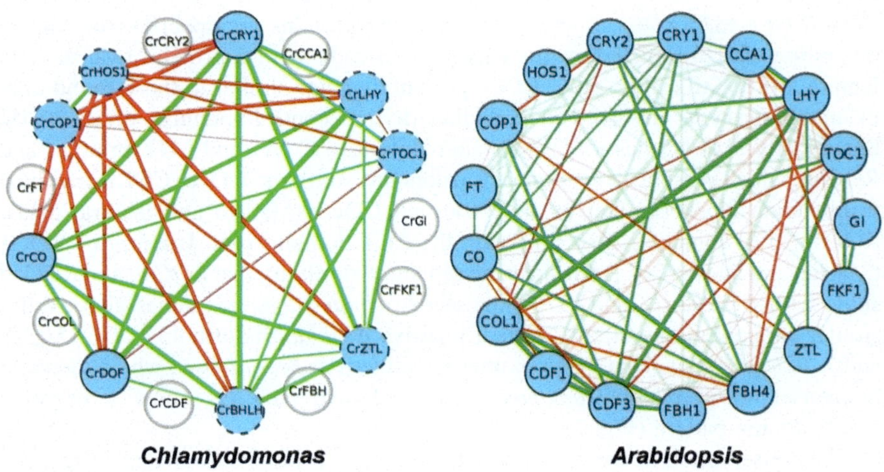

Fig. 2 Co-expression patterns between genes involved in the photoperiod response in *Chlamydomonas* and *Arabidopsis*. The figure represents co-expression relationships (*green*, positive; *red*, negative) between the genes (*blue circles*) involved in the photoperiod response in *Chlamydomonas* and *Arabidopsis*. A conserved co-expression pattern is apparent together with processes of gene duplication as well as specific network rewiring: the circadian clock genes *CCA1*/*LHY* and *TOC1* are negatively co-expressed in both organisms. Processes of gene duplication have produced *Arabidopsis* genes such as *CDF1* and *CDF3* from the *CrDOF*, or *CO* and *COL1* from *CrCO*, from *Chlamydomonas*. Nevertheless, while the positive co-expression between *CrDOF* and *CrCO* in *Chlamydomonas* has been conserved in the *Arabidopsis CDF1* and *COL1*, the co-expression between *CDF1* and *CO* is negative in *Arabidopsis*

phytochromes (PHYs) are involved in morphogenetic, photoperiodic, and circadian mechanisms like flowering.

PHOTs are the principal sensory molecules for light-dependent life cycle control in *Chlamydomonas* and other green algae like *Ostreococcus tauri*. PHOT is a modular protein formed by a light, oxygen, or voltage (LOV) domain, similar to that of the protein family ZTL-FKF1-LKP2, in the amino terminal part of the protein, followed by a carboxy-terminal histidine kinase (HK) domain (LOV-HK). In contrast to PHOTs that are specific of the green linage, the LOV-HK domain is related to the large family of LOV-HK domains found in different kinds of prokaryotes (Djouani-Tahri et al. 2011).

Chlamydomonas PHOTOLIASE HOMOLOG 1 (*CPH1*) encodes a protein with a significant sequence similarity with two plant-specific CRYs (CRY1 and CRY2) involved in the photoperiodic pathway. CPH1 levels are controlled by blue and red light, which induce the instability of the protein (Reisdorph and Small 2004). In this text we refer to *CPH1* as *CrCRY1*. Specific co-expression patterns such as the positive co-expression between *CrCRY1* and *CrCO* seem to be conserved in *Arabidopsis* between the genes *CRY1* and *CO*.

Moreover, in *Chlamydomonas* aCRY (animallike CRY) and DASH-CRYs (*Drosophila*, *Arabidopsis*, *Synechocystis*, and *Homo*-like CRY) (Beel et al. 2013) have been described, indicating that the evolutionary origin of cryptochromes precedes the green eukaryote lineage separation.

PHY-related proteins are a conserved multidomain protein found in bacteria (including cyanobacteria), fungi, and many eukaryotic algae like prasinophytes (green algae), heterokonts (diatoms and brown algae), and glaucophytes. All PHYs use bilin chromophores to sense light. Nevertheless, in algae unlike plants, PHY can sense orange, green, and blue light. In *Chlamydomonas*, in spite of retaining the ability to synthesize bilin, no protein with a significant sequence similarity with any PHY has been identified (Rockwell et al. 2014).

Flowering bHLH Homologs

FBH proteins are part of the large family of eukaryotic basic helix–loop–helix (bHLH)-type transcription factors. bHLHs present a wide diversity and a great number of genes in plant and mosses; in contrast, there is a small family in green and red algae. Particularly, in the *Chlamydomonas* genome only four bHLH genes have been identified (Riaño-Pachón et al. 2008; Carretero-Paulet et al. 2010; Pires and Dolan 2010). Only one of these genes presents significant similarity with bHLH genes present in higher plants such as *Arabidopsis*. We will refer to this gene as *CrbHLH*. The rest of *bHLH* genes seem to be specific of the Chlorophyceae. Additionally, *CrbHLH* exhibits positive co-expression patterns with genes such as *CrCO* and *CrCRY1*. These patterns are conserved in *Arabidopsis* between the genes *FBH4*, *CO*, and *CRY2*.

Constitutive Photomorphogenic 1 and High Expression of Osmotically Responsive Genes 1

COP1 and HOS1 are members of the E3 ubiquitin ligase family with a Ring-finger domain. Up to now, COP1 has been identified in plants and red algae like *Cyanidioschyzon merolae*, whereas HOS1 has been found only in plants (Riaño-Pachón et al. 2008). Nevertheless, recent updates of the web portal for plant comparative genomics Phytozome include potential *Chlamydomonas* orthologs for both genes. These genes have been identified using automatic bioinformatic tools such as the bidirectional best hit method. The conservation of certain co-expression patterns involving these genes supports their consideration as potential orthologs.

2.2.3 Unidentified Genes in Algae

Up to now, no *GI* and *FT* homologs have been identified in any alga species (Corellou et al. 2009; Piñeiro and Jarillo 2013). Therefore, these proteins may have been acquired later in evolution. In fact, the first evidence of a GI binding site in a DOF protein has been found in *Physcomitrella patens* (Lucas-Reina et al. 2015).

3 Overcoming Temperature Changes

Temperature is a key environmental variable that exerts a strong influence on the floral transition. Plants adapted to temperate climates are exposed to annual cold cycles but also to fluctuations of temperature within the different seasons; consequently, they need to differentiate the timing and interval of cold to bloom at the right time in order to increase their reproductive success (Preston and Sandve 2013). Many species from temperate climates require a prolonged exposure to cold in order to become competent to flower (Chouard 1960); this period is known as vernalization. The requirement for vernalization delays reproductive growth during winter minimizing the risk of frost damage to cold-sensitive reproductive organs and ensures that reproductive development and seed production occur in spring and summer (Amasino 2004, 2010; Kim et al. 2009). In addition, most plants in temperate regions face fluctuations in temperatures within the ambient range (above 10 °C) and should be able to perceive and integrate these signals (Samach and Wigge 2005). These non-stressful temperatures have been shown to strongly influence flowering time, causing either a delay or an acceleration of flowering (Westerman and Lawrence 1970; Blazquez et al. 2003). Interestingly, recent reports indicate that the ambient temperature changes are sensed and transduced differently than extreme temperature changes. Here, we will discuss the current knowledge at the molecular level on the mechanisms that control flowering time in response to cold and non-stressful temperatures in different plant species, which will help us to understand the evolution of alternative mechanisms.

3.1 Vernalization

Vernalization responsiveness has evolved independently on multiple occasions (Greenup et al. 2011; Oliver et al. 2013); accordingly, genes controlling vernalization have been identified in different plant lineages (Danyluk et al. 1998; Michaels and Amasino 1999; Sheldon et al. 1999; Izawa et al. 2003; Trevaskis et al. 2003; Pin et al. 2010).

In *A. thaliana*, two genes, *FRIGIDA* (*FRI*) and *FLOWERING LOCUS C* (*FLC*), are the major natural determinants for the vernalization response (Shindo et al. 2005; Lovell et al. 2013; Li et al. 2014). The role of the single-copy gene *FRI* is to activate the expression of *FLC*, which is a MADS-box-type repressor that prevents flowering. Downregulation of *FLC* expression requires a long exposure to cold (Michaels and Amasino 1999). FRI induces *FLC* expression through direct interaction with the nuclear cap-binding complex (Geraldo et al. 2009; Crevillen and Dean 2011). In addition, recent studies have demonstrated that FRI-mediated upregulation of *FLC* is associated with epigenetic modifications, primarily to a marked increase in the histone H3 lysine 4 trimethylation (H3K4me3) pattern (Bastow et al. 2004; Sung and Amasino 2004a; Finnegan and Dennis 2007). The repression of *FLC* by cold involves different mechanisms (Song et al. 2012a). Briefly, an antisense transcript called *COOLAIR* is upregulated after 2–3 weeks of cold leading to the downregulation of *FLC* transcription (Swiezewski et al. 2009). In addition, a sense noncoding RNA (ncRNA) transcript, called *COLDAIR* (Heo and Sung 2011), is also induced by cold but later than *COOLAIR*. *COLDAIR* recruits the polycomb group complex VRN-PRC2 to *FLC* chromatin to mediate gene silencing through the incorporation of histone 3 lysine 27 trimethyl (H3K27me3) marks (De Lucia et al. 2008; Heo and Sung 2011; Crevillen et al. 2013; Kim and Sung 2014). Components of VRN-PRC2 complex are the VEFS domain containing protein VERNALIZATION 2 (VRN2), the SET-domain catalytic subunit CURLY LEAF (CLF) or SWINGER (SWN), FERTILIZATION-INDEPENDENT ENDOSPERM (FIE), and MULTICOPY SUPRESSOR OF IRA1 (MSI1) (Kim and Sung 2014). Additional components of the VRN-PRC2-mediated repression are the plant-specific B3 DNA-binding protein VRN1, and the plant homeodomain (PHD) motif containing proteins VERNALIZATION INSENSITIVE 3 (VIN3), VIN3-LIKE 1 (VIL1), and VERNALIZATION 5 (VRN5), which are nonredundantly necessary for the repression (Levy et al. 2002; Sung and Amasino 2004b; Sung et al. 2006; Greb et al. 2007). *VRN1*, *VRN2*, and *VIL1/VRN5* are constitutively expressed regardless of vernalization. In contrast, *VIN3* is only induced when plants are kept under prolonged periods of cold temperature and quickly decreases when plants are returned to warm growth temperatures. Therefore, *VIN3* is a cold-specific component of the vernalization pathway in *A. thaliana*

Table 2 *Arabidopsis* vernalization orthologs in monocots and brassicas

Arabidopsis thaliana	*T. aestivum H. vulgare*	*Brassica oleracea*	*Beta vulgaris*	*Arabis alpina*
AP1	*VRN1*	*AP1*	–	–
FLC	*VRN2* (*COL* family)	*FLC1*, *FLC2*, *FLC3*, *FLC4*, *FLC5*	*FL1*	*PEP1*
FT	*VRN3*	*FT*	*FT1*, *FT2*	–

(Sung and Amasino 2004b; Kim and Sung 2013, 2014). Nevertheless, the promotion of flowering by vernalization is not exclusively caused by the repression of *FLC*, as plants with a null allele of *FLC* maintain some response to vernalization (Michaels and Amasino 2001), suggesting that other genes are involved. The *MADS AFFECTING FLOWERING 1–5* (*MAF1–5*), which are *FLC* homologs (Ratcliffe et al. 2001, 2003; Scortecci et al. 2001), have been proposed to play a role in the vernalization response; however, their molecular mechanism of action is unknown (Ratcliffe et al. 2003).

Interestingly, several data showed that the extensive allelic heterogeneity at both *FRI* and *FLC* can account for a major fraction of the natural variation in vernalization rate in different *A. thaliana* ecotypes (Johanson et al. 2000; Gazzani et al. 2003; Shindo et al. 2005; Geraldo et al. 2009; Li et al. 2014). *FRI*-like genes with a similar function to *A. thaliana* FRI have been identified in many species, such as *Brassica oleracea*, *A. lyrata*, *Capsella* sp., *Thellungiella halophila*, *Medicago truncatula*, *Lotus japonicus*, *Vitis vinifera*, *Populus balsamífera*, *and Oryza sativa* (Goff et al. 2002; Fang et al. 2008; Kuittinen et al. 2008; Slotte et al. 2008; Risk et al. 2010; Keller et al. 2011; Irwin et al. 2012). Variations in the vernalization responsiveness have been also shown in many of these species (Irwin et al. 2012), suggesting its functional conservation throughout plant evolution. Conversely, *FLC*-like genes as temperature-controlled floral repressors have been identified only in *Arabidopsis*, *Brassica*, *Arabis*, sugar beet (*Beta vulgaris*), and *Petunia* (Michaels and Amasino 1999; Tadege et al. 2001; Schranz et al. 2002; Vandenbussche et al. 2003; Reeves et al. 2007; Wang et al. 2009) (Table 2).

Arabis alpina, a perennial relative of *Arabidopsis*, resumes vegetative growth in fall and repeatedly undergoes vernalization. An *FLC* ortholog [*PERPETUAL FLOWERING 1* (*PEP1*)] acts as a major floral repressor in *Arabis* (Wang et al. 2009). *PEP1* is repressed by vernalizing cold and thus allows plants to bloom. Unlike *Arabidopsis*, *PEP1* is reactivated when plants are returned to warm growth temperature (Kim and Sung 2014). In sugar beet, a pair of *FT* homologs (*BvFT1* and *BvFT2*) acts antagonistically in the floral transition. *BvFT1* acts as a floral repressor whereas *BvFT2* promotes flowering (Pin et al. 2010). Vernalization results in downregulation of *BvFT1*. Vernalization-induced repression of Bv*FT1* is stably maintained even after plants are returned to warm growth temperatures, indicating that *BvFT1* functions similarly to *FLC*. Vernalization requirement in sugar beet is mainly conferred by a dominant allele named *BvBTC1* through its regulation of *BvFT1* and *BvFT2* (Pin et al. 2010). Annual sugar beet plants with a dominant *BvBTC1* allele do not need vernalization for early flowering. In contrast, biennial sugar beet plants carry a partial loss-of-function allele of *Bvbtc1*. *Bvbtc1* is not significantly induced even under LD without vernalization treatment. *Bvbtc1* allele can be gradually activated by vernalization treatment to the level sufficient to repress *BvFT1* and activate *BvFT2* (Kim and Sung 2014).

Recent studies have revealed that the vernalization pathway emerged from a convergent evolution in dicots and monocots (Amasino and Michaels 2010; Greenup et al. 2011; Ream et al. 2012). In cereals, like wheat or barley, flowering is accelerated by vernalization (by a gene resembling *CONSTANS*), as the change in photoperiod in winter time is a stronger floral determinant than temperature (Dubcovsky et al. 2006). In fact, in rice the flowering pathway is regulated mainly by photoperiod, as it does not present a vernalization requirement (Song et al. 2012b).

Genetic analyses in the temperate cereals wheat and barley have shown that three genes determine the vernalization responsiveness: *VRN1*, *VRN2*, and *VRN3* (Pugsley 1971; Yan et al. 2006). They are, nevertheless, different genes than those with the same name in A. thaliana (Table 2). *VRN1* encodes an *APETALA1*-like MADS-box transcription factor with high similarity to the *A. thaliana* meristem identity genes *APETALA1* (*AP1*), *CAULIFLOWER* (*CAL*), and *FRUITFULL* (*FUL*). *VRN1* is induced after vernalization (Trevaskis et al. 2003; Yan et al. 2003; Oliver et al. 2009; Xiao et al. 2014). *VRN2* is the *A. thaliana FLC* functional analogue, although it belongs to the *COL* gene family (Yan et al. 2004; Higgins et al. 2010). VRN2 is a floral repressor that represses *VRN3*, the ortholog of *A. thaliana FT*, under LD conditions. *VRN2* expression is downregulated after vernalization (Trevaskis et al. 2007). Hence, after vernalization the expression of *VRN1* increases, while *VRN2* expression decreases (Yan et al. 2004). On the other hand, *VRN3* induces *VRN1* in LD conditions (Wigge et al. 2005; Yan et al. 2006). The three genes thus form a regulatory loop. Interestingly, *Arabidopsis* and wheat have different genes, *FLC* and *VRN2*, with the same function. However, vernalization in wheat does not result in significant changes in histone modifications at *VRN2*, suggesting that changes of chromatin structure at *VRN2* locus do not occur. Conversely, induction of *VRN1* in barley is epigenetic; however, the epigenetic changes are the opposite of those in *FLC*. In *VRN1* there is a decrease in H3K27me3, the mark of a transcriptionally inactive gene, and an increase in H3K4me3, a mark of an active gene. Activation of VRN1 is quantitative, with longer cold treatments inducing higher levels of expression (Distelfeld et al. 2009; Oliver et al. 2009, 2013). On the other hand, *Brachypodium* spp. have an ortholog of *VRN1* similar to both wheat and barley that promotes flowering; however, *VRN2* is not conserved in this plant (Ream et al. 2014). Surprisingly, a recent report suggested that an *FLC-like* gene is present in monocots, although its function remains to be investigated (Ruelens et al. 2013).

The epigenetic memory of vernalization is maintained by the PcG proteins in *Arabidopsis*. PcG proteins evolved early in evolution, probably in the common ancestor of animals and plants. As evidenced from the variable copy number of homologs in plants, diversification of PRC2 subunits occurred only recently in evolution, mostly after the split of monocots and dicots. There are three VEFS domain containing proteins in *A. thaliana*, EMBRYONIC FLOWER2 (EMF2), VRN2, and FERTILIZATION-INDEPENDENT SEED 2 (FIS2), that bestow partially specialized functions on the corresponding PRC2 complexes. In general, there are several copies of *VEF* genes in dicots as well as in monocots; however, the absence of a *VRN2* ortholog in other species (Luo et al. 2009) suggests that PcG function in the regulation of vernalization response evolved especially in Brassicaceae (Derkacheva and Hennig 2014). Nevertheless, it might be possible that a different VEFS gene participates in the vernalization response in other species. Interestingly, three *VIL* homologs have been identified in the einkorn wheat (*Triticum monococcum* L.) (Fu et al. 2007) and in its wild relative *Aegilops tauschii* (Koyama et al. 2012). Of the three *AetVIL* genes, *AetVIL2* was upregulated after 1 week of low-temperature treatment, and its expression pattern was distinct for winter and spring habit accessions. These observations strongly suggest that *AetVIL2* is associated with the vernalization-responsive pathway in *A. tauschii* (Koyama et al. 2012).

3.2 Ambient Temperature

Recent works in *Arabidopsis* have shed some light in the molecular mechanisms underlying the effect of ambient temperatures on flowering time (Verhage et al. 2014). Warm temperature induces flowering in *Arabidopsis* by upregulation of *FT* expression (Halliday et al. 2003; Balasubramanian and Weigel 2006). The acceleration of flowering in response to high temperature requires the activity of PHYTOCHROME INTERACTING FACTOR4 (PIF4) that directly binds to the *FT* promoter in a temperature-dependent manner (Kumar et al. 2012). The PIF4 binding site in the *FT* promoter is occupied by the histone H2A variant H2A.Z, inhibiting its transcription. *FT* expression increases as H2A.Z-containing nucleosomes are evicted in response to high temperatures (Kumar and Wigge 2010; Kumar et al. 2012). Accordingly, mutations of *ACTIN-RELATED PROTEIN6* (*ARP6*) that compromise H2A.Z occupancy cause the warm temperature transcriptome to be constitutively expressed (Kumar and Wigge 2010). However, other plant species respond in an opposite manner to an increase in the ambient temperature or stay largely independent. Therefore, it is important to determine the evolution of these genes and mechanisms to understand plant response to temperature fluctuations. Recent analysis of the genome of *Brassica rapa* revealed the presence of three orthologs of *PIF4* (Song et al. 2014), while two close orthologs of *PIF4* and *PIF5* exist in rice (*Oryza sativa*) (Nakamura et al. 2007), indicating that *PIF4* might be conserved. However, whether there is also a functional conservation cannot be inferred from these genomic data. On the other hand, histone variant H2A.Z is conserved among eukaryotes and has been proposed to mediate warm temperature signals in budding yeast (*Saccharomyces cerevisiae*) as in *Arabidopsis*

(Kumar and Wigge 2010). Therefore, concerning the conservation of the H2A.Z–PIF4 mechanism, H2A.Z is likely not to be the variable factor. As H2A.Z depletion functions as an enabler, rather than an activator of the higher temperature response, transcription factors can differentially regulate gene expression when shifted to a higher temperature. The fact that H2A.Z depletion only provides access to their targets might explain why plants have evolved a different response to increasing ambient temperatures.

Conversely, the MADS-domain proteins FLM and SVP (SHORT VEGETATIVE PHASE) are involved in the suppression of flowering at low ambient temperatures in *Arabidopsis* (Hartmann et al. 2000; Ratcliffe et al. 2001; Scortecci et al. 2001; Werner et al. 2005; Balasubramanian and Weigel 2006; Lee et al. 2007, 2013; Pose et al. 2013). *FLM* (also known as *MAF1*) is a transcription factor that belongs to the *FLC* clade. Interestingly, *FLM* is alternatively spliced under different ambient temperatures. The two main splice forms function antagonistically through interaction with SVP (Balasubramanian and Weigel 2006; Pose et al. 2013). Low ambient temperatures favor the production of the *FLMβ* splice form, whereas more of the *FLMδ* splice form is produced at high ambient temperatures. Both FLMβ and FLMδ interact with SVP. FLMβ–SVP complex binds to DNA as a repressor of flowering. However, the interaction between SVP and FLMδ results in a functionally ineffective complex, leading to the formation of less repressive FLMβ–SVP complexes. In addition, FLMβ–SVP complex is regulated through protein stability of SVP (Lee et al. 2013). SVP protein becomes gradually less abundant as temperature increases from 16 to 27 °C. Decrease in SVP protein leads to a lower abundance of the repressing FLMβ–SVP complex. Therefore, the regulation of FLM isoforms together with the regulation of SVP protein abundance contributes to repress flowering under low ambient temperatures. Interestingly, all *FLC* clade members (*FLM*/*MAF1*, *MAF2*, *MAF3*, *MAF4*, and *MAF5*) are alternatively spliced. However, it seems that *MAF2*–*MAF4* have evolved different temperature sensitivities (Verhage et al. 2014).

Little is known about the implication of these MADS-box genes in the regulation of flowering time in response to ambient temperature in other species. *FLC*-like genes have been mainly identified as temperature-controlled floral repressors in *Arabidopsis*, *Brassica*, and sugar beet (*Beta vulgaris*) (Michaels and Amasino 1999; Tadege et al. 2001; Schranz et al. 2002; Reeves et al. 2007). Many MADS-box genes have conserved functions across the flowering plants; however, some have acquired novel functions in specific species during evolution. Particularly, the evolution of MADS-box gene subfamilies that control the vegetative-to-floral transition appears to be highly dynamic and linked to the enormous complexity of life history strategies in flowering plants ranging from ephemeral annuals to long-lived trees (Smaczniak et al. 2012a). Future research in other plant species will help to determine whether the orthologs of these or other MADS-box genes have been recruited to this function in other species.

Finally, miR156 and miR172 have been also proposed to regulate floral timing by ambient temperature. Besides timing of the juvenile phase, these two miRNAs have a role in the timing of the phase change from vegetative to reproductive

(Aukerman and Sakai 2003; Wu and Poethig 2006; Verhage et al. 2014). Interestingly, it has been recently shown that miR156-SQUAMOSA PROMOTER BINDING PROTEIN-LIKE 3 (SPL3) module directly regulates *FT* expression in the leaf to control ambient temperature response to flowering. Overexpression of miR156 leads to more delayed flowering at a lower ambient temperature (16 °C), which has been associated with downregulation of *FT* and *FUL* expression. Among miR156 target genes, *SPL3* mRNA levels are significantly reduced at 16 °C. Overexpression of miR156-resistant *SPL3* causes early flowering, regardless of the ambient temperature. Furthermore, SPL3 protein directly binds to GTAC motifs within the *FT* promoter. These data suggest that the interaction between the *miR156–SPL3* module and *FT* is part of the regulatory mechanism controlling flowering time in response to ambient temperature (Kim et al. 2012). Conversely, a higher miR172 expression was observed at 23 °C than at 16 °C (Lee et al. 2010). Both miR156 and miR172 belong to a subset of evolutionary conserved miRNAs that are present throughout the angiosperms (Axtell and Bowman 2008; Cuperus et al. 2011). Results obtained in different dicots and monocots indicate that these miRNAs are not only conserved in sequence but also in their role in regulating phase transition. In addition, mature miRNA has been detected in various mosses, ferns, and gymnosperms (Arazi et al. 2005; Zhang et al. 2006; Axtell and Bowman 2008; Cuperus et al. 2011). In contrast to miR156, miR172 appears to be angiosperm specific, and it has not been cloned from other land plants (Axtell and Bowman 2008; Cuperus et al. 2011), even though the expression of miR172 has been detected by microarrays of RNA extracted from ferns (Axtell and Bowman 2008) and has been computationally predicted in *Physcomitrella* (Fattash et al. 2007). However, whether these miRNAs have a role in controlling thermosensory flowering time in other plants remains to be investigated.

4 Nutrients Signaling to Flowering

Sugars are the main source of carbon and energy for most cell types. For that reason, sugars have been recruited as key regulators of metabolic processes, but they are also involved in the regulation of many other physiological and developmental processes. Its widespread function has contributed to the increase in diversification and plasticity of higher eukaryotes, a phenomenon that acquires an enormous importance in photosynthetic and sessile organism like plants. Therefore, plants have developed more complex and flexible regulatory mechanisms than the rest of higher eukaryotes, and one of such processes is flowering (Rolland et al. 2006). In unicellular algae, routes controlled by sugars are poorly known, and sugar sensing has been involved in metabolic processes such as amino acid transport and astaxanthin biosynthesis in *Chlorella* (Kato and Imamura 2008; Li et al. 2008).

While temperature and photoperiodic signals are key external factors in the *Arabidopsis* floral transition, internal factors such as hormones, nutrients, or plant age have also a strong influence on flowering time (Amasino 2010; Fornara et al. 2010). However, the connection between carbohydrates and flowering is not entirely understood. There are numerous physiological studies showing the effect of sugars in flowering time in different species (Bernier et al. 1993; Lebon et al. 2008), although it is not clear whether they act to promote flowering (Corbesier et al. 1998; Roldan et al. 1999; Wahl et al. 2013) or as floral inhibitors (Zhou et al. 1998; Ohto et al. 2001). The induction of flowering is also associated with the mobilization of starch reserves and a transient increase in carbohydrate transport to the shoot apical meristem (SAM) during the floral transition (Corbesier et al. 1998). Recent studies have shown that this mechanism is controlled by CO, the central photoperiod regulator (Ortiz-Marchena et al. 2014). Interestingly, this process seems to be conserved throughout evolution, as the ancestral CO homolog, CrCO, is also involved in the photoperiodic control of starch accumulation in *Chlamydomonas* (Serrano et al. 2009; Romero and Valverde 2009; Valverde 2011).

It has been shown that trehalose-6-phosphate (T6P) affects flowering in *Arabidopsis* WT plants, so that an increase in sucrose during the floral transition would be signaled by an increase in T6P (Wahl et al. 2013). Plants with abnormal levels of T6P have altered flowering time. Thereby, high levels of T6P would induce the floral transition and vice versa (Schluepmann et al. 2003; Wahl et al. 2013). *FT* expression is reduced in plants with low amount of T6P, so it could be possible that T6P promotes flowering through activation of the florigen (Wahl et al. 2013). Therefore, it has been suggested that T6P promotes flowering when carbohydrate levels are high, influencing the photoperiod pathway (Tsai and Gazzarrini 2014). In this sense, T6P signal could affect flowering through miR156 and SPL (Matsoukas et al. 2012), so that T6P inhibits *miRNA156* expression and SPL is then able to promote the floral transition (Wahl et al. 2013). Although in green algae T6P regulatory function is unknown, its biosynthetic mechanism is conserved in all algae and even in bacteria (Avonce et al. 2010; Michel et al. 2010; Deng et al. 2014; Pade et al. 2014).

In plants, transcriptional regulation by sugars interacts with signaling pathways mediated by hormones, although the mechanism by which this occurs is unknown. Evidence suggests that it is probably due to direct interactions between protein components of both routes in complexes, although there may also be indirect interactions (Gibson 2004; Jossier et al. 2009). Hexose levels, such as glucose and fructose, for example, are sensed by HEXOKINASE1 (HXK1). HXK1 is a glucose-phosphorylating enzyme that exerts a dual function as sugar sensor and hexose kinase. Both functions are independent, so that the metabolism of the hexose phosphate is not involved in the signaling function (Loreti et al. 2000; Moore et al. 2003; Valverde et al. 2005). The conservation of some steps in the signal cascade of sugar sensing is still in controversy. However, HXK is considered a conserved glucose sensor among algae, yeast, plants and animals (Pego et al. 2000; Li et al. 2008; Oesterhelt and Gross 2014).

Two other important systems regulate sugar signaling in plants, the Snf1-related kinase 1 (SnRK1) and the target of rapamycin (TOR) kinase. Both of them are central regulators that sense nutrient levels and promote or inhibit growth in an antagonistic way: low sugar levels promote *SnRK1* expression and high sugar levels upregulate TOR activity (Deprost et al. 2007; Smeekens et al. 2010; Robaglia et al. 2012). Although there are two possible orthologs of *SnRK1* annotated in the *Chlamydomonas* genome, there is no evidence about its functions. However, TOR is a central regulator of cell growth in all eukaryotes (Crespo 2012), and *Chlamydomonas* is no exception as TOR is regulated by nutrients (Crespo et al. 2005). Recently, T6P has been shown to inhibit SnRK1 activity in *Arabidopsis* (Zhang et al. 2009). T6P seems to have this function also in monocots, indicating a conserved role for this sugar (Zhang et al. 2009; Wu and Birch 2010; Debast et al. 2011; Martinez-Barajas et al. 2011; Nunes et al. 2013; Lawlor and Paul 2014). Both T6P and SnRK1 have opposite functions as major regulators of gene expression related to growth and energy (Baena-González and Sheen 2008; Zhang et al. 2009).

It has also been reported in *Arabidopsis* that *EXORDIUM* (*EXO*) and *EXO-LIKE* genes control growth on different environmental conditions through the response to brassinosteroids (Schroder et al. 2009). EXO proteins seem to modify the response to sugars in seedlings and to control general gene expression by sugars and the accumulation of starch mediated by sugars, ABA, and anthocyanins. Therefore, EXO protein would establish a balance between the levels of external carbon available for plant and the cell status (Lisso et al. 2013). In green algae, it has been shown that brassinosteroids and auxins work synergistically in the control of growth and metabolism (Bajguz and Piotrowska-Niczyporuk 2013), but until now, no *EXO* homolog has been described in any algal genome.

All these premises suggest that sugar sensing is an ancient, flexible regulatory mechanism that evolved, using ancestral elements, according to the needs of each organism.

Although sugars play an important role in the floral transition, nitrogen (N) availability also influences flowering time (Frink et al. 1999). N is an essential macronutrient and specifically N deprivation induces early flowering in different plants including *Arabidopsis* (Dickens and Staden 1988; Bernier et al. 1993; Loeppky and Coulman 2001; Castro Marin et al. 2011; Kant et al. 2011; Liu et al. 2013). Under N deprivation, the flowering integrators *FT*, *AP1*, and *LEAFY* (*LFY*) are induced (Kant et al. 2011). Also, *CO* expression is induced in low nitrate conditions and is repressed by high nitrate levels (Liu et al. 2013). On the other hand, spray of nitrate to stem and leaves induces flowering formation in mango trees in the tropics (Núñez-Elisea and Caldeira 1988). N also governs many processes in algae. In *Chlamydomonas*, N controls sexual life cycle (Goodenough et al. 2007), photosynthesis (Grossman 2000), and lipid induction (Sharma et al. 2012), among other processes. Nevertheless, the general regulatory mechanisms that connect N metabolism to developmental responses are widely unknown.

5 Flower Development

Floral organogenesis is a natural extension of the floral transition process and shares many early genes involved in SAM differentiation and tissue organization. Floral integrators such as *FT*, *AP1*, and *LFY* have a significant role in the early stages of floral tissue formation, and their mutation aborts the early differentiation process of the vegetative apical meristem into a reproductive meristem. In fact, flower appearance is extremely variable among species in size, shape, symmetry, and pigmentation, although the different whorls of organs originate from the floral meristem, a small group of undifferentiated cells. Typical angiosperm flowers consist of four organ types arranged in four concentric whorls at the tip of a floral shoot. From the outside to the inside of the flower, these organs are leaflike green sepals (whorl 1), generally colored petals (whorl 2), the male reproductive organs or stamens (whorl 3), and carpels (whorl 4), the female reproductive organs. During their life cycle, plants undergo several phase transitions in which miR156 and miR172 play an important role (Huijser and Schmid 2011; Poethig 2013; Wu and Poethig 2006). Among them, the vegetative-to-reproductive phase transition ends up with the formation of the flower. During this transition, the SAM changes to an inflorescence meristem (IM). The IM can be converted in a floral meristem (FM) or produce lateral meristems that will be, in turn, converted in a FM. The FM undergoes an early growth phase before the identity of the floral organs is established (McKim and Hay 2010). The characterization in *Arabidopsis thaliana* and *Antirrhinum majus* of different homeotic mutants in which the identity of floral organs was altered leads to the proposal of the ABC model for flower development (Haughn and Somerville 1988; Sommer et al. 1990; Coen et al. 1990, 1991; Yanofsky et al. 1990; Coen 1991; Carpenter and Coen 1990; Coen and Meyerowitz 1991; Schwarz-Sommer et al. 1990; Bowman et al. 1991). These homeotic mutants defined three overlapping functions, A, B, and C (Fig. 3), each operating in two adjacent whorls that specify the identity of the four floral organ types (Coen and Meyerowitz 1991). A-function mutants display carpels in the first whorl and stamens in the second whorl instead of sepal and petals, respectively. B-function mutants have sepals in the second whorl and carpels in the third whorl rather than petals and stamens. Finally, in C-function mutants petals substitute stamens in the third whorl and sepals carpels in the fourth whorl. Besides, C-function mutants are indeterminate and produce floral organs inside the fourth whorl. The A function acts alone in the outermost whorl (whorl 1) to specify sepal identity. A and B functions act in the second whorl to specify petals. The reproductive organs are specified by the action of B and C functions. Thus, stamens are determined by the joint action of B and C functions in the third whorl. At the center of the flower, in whorl 4, the C function acts alone to initiate carpel development and to terminate further development of the floral meristem. The ABC model also proposes that activity of C and A functions is mutually exclusive and C function is restricted to the third and fourth whorls by A function and vice versa (Fig. 3) (Coen and Meyerowitz 1991). Most floral homeotic genes controlling floral organ identity

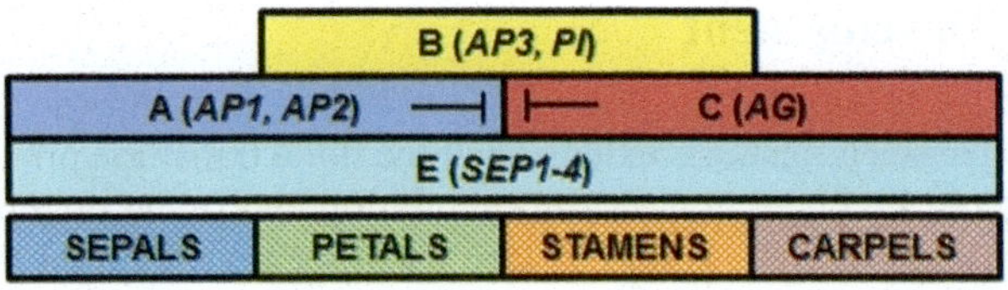

Fig. 3 Specification of floral organ identity. The combination of A, B, C, and E functions originates the specification of the four organ types. *Arabidopsis* genes responsible for the corresponding functions are indicated inside the expression domains of each function in a color-coded pattern

encode MADS-box transcription factors (Meyerowitz 1997; Ng and Yanofsky 2001; Theissen 2001; Schwarz-Sommer et al. 1990; Krizek and Fletcher 2005; Lohmann and Weigel 2002; Jack 2001). MADS is an acronym for *MCM1* (yeast), *AGAMOUS* (*Arabidopsis*), *DEFICIENS* (*Antirrhinum*), and *SRF* (human) on which the definition of this gene family was based (Schwarz-Sommer et al. 1990).

5.1 Floral Identity Determination

Plant floral meristem identity genes control floral meristem versus shoot/inflorescence fate (Bartlett et al. 2008). The meristem identity genes *LFY* and *AP1* in *Arabidopsis* and *FLORICAULA* (*FLO*) and *SQUAMOSA* (*SQUA*) in *Antirrhinum* induce flower development, whereas *TERMINAL FLOWER1* (*TFL1*) in *Arabidopsis* and *CENTRORADIALIS* (*CEN*) in *Antirrhinum* promote inflorescence development (Blazquez et al. 2006; Bradley et al. 1996; Alvarez et al. 1992; Coen et al. 1990; Huijser et al. 1992; Weigel et al. 1992; Mandel et al. 1992). Meristem identity genes are responsible for the determination of the floral meristem at the SAM for the control of the floral organ identity functions (mainly *MADS*-box genes). This transition represents the first step specific to floral development and is driven by the *FLO*/*LFY* genes. *flo* and *lfy* mutants produce proliferating inflorescence shoots instead of flowers (Coen et al. 1990; Schultz and Haughn 1991; Weigel et al. 1992). Homologs to *FLO*/*LFY* have been identified in many different plants and are present in most of the terrestrial plants analyzed, including mosses, ferns, gymnosperms, and angiosperms (Maizel et al. 2005).

The flowering signaling pathways responding to environmental, autonomous, and endogenous signals converge in the so-called floral integrators. *FT–FD* complex at the SAM induces flowering by activating *SUPRESSOR OF OVEREXPRESSION OF CONSTANS 1* (*SOC1*), which in combination with *AGAMOUS-LIKE 24* (*AGL24*) promotes the expression of the floral meristem identity gene *LFY* (Lee et al. 2008), which in turn will directly induce the expression of *AP1* (Mandel and Yanofsky 1995; Parcy et al. 1998; Wagner et al. 1999). The *FT–FD* complex also directly activates *AP1* originating a feed-forward loop. Induction of *LFY* is also mediated by a set of different genes as *SHOOT*

MERISTEMLESS (*STM*), *PENNYWISE* (*PNY*), *POUND-FOOLISH* (*PNF*), and *SPL3* that activate *LFY* and thus the transition to FM (Yamaguchi et al. 2009; Lee et al. 2008; Kanrar et al. 2008; Smith et al. 2011; Pose et al. 2012; Wigge et al. 2005).

Although *LFY* is considered to be the main actor of this transition, other transcription factors from the MADS-box family as *FUL* and *AP1* are also necessary (Ferrandiz et al. 2000; Melzer et al. 2008; Bowman et al. 1993; Mandel and Yanofsky 1995; Weigel and Nilsson 1995) and are co-regulated with *LFY* by *SPL3* (Yamaguchi et al. 2009; Huijser and Schmid 2011). *LFY* and *AP1* control the whole floral network regulating genes involved in the determinacy of the floral meristem and floral organ primordia (Coen et al. 1990; Benlloch et al. 2007; Weigel et al. 1992; Moyroud et al. 2009, 2010; Liu et al. 2009; Irish 2010) and constitute hubs that coordinate multiple processes and developmental pathways (O'Maoileidigh et al. 2014). *LFY* codes for a plant-specific transcription factor that is present as a single-copy gene in most angiosperms and binds to the regulatory regions of its target genes as a dimer with a DNA-binding domain structurally similar to the helix–turn–helix domain (Maizel et al. 2005; Benlloch et al. 2007; Hames et al. 2008; Parcy et al. 1998; Busch et al. 1999; Lamb et al. 2002; Lohmann et al. 2001; Moyroud et al. 2009). *LFY* is expressed at low levels in vegetative tissues, is upregulated in response to the flowering signals, and is expressed in the floral organ primordial where it participates in establishing specific gene expression patterns in the floral organ primordia.

Angiosperms evolved from gymnosperm ancestors at least 130–136 MYA, as evidenced by the earliest fossilized record of pollen from an apparent angiosperm known to date (Frohlich 2006). During plant evolution, several genome duplication events have occurred. However, as indicated before, *LFY* in angiosperms is a single-copy gene in most species with the exception of maize and Lamiales (Aagaard et al. 2006; Bomblies et al. 2003); thus, *LFY* can provide evidences on the evolutionary pace of plants. Some species exhibit various *LFY*-like genes that have been shown to be paralogs acquired recently by polyploidy as in *Nicotiana tabacum* or from small-scale duplication events (Moyroud et al. 2009). On the other hand, gymnosperms usually present two paralogs, *LFY* and *NEEDLY* (*NDLY*) (Mellerowicz et al. 1998; Mouradov et al. 1998), originated in a gymnosperm-specific duplication, with the *NDLY* lineage being lost in angiosperms (Frohlich and Estabrook 2000; Maizel et al. 2005; Himi et al. 2001; Frohlich 2003). Gymnosperm *LFY* homologs are mainly expressed in reproductive meristems and are able to complement *Arabidopsis lfy* mutants, indicating that LFY function is conserved between gymnosperms and angiosperms (Mouradov et al. 1998; Shindo et al. 2001; Maizel et al. 2005). Homologs of *LFY* have also been identified in ferns, mosses, and thallophytic green algae (Himi et al. 2001; Tanahashi et al. 2005; Sayou et al. 2014). Fern *LFY* homolog *CrLFY2* can partially rescue the *Arabidopsis lfy* phenotype (Maizel et al. 2005). In the moss *Physcomitrella patens*, two *LFY* homologs have been identified (*PpLF1*, 2) that have been shown to regulate cell division in the zygote (Tanahashi et al. 2005). PpLFY1 is unable to bind the sequence recognized by *Arabidopsis* LFY, although one amino acid substitution is sufficient for binding to a canonical LFY binding site (Maizel et al. 2005).

By analyzing the binding specificity of LFY homologs from different groups of plants, including green algae, it has been suggested that during evolution LFY modified its DNA binding specificity even though plant genomes generally contain a single *LFY* copy (Sayou et al. 2014). Gene duplication followed by sub-functionalization is a common mechanism in evolution. Duplicated genes loose the obligation to maintain its original function and can evolve to acquire new functions through mutations in their regulatory or coding regions. However, in the case of *LFY*, the acquisition of the floral function seems to be related to changes in its DNA-binding domain (and probably in the cis-regulatory elements of its target genes) through an intermediate showing various binding specificities, thus avoiding deleterious effects (Sayou et al. 2014; Maizel et al. 2005; Della Pina et al. 2014; Kovach and Lamb 2014). The fact that *LFY* is present in multicellular and not in unicellular algae and that it is related to meristem organization suggests that *LFY* is associated to multicellularity, in contrast to *COLs*, *DOFs*, *bHLHs*, and other families of regulatory genes that originated in unicellular algae (Serrano et al. 2009; Romero-Campero et al. 2013).

5.2 *Floral Organ Identity Determination*

As indicated above, the floral meristem identity genes control the floral organ identity genes, whose mutation induces homeotic transformation of one organ into another. Genes that contribute to the A, B, and C functions are transcription factors and are known in different plants. In the case of *Arabidopsis*, AP1 and *APETALA2* (*AP2*) are A-function genes, *APETALA3* (*AP3*) and *PISTILLATA* (*PI*) are B-function genes, and AGAMOUS (AG) is a C-function gene (Fig. 3) (Theissen 2001). The ABC function genes belong to the MADS-box family of transcription factors, with the exception of AP2, which belong to the AP2/ERF family (Jofuku et al. 1994; Weigel 1995; Okamuro et al. 1997; Riechmann and Meyerowitz 1998). The ABC model has been implemented by the identification and characterization of four MADS-box SEPALLATA genes (SEP1–4), which act redundantly and are required for the A, B, and C functions (Pelaz et al. 2000; Ditta et al. 2004), giving rise to the ABCE model for flower development (Wellmer et al. 2014; Theissen 2001). The ABCE functions would act in a combinatorial manner to specify each of the four floral organs. Thus, class A and E genes are necessary to specify sepals; class B and E genes are necessary to specify petals; class B, C, and E genes specify stamens; and finally class C and E genes specify carpels (Fig. 3) (Theissen 2001; Ditta et al. 2004; Theissen and Melzer 2007a).

According to the ABCE model, floral organ determination is accomplished by the formation of multimeric complexes of floral organ identity proteins that bind to two CArG boxes with a consensus sequence $CC(A/T)_6GG$ (Wynne and Treisman 1992; Honma and Goto 2001). Analysis of the interaction between DEFICIENS

(DEF), GLOBOSA (GLO), and SQUAMOSA (SQUA) from *Antirrhinum majus* provided the first evidences on the establishment of tetramers composed of a heterodimer DEF–GLO and a homodimer SQUA–SQUA (Egea-Cortines et al. 1999). *DEF*, *GLO*, and *SQUA* are the orthologs of *Arabidopsis AP3*, *PI*, and *AP1*, respectively (Becker and Theissen 2003). Based on the observation that the SEP genes are also involved in the formation of petals, stamens, and carpels (Pelaz et al. 2000) and act as mediators of higher-order complex formation, the floral quartet model was coined as a mechanistic model for the determination of floral organs (Theissen and Saedler 2001; Honma and Goto 2001; Wellmer et al. 2014; Theissen and Melzer 2007b; Melzer and Theissen 2009; Erdmann et al. 2010; Melzer et al. 2009; Jetha et al. 2015). The floral quartet model indicates that specification of floral organs is mediated by the combinatorial formation of tetramers of MADS-domain proteins, although it has also been shown that floral organ identity MADS-box proteins interact with other types of proteins as chromatin-associated proteins and other transcription factors to establish higher-order complexes (Smaczniak et al. 2012a, b; Wellmer et al. 2014; O'Maoileidigh et al. 2014; Simonini et al. 2012; Liu et al. 2009).

MADS-box genes constitute a large family that has been divided in two main lineages, type I and type II, which are present in plants, animals, and fungi (Alvarez-Buylla et al. 2000a). Members of the MADS-box transcription family are characterized for the presence of a highly conserved MADS-box with a length of about 180 nucleotides that codes for the DNA binding to the CArG box (Alvarez-Buylla et al. 2000b; Theissen et al. 2000; Riechmann and Meyerowitz 1997). The MADS-box genes in plants, with more than 100 members, were initially implicated in floral organ specification, although it has been shown to participate in many different developmental processes during the life cycle of plants (Smaczniak et al. 2012a; De Bodt et al. 2005). The family of MADS-box genes increased considerably during evolution by duplication-divergence-specialization of individual paralogs. Type I MADS-box genes form a heterogeneous group that just share the MADS domain (Kofuji et al. 2003; Parenicova et al. 2003; De Bodt et al. 2003). Type I and II MADS-box genes have been identified in all land plant lineages, from bryophytes to angiosperms. Their number and their functional diversity increased considerably during evolution (Becker and Theissen 2003; Kramer and Hall 2005; Kaufmann et al. 2005; Gramzow and Theissen 2010). Recently, several type I MADS-box genes have been shown to have regulatory roles in different aspects of plant reproduction as female gametogenesis and seed development (Masiero et al. 2011; Portereiko et al. 2006; Steffen et al. 2008; Kang et al. 2008). It has also been suggested that type I MADS-box proteins form heteromeric complexes (de Folter et al. 2005). The MADS-box type II lineage includes the floral homeotic genes as well as genes participating in embryogenesis, flowering time, and fruit development, among others (Smaczniak et al. 2012a). Type II MADS-box genes are characterized for having an N-terminal MADS domain, an intervening domain (I) and a keratin-like domain (K) that are essential for protein–protein interaction, and a very variable C-terminal domain, thus named MIKC-type MADS-box (Kaufmann et al. 2005; Smaczniak et al. 2012a). MIKC-type has been subdivided in two groups, MIKCc and MIKC*, the latter generally having a longer K domain (Henschel et al. 2002; Kwantes et al. 2012; Smaczniak et al. 2012a), that have been characterized in seed plants, pteridophytes, and mosses, indicating that the two groups diverged before the separation of mosses and land plants. In the unicellular green and red algae *Chlamydomonas reinhardtii* and *Cyanidioschyzon merolae*, respectively, a single MADS-box gene, lacking the I, C, and K domains, has been identified (Tanabe et al. 2005). However, MIKC-type MADS-box genes have been characterized in charophycean green algae, having a role in haploid reproductive development during the gametophytic phase (Tanabe et al. 2005). Land plants originated from multicellular charophycean algae about 500 MYA (Graham et al. 2000); thus, MIKC-type MADS-box genes might be recruited to form higher-order complexes before the origin of land plants. The fact that all the charophycean algae MADS-box genes characterized belong to the MIKCc type indicates that they are ancestral to the MIKC* type (Tanabe et al. 2005) and that MICK*-type genes evolved in the charophycean–land plant lineage after its divergence from *Chlamydomonas*. Considering that mosses and club moss (lycophyte) (Henschel et al. 2002), and the rest of land plant lineages, have both types of MIKC genes, it can be assumed that the last common ancestor of mosses and land plants (about 450 MYA) already had both types of MIKC MADS-box genes.

MADS-box genes are generally associated with the development of reproduction in extant land plant, mosses, and green algae relatives. However, extensive duplication events followed by specialization gave rise to a plethora of MADS-box genes involved in many different aspects of plant life cycle other than reproductive processes (Smaczniak et al. 2012a). Many different target genes involved in transcriptional and cellular signaling have been identified for FLC, SEP3, and AP1 (Deng et al. 2011; Kaufmann et al. 2009, 2010; Ito 2011; Dornelas et al. 2011), so the complexity of MADS-box transcription factors at the level of number of members, functions, spatiotemporal expression, posttranscriptional regulation, establishment of high-order complexes, and their putative role in more than organ or developmental stage will require the use of massive analysis techniques to generate a global framework to understand the evolution of this transcription factor family. Besides, the characterization of gene regulatory networks (GRN) will also provide primordial information to the study of MADs-box genes (Espinosa-Soto et al. 2004; van Mourik et al. 2010).

6 Conclusions

The study of the flowering pathways during the evolutionary history of plants unveils regulatory aspects that cannot be deduced from the study of single stories within the same species. We have learned that some of these regulatory pathways are conformed by a set of evolutionarily conserved genes that share even the same hierarchical regulatory mechanisms and modules. These "toolkits" were present as simple, short pathways in unicellular algae and evolved to long, complex ones in angiosperms. The addition of gene copies and new regulatory modules seem to have been a constant in many of the flowering pathways that allowed modern plant to respond with high efficiency to changing environmental conditions. This plasticity is essential to assure that flowering, and thus seed release, will be planned ahead and triggered at the moment of the year that guarantees a successful offspring for the species. This is of course intertwined with other signals such as the synchronicity with pollinator's signals and competing species that are too complex to discover in a direct analysis, but perhaps will become easier to understand if we learn to identify the gene toolkits and basic mechanism that rule these transitions.

The advent of massive analysis techniques is allowing us the rigorous and systematic study of non-model plant species. This information is being fed to computational analysis built upon the regulatory pathways constructed in model species. Surprisingly, these analyses have revealed a lot of homogeneity in the flowering pathways even among very different plant families. Therefore, it seems plausible to believe that these signaling mechanisms were mastered in the early flowering plants, were recruited from mechanisms that triggered developmental decisions in primitive plants, and have thus remained relatively unchanged during evolution due to their importance. This evolution and development perspective could allow us to better understand the response of plants to the incoming changing environmental conditions, intensified by human activity, and develop strategies to make plants flower at the correct time of the year in order to better perpetuate their species and ours.

Acknowledgments The authors would like to thank the help from the coordinated projects BIO2011-28847-C02-00 and BIO2014-52425-P to FV and JMR, the BIO2013-44078-P project to MC, and the TRANSPLANTA Consolider 28317 project from Spanish Ministry of Economy and Innovation (MINECO). The help from the Marie Curie Grant ID333748 to MC and from the Excellence Project P08-AGR-03582 and CVI-281 to FV from the Andalusian Government is also acknowledged. The authors regret the exclusion, due to lack of space, of excellent works from other colleagues that could not be referred in this review.

References

Aagaard JE, Willis JH, Phillips PC (2006) Relaxed selection among duplicate floral regulatory genes in Lamiales. J Mol Evol 63:493–503

Alvarez J, Guli CL, Yu X-H, Smyth DR (1992) Terminal flower: a gene affecting inflorescence development in Arabidopsis thaliana. Plant J 2:103–116

Alvarez-Buylla ER, Liljegren SJ, Pelaz S, Gold SE, Burgeff C, Ditta GS, Vergara-Silva F, Yanofsky MF (2000a) MADS-box gene evolution beyond flowers: expression in pollen, endosperm, guard cells, roots and trichomes. Plant J 24:457–466

Alvarez-Buylla ER, Pelaz S, Liljegren SJ, Gold SE, Burgeff C, Ditta GS, Ribas de Pouplana L, Martinez-Castilla L, Yanofsky MF (2000b) An ancestral MADS-box gene duplication occurred before the divergence of plants and animals. Proc Natl Acad Sci USA 97:5328–5333

Amasino R (2004) Vernalization, competence, and the epigenetic memory of winter. Plant Cell 16:2553–2559

Amasino R (2010) Seasonal and developmental timing of flowering. Plant J 61:1001–1013

Amasino RM, Michaels SD (2010) The timing of flowering. Plant Physiol 154:516–520
Andrés F, Coupland G (2012) The genetic basis of flowering responses to seasonal cues. Nat Rev Genet 13:627–639
Arazi T, Talmor-Neiman M, Stav R, Riese M, Huijser P, Baulcombe DC (2005) Cloning and characterization of micro-RNAs from moss. Plant J 43:837–848
Aukerman MJ, Sakai H (2003) Regulation of flowering time and floral organ identity by a MicroRNA and its APETALA2-like target genes. Plant Cell 15:2730–2741
Avonce N, Wuyts J, Verschooten K, Vandesteene L, Van Dijck P (2010) The Cytophaga hutchinsonii ChTPSP: first characterized bifunctional TPS-TPP protein as putative ancestor of all eukaryotic trehalose biosynthesis proteins. Mol Biol Evol 27:359–369
Axtell MJ, Bowman JL (2008) Evolution of plant microRNAs and their targets. Trends Plant Sci 13:343–349
Baena-González E, Sheen J (2008) Convergent energy and stress signaling. Trends Plant Sci 13:474–482
Bajguz A, Piotrowska-Niczyporuk A (2013) Synergistic effect of auxins and brassinosteroids on the growth and regulation of metabolite content in the green alga Chlorella vulgaris (Trebouxiophyceae). Plant Physiol Biochem 71:290–297
Balasubramanian S, Weigel D (2006) Temperature induced flowering in Arabidopsis thaliana. Plant Signal Behav 1:227–228
Bartlett ME, Kirchoff BK, Specht CD (2008) Epi-illumination microscopy coupled to in situ hybridization and its utility in the study of evolution and development in non-model species. Dev Genes Evol 218:273–279
Bastow R, Mylne JS, Lister C, Lippman Z, Martienssen RA, Dean C (2004) Vernalization requires epigenetic silencing of FLC by histone methylation. Nature 427:164–167
Becker A, Theissen G (2003) The major clades of MADS-box genes and their role in the development and evolution of flowering plants. Mol Phylogenet Evol 29:464–489
Beel B, Müller N, Kottke T, Mittag M (2013) News about cryptochrome photoreceptors in algae. Plant Signal Behav 8:e22870
Benlloch R, Berbel A, Serrano-Mislata A, Madueño F (2007) Floral initiation and inflorescence architecture: a comparative view. Ann Bot 100:659–676
Bernier G, Havelange A, Houssa C, Petitjean A, Lejeune P (1993) Physiological signals that induce flowering. Plant Cell 5:1147–1155
Blazquez MA, Ahn JH, Weigel D (2003) A thermosensory pathway controlling flowering time in Arabidopsis thaliana. Nat Genet 33:168–171
Blazquez MA, Ferrándiz C, Madueño F, Parcy F (2006) How floral meristems are built. Plant Mol Biol 60:855–870
Böhlenius H, Huang T, Charbonnel-Campaa L, Brunner AM, Jansson S, Strauss SH, Nilsson O (2006) CO/FT regulatory module controls timing of flowering and seasonal growth cessation in trees. Science 312:1040–1043
Bomblies K, Wang RL, Ambrose BA, Schmidt RJ, Meeley RB, Doebley J (2003) Duplicate FLORICAULA/LEAFY homologs zfl1 and zfl2 control inflorescence architecture and flower patterning in maize. Development 130:2385–2395
Bouget FY, Lefranc M, Thommen Q, Pfeuty B, Lozano JC, Schatt P, Botebol H, Vergé V (2014) Transcriptional versus non-transcriptional clocks: a case study in Ostreococcus. Mar Genomics 14:1–6
Bowman JL, Smyth DR, Meyerowitz EM (1991) Genetic interactions among floral homeotic genes of Arabidopsis. Development 112:1–20
Bowman JL, Alvarez J, Weigel D, Meyerowitz EM, Smyth DR (1993) Control of flower development in Arabidopsis thaliana by APETALA1 and interacting genes. Development 119:721–743
Bradley D, Carpenter R, Copsey L, Vincent C, Rothstein S, Coen E (1996) Control of inflorescence architecture in Antirrhinum. Nature 379:791–797

Bradshaw WE, Holzapfel CM (2007) Evolution of animal photoperiodism. Annu Rev Ecol Evol Syst 38:1–25
Busch MA, Bomblies K, Weigel D (1999) Activation of a floral homeotic gene in Arabidopsis. Science 285:585–587
Carpenter R, Coen ES (1990) Floral homeotic mutations produced by transposon-mutagenesis in Antirrhinum majus. Genes Dev 4:1483–1493
Carretero-Paulet L, Galstyan A, Roig-Villanova I, Martínez-García JF, Bilbao-Castro JR, Robertson DL (2010) Genome-wide classification and evolutionary analysis of the bHLH family of transcription factors in Arabidopsis, poplar, rice, moss, and algae. Plant Physiol 153:1398–1412
Casal JJ, Fankhauser C, Coupland G, Blázquez MA (2004) Signalling for developmental plasticity. Trends Plant Sci 9:309–314
Castro Marin I, Loef I, Bartetzko L, Searle I, Coupland G, Stitt M, Osuna D (2011) Nitrate regulates floral induction in Arabidopsis, acting independently of light, gibberellin and autonomous pathways. Planta 233:539–552
Chouard P (1960) Vernalization and its relations to dormancy. Annu Rev Plant Physiol 11:47
Coen ES (1991) The role of homeotic genes in flower development and evolution. Annu Rev Plant Physiol Plant Mol Biol 42:241–279
Coen ES, Meyerowitz EM (1991) The war of the whorls: genetic interactions controlling flower development. Nature 353:31–37
Coen ES, Romero JM, Doyle S, Elliott R, Murphy G, Carpenter R (1990) Floricaula: a homeotic gene required for flower development in antirrhinum majus. Cell 63:1311–1322
Coen ES, Doyle S, Romero JM, Elliott R, Magrath R, Carpenter R (1991) Homeotic genes controlling flower development in Antirrhinum. Development 113:149–155
Corbesier L, Lejeune P, Bernier G (1998) The role of carbohydrates in the induction of flowering in Arabidopsis thaliana: comparison between the wild type and a starchless mutant. Planta 206:131–137
Corellou F, Schwartz C, Motta JP, Djouani-Tahri EB, Sanchez F, Bouget FY (2009) Clocks in the green lineage: comparative functional analysis of the circadian architecture of the picoeukaryote ostreococcus. Plant Cell 21:3436–3449
Corrales AR, Nebauer SG, Carrillo L, Fernández-Nohales P, Marqués J, Renau-Morata B, Granell A, Pollmann S, Vicente-Carbajosa J, Molina RV, Medina J (2014) Characterization of tomato Cycling Dof Factors reveals conserved and new functions in the control of flowering time and abiotic stress responses. J Exp Bot 65:995–1012
Crespo JL (2012) BiP links TOR signaling to ER stress in Chlamydomonas. Plant Signal Behav 7:273–275
Crespo L, Díaz-Troya S, Florencio FJ (2005) Inhibition of target of rapamycin signaling by rapamycin in the unicellular green alga. Plant Physiol 139:1736–1749
Crevillen P, Dean C (2011) Regulation of the floral repressor gene FLC: the complexity of transcription in a chromatin context. Curr Opin Plant Biol 14:38–44
Crevillen P, Sonmez C, Wu Z, Dean C (2013) A gene loop containing the floral repressor FLC is disrupted in the early phase of vernalization. EMBO J 32:140–148
Cuperus JT, Fahlgren N, Carrington JC (2011) Evolution and functional diversification of MIRNA genes. Plant Cell 23:431–442
Danyluk J, Perron A, Houde M, Limin A, Fowler B, Benhamou N, Sarhan F (1998) Accumulation of an acidic dehydrin in the vicinity of the plasma membrane during cold acclimation of wheat. Plant Cell 10:623–638
De Bodt S, Raes J, Florquin K, Rombauts S, Rouze P, Theissen G, Van de Peer Y (2003) Genome-wide structural annotation and evolutionary analysis of the type I MADS-box genes in plants. J Mol Evol 56:573–586
De Bodt S, Maere S, Van de Peer Y (2005) Genome duplication and the origin of angiosperms. Trends Ecol Evol 20:591–597

de Folter S, Immink RG, Kieffer M, Parenicova L, Henz SR, Weigel D, Busscher M, Kooiker M, Colombo L, Kater MM, Davies B, Angenent GC (2005) Comprehensive interaction map of the Arabidopsis MADS Box transcription factors. Plant Cell 17:1424–1433

De Lucia F, Crevillen P, Jones AM, Greb T, Dean C (2008) A PHD-polycomb repressive complex 2 triggers the epigenetic silencing of FLC during vernalization. Proc Natl Acad Sci USA 105:16831–16836

Debast S, Nunes-Nesi A, Hajirezaei MR, Hofmann J, Sonnewald U, Fernie AR, Bornke F (2011) Altering trehalose-6-phosphate content in transgenic potato tubers affects tuber growth and alters responsiveness to hormones during sprouting. Plant Physiol 156:1754–1771

Della Pina S, Souer E, Koes R (2014) Arguments in the evo-devo debate: say it with flowers! J Exp Bot 65:2231–2242

Deng W, Ying H, Helliwell CA, Taylor JM, Peacock WJ, Dennis ES (2011) FLOWERING LOCUS C (FLC) regulates development pathways throughout the life cycle of Arabidopsis. Proc Natl Acad Sci USA 108:6680–6685

Deng Y, Wang X, Guo H, Duan D (2014) A trehalose-6-phosphate synthase gene from Saccharina japonica (Laminariales, Phaeophyceae). Mol Biol Rep 41:529–536

Deprost D, Yao L, Sormani R, Moreau M, Leterreus G, Nicolai M, Bedu M, Robaglia C, Meyer C (2007) The Arabidopsis TOR kinase links plant growth, yield, stress resistance and mRNA translation. EMBO Rep 8:864–870

Derkacheva M, Hennig L (2014) Variations on a theme: polycomb group proteins in plants. J Exp Bot 65:2769–2784

Dickens CWS, Staden JV (1988) The in vitro flowering of Kalanchöe blossfeldiana Poellnitz: I. Role of culture conditions and nutrients. J Exp Bot 39:461–471

Distelfeld A, Li C, Dubcovsky J (2009) Regulation of flowering in temperate cereals. Curr Opin Plant Biol 12:178–184

Ditta G, Pinyopich A, Robles P, Pelaz S, Yanofsky MF (2004) The SEP4 gene of Arabidopsis thaliana functions in floral organ and meristem identity. Curr Biol 14:1935–1940

Djouani-Tahri EB, Christie JM, Sanchez-Ferandin S, Sanchez F, Bouget FY, Corellou F (2011) A eukaryotic LOV-histidine kinase with circadian clock function in the picoalga Ostreococcus. Plant J 65:578–588

Doi K, Izawa T, Fuse T, Yamanouchi U, Kubo T, Shimatani Z, Yano M, Yoshimura A (2004) Ehd1, a B-type response regulator in rice, confers short-day promotion of flowering and controls FT-like gene expression independently of Hd1. Genes Dev 2:926–936

Dornelas MC, Patreze CM, Angenent GC, Immink RG (2011) MADS: the missing link between identity and growth? Trends Plant Sci 16:89–97

Dubcovsky J, Loukoianov A, Fu D, Valarik M, Sanchez A, Yan L (2006) Effect of photoperiod on the regulation of wheat vernalization genes VRN1 and VRN2. Plant Mol Biol 60:469–480

Egea-Cortines M, Saedler H, Sommer H (1999) Ternary complex formation between the MADS-box proteins SQUAMOSA, DEFICIENS and GLOBOSA is involved in the control of floral architecture in Antirrhinum majus. EMBO J 18:5370–5379

Erdmann R, Gramzow L, Melzer R, Theissen G, Becker A (2010) GORDITA (AGL63) is a young paralog of the Arabidopsis thaliana B(sister) MADS box gene ABS (TT16) that has undergone neofunctionalization. Plant J 63:914–924

Espinosa-Soto C, Padilla-Longoria P, Alvarez-Buylla ER (2004) A gene regulatory network model for cell-fate determination during Arabidopsis thaliana flower development that is robust and recovers experimental gene expression profiles. Plant Cell 16:2923–2939

Fang Q, Liu J, Xu Z, Song R (2008) Cloning and characterization of a flowering time gene from Thellungiella halophila. Acta Biochim Biophys Sin (Shanghai) 40:747–753

Fattash I, Voss B, Reski R, Hess WR, Frank W (2007) Evidence for the rapid expansion of microRNA-mediated regulation in early land plant evolution. BMC Plant Biol 7:13

Ferrandiz C, Gu Q, Martienssen R, Yanofsky MF (2000) Redundant regulation of meristem identity and plant architecture by FRUITFULL, APETALA1 and CAULIFLOWER. Development 127:725–734

Finnegan EJ, Dennis ES (2007) Vernalization-induced trimethylation of histone H3 lysine 27 at FLC is not maintained in mitotically quiescent cells. Curr Biol 17:1978–1983
Fornara F, Panigrahi KCS, Gissot L, Sauerbrunn N, Rühl M, Jarillo JA, Coupland G (2009) Arabidopsis DOF transcription factors act redundantly to reduce CONSTANS expression and are essential for a photoperiodic flowering response. Dev Cell 17:75–86
Fornara F, de Montaigu A, Coupland G (2010) SnapShot: control of flowering in Arabidopsis. Cell 141:551–552
Frink CR, Waggoner PE, Ausubel JH (1999) Nitrogen fertilizer: retrospect and prospect. Proc Natl Acad Sci USA 96:1175–1180
Frohlich MW (2003) An evolutionary scenario for the origin of flowers. Nat Rev Genet 4:559–566
Frohlich MW (2006) Recent developments regarding the evolutionary origin of flowers. In: Soltis DE, Callow JA (eds) Advances in botanical research, vol 44. Academic Press, San Diego, CA, pp 63–127
Frohlich MW, Estabrook GF (2000) Wilkinson support calculated with exact probabilities: an example using Floricaula/LEAFY amino acid sequences that compares three hypotheses involving gene gain/loss in seed plants. Mol Biol Evol 17:1914–1925
Fu D, Dunbar M, Dubcovsky J (2007) Wheat VIN3-like PHD finger genes are up-regulated by vernalization. Mol Genet Genomics 277:301–313
Gazzani S, Gendall AR, Lister C, Dean C (2003) Analysis of the molecular basis of flowering time variation in Arabidopsis accessions. Plant Physiol 132:1107–1114
Geraldo N, Baurle I, Kidou S, Hu X, Dean C (2009) FRIGIDA delays flowering in Arabidopsis via a cotranscriptional mechanism involving direct interaction with the nuclear cap-binding complex. Plant Physiol 150:1611–1618
Gibson SI (2004) Sugar and phytohormone response pathways: navigating a signalling network. J Exp Bot 55:253–264
Goff SA, Ricke D, Lan TH, Presting G, Wang R, Dunn M, Glazebrook J, Sessions A, Oeller P, Varma H et al (2002) A draft sequence of the rice genome (Oryza sativa L. ssp. japonica). Science 296:92–100
González-Schain ND, Díaz-Mendoza M, Zurczak M, Suárez-López P (2012) Potato CONSTANS is involved in photoperiodic tuberization in a graft-transmissible manner. Plant J 70:678–690
Goodenough U, Lin H, Lee JH (2007) Sex determination in Chlamydomonas. Semin Cell Dev Biol 18:350–361
Graham LE, Cook ME, Busse JS (2000) The origin of plants: body plan changes contributing to a major evolutionary radiation. Proc Natl Acad Sci USA 97:4535–4540
Gramzow L, Theissen G (2010) A hitchhiker's guide to the MADS world of plants. Genome Biol 11:214
Greb T, Mylne JS, Crevillen P, Geraldo N, An H, Gendall AR, Dean C (2007) The PHD finger protein VRN5 functions in the epigenetic silencing of Arabidopsis FLC. Curr Biol 17:73–78
Greenup AG, Sasani S, Oliver SN, Walford SA, Millar AA, Trevaskis B (2011) Transcriptome analysis of the vernalization response in barley (Hordeum vulgare) seedlings. PLoS One 6: e17900
Grossman A (2000) Acclimation of *Chlamydomonas reinhardtii* to its nutrient environment. Protist 151:201–224
Halliday KJ, Salter MG, Thingnaes E, Whitelam GC (2003) Phytochrome control of flowering is temperature sensitive and correlates with expression of the floral integrator FT. Plant J 33:875–885
Hames C, Ptchelkine D, Grimm C, Thevenon E, Moyroud E, Gerard F, Martiel JL, Benlloch R, Parcy F, Muller CW (2008) Structural basis for LEAFY floral switch function and similarity with helix-turn-helix proteins. EMBO J 27:2628–2637
Hartmann U, Hohmann S, Nettesheim K, Wisman E, Saedler H, Huijser P (2000) Molecular cloning of SVP: a negative regulator of the floral transition in Arabidopsis. Plant J 21:351–360
Haughn GW, Somerville CR (1988) Genetic control of morphogenesis in Arabidopsis. Dev Genet 9:73–89

Henschel K, Kofuji R, Hasebe M, Saedler H, Munster T, Theissen G (2002) Two ancient classes of MIKC-type MADS-box genes are present in the moss Physcomitrella patens. Mol Biol Evol 19:801–814

Heo JB, Sung S (2011) Vernalization-mediated epigenetic silencing by a long intronic noncoding RNA. Science 331:76–79

Higgins J, Bailey PC, Laurie DA (2010) Comparative genomics of flowering time pathways using Brachypodium distachyon as a model for the temperate grasses. PLoS One 5:e10065

Himi S, Sano R, Nishiyama T, Tanahashi T, Kato M, Ueda K, Hasebe M (2001) Evolution of MADS-box gene induction by FLO/LFY genes. J Mol Evol 53:387–393

Honma T, Goto K (2001) Complexes of MADS-box proteins are sufficient to convert leaves into floral organs. Nature 409:525–529

Huijser P, Schmid M (2011) The control of developmental phase transitions in plants. Development 138:4117–4129

Huijser P, Klein J, Lonnig WE, Meijer H, Saedler H, Sommer H (1992) Bracteomania, an inflorescence anomaly, is caused by the loss of function of the MADS-box gene squamosa in Antirrhinum majus. EMBO J 11:1239–1249

Imaizumi T, Schultz TF, Harmon FG, Ho LA, Kay SA (2005) FKF1 F-box protein mediates cyclic degradation of a repressor of CONSTANS in Arabidopsis. Science 309:293–297

Irish VF (2010) The flowering of Arabidopsis flower development. Plant J 61:1014–1028

Irwin JA, Lister C, Soumpourou E, Zhang Y, Howell EC, Teakle G, Dean C (2012) Functional alleles of the flowering time regulator FRIGIDA in the *Brassica oleracea* genome. BMC Plant Biol 12:21

Ito T (2011) Coordination of flower development by homeotic master regulators. Curr Opin Plant Biol 14:53–59

Ito S, Song YH, Josephson-Day AR, Miller RJ, Breton G, Olmstead RG, Imaizumi T (2012) FLOWERING BHLH transcriptional activators control expression of the photoperiodic flowering regulator CONSTANS in Arabidopsis. Proc Natl Acad Sci USA 109:3582–3587

Izawa T, Takahashi Y, Yano M (2003) Comparative biology comes into bloom: genomic and genetic comparison of flowering pathways in rice and Arabidopsis. Curr Opin Plant Biol 6:113–120

Jack T (2001) Plant development going MADS. Plant Mol Biol 46:515–520

Jang S, Marchal V, Panigrahi KCS, Wenkel S, Soppe W, Deng XW, Valverde F, Coupland G (2008) Arabidopsis COP1 shapes the temporal pattern of CO accumulation conferring a photoperiodic flowering response. EMBO J 27:1277–1288

Jetha K, Theissen G, Melzer R (2015) Arabidopsis SEPALLATA proteins differ in cooperative DNA-binding during the formation of floral quartet-like complexes. Nucleic Acids Res 42:10927–10942

Jofuku KD, den Boer BG, Van Montagu M, Okamuro JK (1994) Control of Arabidopsis flower and seed development by the homeotic gene APETALA2. Plant Cell 6:1211–1225

Johanson U, West J, Lister C, Michaels S, Amasino R, Dean C (2000) Molecular analysis of FRIGIDA, a major determinant of natural variation in Arabidopsis flowering time. Science 290:344–347

Jossier M, Bouly JP, Meimoun P, Arjmand A, Lessard P, Hawley S, Grahame Hardie D, Thomas M (2009) SnRK1 (SNF1-related kinase 1) has a central role in sugar and ABA signalling in Arabidopsis thaliana. Plant J 59:316–328

Kang IH, Steffen JG, Portereiko MF, Lloyd A, Drews GN (2008) The AGL62 MADS domain protein regulates cellularization during endosperm development in Arabidopsis. Plant Cell 20:635–647

Kanrar S, Bhattacharya M, Arthur B, Courtier J, Smith HM (2008) Regulatory networks that function to specify flower meristems require the function of homeobox genes PENNYWISE and POUND-FOOLISH in Arabidopsis. Plant J 54:924–937

Kant S, Peng M, Rothstein SJ (2011) Genetic regulation by NLA and microRNA827 for maintaining nitrate-dependent phosphate homeostasis in Arabidopsis. PLoS Genet 7:e1002021

Kato Y, Imamura N (2008) Effect of sugars on amino acid transport by symbiotic Chlorella. Plant Physiol Biochem 46:911–917

Kaufmann K, Melzer R, Theissen G (2005) MIKC-type MADS-domain proteins: structural modularity, protein interactions and network evolution in land plants. Gene 347:183–198

Kaufmann K, Muino JM, Jauregui R, Airoldi CA, Smaczniak C, Krajewski P, Angenent GC (2009) Target genes of the MADS transcription factor SEPALLATA3: integration of developmental and hormonal pathways in the Arabidopsis flower. PLoS Biol 7:e1000090

Kaufmann K, Wellmer F, Muino JM, Ferrier T, Wuest SE, Kumar V, Serrano-Mislata A, Madueno F, Krajewski P, Meyerowitz EM, Angenent GC, Riechmann JL (2010) Orchestration of floral initiation by APETALA1. Science 328:85–89

Keller SR, Levsen N, Ingvarsson PK, Olson MS, Tiffin P (2011) Local selection across a latitudinal gradient shapes nucleotide diversity in balsam poplar, *Populus balsamifera* L. Genetics 188:941–952

Kim DH, Sung S (2013) Coordination of the vernalization response through a VIN3 and FLC gene family regulatory network in Arabidopsis. Plant Cell 25:454–469

Kim DH, Sung S (2014) Genetic and epigenetic mechanisms underlying vernalization. Arabidopsis Book 12:e0171

Kim DH, Doyle MR, Sung S, Amasino RM (2009) Vernalization: winter and the timing of flowering in plants. Annu Rev Cell Dev Biol 25:277–299

Kim JJ, Lee JH, Kim W, Jung HS, Huijser P, Ahn JH (2012) The microRNA156-SQUAMOSA PROMOTER BINDING PROTEIN-LIKE3 module regulates ambient temperature-responsive flowering via FLOWERING LOCUS T in Arabidopsis. Plant Physiol 159:461–478

Kloosterman B, Abelenda JA, Gomez MDMC, Oortwijn M, de Boer JM, Kowitwanich K, Horvath BM, van Eck HJ, Smaczniak C, Prat S, Visser RGF, Bachem CWB (2013) Naturally occurring allele diversity allows potato cultivation in northern latitudes. Nature 495:246–250

Kofuji R, Sumikawa N, Yamasaki M, Kondo K, Ueda K, Ito M, Hasebe M (2003) Evolution and divergence of the MADS-box gene family based on genome-wide expression analyses. Mol Biol Evol 20:1963–1977

Kovach JD, Lamb RS (2014) There can be only one. Science 343:623–624

Koyama K, Hatano H, Nakamura J, Takumi S (2012) Characterization of three VERNALIZATION INSENSITIVE3-like (VIL) homologs in wild wheat, Aegilops tauschii Coss. Hereditas 149:62–71

Kramer EM, Hall JC (2005) Evolutionary dynamics of genes controlling floral development. Curr Opin Plant Biol 8:13–18

Krizek BA, Fletcher JC (2005) Molecular mechanisms of flower development: an armchair guide. Nat Rev Genet 6:688–698

Kuittinen H, Niittyvuopio A, Rinne P, Savolainen O (2008) Natural variation in Arabidopsis lyrata vernalization requirement conferred by a FRIGIDA indel polymorphism. Mol Biol Evol 25:319–329

Kumar SV, Wigge PA (2010) H2A.Z-containing nucleosomes mediate the thermosensory response in Arabidopsis. Cell 140:136–147

Kumar SV, Lucyshyn D, Jaeger KE, Alos E, Alvey E, Harberd NP, Wigge PA (2012) Transcription factor PIF4 controls the thermosensory activation of flowering. Nature 484:242–245

Kwantes M, Liebsch D, Verelst W (2012) How MIKC* MADS-box genes originated and evidence for their conserved function throughout the evolution of vascular plant gametophytes. Mol Biol Evol 29:293–302

Lamb RS, Hill TA, Tan QK, Irish VF (2002) Regulation of APETALA3 floral homeotic gene expression by meristem identity genes. Development 129:2079–2086

Lawlor DW, Paul MJ (2014) Source/sink interactions underpin crop yield: the case for trehalose 6-phosphate/SnRK1 in improvement of wheat. Front Plant Sci 5:418. doi:10.3389/fpls.2014.00418

Lazaro A, Valverde F, Pineiro M, Jarillo JA (2012) The Arabidopsis E3 Ubiquitin Ligase HOS1 negatively regulates CONSTANS abundance in the photoperiodic control of flowering. Plant Cell 24:982–999

Lebon G, Wojnarowiez G, Holzapfel B, Fontaine F, Vaillant-Gaveau N, Clement C (2008) Sugars and flowering in the grapevine (Vitis vinifera L.). J Exp Bot 59:2565–2578

Ledger S, Strayer C, Ashton F, Sa K, Putterill J (2001) Analysis of the function of two circadian-regulated CONSTANS-LIKE genes. Plant J 26:15–22

Lee JH, Park SH, Lee JS, Ahn JH (2007) A conserved role of SHORT VEGETATIVE PHASE (SVP) in controlling flowering time of Brassica plants. Biochim Biophys Acta 1769:455–461

Lee J, Oh M, Park H, Lee I (2008) SOC1 translocated to the nucleus by interaction with AGL24 directly regulates leafy. Plant J 55:832–843

Lee H, Yoo SJ, Lee JH, Kim W, Yoo SK, Fitzgerald H, Carrington JC, Ahn JH (2010) Genetic framework for flowering-time regulation by ambient temperature-responsive miRNAs in Arabidopsis. Nucleic Acids Res 38:3081–3093

Lee JH, Ryu HS, Chung KS, Pose D, Kim S, Schmid M, Ahn JH (2013) Regulation of temperature-responsive flowering by MADS-box transcription factor repressors. Science 342:628–632

Levy YY, Mesnage S, Mylne JS, Gendall AR, Dean C (2002) Multiple roles of Arabidopsis VRN1 in vernalization and flowering time control. Science 297:243–246

Li Y, Huang J, Sandmann G, Chen F (2008) Glucose sensing and the mitochondrial alternative pathway are involved in the regulation of astaxanthin biosynthesis in the dark-grown Chlorella zofingiensis (Chlorophyceae). Planta 228:735–743

Li D, Yang C, Li X, Gan Q, Zhao X, Zhu L (2009) Functional characterization of rice OsDof12. Planta 229:1159–1169

Li P, Filiault D, Box MS, Kerdaffrec E, van Oosterhout C, Wilczek AM, Schmitt J, McMullan M, Bergelson J, Nordborg M et al (2014) Multiple FLC haplotypes defined by independent cis-regulatory variation underpin life history diversity in Arabidopsis thaliana. Genes Dev 28:1635–1640

Lisso J, Schroder F, Mussig C (2013) EXO modifies sucrose and trehalose responses and connects the extracellular carbon status to growth. Front Plant Sci 4:219

Liu C, Xi W, Shen L, Tan C, Yu H (2009) Regulation of floral patterning by flowering time genes. Dev Cell 16:711–722

Liu T, Li Y, Ren J, Qian Y, Yang X, Duan W, Hou X (2013) Nitrate or NaCl regulates floral induction in Arabidopsis thaliana. Biologia 68:215–222

Loeppky HA, Coulman BE (2001) Residue removal and nitrogen fertilization affects tiller development and flowering in meadow bromegrass. Agron J 93:891–895

Lohmann JU, Weigel D (2002) Building beauty: the genetic control of floral patterning. Dev Cell 2:135–142

Lohmann JU, Hong RL, Hobe M, Busch MA, Parcy F, Simon R, Weigel D (2001) A molecular link between stem cell regulation and floral patterning in Arabidopsis. Cell 105:793–803

Loreti E, Matsukura C, Gubler F, Alpi A, Yamaguchi J, Perata P (2000) Glucose repression of alpha-amylase in barley embryos is independent of GAMYB transcription. Plant Mol Biol 44:85–90

Lovell JT, Juenger TE, Michaels SD, Lasky JR, Platt A, Richards JH, Yu X, Easlon HM, Sen S, McKay JK (2013) Pleiotropy of FRIGIDA enhances the potential for multivariate adaptation. Proc Biol Sci 280:20131043

Lucas-Reina E, Romero-Campero FJ, Romero JM, Valverde F (2015) An evolutionarily conserved DOF-CONSTANS module controls plant photoperiodic signalling. Plant Physiol 168(2):561–574. doi:10.1104/pp.15.00321

Luo M, Platten D, Chaudhury A, Peacock WJ, Dennis ES (2009) Expression, imprinting, and evolution of rice homologs of the polycomb group genes. Mol Plant 2:711–723

Maizel A, Busch MA, Tanahashi T, Perkovic J, Kato M, Hasebe M, Weigel D (2005) The floral regulator LEAFY evolves by substitutions in the DNA binding domain. Science 308:260–263

Mandel MA, Yanofsky MF (1995) A gene triggering flower formation in Arabidopsis. Nature 377:522–524
Mandel MA, Gustafson-Brown C, Savidge B, Yanofsky MF (1992) Molecular characterization of the Arabidopsis floral homeotic gene APETALA1. Nature 360:273–277
Martinez-Barajas E, Delatte T, Schluepmann H, de Jong GJ, Somsen GW, Nunes C, Primavesi LF, Coello P, Mitchel RAC, Paul MJ (2011) Wheat grain development is characterized by remarkable trehalose 6-phosphate accumulation pregrain filling: tissue distribution and relationship to SNF1-related protein kinase1 activity. Plant Physiol 156:373–381
Martínez-García JF, Virgós-Soler A, Prat S (2002) Control of photoperiod-regulated tuberization in potato by the Arabidopsis flowering-time gene CONSTANS. Proc Natl Acad Sci USA 99:15211–15216
Masiero S, Colombo L, Grini PE, Schnittger A, Kater MM (2011) The emerging importance of type I MADS box transcription factors for plant reproduction. Plant Cell 23:865–872
Matsoukas IG, Massiah AJ, Thomas B (2012) Florigenic and antiflorigenic signaling in plants. Plant Cell Physiol 53:1827–1842
Matsuo T, Ishiura M (2011) Chlamydomonas reinhardtii as a new model system for studying the molecular basis of the circadian clock. FEBS Lett 585:1495–1502
McClung CR (2014) Wheels within wheels: new transcriptional feedback loops in the Arabidopsis circadian clock. F1000Prime Rep 6:2
McKim S, Hay A (2010) Patterning and evolution of floral structures—marking time. Curr Opin Genet Dev 20:448–453
Mellerowicz EJ, Horgan K, Walden A, Coker A, Walter C (1998) PRFLL—a Pinus radiata homologue of FLORICAULA and LEAFY is expressed in buds containing vegetative shoot and undifferentiated male cone primordia. Planta 206:619–629
Melzer R, Theissen G (2009) Reconstitution of 'floral quartets' in vitro involving class B and class E floral homeotic proteins. Nucleic Acids Res 37:2723–2736
Melzer S, Lens F, Gennen J, Vanneste S, Rohde A, Beeckman T (2008) Flowering-time genes modulate meristem determinacy and growth form in Arabidopsis thaliana. Nat Genet 40:1489–1492
Melzer R, Verelst W, Theissen G (2009) The class E floral homeotic protein SEPALLATA3 is sufficient to loop DNA in 'floral quartet'-like complexes in vitro. Nucleic Acids Res 37:144–157
Meyerowitz EM (1997) Plants and the logic of development. Genetics 145:5–9
Michaels SD, Amasino RM (1999) The gibberellic acid biosynthesis mutant ga1-3 of Arabidopsis thaliana is responsive to vernalization. Dev Genet 25:194–198
Michaels SD, Amasino RM (2001) Loss of FLOWERING LOCUS C activity eliminates the late-flowering phenotype of FRIGIDA and autonomous pathway mutations but not responsiveness to vernalization. Plant Cell 13:935–941
Michel G, Tonon T, Scornet D, Cock JM, Kloareg B (2010) Central and storage carbon metabolism of the brown alga Ectocarpus siliculosus: insights into the origin and evolution of storage carbohydrates in Eukaryotes. New Phytol 188:67–81
Moore B, Zhou L, Rolland F, Hall Q, Cheng WH, Liu YX, Hwang I, Jones T, Sheen J (2003) Role of the Arabidopsis glucose sensor HXK1 in nutrient, light, and hormonal signaling. Science 300:332–336
Moreno-Risueño MA, Martínez M, Vicente-Carbajosa J, Carbonero P (2007) The family of DOF transcription factors: from green unicellular algae to vascular plants. Mol Genet Genomics 277:379–390
Mouradov A, Glassick T, Hamdorf B, Murphy L, Fowler B, Marla S, Teasdale RD (1998) NEEDLY, a Pinus radiata ortholog of FLORICAULA/LEAFY genes, expressed in both reproductive and vegetative meristems. Proc Natl Acad Sci USA 95:6537–6542
Moyroud E, Tichtinsky G, Parcy F (2009) The LEAFY floral regulators in Angiosperms: conserved proteins with diverse roles. J Plant Biol 52:177–185

Moyroud E, Kusters E, Monniaux M, Koes R, Parcy F (2010) LEAFY blossoms. Trends Plant Sci 15:346–352

Murphy RL, Klein RR, Morishige DT, Brady JA, Rooney WL, Miller FR, Dugas DV, Klein PE, Mullet JE (2011) Coincident light and clock regulation of pseudoresponse regulator protein 37 (PRR37) controls photoperiodic flowering in sorghum. Proc Natl Acad Sci USA 108:16469–16474

Nakamura Y, Kato T, Yamashino T, Murakami M, Mizuno T (2007) Characterization of a set of phytochrome-interacting factor-like bHLH proteins in Oryza sativa. Biosci Biotech Biochem 71:1183–1191

Navarro C, Ja A, Cruz-Oró E, Cuéllar Ca Tamaki S, Silva J, Shimamoto K, Prat S (2011) Control of flowering and storage organ formation in potato by FLOWERING LOCUS T. Nature 478:119–122

Ng M, Yanofsky MF (2001) Function and evolution of the plant MADS-box gene family. Nat Rev Genet 2:186–195

Noguero M, Atif RM, Ochatt S, Thompson RD (2013) The role of the DNA-binding One Zinc Finger (DOF) transcription factor family in plants. Plant Sci 209:32–45

Nunes C, Primavesi LF, Patel MK, Martinez-Barajas E, Powers SJ, Sagar R et al (2013) Inhibition of SnRK1 by metabolites: tissue-dependent effects and cooperative inhibition by glucose1-phosphate in combination with trehalose-6-phosphate. Plant Physiol Biochem 63:89–98

Núñez-Elisea R, Caldeira ML (1988) Induction of flowering in mango (*Mangifera indica* L.) within ammonium nitrate sprays. HortSci 23:883

O'Maoileidigh DS, Graciet E, Wellmer F (2014) Gene networks controlling Arabidopsis thaliana flower development. New Phytol 201:16–30

Oesterhelt C, Gross W (2014) Different sugar kinases are involved in the sugar sensing of Galdieria sulphuraria. Plant Physiol 128:291–299

Ohto M, Onai K, Furukawa Y, Aoki E, Araki T, Nakamura K (2001) Effects of sugar on vegetative development and floral transition in Arabidopsis. Plant Physiol 127:252–261

Okamuro JK, Caster B, Villarroel R, Van Montagu M, Jofuku KD (1997) The AP2 domain of APETALA2 defines a large new family of DNA binding proteins in Arabidopsis. Proc Natl Acad Sci USA 94:7076–7081

Oliver SN, Finnegan EJ, Dennis ES, Peacock WJ, Trevaskis B (2009) Vernalization-induced flowering in cereals is associated with changes in histone methylation at the *VERNALIZATION1* gene. Proc Natl Acad Sci USA 106:8386–8391

Oliver SN, Deng W, Casao MC, Trevaskis B (2013) Low temperatures induce rapid changes in chromatin state and transcript levels of the cereal VERNALIZATION1 gene. J Exp Bot 64:2413–2422

Ortiz-Marchena MI, Albi T, Lucas-Reina E, Said FE, Romero-Campero FJ, Cano B, Ruiz MT, Romero JM, Valverde F (2014) Photoperiodic control of carbon distribution during the floral transition in Arabidopsis thaliana. Plant Cell 26:565–584

Pade N, Linka N, Ruth W, Weber APM, Hagemann M (2014) Floridoside and isofloridoside are synthesized by trehalose 6-phosphate synthase-like enzymes in the red alga Galdieria sulphuraria. New Phytol 205(3):1227–1238. doi:10.1111/nph.13108

Parcy F, Nilsson O, Busch MA, Lee I, Weigel D (1998) A genetic framework for floral patterning. Nature 395:561–566

Parenicova L, de Folter S, Kieffer M, Horner DS, Favalli C, Busscher J, Cook HE, Ingram RM, Kater MM, Davies B, Angenent GC, Colombo L (2003) Molecular and phylogenetic analyses of the complete MADS-box transcription factor family in Arabidopsis: new openings to the MADS world. Plant Cell 15:1538–1551

Pego V, Kortstee AJ, Huijser C, Smeekens SCM (2000) Photosynthesis, sugars and the regulation of gene expression. J Exp Bot 51:407–416

Pelaz S, Ditta GS, Baumann E, Wisman E, Yanofsky MF (2000) B and C floral organ identity functions require SEPALLATA MADS-box genes. Nature 405:200–203

Pin PA, Benlloch R, Bonnet D, Wremerth-Weich E, Kraft T, Gielen JJ, Nilsson O (2010) An antagonistic pair of FT homologs mediates the control of flowering time in sugar beet. Science 330:1397–1400

Piñeiro M, Jarillo JA (2013) Ubiquitination in the control of photoperiodic flowering. Plant Sci 198:98–109

Pires N, Dolan L (2010) Early evolution of bHLH proteins in plants. Plant Signal Behav 5:911–912

Poethig RS (2013) Vegetative phase change and shoot maturation in plants. Curr Top Dev Biol 105:125–152

Portereiko MF, Lloyd A, Steffen JG, Punwani JA, Otsuga D, Drews GN (2006) AGL80 is required for central cell and endosperm development in Arabidopsis. Plant Cell 18:1862–1872

Pose D, Yant L, Schmid M (2012) The end of innocence: flowering networks explode in complexity. Curr Opin Plant Biol 15:45–50

Pose D, Verhage L, Ott F, Yant L, Mathieu J, Angenent GC, Immink RG, Schmid M (2013) Temperature-dependent regulation of flowering by antagonistic FLM variants. Nature 503:414–417

Preston JC, Sandve SR (2013) Adaptation to seasonality and the winter freeze. Front Plant Sci 4:167

Pugsley AT (1971) A genetic analysis of the spring-winter habit of growth in wheat. Aust J Agric Res 22:10

Ratcliffe OJ, Nadzan GC, Reuber TL, Riechmann JL (2001) Regulation of flowering in Arabidopsis by an FLC homologue. Plant Physiol 126:122–132

Ratcliffe OJ, Kumimoto RW, Wong BJ, Riechmann JL (2003) Analysis of the Arabidopsis MADS AFFECTING FLOWERING gene family: MAF2 prevents vernalization by short periods of cold. Plant Cell 15:1159–1169

Ream TS, Woods DP, Amasino RM (2012) The molecular basis of vernalization in different plant groups. Cold Spring Harb Symp Quant Biol 77:105–115

Ream TS, Woods DP, Schwartz CJ, Sanabria CP, Mahoy JA, Walters EM, Kaeppler HF, Amasino RM (2014) Interaction of photoperiod and vernalization determines flowering time of Brachypodium distachyon. Plant Physiol 164:694–709

Reeves PA, He Y, Schmitz RJ, Amasino RM, Panella LW, Richards CM (2007) Evolutionary conservation of the FLOWERING LOCUS C-mediated vernalization response: evidence from the sugar beet (*Beta vulgaris*). Genetics 176:295–307

Reisdorph NA, Small GD (2004) The CPH1 gene of Chlamydomonas reinhardtii encodes two forms of cryptochrome whose levels are controlled by light-induced proteolysis 1 [w]. Plant Physiol 134:1546–1554

Riaño-Pachón DM, Corrêa LGG, Trejos-Espinosa R, Mueller-Roeber B (2008) Green transcription factors: a chlamydomonas overview. Genetics 179:31–39

Riechmann JL, Meyerowitz EM (1997) MADS domain proteins in plant development. Biol Chem 378:1079–1101

Riechmann JL, Meyerowitz EM (1998) The AP2/EREBP family of plant transcription factors. Biol Chem 379:633–646

Risk JM, Laurie RE, Macknight RC, Day CL (2010) FRIGIDA and related proteins have a conserved central domain and family specific N- and C-terminal regions that are functionally important. Plant Mol Biol 73:493–505

Robaglia C, Thomas M, Meyer C (2012) Sensing nutrient and energy status by SnRK1 and TOR kinases. Curr Opin Plant Biol 57:301–307

Rockwell NC, Duanmu D, Martin SS, Bachy C, Price DC, Bhattacharya D, Worden AZ, Lagarias JC (2014) Eukaryotic algal phytochromes span the visible spectrum. Proc Natl Acad Sci USA 111:3871–3876

Roldan M, Gomez-Mena C, Ruiz-Garcia L, Salinas J, Martinez-Zapater JM (1999) Sucrose availability on the aerial part of the plant promotes morphogenesis and flowering of Arabidopsis in the dark. Plant J 20:581–590

Rolland F, Baena-González E, Sheen J (2006) Sugar sensing and signaling in plants: conserved and novel mechanisms. Annu Rev Plant Biol 57:675–709

Romero JM, Valverde F (2009) Evolutionarily conserved photoperiod mechanisms in plants. Plant Signal Behav 4:642–644

Romero-Campero FJ, Lucas-Reina E, Said FE, Romero JM, Valverde F (2013) A contribution to the study of plant development evolution based on gene co-expression networks. Front Plant Sci 4:1–17

Rubio V, Deng XW (2007) Plant science: standing on the shoulders of GIGANTEA. Science 318:206–207

Ruelens P, de Maagd RA, Proost S, Theissen G, Geuten K, Kaufmann K (2013) FLOWERING LOCUS C in monocots and the tandem origin of angiosperm-specific MADS-box genes. Nat Commun 4:2280

Samach A, Wigge PA (2005) Ambient temperature perception in plants. Curr Opin Plant Biol 8:483–486

Sawa M, Kay SA (2011) GIGANTEA directly activates Flowering Locus T in Arabidopsis thaliana. Proc Natl Acad Sci USA 28:11698–11703

Sayou C, Monniaux M, Nanao MH, Moyroud E, Brockington SF, Thevenon E, Chahtane H, Warthmann N, Melkonian M, Zhang Y, Wong GK, Weigel D, Parcy F, Dumas R (2014) A promiscuous intermediate underlies the evolution of LEAFY DNA binding specificity. Science 343:645–648

Schluepmann H, Pellny T, van Dijken A, Smeekens S, Paul M (2003) Trehalose 6-phosphate is indispensable for carbohydrate utilization and growth in Arabidopsis thaliana. Proc Natl Acad Sci USA 100:6849–6854

Schranz ME, Quijada P, Sung SB, Lukens L, Amasino R, Osborn TC (2002) Characterization and effects of the replicated flowering time gene FLC in Brassica rapa. Genetics 162:1457–1468

Schroder F, Lisso J, Lange P, Mussig C (2009) The extracellular EXO protein mediates cell expansion in Arabidopsis leaves. BMC Plant Biol 9:20

Schultz EA, Haughn GW (1991) LEAFY, a homeotic gene that regulates inflorescence development in Arabidopsis. Plant Cell 3:771–781

Schulze T, Prager K, Dathe H, Kelm J, Kiessling P, Mittag M (2010) How the green alga Chlamydomonas reinhardtii keeps time. Protoplasma 244:3–14

Schwarz-Sommer Z, Huijser P, Nacken W, Saedler H, Sommer H (1990) Genetic control of flower development by homeotic genes in Antirrhinum majus. Science 250:931–936

Scortecci KC, Michaels SD, Amasino RM (2001) Identification of a MADS-box gene, FLOWERING LOCUS M, that represses flowering. Plant J 26:229–236

Serrano G, Herrera-palau R, Romero JM, Serrano A, Coupland G, Valverde F (2009) Chlamydomonas CONSTANS and the evolution of plant photoperiodic signaling. Curr Biol 19:359–368

Sharma KK, Schuhmann H, Schenk PM (2012) High lipid induction in microalgae for biodiesel production. Energies 5:1532–1553

Sheldon CC, Burn JE, Perez PP, Metzger J, Edwards JA, Peacock WJ, Dennis ES (1999) The FLF MADS box gene: a repressor of flowering in Arabidopsis regulated by vernalization and methylation. Plant Cell 11:445–458

Shindo S, Sakakibara K, Sano R, Ueda K, Hasebe M (2001) Characterization of a FLORICAULA/LEAFY homologue of Gnetum parvifolium and its implications for the evolution of reproductive organs in seed plants. Int J Plant Sci 162:1199–1209

Shindo C, Aranzana MJ, Lister C, Baxter C, Nicholls C, Nordborg M, Dean C (2005) Role of FRIGIDA and FLOWERING LOCUS C in determining variation in flowering time of Arabidopsis. Plant Physiol 138:1163–1173

Simonini S, Roig-Villanova I, Gregis V, Colombo B, Colombo L, Kater MM (2012) Basic pentacysteine proteins mediate MADS domain complex binding to the DNA for tissue-specific expression of target genes in Arabidopsis. Plant Cell 24:4163–4172

Slotte T, Huang H, Lascoux M, Ceplitis A (2008) Polyploid speciation did not confer instant reproductive isolation in Capsella (Brassicaceae). Mol Biol Evol 25:1472–1481
Smaczniak C, Immink RG, Angenent GC, Kaufmann K (2012a) Developmental and evolutionary diversity of plant MADS-domain factors: insights from recent studies. Development 139:3081–3098
Smaczniak C, Immink RG, Muino JM, Blanvillain R, Busscher M, Busscher-Lange J, Dinh QD, Liu S, Westphal AH, Boeren S, Parcy F, Xu L, Carles CC, Angenent GC, Kaufmann K (2012b) Characterization of MADS-domain transcription factor complexes in Arabidopsis flower development. Proc Natl Acad Sci USA 109:1560–1565
Smeekens S, Ma J, Hanson J, Rolland F (2010) Sugar signals and molecular networks controlling plant growth. Curr Opin Plant Biol 13:274–279
Smith HM, Ung N, Lal S, Courtier J (2011) Specification of reproductive meristems requires the combined function of SHOOT MERISTEMLESS and floral integrators FLOWERING LOCUS T and FD during Arabidopsis inflorescence development. J Exp Bot 62:583–593
Sommer H, Beltran JP, Huijser P, Pape H, Lonnig WE, Saedler H, Schwarz-Sommer Z (1990) Deficiens, a homeotic gene involved in the control of flower morphogenesis in Antirrhinum majus: the protein shows homology to transcription factors. EMBO J 9:605–613
Song J, Angel A, Howard M, Dean C (2012a) Vernalization—a cold-induced epigenetic switch. J Cell Sci 125:3723–3731
Song Y, Gao Z, Luan W (2012b) Interaction between temperature and photoperiod. Sci China Life Sci 55:241–249
Song YH, Smith RW, To BJ, Millar AJ, Imaizumi T (2012c) FKF1 conveys timing information for CONSTANS stabilization in photoperiodic flowering. Science 336:1045–1049
Song XM, Huang ZN, Duan WK, Ren J, Liu TK, Li Y, Hou XL (2014) Genome-wide analysis of the bHLH transcription factor family in Chinese cabbage (Brassica rapa ssp. pekinensis). Mol Genet Genomics 289:77–91
Srikanth A, Schmid M (2011) Regulation of flowering time: all roads lead to Rome. Cell Mol Life Sci 68:2013–2037
Steffen JG, Kang IH, Portereiko MF, Lloyd A, Drews GN (2008) AGL61 interacts with AGL80 and is required for central cell development in Arabidopsis. Plant Physiol 148:259–268
Sung S, Amasino RM (2004a) Vernalization and epigenetics: how plants remember winter. Curr Opin Plant Biol 7:4–10
Sung S, Amasino RM (2004b) Vernalization in Arabidopsis thaliana is mediated by the PHD finger protein VIN3. Nature 427:159–164
Sung S, Schmitz RJ, Amasino RM (2006) A PHD finger protein involved in both the vernalization and photoperiod pathways in Arabidopsis. Genes Dev 20:3244–3248
Swiezewski S, Liu F, Magusin A, Dean C (2009) Cold-induced silencing by long antisense transcripts of an Arabidopsis Polycomb target. Nature 462:799–802
Tadege M, Sheldon CC, Helliwell CA, Stoutjesdijk P, Dennis ES, Peacock WJ (2001) Control of flowering time by FLC orthologues in Brassica napus. Plant J 28:545–553
Tanabe Y, Hasebe M, Sekimoto H, Nishiyama T, Kitani M, Henschel K, Munster T, Theissen G, Nozaki H, Ito M (2005) Characterization of MADS-box genes in charophycean green algae and its implication for the evolution of MADS-box genes. Proc Natl Acad Sci USA 102:2436–2441
Tanahashi T, Sumikawa N, Kato M, Hasebe M (2005) Diversification of gene function: homologs of the floral regulator FLO/LFY control the first zygotic cell division in the moss Physcomitrella patens. Development 132:1727–1736
Theissen G (2001) Development of floral organ identity: stories from the MADS house. Curr Opin Plant Biol 4:75–85
Theissen G, Melzer R (2007a) Combinatorial control of floral organ identity by MADS-domain transcription factors. In: Annual plant reviews, vol 29: regulation of transcription in plants. Blackwell, Oxford, pp 253–265

Theissen G, Melzer R (2007b) Molecular mechanisms underlying origin and diversification of the angiosperm flower. Ann Bot 100:603–619

Theissen G, Saedler H (2001) Plant biology. Floral quartets. Nature 409:469–471

Theissen G, Becker A, Di Rosa A, Kanno A, Kim JT, Munster T, Winter KU, Saedler H (2000) A short history of MADS-box genes in plants. Plant Mol Biol 42:115–149

Trevaskis B, Bagnall DJ, Ellis MH, Peacock WJ, Dennis ES (2003) MADS box genes control vernalization-induced flowering in cereals. Proc Natl Acad Sci USA 100:13099–13104

Trevaskis B, Hemming MN, Dennis ES, Peacock WJ (2007) The molecular basis of vernalization-induced flowering in cereals. Trends Plant Sci 12:352–357

Tsai AYL, Gazzarrini S (2014) Trehalose-6-phosphate and SnRK1 kinases in plant development and signalling: the emerging picture. Front Plant Sci 5:119. doi:10.3389/fpls.2014.00119

Turck F, Fornara F, Coupland G (2008) Regulation and identity of florigen: FLOWERING LOCUS T moves center stage. Annu Rev Plant Biol 59:573–594

Valverde F (2011) CONSTANS and the evolutionary origin of photoperiodic timing of flowering. J Exp Bot 62:2453–2463

Valverde F, Mouradov A, Soppe W, Ravenscroft D, Samach A, Coupland G (2004) Photoreceptor regulation of CONSTANS protein in photoperiodic flowering. Science 303:1003–1006

Valverde F, Ortega JM, Losada M, Serrano A (2005) Sugar-mediated transcriptional regulation of the Gap gene system and concerted photosystem II functional modulation in the microalga Scenedesmus vacuolatus. Planta 221:937–952

van Mourik S, van Dijk A, de Gee M, Immink R, Kaufmann K, Angenent G, van Ham R, Molenaar J (2010) Continuous-time modeling of cell fate determination in Arabidopsis flowers. BMC Syst Biol 4:101

Vandenbussche M, Zethof J, Souer E, Koes R, Tornielli GB, Pezzotti M, Ferrario S, Angenent GC, Gerats T (2003) Toward the analysis of the petunia MADS box gene family by reverse and forward transposon insertion mutagenesis approaches: B, C, and D floral organ identity functions require SEPALLATA-like MADS box genes in petunia. Plant Cell 15:2680–2693

Verhage L, Angenent GC, Immink RG (2014) Research on floral timing by ambient temperature comes into blossom. Trends Plant Sci 9:583–591

Wagner D, Sablowski RW, Meyerowitz EM (1999) Transcriptional activation of APETALA1 by LEAFY. Science 285:582–584

Wahl V, Ponnu J, Schlereth A, Arrivault S, Langenecker T, Franke A, Feil R, Lunn JE, Stitt M, Schmid M (2013) Regulation of flowering by trehalose-6-phosphate signaling in Arabidopsis thaliana. Science 339:704–707

Wang R, Farrona S, Vincent C, Joecker A, Schoof H, Turck F, Alonso-Blanco C, Coupland G, Albani MC (2009) PEP1 regulates perennial flowering in Arabis alpina. Nature 459:423–427

Wang H, Zhang Z, Li H, Zhao X, Liu X, Ortiz M, Lin C, Liu B (2013) CONSTANS-LIKE 7 regulates branching and shade avoidance response in Arabidopsis. J Exp Bot 64:1017–1024

Weigel D (1995) The APETALA2 domain is related to a novel type of DNA binding domain. Plant Cell 7:388–389

Weigel D, Nilsson O (1995) A developmental switch sufficient for flower initiation in diverse plants. Nature 377:495–500

Weigel D, Alvarez J, Smyth DR, Yanofsky MF, Meyerowitz EM (1992) LEAFY controls floral meristem identity in Arabidopsis. Cell 69:843–859

Wellmer F, Graciet E, Riechmann JL (2014) Specification of floral organs in Arabidopsis. J Exp Bot 65:1–9

Werner JD, Borevitz JO, Warthmann N, Trainer GT, Ecker JR, Chory J, Weigel D (2005) Quantitative trait locus mapping and DNA array hybridization identify an FLM deletion as a cause for natural flowering-time variation. Proc Natl Acad Sci USA 102:2460–2465

Westerman JM, Lawrence MJ (1970) Genotype-environment interaction and developmental regulation in Arabidopsis thaliana. I. Inbred lines; description. Heredity 25:18

Wigge PA, Kim MC, Jaeger KE, Busch W, Schmid M, Lohmann JU, Weigel D (2005) Integration of spatial and temporal information during floral induction in Arabidopsis. Science 309:1056–1059

Wu L, Birch RG (2010) Physiological basis for enhanced sucrose accumulation in an engineered sugarcane cell line. Funct Plant Biol 37:1161–1174

Wu G, Poethig RS (2006) Temporal regulation of shoot development in Arabidopsis thaliana by miR156 and its target SPL3. Development 133:3539–3547

Wynne J, Treisman R (1992) SRF and MCM1 have related but distinct DNA binding specificities. Nucleic Acids Res 20:3297–3303

Xiao J, Xu S, Li C, Xu Y, Xing L, Niu Y, Huan Q, Tang Y, Zhao C, Wagner D et al (2014) O-GlcNAc-mediated interaction between VER2 and TaGRP2 elicits TaVRN1 mRNA accumulation during vernalization in winter wheat. Nat Commun 5:4572

Xue W, Xing Y, Weng X, Zhao Y, Tang W, Wang L, Zhou H, Yu S, Xu C, Li X, Zhang Q (2008) Natural variation in Ghd7 is an important regulator of heading date and yield potential in rice. Nat Genet 40:761–767

Yamaguchi A, Wu MF, Yang L, Wu G, Poethig RS, Wagner D (2009) The microRNA-regulated SBP-Box transcription factor SPL3 is a direct upstream activator of LEAFY, FRUITFULL, and APETALA1. Dev Cell 17:268–278

Yan L, Loukoianov A, Tranquilli G, Helguera M, Fahima T, Dubcovsky J (2003) Positional cloning of the wheat vernalization gene VRN1. Proc Natl Acad Sci USA 100:6263–6268

Yan L, Loukoianov A, Blechl A, Tranquilli G, Ramakrishna W, SanMiguel P, Bennetzen JL, Echenique V, Dubcovsky J (2004) The wheat VRN2 gene is a flowering repressor down-regulated by vernalization. Science 303:1640–1644

Yan L, Fu D, Li C, Blechl A, Tranquilli G, Bonafede M, Sanchez A, Valarik M, Yasuda S, Dubcovsky J (2006) The wheat and barley vernalization gene VRN3 is an orthologue of FT. Proc Natl Acad Sci USA 103:19581–19586

Yang J, Yang MF, Zhang WP, Chen F, Shen SH (2011) A putative flowering-time-related Dof transcription factor gene, JcDof3, is controlled by the circadian clock in Jatropha curcas. Plant Sci 181:667–674

Yano M, Katayose Y, Ashikari M, Yamanouchi U, Monna L, Fuse T, Baba T, Yamamoto K, Umehara Y, Nagamura Y, Sasaki T (2000) Hd1, a major photoperiod sensitivity quantitative trait locus in rice, is closely related to the Arabidopsis flowering time gene CONSTANS. Plant Cell 12:2473–2484

Yanofsky MF, Ma H, Bowman JL, Drews GN, Feldmann KA, Meyerowitz EM (1990) The protein encoded by the Arabidopsis homeotic gene agamous resembles transcription factors. Nature 346:35–39

Zhang B, Pan X, Cannon CH, Cobb GP, Anderson TA (2006) Conservation and divergence of plant microRNA genes. Plant J 46:243–259

Zhang Y, Primavesi LF, Jhurreea D, Andraloja PC, Mitchell RA, Powers SJ, Schluepmann H, Delatte T, Wingler A, Paul MJ (2009) Inhibition of SNF1-related protein kinase 1 and regulation of metabolic pathway by trehalose. Plant Physiol 149:1860–1871

Zhou L, Jang JC, Jones TL, Sheen J (1998) Glucose and ethylene signal transduction crosstalk revealed by an Arabidopsis glucose-insensitive mutant. Proc Natl Acad Sci USA 95:10294–10299

Part III
Physiology

Boron Stress and Plant Carbon and Nitrogen Relations

Sasmita Mishra and Scott Heckathorn

Contents

Abstract Boron (B) is an essential plant micronutrient, but our understanding of the effects of B stress (deficiency and toxicity) remains incomplete. Here we summarize and analyze the current literature related to B stress and carbon (C) and nitrogen (N) relations. We conclude that photosynthesis is an early sensitive target for B stress, with many aspects of both light and CO_2-fixation reactions negatively affected, and this decreases C skeletons and energy for other functions. B stress may also decrease normal aerobic respiration and increase fermentation and the pentose phosphate pathway. Levels of nonstructural carbohydrates are altered by B stress, likely via shifts in fluxes among metabolic pathways. B stress also impacts many aspects of N relations, including changing levels of N uptake proteins, decreasing N uptake rates, and affecting levels or activities of N assimilation enzymes, which then change amino-acid composition and %N. B stress also may impact the long-distance transport of C and N between shoots and roots. The negative effects of B stress on C and N relations are likely interrelated. Because effects of B deficiency and toxicity on C and N relations are often similar (especially for C), multifaceted, and associated with both soluble and membrane

S. Mishra • S. Heckathorn (✉)
Department of Environmental Sciences, University of Toledo, Toledo, OH 43606, USA
e-mail: scott.heckathorn@utoledo.edu

U. Lüttge et al. (eds.), *Progress in Botany*, Progress in Botany 77,
DOI 10.1007/978-3-319-25688-7_11

components, it is likely that B-stress effects are caused by imbalances in the interaction of B with multiple molecules (especially those with *cis*-hydroxyl groups).

1 Introduction

Boron (B) was recognized as an essential micronutrient for higher plants in 1923 (Warington 1923). Since the range of B concentration in plant tissues between deficiency and toxicity is relatively narrow (Marschner 1995, Chap. 9), B stress in crops is very common (Nable et al. 1997; Shorrocks 1997). B deficiency has been reported in the field for at least 132 crops from 80 countries (Shorrocks 1997), most commonly in highly-leached sandy soils with low pH and low organic-matter content (Adriano 2001; Havlin et al. 2004). B toxicity in plants is less common than deficiency, and it mostly occurs in only a few regions of the world where soils originate from marine sediments with high B content (Nable et al. 1997) or in regions irrigated with B-rich water (Huang et al. 2014).

To date, the best-documented function of B in plants is that of a structural function in plant cell walls, cross-linking pectic carbohydrates via complexes with *cis*-hydroxyl or diol groups (Bolaños et al. 2004; Brown et al. 2002; Goldbach and Wimmer 2007; O'Neill et al. 2004). Because of this potential for interacting with *cis*-hydroxyls, which are abundant in biomolecules, B may potentially interact with a wide array of cell targets, which likely explains the pleiotropic effects of B stress on plants (Bolaños et al. 2004; Goldbach and Wimmer 2007). For example, B stress has strong effects on root growth and elongation, and B is involved in plant reproduction, which may or may not be related solely to the structural role of B in cell walls, as B stress has effects on many aspects of plant function. Plant aspects affected by B stress include membrane structure or function, and metabolism of nucleic acids, proteins, photosynthesis, and antioxidants (Bolaños et al. 2004; Brown et al. 2002; Goldbach and Wimmer 2007).

Plant roots take up B primarily as uncharged boric acid (H_3BO_3), which can occur largely by diffusion across the plasma membrane at high soil B levels, because of the high permeability of the lipid bilayer to uncharged small-sized boric acid, but occurs mostly via B-transport proteins at low B levels (Brown et al. 2002). Under natural conditions, B availability may vary from sub- to supra-optimal levels, and plants need to strictly control B uptake, transport, and distribution among organs, tissues, and cell compartments or in apoplast vs. symplast (Dannel et al. 2002; Reid 2014). B uptake in plants involves two main families of B-transport proteins which have been widely studied: the BORs and NIPs (Miwa and Fujiwara 2010; Reid 2014; Takano et al. 2008, 2010). The first B transporter to be discovered was BOR1, which is an efflux-type transporter that is localized to the plasma membrane. The other major B transporter, NIP5;1, is a member of the "major intrinsic protein" (MIP) family of membrane proteins, and it belongs to the NOD26-like intrinsic protein (NIP) subfamily of aquaporins, which facilitates uptake of boric acid and water. Expression of both BOR1 and NIP5;1, along with several other closely-related members of these two protein families, is often affected by B stress (e.g., upregulated by B deficiency) (Miwa and Fujiwara 2010; Reid 2014; Takano et al. 2008, 2010).

The last 15 years have witnessed impressive progress in understanding how plants acquire, transport, and utilize B, and this progress has been described in several excellent reviews (Bolaños et al. 2004; Brown et al. 2002; Camacho-Cristóbal et al. 2011; Dannel et al. 2002; Goldbach and Wimmer 2007; Miwa and Fujiwara 2010; Reid 2014; Takano et al. 2008, 2010; Wimmer and Eichert 2013). None of these past reviews have focused on effects of B stress on plant carbon or nutrient relations, though several briefly discussed B stress and photosynthesis or nitrogen assimilation (Bolaños et al. 2004; Brown et al. 2002; Camacho-Cristóbal et al. 2011; Goldbach and Wimmer 2007). Most recently, Wimmer and Eichert (2013), while discussing B deficiency and plant-water relations, speculated on how photosynthesis might be indirectly affected by B stress. In the last few years, a number of new studies, some employing transcriptomic/proteomic/metabolomic approaches, have provided additional novel information on how B stress impacts plant C (mostly photosynthesis) and nutrient (mostly N) relations. In this review, we analyze the available literature, both old and new, relating to B stress and C and N relations, with the goals of (1) summarizing "what we know", (2) proposing some new testable hypotheses regarding B effects, and (3) identifying some needed areas for future inquiry.

2 B Stress and Carbon Relations

2.1 *B Stress and Photosynthesis*

After searching through the existing literature, we compiled a list of published papers that contained results pertaining to the effects of B stress, either deficiency or toxicity, on some aspect of photosynthesis, and though the number of papers is limited, there are several important patterns that emerge from this analysis (Table 1). First, in every paper but one that we could find wherein photosynthesis was measured in some way, B stress had a negative impact on photosynthesis, and this was true for mild or severe B deficiency and toxicity, across diverse species, in hydroponics or solid soil media, and for both short-term (abrupt) and long-term (chronic) B treatments. Second, in nearly every study for which appropriate data were available, the negative effects of B stress on the overall rate of photosynthesis (light and dark reactions together) exceeded the negative impacts of B stress on plant growth (total plant biomass, or leaf mass when total is not available). Third, the effects of B stress on photosynthesis were multifaceted, affecting multiple

Table 1 Summary of results from published studies of effects of B stress on photosynthesis

References	Species	B treatment	B effects on photosynthesis[a]	Other effects
B deficiency				
Dixit et al. (2002)[b]	*Curcuma longa*	0 or 5.6 μM for 120 days (in sand)	↓P_n (48 %), G_s (50 %); ↓chl (11 %)	↓Leaf mass (39 %)
El-Shintinawy (1999)	*Helianthus annus*	0.02–50 μM B for 4–10 days (in sand)	↓P_{et} (isolated chloroplasts) (whole chain 31 %, PSII 23 %); ↓chl (18 %), chl *a:b*	↓Plant mass (66 %); ↑suc, ↓membrane oxidation, ↑K^+ leakage
Kastori et al. (1995)[c]	*Helianthus annus*	25 or 2.5 μM for 23 days; or 1.0 μM for 13 days, then 0 μM for 10 days (in sand)	↓P_{max} (40 %), QY (34 %) (leaf pieces); ↓q_p (14, 21 %, lo, hi light), Φ_{PSII} (15, 34 %, lo, hi light), ↓q_n (lo light, 37 %), ↑q_n (hi light, 4 %) (attached leaves)	↓Plant mass (28 %); ↑R_{leaf}, ↑glu, fru, suc
Mishra et al. (2009)[d]	*Pelargonium x hortorum*	45 μM for 40 days, then 45 or 0 μM for 1–5 days (hydroponics)	Lo light: ↓P_n (18 %), CE (23 %), chl (16 %); Hi light: ↑P_n (12 %), CE (8 %), chl (20 %); Lo and Hi: no Δ F_v/F_m, rubisco/rubisco activase/OEC23 content	Lo light: ↓Plant mass (2 %); Hi light: ↓Plant mass (8 %); no Δ soluble sugars, protein
Sheng et al. (2009)[e]	*Citrus sinensis*	1–25 μM for 183 days (in sand/perlite)	↓P_n (79 %), G_s (38 %); ↑C_i (93 %)	↓Plant mass (16 %)
Zhao and Oosterhuis (2002, 2003)	*Gossypium hirsutum*	23 μM B for 14 days, then 23 or 0 μM for 35 days (in sand)	↓P_n (43 %), G_s (80 %), C_i (11 %); ↑chl (13 %)	↓Plant mass (29 %); ↓glu, fru, suc, starch; ↑ion leakage
B toxicity				
Landi et al. (2013)[f]	*Ocimum basilicum*	20 or 2,000 μM for 20 days (hydroponics)	↓P_n (52 %), G_s (48 %), Φ_{PSII} (21 %); ↑C_i (3 %), q_p (11 %), q_n (33 %)	↑Membrane oxidation
Lovatt and Bates (1984)	*Cucurbita pepo*	1 or 400 μM for 5 days (hydroponics)	↓P_n (31 %), G_s (25 %); ↓chl (83 %)	↓Shoot mass (63 %)

(continued)

Table 1 (continued)

References	Species	B treatment	B effects on photosynthesis[a]	Other effects
Papadakis et al. (2004a, b)[g]	*Citrus sinensis*	25 or 250 μM for 204 days (in sand/perlite)	↓P_n (20 %), G_s (40 %), F_v/F_m (20 %); ↑C_i (9 %); ↓chl (21 %)	↓Sugar, starch
Reid et al. (2004)	*Hordeum vulgare*	12 μM for 14 days, then leaf pieces in 0.01, 50, 100 mM for 3 h (hydroponics)	↓P_{max} (leaf pieces) (23 %)	↓R_{leaf} (60 %)
Sheng et al. (2010)[h]	*Citrus sinensis*	25 or 250 μM for 183 days (in sand/perlite)	↓P_n (49 %), G_s (44 %); ↑C_i (16 %)	↓Plant mass (27 %)
Simón et al. (2013)	*Jatropha curcas*	25–700 μM for 70 days (peat moss/perlite)	↓P_n (63 %), G_s (59 %), Φ_{PSII} (37 %); no Δ q_P; ↑C_i (25 %), q_n (62 %)	↓Plant mass (29 %); ↓sugar, starch; ↑membrane oxidation
B deficiency and toxicity				
Chen et al. (2014)	*Arabidopsis thaliana*	30 μM for 30 days, then 0, 30, or 3,000 μM for 60 h (hydroponics)	↓ (Lo/Hi B): rubisco activase (31/47 %), FBP aldolase (78/83 %), ATPase-δ (66/79 %), OEC23 (62/96 %), PSI-2.1 (61/81 %) content; no Δ chl (Lo and Hi B)	↓Leaf mass (3/9 %); no Δ protein
Han et al. (2009)	*Citrus grandis*	10 μM for 15 weeks, then 0, 10, or 500 μM for 15 weeks (sand)	↓ (Lo/Hi B): P_n (79/68 %), G_s (56/54 %); F_v/F_m (23/48 %); chl (63/46 %), rubisco (73/81 %) and stromal FBP phosphatase activity (63/46 %); ↑C_i (66/38 %)	NSC, starch: ↑Lo B, ↓Hi B; Lo and Hi B: ↑membrane oxidation, ↓protein (25/45 %)

(continued)

Table 1 (continued)

References	Species	B treatment	B effects on photosynthesis[a]	Other effects

ATPase-δ H$^+$-ATPase-δ subunit, *chl* leaf chlorophyll concentration, *CE* carboxylation efficiency (of leaves), C_i leaf sub-stomatal internal [CO_2], *glu* glucose, G_s stomatal conductance to water vapor, *FBP* fructose bisphosphate, *fru* fructose, F_v/F_m photochemical efficiency of dark-adapted PSII, *NSC* total nonstructural carbohydrates, *OEC23* 23-kD oxygen-evolving-complex protein of PSII, P_{et} photosynthetic electron transport, P_n net photosynthesis in attached leaves, P_{max} maximum net photosynthesis at light and CO_2 saturation, *PSI-2.1* photosystem I reaction center subunit 2.1, *PSII* photosystem II, q_n non-photochemical quenching, q_p photochemical quenching, *QY* quantum yield of P_n at limiting light, Φ_{PSII} quantum efficiency of light-adapted photosystem II, R_{leaf} leaf dark respiration, *suc* sucrose

[a]Photosynthesis measured as net CO_2 uptake in attached leaves, unless noted

[b]Plants started from rhizomes. Results calculated for "leaf position 3" (results for other leaves similar)

[c]Photosynthetic measurements at 100 or 1,000 μmol m^{-2} s^{-1} PAR

[d]Results for photosynthesis shown only for day 1, but day 5 for biomass; plants grown and measured at 100 or 300 μmol m^{-2} s^{-1} PAR

[e]Results for trifoliate-orange rootstock (other genotypes were similar)

[f]Results for cultivar Greco a Palla

[g]Results for cultivars Navelina and Clementine with Sour-orange rootstock (other genotypes were similar)

[h]Results for cultivar Newhall with Carrizo rootstock (other genotypes were similar)

components of both the light and dark (CO_2 fixation or Calvin Cycle) reactions. Fourth, B stress often decreased CO_2 fixation to a greater extent than the light reactions. Fifth, decreases in net photosynthesis with B stress were nearly always non-stomatal in nature (i.e., were caused instead by effects on primary photosynthetic metabolism). Sixth, B effects on photosynthesis can be rapid, occurring within hours and before visible symptoms are apparent or before growth is affected. Seventh, negative effects of B stress on photosynthesis were not always accompanied by increases in general oxidative damage or photoinhibition. Eighth, the effects of B deficiency on photosynthesis were similar to the effects of B toxicity.

Decreases in the overall rate of photosynthesis with both B deficiency and toxicity that exceeded decreases in plant growth occurred for both in situ net photosynthesis in intact leaves (P_n) (Dixit et al. 2002; Han et al. 2009; Mishra et al. 2009 (low light); Sheng et al. 2009, 2010; Simón et al. 2013; Zhao and Oosterhuis 2003; vs. Lovatt and Bates 1984) and maximum potential photosynthesis under light and CO_2 saturation (P_{max}) (Kastori et al. 1995), but did not hold true for the rate of photosynthetic electron transport (P_{et}) measured alone (El-Shintinawy 1999). Consistent with the above, Chen et al. (2014) observed decreases in the concentration of photosynthetic proteins that exceeded decreases in leaf mass. Importantly, the magnitude of decreases in photosynthesis or photosynthetic proteins with B stress typically exceeded decreases in total protein and/or chlorophyll concentration in leaves (Chen et al. 2014; Dixit et al. 2002; El-Shintinawy 1999; Han et al. 2009; Mishra et al. 2009 (protein); Zhao and Oosterhuis 2003; vs. Lovatt and Bates 1984; Papadakis et al. 2004a, b); hence, negative effects of B stress on photosynthesis are not caused by simple dilution effects on photosynthetic machinery.

Boron deficiency and toxicity both had wide-ranging effects on both the light and CO_2-fixation reactions. For example, El-Shintinawy (1999) observed greater decreases with B deficiency on whole-chain P_{et} than photosystem II (PSII) electron transport, indicating that electron transport downstream from PSII was also negatively affected by B stress (to a larger extent, in fact). Kastori et al. (1995) observed decreases in the quantum efficiency of light-adapted PSII electron transport (Φ_{PSII}) and photochemical quenching (q_p = the fraction of open or oxidized PSII), the latter of which is dependent on, and thus indicative of, electron transport downstream from PSII. Chen et al. (2014) observed preferential decreases in specific PSII, PSI, and chloroplast ATPase proteins with B deficiency and toxicity. Note that neither Landi et al. (2013) or Simón et al. (2013) observed decreases in electron transport downstream from PSII (i.e., decreases in q_p), so PSII may often be the weakest link within electron transport to B stress. In addition, in many studies, B stress caused a decrease in chlorophyll concentration too. For the CO_2-fixation reactions, B stress can preferentially decrease the content of rubisco activase and fructose bisphosphate (FBP) aldolase (Chen et al. 2014) and decrease the activities of rubisco and FBP phosphatase (Mishra et al. 2009; Han et al. 2009). Within a study, the relative effects of B stress are often larger on CO_2 fixation than on light-reaction activity or components, e.g., P_{max} vs. quantum yield at limiting light (which reflects P_{et}) (Kastori et al. 1995), P_n vs. Φ_{PSII} (Landi et al. 2013; Simón et al. 2013), carboxylation efficiency (CE) vs. PSII capacity (F_v/F_m) (Mishra et al. 2009), and rubisco and FBP phosphatase activity vs. PSII capacity (F_v/F_m) (Han et al. 2009).

In most of the past studies wherein leaf internal CO_2 concentration (C_i) was determined (or could be calculated), C_i increased with B deficiency and toxicity (Han et al. 2009; Papadakis et al. 2004a, b; Sheng et al. 2009, 2010; Simón et al. 2013) or did not change significantly (Dixit et al. 2002; Landi et al. 2013; Mishra et al. 2012); only in one study did C_i decrease (Zhao and Oosterhuis 2003). Increases in C_i indicate that the relative limitation of photosynthetic metabolism vs. stomatal conductance to net photosynthesis has increased (i.e., the rate of photosynthetic consumption of CO_2 has decreased relative to stomatal opening, so internal CO_2 concentration increases); when C_i decreases, the reverse is true (i.e., the relative limitation of stomates to photosynthesis has increased). In a recent review of the effects of B deficiency on plant-water relations, Wimmer and Eichert (2013) hypothesize that B deficiency may cause stomatal closure due to disruption of stomatal function, reduced photosynthesis, or decreases in leaf water status (due to either thinner cuticle or decreased water uptake by roots or xylem hydraulic conductivity). The C_i results discussed above indicate that stomatal closure with B stress is typically associated with reductions in photosynthesis, rather than with increases in water stress in leaves (which would result in decreases in C_i) or disruption of stomatal function (which would result in decreases in G_s, but no decreases in CE, P_{et}, etc.).

Several previous studies have illustrated that B stress can have rapid effects on photosynthesis or affect photosynthesis prior to effects on plant growth. For example, Chen et al. (2014) observed effects of both B toxicity and deficiency on multiple photosynthetic proteins within 60 h of transferring arabidopsis plants from normal B (30 μM) to 0 μM B and prior to effects on concentration of total protein or chlorophyll in leaves. Mishra et al. (2009) observed decreases in P_n and CE, but not F_v/F_m, within 24 h of shifting geranium plants from normal B (45 μM) to 0 μM B. Zhao and Oosterhuis (2003) observed decreases in P_n in cotton plants within 2 weeks of B removal, but effects on plant growth were not observed until after 3 weeks of B withdrawal. Lastly, in Lovatt and Bates (1984), photosynthesis decreased by 51 % within 48 h of the transfer of squash plants to B toxicity treatments, yet shoot mass was not yet affected and root mass had decreased by <5 %.

Decreases in photosynthesis during B deficiency and toxicity are often accompanied by increases in general cell damage, e.g., general oxidative stress as indicated by increases in membrane oxidation (Han et al. 2009; Landi et al. 2013; Simón et al. 2013) or general membrane damage as indicated by increased membrane ion leakage (El-Shintinawy 1999; Zhao and Oosterhuis 2003). On the other hand, several studies have observed decreases in photosynthesis or photosynthetic proteins in the absence of evidence for oxidative damage (El-Shintinawy 1999) or upregulation of antioxidants (Chen et al. 2014; Mishra et al. 2009). Similarly, B stress is often accompanied by photoinhibition, most commonly indicated by decreases in the quantum efficiency of PSII (e.g., decreases in F_v/F_m, Φ_{PSII}) or by decreases in photosynthesis or growth at high vs. low light. For example, Kastori et al. (1995) observed larger decreases in photosynthesis with B deficiency in sunflower when measurements were made at 1,000 vs. 100 μmol m^{-2} s^{-1} PAR, and Mishra et al. (2014) observed decreases in photosynthesis with B toxicity in geranium plants grown and measured at 500 vs. 300 or 100 μmol m^{-2} s^{-1} PAR (no photoinhibition was observed for B deficiency). Consistent with effects of high light and B stress on photosynthesis, Cakmak et al. (1995) found higher growth light (100, 250, 500 μmol m^{-2} s^{-1} PAR) exacerbated effects of B deficiency on growth of sunflower. However, Mishra et al. (2009) observed a protective effect of higher light on photosynthesis during B withdrawal in plants grown and measured under non-photoinhibitory low-light conditions (100 vs. 300 μmol m^{-2} s^{-1} PAR). Consistent with this, a protective effect of higher light on plant biomass (but not photosynthesis) was observed during chronic B deficiency and toxicity in geranium grown and measured at low-to-medium light (100, 300, 500 μmol m^{-2} s^{-1} PAR; Mishra et al. 2014) and in *Lemna pausicostata* during B toxicity (1,000–5,500 lux; Tanaka 1966). Thus, general oxidative stress and membrane damage, or photoinhibition, commonly occur during B stress, especially at medium to high light levels, but do not necessarily occur under lower light conditions when decreases in photosynthesis with B stress may still be evident. Hence, decreases in photosynthesis with B stress are exacerbated by high light, but are not always caused by general cell damage or photoinhibition.

Together, the above studies and observations indicate that B stress likely affects photosynthesis initially via widespread multifaceted disruption of photosynthetic metabolism, which later may be exacerbated by general oxidative damage or photoinhibition, as suggested for toxicity by Reid et al. (2004). Given that many aspects of photosynthesis are affected by B stress, it is unlikely that most B effects are direct, even though they may be rapid. Also, given that both soluble and membrane-associated components of photosynthesis are affected by B stress, it is unlikely that B effects on membranes are the sole mechanism, even though B is strongly implicated in a physical role in membrane function (Brown et al. 2002; Goldbach and Wimmer 2007). Notably, B deficiency and toxicity have similar effects on photosynthesis, suggesting that initially it is the imbalance in cellular B concentration that leads to disruption of photosynthesis. Such imbalance might, in part, be related to binding of B to ATP or NADPH and related compounds (Goldbach and Wimmer 2007; Reid et al. 2004), which participate in many aspects of photosynthesis, including rubisco activase function (Buchanan et al. 2000, Chap. 12). Rubisco activase participates in activation of rubisco, via the ATP-dependent removal of ribulose bisphosphate from inactivated rubisco, allowing rubisco to fix CO_2 in the first step of the Calvin Cycle (Buchanan et al. 2000, Chap. 12).

2.2 B Stress and Respiration and Soluble Carbohydrates

Few studies have examined effects of B stress on respiration. As noted in Table 1, Kastori et al. (1995) found that B deficiency increased dark respiration in sunflower leaves, while Krueger et al. (1987) saw no effect of B withdrawal on respiration in root tips of squash. In contrast, Reid et al. (2004) observed decreases in leaf respiration in barley with B toxicity. Hence, there are insufficient results to permit generalizations regarding B-stress effects on plant respiration. However, multiple studies have observed effects of B stress on levels of specific enzymes involved in carbohydrate metabolism (aside from photosynthetic effects discussed above). For example, Wang et al. (2010, 2011) employed a proteomics approach to investigate effects of B deficiency on root proteins in *Brassica napus*, and they found that levels (per unit total protein) of several enzymes involved in glycolysis and the tricarboxylic acid (TCA or Krebs) cycle decreased, and several enzymes in the pentose phosphate pathway (PPP) increased. Interestingly, many of the affected enzymes catalyze reactions involving NADH, both enzymes that increased and others that decreased; hence, it is unlikely that B deficiency is causing these enzymes to decrease because they are malfunctioning. In addition, even though the method used to extract total root protein from *B. napus* roots was able to extract both soluble and membrane proteins, no membrane proteins involved in mitochondrial electron transport were found to be affected by B deficiency, but Koshiba et al. (2010) observed an increase in transcript levels of mitochondrial alternative oxidase in BY-2 cells during B withdrawal, suggesting a possible increase in non-ATP vs. ATP-generating electron transport. Choi et al. (2007) showed that in barley, B toxicity increases invertase levels in leaves, but decreases levels in roots (notably: invertase levels were not strongly correlated with glucose, fructose, and sucrose levels). In contrast, B toxicity increased sucrose phosphate synthase activity in tobacco leaves, but decreased sucrose synthase and amylase activities, resulting in increases in sucrose, glucose, fructose, and starch levels in leaves (Shi et al. 2012).

Both B deficiency and toxicity tend to increase the concentration of soluble nonstructural carbohydrates in plant tissues, but this is not universal among studies and can vary among tissues and cultivars within a study. For example, B deficiency increased levels of glucose, fructose, and sucrose levels in sunflower and tobacco (Camacho-Cristóbal and González-Fontes 1999; El-Shintinawy 1999; Kastori et al. 1995), decreased sugars and starch in cotton (Zhao and Oosterhuis 2002), and had no effect on total soluble sugars in geranium (Mishra et al. 2009). With B toxicity, levels of soluble sugars or starch can increase (glucose, fructose, sucrose in tomato, Cervilla et al. 2007; tobacco, Shi et al. 2012) or decrease (total soluble sugars and starch in citrus, Papadakis et al. 2004a, b; sugars and starch in *Jatropha*, Simón et al. 2013). Further, effects of B toxicity on sugar levels differed between high-B-tolerant and B-intolerant barley cultivars and between roots and shoots within a cultivar with no consistent pattern across studies (Choi et al. 2007; Roessner et al. 2006).

As with photosynthesis, the limited results indicate that B stress typically affects respiration and carbon metabolism, with widespread effects, often on enzymes that involve reactions with ATP or NADH. Further, the limited results suggest the possibility that B stress causes metabolic shifts from one biochemical pathway to another (e.g., from glycolysis and Krebs Cycle to the pentose phosphate pathway, from starch to sucrose accumulation). If true, such metabolic shifts may result from alterations in resource acquisition vs. utilization or source-sink relationships caused by B stress, such as differential sensitivity to B stress of photosynthesis vs. leaf growth or roots vs. shoots.

Drawing on the above results pertaining to photosynthesis, respiration, and other aspects of carbon metabolism, we propose a tentative model of how B stress affects plant carbon relations (Fig. 1). First, there is reasonably strong and sufficient evidence to indicate that photosynthesis is an early and sensitive target of B stress and that B deficiency and toxicity typically have similar effects on photosynthesis. Though less well supported, the available evidence suggests that the CO_2-fixation reactions are likely somewhat more sensitive than the light reactions. A reduction in the CO_2-fixation reactions will initially decrease availability of carbon skeletons for other cell metabolism and for transport to non-photosynthetic tissues, and a reduction in the light reactions will decrease ATP and NADPH availability. Decreases in photosynthesis may then be followed by (or are possibly concurrent with) decreases in aerobic respiration, which will decrease ATP and NADH availability in roots or further decrease it in leaves. It is possible that fermentation and alternative oxidase (AOx) activity increases during B stress, in order to maintain cell redox state, but this is unconfirmed speculation based on single reports of increases in levels of pyruvate decarboxylase (Wang et al. 2011) and AOx proteins (Koshiba et al. 2010)

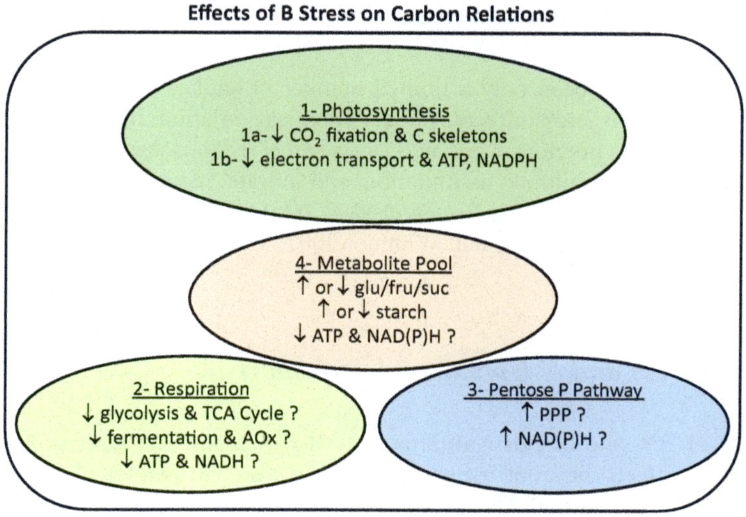

Fig. 1 Tentative model of the effects of B stress on plant carbon (C) relations. At present, evidence indicates that B deficiency and toxicity have similar effects on photosynthesis, but there is insufficient evidence to indicate if this holds true for non-photosynthetic C metabolism; hence, no distinction is made in the present model between B deficiency and toxicity. It is hypothesized that B stress decreases photosynthesis, then respiration, which results in an increase in the oxidative pentose phosphate pathway to generate needed reductant

with B stress. Reductions in the production of ATP and NAD(P)H might then lead to increases in the activity of the oxidative pentose phosphate pathway, in order to generate reductant necessary for normal cell function (Buchanan et al. 2000, Chap. 13), as well as protective responses to B stress (e.g., increases in antioxidants, which are common with B stress and which require reductant). Changes in the cellular pool sizes of nonstructural carbohydrates like sugars and starch will typically occur, but the magnitude and direction of changes will depend on the net balance between anabolic and catabolic pathways, as well as carbon assimilation in leaves vs. its export to and consumption in non-photosynthetic tissue. It should be emphasized that few studies have measured effects of B stress on respiration and other non-photosynthetic aspects of carbon metabolism, and our predicted effects of B stress on non-photosynthetic carbon metabolism derive often from changes in levels of respiratory proteins that have been investigated in only one or two studies. Hence, our proposed model is intended to serve as a "starting point for discussion" and to highlight the need for additional research on effects of B stress on plant carbon metabolism.

3 B Stress and Nitrogen Relations

To date, there have been only a limited number of studies (e.g., <20) that have investigated how B stress affects N acquisition, assimilation, transport, or utilization in plants. Most previous work on B stress and N relations has focused on effects of B deficiency on nitrate assimilation and nitrate reductase activity (recent reviews: Bolaños et al. 2004; Brown et al. 2002; Camacho-Cristóbal et al. 2011; Dannel et al. 2002; Goldbach and Wimmer 2007; Wimmer and Eichert 2013).

3.1 B Stress and N Uptake and Transport

Because of B's role in cell wall structure, B deficiency and toxicity both have deleterious effects on root growth and elongation (Brown et al. 2002; Reid et al. 2004; Wimmer and Eichert 2013). Decreases in root growth with B stress will decrease water and nutrient acquisition by roots, simply by decreasing root absorptive surface area and mass, especially root tips, wherein most nutrient uptake occurs (Marschner 1995, Chaps. 2, 13, and 14; Wimmer and Eichert 2013). In addition, a few studies have provided evidence to indicate that the rate of nutrient (i.e., N) uptake per g (or cm^2) of root (or root-specific uptake rate) is also decreased by B stress, which may change either because of change in the concentration of nutrient transporters per g or cm^2 root or in activity per transporter. In general, N uptake by roots mostly involves three major protein transporters: NRT1 (low affinity for NO_3^-), NRT2 (high affinity for NO_3^-), and AMT1 (high affinity for NH_4^+) (Tsay et al. 2011). Boron deficiency decreased the rate of NO_3^- uptake per g root within 5 days in tobacco roots by 37 %, as well as decreased the level of NRT2 transcript by 40 % (Camacho-Cristóbal and González-Fontes 2007), suggesting that the effect of B stress on N uptake was determined, at least partly, by the effect on the concentration of N uptake proteins in roots. Camacho-Cristóbal and González-Fontes (2007) also measured the effect of B deficiency on transcript level of the plasmalemma H^+-ATPase, since it is responsible for maintaining the electrochemical gradients for cotransport of nitrate and protons, and found that B deficiency decreased it also. In contrast, in our own study on barley, exposed to suboptimal (0.02 μM) and supraoptimal (2,000 μM) B concentrations (vs. 20 μM = optimal), we observed an increase in the concentration of N transporters (NRT1, NRT2, and AMT1) per unit total root protein (Fig. 2a, c, e) or per g root (not shown) with suboptimal B and a change (small decrease) in only NRT2 with supraoptimal B; interestingly, the concentration of the main phosphate uptake protein, PHT1, was relatively insensitive to B levels (note that N uptake rate was not determined) (Fig. 2g). Reid et al. (2004) observed a decrease in the uptake rate of glycine into barley roots with B toxicity.

Once nutrients have been taken up by roots, they then need to be transported to shoots (before or after assimilation for some nutrients). There is evidence that B

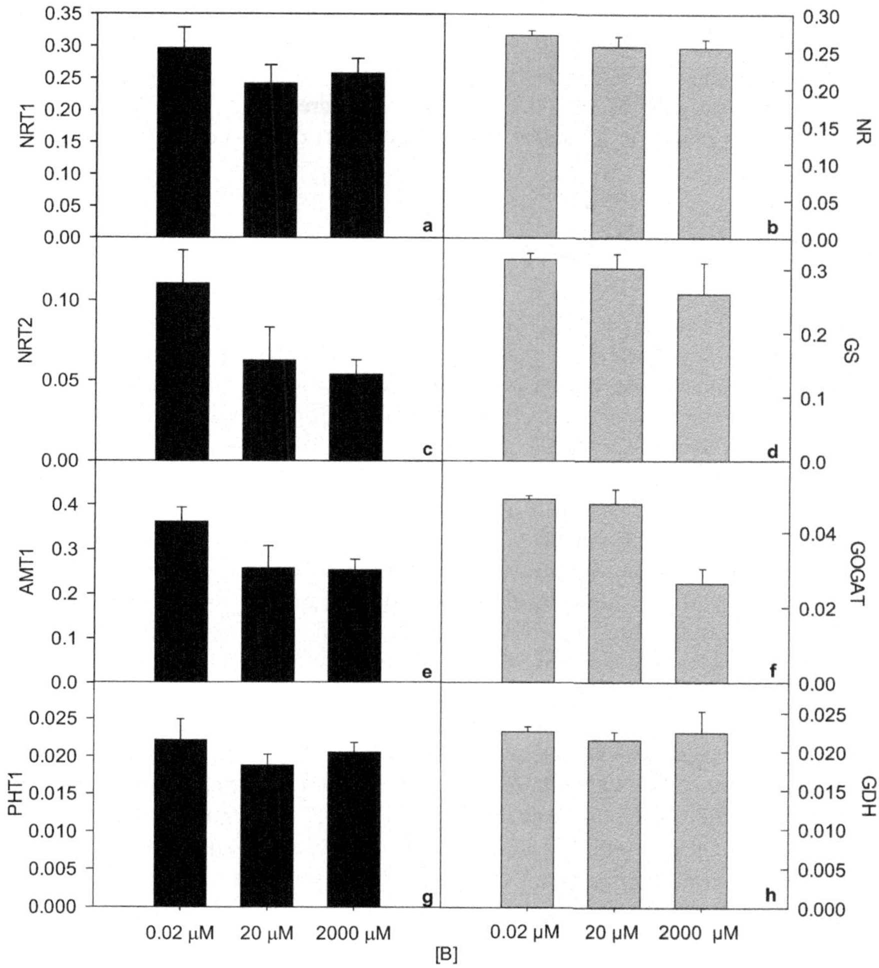

Fig. 2 Effects of B availability on the relative concentration of nitrogen (N) uptake and assimilation proteins in roots of barley (*Hordeum vulgare* cv. Schooner). *NRT1* low-affinity nitrate transporter, *NRT2* high-affinity nitrate transporter, *AMT1* high-affinity ammonium transporter, *PHT1* high-affinity phosphate transporter, *NR* nitrate reductase, *GS* glutamine synthetase, *GOGAT* glutamine oxoglutarate aminotransferase (or glutamate synthase), *GDH* glutamate dehydrogenase. Plants were grown hydroponically for 7 days in complete nutrient solutions containing 9.5 mM NO_3^- and 0.5 mM NH_4^+, but with different B concentrations (0.2, 20, 2,000 μM). Levels of each protein per unit total root protein, relativized to a standard extract, were measured by ELISA using protein-specific antibodies. Each bar represents the mean + 1 SE ($n = 5$). Similar results were obtained per g root (not shown)

deficiency can damage xylem and phloem and potentially affect long-distance transport (reviewed by Wimmer and Eichert 2013), and recently, phloem structure was found to be relatively sensitive to B toxicity in citrus (Huang et al. 2014).

The above studies are not sufficient in number to reach firm conclusions regarding effects of B stress on the uptake of nutrients by roots and transport of nutrients between roots and shoots. While the bulk of the evidence points to mostly negative effects of B stress on nutrient uptake by roots, via decreases in both root growth and uptake rate per unit root, it is clear that additional work is needed in this area.

3.2 B Stress and N Assimilation

In contrast to B stress and N uptake and transport, more is known regarding effects of B stress on N assimilation, particularly on NO_3^- reduction. For example, in tobacco, B deficiency decreased nitrate reductase (NR) activity, but not NR activation state, in both leaves and roots (Camacho-Cristóbal and González-Fontes 1999, 2007; Matas et al. 2009), and in roots, the decrease in NR activity was accompanied by a larger decrease in NO_3^- uptake, leading to a decrease in NO_3^- content. In one of these studies (Camacho-Cristóbal and González-Fontes 2007), B deficiency increased NH_4^+ levels in roots, as well as levels of asparagine, glutamine, and total amino acids (with no decrease in total protein); however, since the plants were provided N solely as NO_3^-, increases in NH_4^+ were likely the result of increases in deamination reactions. Similar results were obtained in tomato, wherein B deficiency decreased NR activity and NO_3^- levels in leaves and roots and increased NH_4^+ levels in roots (Ramón et al. 1989; plant N source not provided). In alfalfa, NO_3^- concentration also decreased with B deficiency (Scripture and McHargue 1943). Interestingly, a 24-h B-deficiency treatment did not change NO_3^- or NH_4^+ concentration in tobacco plants receiving only NO_3^- (Beato et al. 2011), indicating that effects of B deficiency may not be immediate. In contrast to the above studies, in sugar beet (Bonilla et al. 1980), sunflower (Kastori and Petrović 1989), oilseed rape (Shen et al. 1993), and tobacco (Ruiz et al. 1998), NO_3^- levels increased in B-deficient plants, accompanied by decreases in activities of nitrate reductase, suggesting that B deficiency might be suppressing nitrate reductase enzyme activity more than NO_3^- uptake, perhaps due to decreased demand for assimilated N as a result of decreased growth. Similar to deficiency, B toxicity can also decrease NR activity. For example, in leaves of two different tomato cultivars, NR activity decreased by 71–92 % during 14 days of exposure to B toxicity (2 mM), while nitrite reductase (NiR) activity decreased by 27–34 % (Cervilla et al. 2009). In contrast, in tomato and pepper plants experiencing B toxicity, Eraslan et al. (2007) observed an increase in NR activity, and NR activity increased with B toxicity in sunflower also (Kastori and Petrović 1989). Combining together results from all these studies, we can conclude that NR and NiR activities are sensitive to B deficiency and toxicity, most often decreasing with B stress. Further, the decreases in NR and NiR activities may partly result from decreases in NO_3^- uptake and concentration, but given that decreases in NR activity are often accompanied by increases in NO_3^- levels, then decreases in NR must be due sometimes to decreases in NR concentration or activity per NR molecule.

As with enzymes involved in C metabolism, there is no evidence that B interacts directly with NR or other N-assimilation enzymes (Brown et al. 2002; Goldbach and Wimmer 2007).

As with NO_3^- assimilation, NH_4^+ assimilation may also be affected by B stress. Assimilation of NH_4^+ occurs primarily via the activity of the GS-GOGAT cycle, wherein NH_4^+ is first fixed by ATP-dependent glutamine synthetase (GS), yielding glutamine (gln) from glutamate (glu), and then NAD(P)H-dependent glutamate synthase (GOGAT) produces two glutamates from glutamine and α-ketoglutarate (Buchanan et al. 2000). Assimilation of NH_4^+ may also occur via other pathways, especially through activity of NAD(P)H-dependent glutamate dehydrogenase (GDH) and ATP-dependent asparagine synthetase when NH_4^+ levels are high (Buchanan et al. 2000; Gaufichon et al. 2010), though recent studies indicate that in vivo, GDH functions primarily in the reverse direction to produce α-ketoglutarate for the Krebs Cycle, thus liberating NH_4^+ (Fontaine et al. 2012). As noted above for tobacco provided only NO_3^- (Camacho-Cristóbal and González-Fontes 2007), though B deficiency decreased NO_3^- assimilation in roots, it increased NH_4^+ levels. Interestingly, in this study, root GS activity decreased by up to 50 % with B deficiency, but there was almost no change in leaf GS activity, and root asparagine and glutamine levels increased, as did total amino acids, suggesting that B deficiency increased NH_4^+ levels via deaminating activity of GDH and NH_4^+ re-assimilation via asparagine synthetase (AS) activity (Camacho-Cristóbal and González-Fontes 2007). In agreement with this, Beato et al. reported increases in AS and GDH transcript levels with B deficiency in roots of tobacco provided either NH_4^+ or NO_3^-, along with an increase in levels of amino acids related to N assimilation, such as asparagine, glutamine, and glutamate (Beato et al. 2010, 2011, 2014). Interestingly, Ruiz et al. (1998) also observed decreases in the activity of GS and GOGAT with B deficiency in tobacco provided only NO_3^-, while Beato et al. (2014) observed increases in cytosolic GS transcript levels in tobacco receiving 3 mM NO_3^- and 3 mM NH_4^+. In contrast to B deficiency, in NO_3^--fed tomato experiencing B toxicity, GS, GOGAT, and GDH activity increased in leaves (125 %, 43 %, and 40 %, respectively) within 14 days of treatment (Cervilla et al. 2009). In barley roots receiving NO_3^- and NH_4^+, we observed small statistically insignificant increases in the concentration of NR, GS, GOGAT, and GDH proteins under B deficiency, while the levels of GOGAT and GS decreased with B toxicity (with no change in GDH and NR) (Fig. 2b, d, f, h).

Together, these results indicate that in general, B deficiency decreases NO_3^- assimilation and, as first hypothesized by Beato et al. (2011, 2014), probably increases internal recycling of NH_4^+ derived from GDH deamination activity followed by NH_4^+ re-assimilation by AS or GS, likely in compensation for decreased NO_3^- uptake and assimilation. Based on the discussion above, we propose that during B deficiency, NH_4^+ assimilation via AS likely occurs when plants are provided only NO_3^-, but that assimilation of NH_4^+ occurs by GS or AS when plants are provided both NO_3^- and NH_4^+. Decreases in NO_3^- assimilation appear to be caused by some combination of (1) decreases in the activity per molecule or concentration of the major N-assimilation enzymes (NR, NiR, GS,

GOGAT), (2) decreases in NO_3^- uptake, or (3) decreases in NO_3^- use due to effects of B stress on photosynthesis, sugar levels, or growth. Importantly, insufficient results exist for B toxicity to draw any conclusions regarding its effects on N assimilation.

It should also be noted that B deficiency and toxicity are known to affect N fixation and nitrogenase activity in legumes and cyanobacteria (reviewed in Brown et al. 2002). Effects of B stress on N fixation are thought to be related to effects on membrane structure or function in cyanobacterial heterocysts or legume nodules, causing increased diffusion of O_2 across membranes, which then poisons nitrogenase.

3.3 B Stress and Amino Acids and Protein Content

Since B stress (both deficiency and toxicity) affects N uptake and assimilation, as well as N demand, one might expect that B stress would affect the total content or pool size of amino acids or protein in plants, or that B stress might affect the composition of the amino-acid or protein pool. For example, in four different studies with tobacco, B deficiency often, but not always, decreased leaf total amino-acid concentration in leaves and roots, but did not affect total protein significantly (Camacho-Cristóbal and González-Fontes 1999, 2007; Matas et al. 2009; Ruiz et al. 1998). Similarly, no effect of B deficiency was observed on total protein of leaves in geranium (Mishra et al. 2009) or arabidopsis (Chen et al. 2014), but it decreased total soluble protein concentration in citrus (Han et al. 2009). In contrast, B toxicity increased the concentration of total amino acids, but decreased soluble protein levels, in tomato leaves, which the authors speculated was caused by enhanced remobilization of protein N to meristematic tissues (Cervilla et al. 2009). However, B toxicity did not affect leaf protein concentration in arabidopsis (Chen et al. 2014) and decreased it in citrus (Han et al. 2009). Even in the absence of changes in total amino-acid concentration, B stress alters amino-acid composition in tissue. For example, in tobacco roots, asparagine concentration increased 6.75-fold within 24 h (Beato et al. 2011) and by 225 % within 5 days of B deprivation (with smaller changes in gln and no change in asp or glu) (Camacho-Cristóbal and González-Fontes 2007). Using a metabolomic approach, Roessner et al. (2006) found that B toxicity increased levels of many amino acids in barley roots, but decreased others. Interestingly, in this study, B toxicity decreased levels of glutamate, while increasing α-ketoglutarate, which is consistent with the previous discussion regarding the sensitivity of GOGAT to B stress. In addition, effects of B stress on amino-acid composition can vary between leaves and roots (Camacho-Cristóbal and González-Fontes 2007; Roessner et al. 2006) and between B-toxicity-tolerant and B-toxicity-sensitive cultivars of a species (Roessner et al. 2006). Hence, though the effects of B deficiency and toxicity on total amino-acid and protein concentration vary across studies and species, it is apparent that in most cases, both B deficiency and toxicity affect amino-acid composition often in a way to suggest impairment of GS-GOGAT but possible upregulation of NH_4 re-assimilation via increased flux through GDH and asparagine synthetase (AS).

3.4 B Stress and Concentration of N and Other Mineral Nutrients

Several previous studies have examined the effect of B deficiency and/or toxicity on the concentration of mineral nutrients in plant tissues, and it is evident from these studies that B stress often affects tissue nutrient concentrations, but there are few consistent patterns that emerge from these studies. For example, B deficiency can increase or decrease or have no effect on Ca, Fe, K, P, Mg, and Zn concentrations, depending on species within a study (Krug et al. 2009), study within the same species (Mishra et al. 2009, 2014), genotypes within a single species within a study (Mozafar 1989), and tissue within a species within a study (Lovatt and Bates 1984; Mishra et al. 2014; Simón et al. 2013). There is no apparent difference between B deficiency and toxicity in terms of effects on nutrient concentrations within a tissue, even within the same species, across or among studies (deficiency: Camacho-Cristóbal and González-Fontes 1999; Davis et al. 2003; Krug et al. 2009; López-Lefebre et al. 2002; Mishra et al. 2009, 2014; Mozafar 1989) (toxicity: Eraslan et al. 2007; Lovatt and Bates 1984; Mishra et al. 2014; Simón et al. 2013). Within a single study, B stress can decrease the concentration of some nutrients but increase the concentration of others. The effects of B stress on a given nutrient appears unrelated to it occurring as a cation or anion, to use of hydroponics vs. solid media (+ nutrient solution), or to growing plants in a greenhouse vs. growth chamber (the latter likely providing lower light levels on average). However, a potential pattern is evident regarding the effects of B stress on the concentration of N (%N) in plant tissues, as B deficiency decreased %N in leaves of the three species (corn, tobacco, tomato) in which it was measured (Davis et al. 2003; López-Lefebre et al. 2002; Mozafar 1989), but B toxicity increased %N in the three species (pepper, tomato, squash) in which it was measured (Eraslan et al. 2007; Lovatt and Bates 1984). Also, though B stress did not always affect Ca concentration, it did decrease %Ca in 8 of 11 studies (7, deficiency; 4, toxicity).

As with photosynthesis, we are proposing a tentative model for how B stress affects N relations in roots, though in this case, due to the paucity of information on B-toxicity effects, we restrict our model to B deficiency (Fig. 3). First, though limited, the available evidence indicates that B deficiency often decreases uptake of NO_3^- and organic N; effects on NH_4^+ uptake are unknown. Second, B deficiency typically decreases NO_3^- assimilation by decreasing the activities of nitrate reductase and nitrite reductase, and when NO_3^- is the sole or predominant N source, of glutamine synthetase and glutamate synthase. Perhaps to compensate for decreases in NO_3^- assimilation, it appears that NH_4^+ recycling may increase via the deaminating activity of glutamate dehydrogenase, followed by re-assimilation of NH_4^+

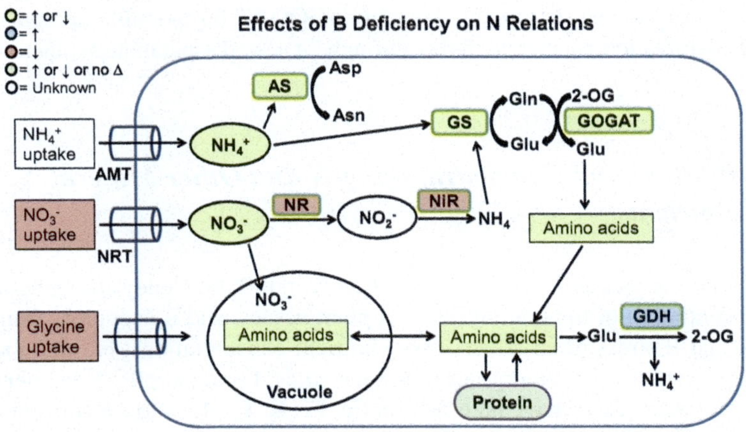

Fig. 3 Tentative model of the effects of B deficiency on root nitrogen (N) relations. At present, insufficient evidence exists to speculate regarding effects of B toxicity on N relations. It is hypothesized that B deficiency typically decreases (*down arrow*) NO_3^- uptake and assimilation, but increases (*up arrow*) NH_4^+ re-assimilation via increases in GDH activity, with effects on NH_4^+ uptake unknown; effects on N pool sizes are variable (*down arrow* or *up arrow*; *down arrow* or *up arrow* or *no change*). *AMT* ammonium transporter, *AS* asparagine synthetase, *Asn* asparagine, *Asp* aspartate, *NRT* nitrate transporters, *NR* nitrate reductase, *NiR* nitrite reductase, *GS* glutamine synthetase, *GOGAT* glutamine-2-oxoglutarate aminotransferase (or glutamate synthase), *GDH* glutamate dehydrogenase, *gln* glutamine, *glu* glutamate, *2-OG* 2-oxoglutarate (or α-ketoglutarate)

by either aspartate synthetase (when NO_3^- is the sole N source) or GS (when both NO_3^- and NH_4^+ are available). Depending on the relative decrease in NO_3^- and NH_4^+ uptake vs. assimilation, NO_3^- and NH_4^+ concentrations may increase or decrease. Alterations in the balance between NO_3^- and NH_4^+ assimilation likely lead or contribute to changes in the composition and pool size of amino acids. Total protein content may or may not be affected, but total N likely often decreases.

4 Conclusions

Boron deficiency and toxicity affect many aspects of plant C and N relations (Fig. 4). Our analysis indicates that photosynthesis is an early and sensitive target for both B deficiency and toxicity, with many aspects of both light and CO_2-fixation reactions (membrane and soluble phases, respectively) affected by B stress, though there is no evidence for direct interaction of B with C (or N)-relations enzymes. The multifaceted effects of B stress on photosynthesis are likely caused initially by metabolic imbalances in B, which might then affect ATP or NAD(P)H metabolism, rather than caused by general oxidative damage to proteins or membranes. Decreases in photosynthesis with B stress typically exceed those for plant growth, indicating that photosynthesis is not necessarily the primary limitation to growth

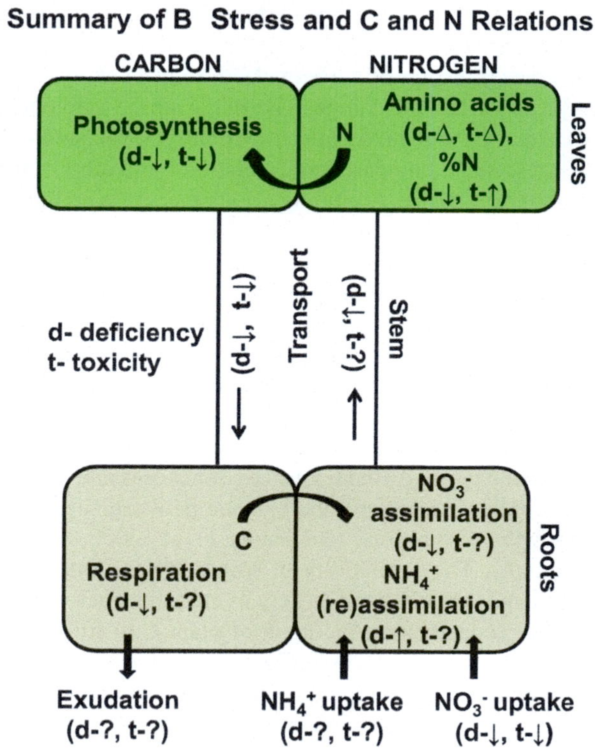

Fig. 4 Summary model of the effects of B deficiency (d) and toxicity (t) on plant carbon (C) and nitrogen (N) relations. Increase = *up arrow*, decrease = *down arrow*, *up arrow* or *down arrow* = *open triangle*, *question mark* = unknown. See Sect. 4 for additional details

under B stress. Nevertheless, decreases in photosynthesis should ultimately cause decreases in available ATP, NADPH, and carbon skeletons, which can be temporarily compensated for by depleting C stores or decreasing C allocation to storage, defense, etc. Also, it is likely that B stress (especially B deficiency) also decreases mitochondrial respiration, via impacts on several glycolytic and Krebs Cycle enzymes. Decreases in photosynthesis and respiration with B stress may cause compensatory increases in the activity of the oxidative pentose phosphate pathway (PPP). Alterations in source-sink metabolism likely explain why B stress typically affects the composition and size of soluble carbohydrate pools. Lastly, it appears that phloem tissue is damaged by both B deficiency and toxicity. Hence, B stress, especially chronic and long-term, is likely to impact the synthesis and subsequent translocation of C to non-photosynthetic tissues (e.g., roots). Importantly, excluding photosynthesis, predictions of effects of B stress on C relations are based on only a few studies, indicating that additional research is needed to complete our understanding of how B deficiency vs. toxicity affects C relations (e.g., especially predictions regarding respiration, PPP, and translocation).

In the case of B stress and N relations, available results suggest that B deficiency likely decreases NO_3^- uptake and assimilation, but effects on NH_4^+ uptake and assimilation are not clear, though there is evidence to suggest that B deficiency increases NH_4^+ recycling, perhaps because N uptake decreases. It appears that B deficiency decreases xylem hydraulic conductivity, which might affect the long-distance transport of nutrients from roots to shoots. Both B deficiency and toxicity typically affect the composition of the amino-acid pool, as well as the concentration of N and Ca in plant tissues. Importantly, fewer studies have investigated impacts of B stress on N, compared to C, relations, and most of this research has focused on B deficiency. Hence, excluding nitrate reductase activity, our understanding of how B stress impacts N relations is woefully incomplete, and much additional research is needed to complete our understanding of B-stress effects on N relations and to test the validity of the predictions that we have made here.

As plant C and N relations are intimately interrelated, so too will be the effects of B stress on C and N. For example, N uptake and assimilation are energetically very expensive (Buchanan et al. 2000, Chap. 16; Lambers et al. 2008, Chap. 6), so root growth and N uptake will be decreased by reductions in photosynthesis, translocation of C from leaves to roots, or root respiration. The uptake of NO_3^- and NH_4^+ is regulated partly by the pool sizes of hexoses and amino acids, which are affected by metabolic demand and the balance between supply (photosynthesis) and demand (growth, respiration) (Buchanan et al. 2000, Chap. 16; Lambers et al. 2008, Chap. 6). The availability of soil N (NO_3^-, NH_4^+, amino acids, etc.) for uptake by plants is determined largely by microbial activity (i.e., mineralization), which is highly dependent on the input of plant C to soil via exudation, root turnover, and litter (Lambers et al. 2008, Chap. 6), but the effects of B stress on plant C loss to soil have not been examined. And, photosynthetic enzymes and chlorophyll are N costly and dependent on organic N provided by the acquisition of N by roots, followed by its assimilation.

In closing, we note that nearly all B stress research to date has been conducted under greenhouse or growth-chamber conditions, yet plant performance in the field does not always mirror that under controlled conditions. Therefore, we strongly advocate that future research efforts on B deficiency and toxicity include field components when possible. We also note that little B research has examined wild species, B effects on plant-soil links, or ecophysiological aspects of B nutrition, so these are potentially fertile areas for future research.

Acknowledgments The authors thank the US Department of Agriculture for financial support.

References

Adriano CD (2001) Trace elements in terrestrial environments: biogeochemistry, bioavailability, and risks of metals, 2nd edn. Springer, New York, NY

Beato VM, Navarro-Gochicoa MT, Rexach J, Herrera-Rodríguez MB, Camacho-Cristóbal JJ, Kempa S, Weckwerth W, González-Fontes A (2010) A tobacco asparagine synthetase gene responds to carbon and nitrogen status and its root expression is affected under boron stress. Plant Sci 178:289–298

Beato VM, Navarro-Gochicoa MT, Rexach J, Herrera-Rodríguez MB, Camacho-Cristóbal JJ, Kempa S, Weckwerth W, González-Fontes A (2011) Expression of root glutamate dehydrogenase genes in tobacco plants subjected to boron deprivation. Plant Physiol Biochem 49:1350–1354

Beato VM, Rexach J, Navarro-Gochicoa MT, Camacho-Cristóbal JJ, Herrera-Rodríguez MB, González-Fontes A (2014) Boron deficiency increases expressions of asparagine synthetase, glutamate dehydrogenase and glutamine synthetase genes in tobacco roots irrespective of the nitrogen source. Soil Sci Plant Nutr 60:1–11

Bolaños L, Lukaszewski K, Bonilla I, Blevins D (2004) Why boron? Plant Physiol Biochem 42:907–912

Bonilla I, Cadahía C, Carpena O, Hernando V (1980) Effects of boron on nitrogen metabolism and sugar levels of sugar beet. Plant Soil 57:3–9

Brown PH, Bellaloui N, Wimmer MA, Bassil ES, Ruiz J, Hu H, Pfeffer H, Dannel F, Römheld V (2002) Boron in plant biology. Plant Biol 4:205–223

Buchanan BB, Gruissem W, Jones RL (2000) Biochemistry and molecular biology of plants. American Society of Plant Physiologists, Maryland

Cakmak I, Kurz H, Marschner H (1995) Short-term effects of boron, germanium and high light intensity on membrane permeability in boron deficient leaves of sunflower. Physiol Plant 95:11–18

Camacho-Cristóbal JJ, González-Fontes A (1999) Boron deficiency causes a drastic decrease in nitrate content and nitrate reductase activity, and increases the content of carbohydrates in leaves from tobacco plants. Planta 209:528–536

Camacho-Cristóbal JJ, González-Fontes A (2007) Boron deficiency decreases plasmalemma H^+-ATPase expression and nitrate uptake, and promotes ammonium assimilation into asparagine in tobacco roots. Planta 226:443–451

Camacho-Cristóbal JJ, Rexach J, Herrera-Rodríguez MB, Navarro-Gochicoa MT, González-Fontes A (2011) Boron deficiency and transcript level changes. Plant Sci 181:85–89

Cervilla LM, Blasco B, Ríos JJ, Romero L, Ruiz JM (2007) Oxidative stress and antioxidants in tomato (*Solanum lycopersicum*) plants subjected to boron toxicity. Ann Bot 100:747–756

Cervilla LM, Blasco B, Ríos JJ, Rosales MA, Rubio-Wilhelmi MM, Sánchez-Rodríguez E, Romero L, Ruiz JM (2009) Response of nitrogen metabolism to boron toxicity in tomato plants. Plant Biol 11:671–677

Chen M, Mishra S, Heckathorn S, Frantz J (2014) Proteomic analysis of *Arabidopsis thaliana* leaves in response to acute boron deficiency and toxicity reveals effects on photosynthesis, carbohydrate metabolism, and protein synthesis. J Plant Physiol 171:235–242

Choi EY, Kolesik P, McNeill A, Collins H, Zhang Q, Huynh BL, Graham R, Stangoulis J (2007) The mechanism of boron tolerance for maintenance of root growth in barley (*Hordeum vulgare* L.). Plant Cell Environ 30:984–993

Dannel F, Pfeffer H, Römheld V (2002) Update on boron in higher plants—uptake, primary translocation and compartmentation. Plant Biol 4:193–204

Davis JM, Sanders DC, Nelson PV, Lengnick L, Sperry WJ (2003) Boron improves growth, yield, quality, and nutrient content of tomato. J Am Soc Horticult Sci 128:441–446

Dixit D, Srivastava NK, Sharma S (2002) Boron deficiency induced changes in translocation of $^{14}CO_2$-photosynthate into primary metabolites in relation to essential oil and curcumin accumulation in turmeric (*Curcuma longa* L.). Photosynthetica 40:109–113

El-Shintinawy F (1999) Structure and functional damage caused by boron deficiency in sunflower leaves. Photosynthetica 36:565–573

Eraslan F, Inal A, Gunes A, Alpaslan M (2007) Boron toxicity alters nitrate reductase activity, proline accumulation, membrane permeability, and mineral constituents of tomato and pepper plants. J Plant Nutr 30:981–994

Fontaine JX, Tercé-Laforgue T, Armengaud P, Clément G et al (2012) Characterization of a NADH-dependent glutamate dehydrogenase mutant of *Arabidopsis* demonstrates the key role of this enzyme in root carbon and nitrogen metabolism. Plant Cell 24:4044–4065

Gaufichon L, Reisdorf-Cren M, Rothstein SJ, Chardon F, Suzuki A (2010) Biological functions of asparagine synthetase in plants. Plant Sci 179:141–153

Goldbach HE, Wimmer MA (2007) Boron in plants and animals: is there a role beyond cell-wall structure? J Plant Nutr Soil Sci 170:39–48

Han S, Tang N, Jiang HX, Yang LT, Li Y, Chen LS (2009) CO_2 assimilation, photosystem II photochemistry, carbohydrate metabolism and antioxidant system of citrus leaves in response to boron stress. Plant Sci 176:143–153

Havlin LJ, Tisdale LS, Nelson LW, Beaton DJ (2004) Soil fertility and fertilizers: an introduction to nutrient management. Prentice Hall, New Jersey

Huang JH, Cai ZJ, Wen SX, Guo P, Ye X, Lin GZ, Chen LS (2014) Effects of boron toxicity on root and leaf anatomy in two *Citrus* species differing in boron tolerance. Trees. doi:10.1007/s00468-014-1075-1

Kastori R, Petrović N (1989) Effects of boron on nitrate reductase activity in young sunflower plants. J Plant Nutr 12:621–632

Kastori R, Plesničar M, Panković D, Sakač Z (1995) Photosynthesis, chlorophyll fluorescence and soluble carbohydrates in sunflower leaves as affected by boron deficiency. J Plant Nutr 18:1751–1763

Koshiba T, Kobayashi M, Ishihara A, Matoh T (2010) Boron nutrition of cultured tobacco BY-2 cells. VI. Calcium is involved in early responses to boron deprivation. Plant Cell Physiol 51:323–327

Krueger RW, Lovatt CL, Albert LS (1987) Metabolic requirement of *Cucurbita pepo* for boron. Plant Physiol 83:254–258

Krug BA, Whipker BE, Frantz J, McCall I (2009) Characterization of calcium and boron deficiency and the effects of temporal disruption of calcium and boron supply on pansy, petunia, and gerbera plugs. HortScience 44:1566–1572

Lambers H, Chapin FS, Pons T (2008) Plant physiological ecology, 2nd edn. Springer, New York, NY

Landi M, Remorini D, Pardossi A, Guidi L (2013) Purple *versus* green-leafed *Ocimum basilicum*: which differences occur with regards to photosynthesis under boron toxicity? J Plant Nutr Soil Sci 176:942–951

López-Lefebre LR, Rivero RM, García PC, Sánchez E, Ruiz JM, Romero L (2002) Boron effect on mineral nutrients of tobacco. J Plant Nutr 25:509–522

Lovatt C, Bates LM (1984) Early effects of excess boron on photosynthesis and growth of *Cucurbita pepo*. J Exp Bot 35:297–305

Marschner H (1995) Mineral nutrition of higher plants, 2nd edn. Academic, London

Matas MA, González-Fontes A, Camacho-Cristóbal JJ (2009) Effect of boron supply on nitrate concentration and its reduction in roots and leaves of tobacco plants. Biol Plant 53:120–124

Mishra S, Heckathorn S, Frantz J, Yu F, Gray J (2009) Effects of boron deficiency on geranium grown under different nonphotoinhibitory light levels. J Am Soc Horticult Sci 134:183–193

Mishra S, Heckathorn S, Frantz J (2012) Elevated CO_2 affects plant responses to variation in boron availability. Plant Soil 350:117–130

Mishra S, Heckathorn S, Frantz J, Krause C (2014) Effects of irradiance during growth on tolerance of geranium to sub- and supra-optimal boron supply. Biol Plant 58:582–588

Miwa K, Fujiwara T (2010) Boron transport in plants: co-ordinated regulation of transporters. Ann Bot 105:1103–1108

Mozafar A (1989) Boron effect on mineral nutrients of maize. Agron J 81:285–290

Nable RO, Bañuelos GS, Paull JG (1997) Boron toxicity. Plant Soil 193:181–198

O'Neill MA, Ishii T, Albersheim P, Darvill AG (2004) Rhamnogalacturonan II: structure and function of a borate cross-linked cell wall pectic polysaccharide. Annu Rev Plant Biol 55:109–139

Papadakis IE, Dimassi KN, Bosabalidis AM, Therios IN, Patakas A, Giannakoula A (2004a) Boron toxicity in 'Clementine' mandarin plants grafted on two rootstocks. Plant Sci 166:539–547

Papadakis IE, Dimassi KN, Bosabalidis AM, Therios IN, Patakas A, Giannakoula A (2004b) Effects of B excess on some physiological and anatomical parameters of 'Navelina' orange plants grafted on two rootstocks. Environ Exp Bot 51:247–257

Ramón AM, Ruiz ROC, Gárate A (1989) *In vitro* stabilization and distribution of nitrate reductase in tomato plants. Incidence of boron deficiency. J Plant Physiol 135:126–128

Reid R (2014) Understanding the boron transport network in plants. Plant Soil 385:1–13

Reid RJ, Hayes JE, Post A, Stangoulis JCR, Graham RD (2004) A critical analysis of the causes of boron toxicity in plants. Plant Cell Environ 27:1405–1414

Roessner U, Patterson JH, Forbes MG, Fincher GB, Langridge P, Bacic A (2006) An investigation of boron toxicity in barley using metabolomics. Plant Physiol 142:1087–1101

Ruiz JM, Baghour M, Bretones G, Belakbir A, Romero L (1998) Nitrogen metabolism in tobacco plants (*Nicotiana tabacum* L.): role of boron as a possible regulatory factor. Int J Plant Sci 159:121–126

Scripture PN, McHargue JS (1943) Effect of boron deficiency on the soluble nitrogen and carbohydrate content of alfalfa. J Am Soc Agron 35:988–992

Shen Z, Liang Y, Shen K (1993) Effects of boron on the nitrate reductase activity in oilseed rape plants. J Plant Nutr 16:1229–1239

Sheng O, Song S, Peng S, Deng X (2009) The effects of low boron on growth, gas exchange, boron concentration and distribution of 'Newhall' navel orange (*Citrus sinensis* Osb.) plants grafted on two rootstocks. Sci Hortic 121:278–283

Sheng O, Zhou G, Wei Q, Peng S, Deng X (2010) Effects of excess boron on growth, gas exchange, and boron status of four orange scion-rootstock combinations. J Plant Nutr Soil Sci 173:469–476

Shi YC, Sun B, Liu WQ (2012) Sucrose phosphate synthase plays a key role in boron-promoted sucrose synthesis in tobacco leaves. J Plant Nutr 175:854–859

Shorrocks VM (1997) The occurrence and correction of boron deficiency. Plant Soil 193:121–148

Simón I, Díaz-López L, Gimeno V, Nieves M, Pereira WE, Martínez V, Lidon V, Garciá-Sánchez F (2013) Effects of boron excess in nutrient solution on growth, mineral nutrition, and physiological parameters of *Jatropha curcas* seedlings. J Plant Nutr Soil Sci 176:165–174

Takano J, Miwa K, Fujiwara T (2008) Boron transport mechanisms: collaboration of channels and transporters. Trends Plant Sci 13:451–457

Takano J, Tanaka M, Toyoda A, Miwa K, Kasai K, Fuji K, Onouchi H, Naito S, Fujiwara T (2010) Polar localization and degradation of *Arabidopsis* boron transporters through distinct trafficking pathways. Proc Natl Acad Sci USA 107:5220–5225

Tanaka H (1966) Response of *Lemna pausicostata* to boron as affected by light intensity. Plant Soil 25:425–434

Tsay YF, Ho CH, Chen HY, Lin SH (2011) Integration of nitrogen and potassium signaling. Annu Rev Plant Biol 62:207–226

Wang Z, Wang Z, Shi L, Wang L, Xu F (2010) Proteomic alterations of *Brassica napus* root in response to boron deficiency. Plant Mol Biol 74:265–278

Wang Z, Wang Z, Chen S, Shi L, Xu F (2011) Proteomics reveals the adaptability mechanism of *Brassica napus* to short-term boron deprivation. Plant Soil 347:195–210

Warington K (1923) The effect of boric acid and borax on the broad bean and certain other plants. Ann Bot 37:629–672

Wimmer MA, Eichert T (2013) Review: mechanisms for boron deficiency-mediated changes in plant water relations. Plant Sci 203–204:25–32

Zhao D, Oosterhuis DM (2002) Cotton carbon exchange, nonstructural carbohydrates, and boron distribution in tissues during development of boron deficiency. Field Crop Res 78:75–87

Zhao D, Oosterhuis DM (2003) Cotton growth and physiological responses to boron deficiency. J Plant Nutr 26:855–867

Diversity and Evolution of Sexual Strategies in *Silene*: A Review

Inés Casimiro-Soriguer, Eduardo Narbona, and M Luisa Buide

Contents

Abstract The variety and evolution of reproductive strategies in plants have attracted the attention of scientists for a long time. The genus *Silene* has been the focus of several studies related to the diversity and evolution of sexual systems. This review will summarize the huge amount of knowledge on sexual strategies in *Silene* species. Hermaphroditism is the most frequent condition in *Silene*; however, there is a relatively high frequency of gynodioecy and dioecy compared to angiosperms and dicotyledons. In some gynodioecious species, gynomonoecious individuals are common, forming a gynodioecious-gynomonoecious sexual system that is rare among angiosperms. Dioecy has independently evolved in the two

I. Casimiro-Soriguer (✉)
Área de Botánica, Departamento de Biología Molecular e Ingeniería Bioquímica, Universidad Pablo de Olavide, Ctra. de Utrera, km 1, 41013 Sevilla, Spain

Área de Botánica, Departamento de Biología Vegetal y Ecología, Universidad de Sevilla, Avenida Reina Mercedes s/n, 41012 Sevilla, Spain
e-mail: inessoriguer@gmail.com

E. Narbona • M.L. Buide
Área de Botánica, Departamento de Biología Molecular e Ingeniería Bioquímica, Universidad Pablo de Olavide, Ctra. de Utrera, km 1, 41013 Sevilla, Spain
e-mail: enarfer@upo.es; mlbuirea@upo.es

U. Lüttge et al. (eds.), *Progress in Botany*, Progress in Botany 77,
DOI 10.1007/978-3-319-25688-7_12

phylogenetically supported subgenera of *Silene*, with a probable origin down the "gynodioecious pathway." Heterogametic sex chromosomes have made *S. latifolia* and other dioecious species of the genus important models for the evolution of sex determination. In *Silene* species, studies on sexual expression at the plant and population level suggest that it is highly variable. Sexual dimorphism in reproductive and vegetative characters of dioecious species showed patterns that generally fit those found in other species. Compared with other genera of angiosperms, *Silene* presents a unique opportunity to evaluate the evolution of the different sexual systems and sex chromosomes (being of the few angiosperm genera with female heterogamety), the maintenance of gynodioecious and gynodioecious-gynomonoecious sexual systems, and the evolutionary implications of sexual dimorphism.

1 Introduction

The great variety of breeding systems in plants has been deeply interesting to scientists since the nineteenth century (Darwin 1877; Müller 1883). In fact, studies of sexual systems in contemporary scientific literature are on the increase, particularly from 1995 on (Fig. 1). How this genetic and morphological diversity has emerged from the ancestral condition of hermaphroditism is an important question, which has been widely discussed (Charlesworth and Charlesworth 1978; Barrett 2013; Crossman and Charlesworth 2014). Many intermediate steps between hermaphroditism and dioecy can be found in nature (e.g., gynodioecy or monoecy) and their study has attracted the attention of researchers over the years (Spigler and Ashman 2012; Golenberg and West 2013; Dufay et al. 2014). One of the main questions is why the separation of sex may evolve and how it may be maintained. This question has been tested in gynodioecious species: females have the advantage of avoiding inbreeding, whereas hermaphrodites may represent a bet-hedging strategy (Pettersson 1992).

Silene is the largest and most diverse genus of Caryophyllaceae (Mabberley 2008), with about 700 species (Mabberley 2008; Oxelman et al. 2013). It is a model system for studies in ecology and evolution, especially the ecology of biotic interactions and the evolution of sex chromosomes in plants (Bernasconi et al. 2009). Species of *Silene* show a high diversity of sexual systems (Desfeux et al. 1996; Jürgens et al. 2002; Casimiro-Soriguer et al. 2015), including dioecious species with sex chromosomes (e.g. *S. latifolia*). All species are self-compatible in *Silene*, and some of them are even cleistogamous (Jürgens et al. 2002; Witt et al. 2013, and references therein). Individuals or flowers may show spatial (e.g., dioecy, gynodioecy, or gynomonoecy) and temporal separation (e.g., protandry in hermaphroditic flowers) of the male and female functions, which could help avoid the effects of inbreeding depression (Charlesworth 1999; but see Davis and Delph

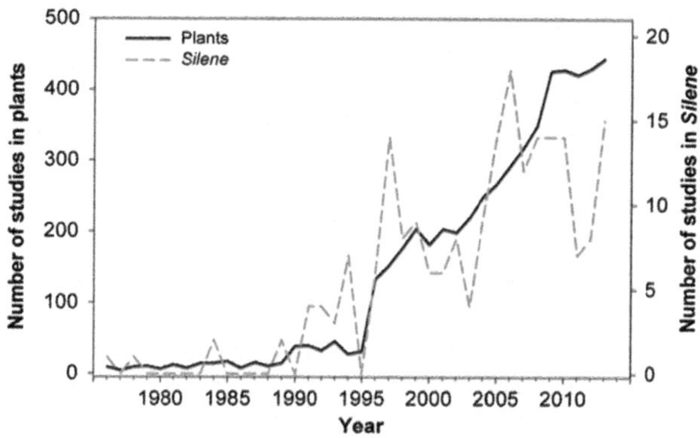

Fig. 1 Studies of sexual systems in plants (*black line*) and in the genus *Silene* (*gray dashed lines*). We performed a literature search for scientific papers in SCOPUS from 1976 to 2013 including at least one of the following words: sexual system, breeding system, hermaphroditic, hermaphrodite, hermaphroditism, dioecious, dioecy, gynodioecious, gynodioecy, gynomonoecious, gynomonoecy, androdioecious, androdioecy, andromonoecious, andromonoecy, monoecious, and monoecy. We limited the results to the area of life sciences restricted to the subject areas of agriculture and biological sciences, biochemistry, genetics and molecular biology, environmental sciences, and multidisciplinary and earth and planetary sciences. We found a total of 5,645 papers on plants as a whole and 210 when restricting the search to genus *Silene*

2005 and Reynolds et al. 2009) or allocating different resources to female and male functions (Lloyd 1979; Pettersson 1992; Delph 2003).

Here we will review the variety of sexual systems in *Silene* and the possible evolutionary pathways that could lead to this diversity. We summarize the advances in sex determination in dioecious and non-dioecious species. The variations in sex and gender expression will then be analyzed at the flower, individual, and population levels. Finally, sexual dimorphism in secondary sexual characters will be addressed for both reproductive and vegetative traits in species with two types of flowers. This review attempts to collate the huge amount of data on the evolution of gender and sexual systems in *Silene*. This may provide a valuable opportunity to generate and test hypotheses regarding the evolution and maintenance of certain sexual strategies that have repeatedly evolved in plants.

2 Diversity and Definition of Sexual Systems

Over time, the studies on plant reproductive strategies and sexual systems have generated a variety of terms to describe reproductive systems, but they are often inconsistently used. To avoid confusion, we therefore start this review by defining the most important terms relating to sexual systems.

Breeding, mating, and sexual systems have been inconsistently defined in the literature (Neal and Anderson 2005). Neal and Anderson (2005) suggest the use of the term "breeding system" for the distribution of sex within and among plants (e.g., gynodioecy), for physiological aspects (e.g., self-incompatibility), and for morphological differences within populations (e.g., heterostyly) and "mating system" when the genetic relatedness and pairings between individuals are examined (e.g., selfing rate). We accept their definition of breeding system, but consider the term "sexual system" more appropriate for the distribution of sex within and among plants. Therefore, "breeding system" should not be considered a synonym for sexual system, as some authors have defined (e.g., Lloyd 1979; Harder and Barrett 2006).

The distribution of sexual organs can be considered on different levels: (1) flower, (2) plant, and (3) population. (1) At the flower level, female and male sexual organs can be located in the same (hermaphroditic or perfect flower) or different flowers (female or pistillate and male or staminate flowers). (2) At the plant level, an individual is considered hermaphrodite when it bears only hermaphroditic flowers. When there are different types of flowers on the same individual, the suffix—monoecious or—monoecy is used. Monoecy itself occurs when the same plant produces both pistillate and staminate flowers. The term andro(gyno) monoecy or trimonoecy is applied to plants bearing perfect plus staminate(pistillate) flowers or all three types, respectively. When different sexes of flowers are distributed among different individuals, the suffix—dioecious or—dioecy is used. (3) On the population level, we can define populations as monomorphic, dimorphic, or polymorphic. Monomorphic populations have only one type of plants (i.e., hermaphroditic, monoecious, andromonoecious, or gynomonoecious). In dimorphic populations, there are two types of plants: e.g., males and females (dioecy), females and hermaphrodites (gynodioecy), or males and hermaphrodites (androdioecy). In polymorphic populations there are more than two types of plants, e.g., males, females, and hermaphrodites (trioecy, Fleming et al. 1994; Cruden and Lloyd 1995). Please note that Harder and Barrett (2006) also applied the term subdioecy to trioecious species. We advise against this and prefer to follow Bailey and Delph (2007a), who consider subdioecy close to dioecy, with a high frequency of females and hermaphrodites achieving nearly all their fitness via pollen, but retaining the possibility of reproducing through seeds. On the other hand, some sexual systems that have been considered dimorphic at the population level show other intermediate morphs. Thus, some gynodioecious populations include hermaphroditic individuals bearing also female flowers (i.e., gynomonoecious individuals). The correct term for this sexual system is gynodioecy-gynomonoecy, in which females, hermaphrodites, and gynomonoecious plants coexist within a population (Desfeux et al. 1996).

3 Sexual Systems in *Silene*

As in the case of studies of sexual systems in plants as a whole, studies of sexual systems in *Silene* have shown an abrupt increase in the last 20 years, with nearly 60 % thereof published in the last decade (Fig. 1). Interestingly, from the 210 studies of sexual systems in *Silene* that we found in our search (see details in Fig. 1), 87 % were focused on three species only: the dioecious *S. latifolia* (63 %), the gynodioecious *S. vulgaris* (14 %), and the sexually variable *S. acaulis* (10 %), in which gynodioecious, trioecious, subdioecious, and dioecious populations have been described (Maurice et al. 1999; Alatalo and Molau 2001). The remaining 13 % of the studies included eight other non-dioecious species (9 %) and another three dioecious species (*S. dioica*, *S. otites*, or *S. diclinis*; 4 %). Overall, studies in *Silene* are skewed toward dioecious species and especially in *S. latifolia*; however, dioecy is the third most common sexual system in the genus after hermaphroditism and gynodioecy (see below). This result is not surprising considering the importance of *S. latifolia* in the study of the evolution of sex chromosomes (see Sect. 5) and sexual dimorphism (see Sect. 7), among other aspects (Bernasconi et al. 2009).

Hermaphroditic, gynodioecious, gynomonoecious, dioecious, trioecious, and even andromonoecious species have been reported in the genus *Silene* (revised in Jürgens et al. 2002 and Casimiro-Soriguer et al. 2015). Despite *Silene* is a large genus (~700 species), there is a relatively important number of species in which the sexual system has been specifically studied or described (98 species; Casimiro-Soriguer et al. 2015). Based on these data, the most frequent sexual system in *Silene* is hermaphroditism (58.2 %; Table 1). However, the proportion of hermaphrodites is clearly lower than the estimate for all angiosperms and dicotyledons. In comparison with floras with available frequencies of the different sexual systems, the frequency of hermaphroditism in *Silene* is lower than those of the species of Levant flora (Israel) and similar to the Hawaiian flora. Gynodioecy (25.5 %) is the second most frequent sexual system, more than three times higher than for angiosperms and dicotyledons as a whole and higher than in the Levant local floras. In the genus *Silene*, gynodioecy-gynomonoecy occurs frequently (Talavera et al. 1996; Desfeux et al. 1996; Jürgens et al. 2002; Dufay et al. 2010; Casimiro-Soriguer et al. 2013), although it is sometimes classified as gynodioecy (e.g., Lafuma and Maurice 2006). In Table 1, we categorized all those species with gynodioecious-gynomonoecious sexual systems as gynodioecious for simplicity (~50 % of the cases, Casimiro-Soriguer et al. 2015). The frequency of dioecy also was more than three times higher than angiosperms and dicotyledons as whole, although very similar to the Hawaiian flora. To our knowledge, monoecious species do not occur in *Silene* (Casimiro-Soriguer et al. 2015); however, monoecy has a relative importance in the different datasets (between 3.6 % and 7.6 %). Although gynomonoecy and andromonoecy are present in the genus, the frequency of these sexual systems is very low (Table 1). The only gynomonoecious species was *S. noctiflora* (Folke and Delph 1997; Davis and Delph 2005), although populations with hermaphrodites and gynomonoecious plants have also been described (Jürgens et al. 2002). Gynomonoecy in *Silene* is to be expected, due to its combination with other sexual systems such as gynodioecy (i.e., gynodioecious-gynomonoecious species) given the possible related genetic determination system (Garraud et al. 2011). Andromonoecy has only been reported in *S. tibetica* (Oxelman et al. 2001), but there are no more studies that confirm the sexual system of this species.

Table 1 Percentage of sexual systems in species *Silene*, angiosperms and dicotyledons in general, and Hawaiian and Levant floras

Sexual system	*Silene*[a]	Angiosperms[b]	Dicotyledons[c]	Hawaiian flora[d]	Levant flora[e]
Hermaphroditism	58.2	72	71	62.4	86.6
Dioecy	14.3	4	4	14.7	2.2
Gynodioecy	25.5	7	*	3.8	0.3
Monoecy	–	5	4	7.6	3.6
Gynomonoecy	1.0	3	7	3.9	0.4
Andromonoecy	1.0	*	**	4.5	5.7
Others	–	9	14	–	1.1

*Sexual system included in the group "others"
**Sexual system included in the group "gynomonoecy"
[a]Percentage of *Silene* species calculated from data of 98 species revised by Casimiro-Soriguer et al. (2015). In the case of species with variable sexual system within or among populations, the most frequent sexual system was used [for a similar sexual system classification, see Jürgens et al. (2002)]. The available percentages data of other groups were taken from Richards (1997)[b], Yampolsky and Yampolsky (1922)[c], Sakai et al. (1995)[d], and de Jong et al. (2008)[e]

Silene was early known for its variety of sexual systems (Müller 1883; Knuth 1908); nevertheless, different sexual systems such as hermaphroditism or dioecy (as gynodioecy or gynomonoecy) are rarely described in checklists, national or local floras, or scientific papers. In fact, Desfeux et al. (1996) already noted that gynodioecy was usually not considered in floras. Thus it is reasonable to think that a non-negligible number of *Silene* species, usually described as hermaphrodites in floras, have other sexual systems. Accordingly, gynodioecy or gynomonoecy may be underestimated in the percentages shown in Table 1, because of the relatively low percentage of species with well-known sexual systems (~14 %; Casimiro-Soriguer et al. 2015). Conversely, most dioecious species are probably already described due to their easy identification; therefore, the percentage would decrease when more species were included in the dataset.

4 Evolutionary Pathways of Sexual System and Sex Determination in *Silene*

Although hermaphroditism is the commonest sexual system in angiosperms, dioecy has evolved many times in angiosperm phylogeny (Weiblen et al. 2000; Dufay et al. 2014). The two main evolutionary pathways proposed to explain the origin of dioecy are gynodioecy and monoecy (Charlesworth and Charlesworth 1978; Barrett 2002; Golenberg and West 2013). Dioecy may also be a transition to the rare androdioecy, as has been hypothesized for *Mercurialis annua* (Pannell 1997; Pannell et al. 2004). There is another less-considered pathway that renders dioecy from distyly (Muenchow and Grebus 1989). Although dioecy has been suggested to have evolved through monoecy in different groups (Renner and Won 2001; Torices et al. 2011), the gynodioecy pathway is generally more supported given the appreciable number of gynodioecious species that are related to dioecious species (Charlesworth and Charlesworth 1978; Maurice et al. 1993; Dufay et al. 2014).

Although *Silene* is a large genus without a complete resolved phylogeny, it has been divided into the well-supported phylogenetic subgenus *Behenantha* and subgenus *Silene* (Popp and Oxelman 2004; Rautenberg et al. 2010). Dioecious species have been placed in these two clades: *S. latifolia* and dioecious relatives in subgenus *Behenantha* and *S. otites* and relatives in subgenus *Silene* (Desfeux et al. 1996; Marais et al. 2011; Slancarova et al. 2013). *Silene* also has many gynodioecious species (Table 1), which is uncommon in other genera of the Caryophyllaceae (Matsunaga et al. 2003). Desfeux et al. (1996) mapped the evolution of sexual systems in a phylogeny of 22 species of *Silene* and suggested that gynodioecy was the ancestral condition of this genus. Fifteen years later, using high-resolution molecular tools, Marais et al. (2011) found that the most probable ancestral condition is either gynodioecy or hermaphroditism. Independently of the ancestral sexual system, it seems more likely that dioecy evolved via gynodioecy (Marais et al. 2011). Recently, it has been found that dioecy, hermaphroditism, gynodioecy, and gynodioecy-gynomonoecy are present with a similar frequency in both subgenera (Casimiro-Soriguer et al. 2015). Their presence in both subgenera is consistent with multiple and independent origins of these sexual systems in *Silene*.

In the gynodioecious pathway, the first step is the invasion of a male-sterile mutant in a population with hermaphrodite plants. In the second step, hermaphrodites gradually become functionally male until a female sterile mutant establishes itself, making the population dioecious (Charlesworth and Charlesworth 1978). During this last step, subdioecy occurs because hermaphrodites increase their male fertility but continue to produce some fruits (Bailey and Delph 2007a; Spigler and Ashman 2012). It is suggested that gender plasticity related to environmental conditions may help stabilize a subdioecious population (Delph and Wolf 2005). With regard to the genetic bases involved in the gynodioecy, in most species the male sterility alleles necessary to produce female individuals are located in the mitochondria (cytoplasmic male sterility factors) and they interact with nuclear restorers of male fertility (Bailey and Delph 2007b). *Silene vulgaris* and *S. nutans* are examples with multiple cytoplasmic male sterility and nuclear restorer loci involved in the expression of the gynodioecious sexual system (Charlesworth and Laporte 1998; Taylor et al. 2001; Bailey and McCauley 2005; Garraud et al. 2011). The fact that these genes are maternally inherited facilitates the spread of mutants in the population, decreasing the magnitude of female advantage needed to establish females in cosexual populations (reviewed in Dufay and Billard 2012). On the other hand, some models may explain the occurrence of gynomonoecious individuals in gynodioecious-gynomonoecious species by an incomplete male fertility restoration (Koelewijn and van Damme 1996; Ehlers and Thompson 2004). However, there are insufficient experimental studies to test these models (see Garraud et al. 2011).

5 Evolution of Sex Chromosomes in Dioecious Species of *Silene*

One of the most fascinating aspects of sexual systems in *Silene* is the acquisition of dioecy in different lineages and possibly at different times (Marais et al. 2011; Slancarova et al. 2013). In the subgenus *Behenantha,* all the species of section *Melandrium* are dioecious (Desfeux et al. 1996; Marais et al. 2011); in the subgenus *Silene*, the group of *S. otites* and relatives include dioecious and non-dioecious species, although the subsection *Otites* has only dioecious members (Slancarova et al. 2013). According to different authors (Marais et al. 2011; Käfer et al. 2013), dioecy is ancestral in *S. latifolia* and its close dioecious relatives (section *Melandrium*). However, according to Slancarova et al. (2013), it evolved more recently in the group *S. otites* and relatives.

Dioecious species of *Silene* show different sex determination types. Most dioecious species have male heterogamety (e.g., *S. latifolia*); however *S. otites* has female heterogamety, which is very rare in plants and has been reported in only a few angiosperms (reviewed in Slancarova et al. 2013). Moreover, in *S. diclinis*, neo-sex chromosomes have been reported (Weingartner and Delph 2014).

Silene latifolia has X and Y chromosomes (females XX/males XY), whose origin has been suggested in the ancestral lineage of section *Melandrium* (Marais et al. 2011). Compared to other organisms, this is a recent origin (10–20 million years or even much later, Filatov 2005; Slancarova et al. 2013), which allows the study of the early stages of the evolution of sex chromosomes (Ming and Moore 2007; Bergero and Charlesworth 2009; Qiu et al. 2011). It has been suggested that *S. latifolia* sex chromosomes have evolved from a single pair of autosomes through the formation and expansion of a large nonrecombining region on the Y chromosome (Filatov 2005; Bergero and Charlesworth 2009). On the other hand, the dioecious *S. diclinis*, included in the same section as *S. latifolia*, has a neo-sex chromosome (a region added to the nonrecombining part of a sex chromosome through a translocation event), which appears to have evolved from the ancestral XY chromosomes present in *S. latifolia* (Weingartner and Delph 2014).

Although Sansome's classic paper (1938) on *S. otites* had already suggested that females were the heterogametic sex, it was not until recently that there have been detailed analyses with molecular markers (Slancarova et al. 2013). They analyzed *S. otites*, showing that females are the heterogametic sex (females ZW/males ZZ). Even more interesting is the fact that *S. colpophylla*, a male heterogametic species, is closely related to *S. otites* and is placed in the same monophyletic group, which supposes an interesting change in heterogamety (Mrackova et al. 2008). On the other hand, the sex-determining system in the subdioecious *S. roemeri* and *S. acaulis* may have a common origin (Slancarova et al. 2013), although the authors suggest the study of more species to make definite conclusions.

6 Sex and Gender Expression

Information about sex expression can be gained from simple morphological observations to detailed studies of the genetics and ecology of the species (Sakai and Weller 1999; Elle and Meagher 2000; Delph and Wolf 2005). Changes in sex expression may occur at different levels of organization: (1) in flowers or inflorescences, for example, in plants with intra- or inter-floral dichogamy (Bertin and Newman 1993; Narbona et al. 2005); (2) in individuals, for example, changes in the proportion of male and hermaphroditic flowers in andromonoecious plants (Narbona et al. 2011 and references therein); or (3) in populations, for example, sex ratio variations in dioecious populations (Barrett et al. 2010).

Sex expression in *Silene* has been studied in gynomonoecious, gynodioecious-gynomonoecious, and dioecious species. For instance, in the gynomonoecious *S. noctiflora*, the frequency of female relative to hermaphroditic flowers within a plant increased due to the effect of high temperature and ethylene (Folke and Delph 1997). In *S. nutans*, the proportion of female flowers in gynomonoecious individuals varied from 0.03 to 0.9 (Dufay et al. 2010). More complexity in sex expression exists when the population level is analyzed. In the dioecious *S. latifolia*, the sex ratio is variable and often female biased (Carroll and Mulcahy 1993; Austerlitz et al. 2012); in fact, it is affected by environmental variables such as soil moisture and density (Lovett Doust et al. 1987). Even more complexity in sex expression can be found in gynodioecious-gynomonoecious species as a result of the changing frequency of the three possible morphs, as is found in *S. italica* (Maurice 1999), *S. littorea* (Guitián and Medrano 2000; Casimiro-Soriguer et al. 2013), *S. nutans* (Dufay et al. 2010), and *S. stockenii* (Talavera et al. 1996). All these findings suggest that sexual expression in *Silene* species is plastic, which may have consequences in the maintenance of the sexual system (Pannell et al. 2008; Delph and Wolf 2005).

Sometimes the morphology of the flower does not reflect its function as donor of male or female gametes. A hermaphroditic flower may function as exclusively male or female if its fitness is obtained only through pollen (or seed-siring success) or ovules (or seed production), respectively. Lloyd and Bawa (1984) emphasized consideration not only of the sex of a plant (i.e., morphology) but also the description of gender based on function: its femaleness or maleness as a parent. Plant gender measures allow evolutionary biologists to test hypotheses about sexual system evolution and their environmental relationships (Lloyd 1976; Bawa 1980; Delph and Wolf 2005). Based on the studies of Lloyd (1979, 1980), Lloyd and Bawa (1984) proposed two measurements for plant gender (for clarification, see Barrett and Harder 2006). The phenotypic gender, also called standardized phenotypic gender, quantifies the investment of parental resources (e.g., pollen, seeds, male or female flowers) in relation to other plants of the population. The phenotypic gender of a plant uses estimates of female investment (number of ovules, often estimated as number of female flowers) and male investment (number of pollen grains, male flowers) and includes the equivalence factor (E) which estimates the ratio of investments in maternal and paternal functions in the population as a whole. Values of phenotypic femaleness are defined between zero (a plant that only produces male flowers) and one (only produces female flowers). On the other hand, functional gender estimates the success of a plant as male or female parent and is calculated as the proportion of a plant's fitness transmitted through ovules or pollen. Accurate estimates of functional gender require information about seed production, pollen availability and dispersal, frequency of self- and cross-fertilization of the population, etc. (Lloyd 1980). The maternal expenditure may be easy to measure as seed production, but the estimation of paternal success is a complicated task. The development of molecular tools has facilitated estimation of male success through paternity analysis (Elle and Meagher 2000; Verdú et al. 2004; Gleiser et al. 2008).

The phenotypic and functional gender of a plant may be similar, but they are not necessarily the same (Primack and Lloyd 1980; Lloyd and Bawa 1984). In fact, study cases showed that both estimates of gender are poorly related, probably due to the complexity of functional gender estimation (Devlin and Stephenson 1987; Méndez 1998; Austen and Weis 2014). In addition, phenotypic gender may also be substantially influenced by the estimates used in its calculation (Thomson and Barrett 1981; Lloyd and Bawa 1984). For instance, the phenotypic gender of the gynodioecious-gynomonoecious *S. littorea* was calculated using two different estimates of female investment: the number of flowers bearing ovules (females and hermaphrodites) and the number of fruits set (Fig. 2; see Casimiro-Soriguer et al. 2013 for details of the study system). Phenotypic femaleness estimated with flowers showed that most plants in the populations invested approximately the same in female and male functions (values near 0.5). However, when phenotypic gender was estimated with fruits, there was a decrease in femaleness in most plants of both populations. This means that plants invested more in the male than the female function when fruit estimate was used. It is remarkable that completely female plants (plants that only produced female flowers) of one population showed the same value for both female investment measures (Fig. 2a). However, a completely male plant appeared in the other population when the fruit estimate was used (Fig. 2b), due to the fact that this plant did not produce fruits.

Temporal variation in functional and phenotypic gender of the plants in a population during the flowering season may also occur (Thomson and Barrett 1981; Méndez 1998; Casimiro-Soriguer et al. 2013). The success of a plant as a male may vary over time due to variations in the mating environment, i.e., the amount of pollen (or male-phase flowers) relative to the number of ovules (or -female-phase flowers) in the population (Brunet and Charlesworth 1995). Thus, in species with dichogamy and sequential blooming, variations in the mating environment and subsequent variation in individual phenotypic or functional gender are expected (Cruden and Hermann-Parker 1977; Brunet and Charlesworth 1995). Even more complexity can be found when the species present andro(gyno)monoecious individuals that are able to produce different proportions of male(female) and hermaphroditic flowers. In fact, extremely high fluctuations in mating environment are found in some species with andro(gyno)monoecious species, dichogamy, and synchronous blooming patterns (Thomson and Barrett 1981; Narbona et al. 2011).

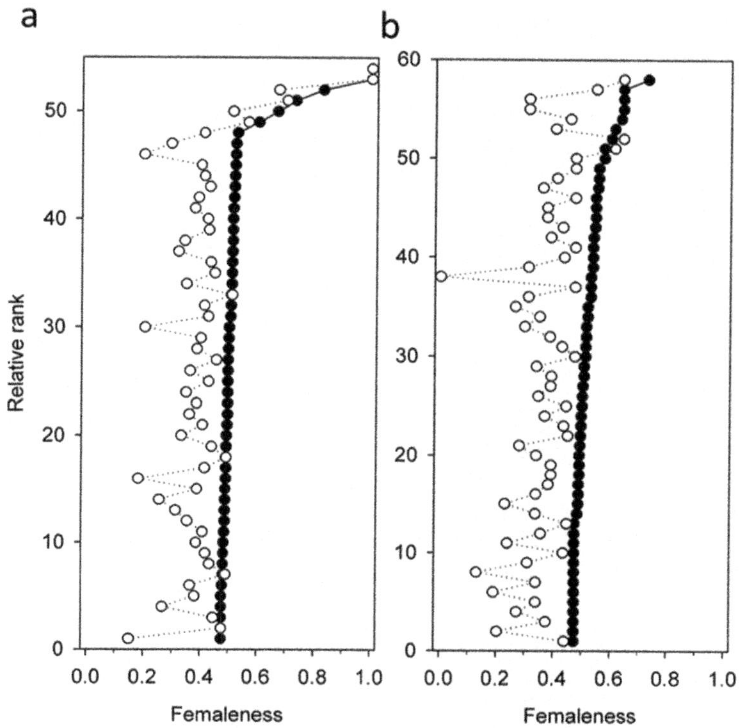

Fig. 2 Phenotypic gender in *Silene littorea*. The phenotypic gender has been calculated in two populations (**a**, **b**) of *Silene littorea* from southern Spain, using two different estimates of female investment: number of flowers (*black circles*) and number of fruits (*white circles*). Plants are ordered by relative value of phenotypic gender calculated by the number of flowers. See Casimiro-Soriguer et al. (2013) for details of the study system

Although the potential number of species of *Silene* with these characteristics is high, the number of cases studied is limited. In *S. littorea*, the mating environment fluctuated relatively little throughout the flowering season, but fluctuations were higher in a population with low flower production (Casimiro-Soriguer et al. 2013). This could be important for the conservation of the species since *S. littorea* usually has small populations and presents problems related to inbreeding depression (Vilas et al. 2006).

7 Sexual Dimorphism in Secondary Sex Characters

The term "sexual dimorphism" in plants is used to describe differences between males and females (plants or flowers) in primary and secondary sex characters (Sakai and Weller 1999; Barrett and Hough 2013). Primary sex characters are those that directly refer to the sexual organs (gynoecium or androecium). Secondary sex characters are those related to differences in other structures of the flower (e.g., nectaries, color and size of petals or sepals, or nectar composition; Eckhart 1999) or vegetative traits (e.g., morphology or physiology of vegetative parts, age of first reproduction, longevity, and growth; Delph 1999). In this section we will focus on secondary sex characters, differentiating between those affected by reproductive and vegetative traits.

7.1 Reproductive Traits

In dioecious species, it is expected that evolution drives the differentiation of male and female flowers in response to the different reproductive functions of the two sexes. In fact, many of the differences found between males and females correspond to reproductive characters, particularly flower number and size (Meagher 1992). The prediction based on Bateman's principle (Bateman 1948; Willson 1994) is that males produce larger flowers than females. Contrary to expectations, females of *S. latifolia* have heavier flowers due to sepals and pedicels (Carroll and Delph 1996), although males are able to produce a higher number of flowers than females (Laporte and Delph 1996; Delph et al. 2002) and invest more biomass overall in flowers than females (Carroll and Delph 1996). In this species, the relative number of flowers/flower size has been found to be an important trade-off with other reproductive and vegetative traits (Delph and Meagher 1995; Delph et al. 2004). Another important aspect analyzed in *S. latifolia* is the quantitative genetic basis of sexual dimorphism (revised in Meagher 1999), and the sex variation in DNA content (more male than female), and its negative correlation to flower size (Meagher and Costich 1994; Meagher et al. 2005). The same significant negative correlation of DNA content and calyx diameter was found in *S. diclinis*, *S. dioica*, and *S. latifolia*, but not in *S. marizii* (Meagher and Costich 2004). As in *S. latifolia*, males of the close relative *S. dioica* produced more flowers than females; however, male flowers were larger than females (Kay et al. 1984). On the other hand, Wright and Meagher (2004) found variations between selection in male and female flowers, and this also varied widely between sites and years.

Although the patterns of variation in dioecious species do not always follow the same path, in gynodioecious species the pattern is much clearer, with hermaphroditic flowers with larger corollas than females (Eckhart 1999). This pattern is followed in the gynodioecious species of *Silene*, such as *S. acaulis* (Delph and Carroll 2001) and *S. vulgaris* (Dykstra et al. 2009), but also in the gynodioecious-gynomonoecious species *S. stockenii* (Talavera et al. 1996), *S. nutans* (Dufay et al. 2010), and four other species of the section *Psammophilae* (Casimiro-Soriguer 2015).

With respect to sexual dimorphism in nectar production, female flowers of the dioecious *S. latifolia* and *S. dioica* produce higher volumes than males, but sugar concentration is higher in males (Kay et al. 1984; Carroll and Delph 1996; Witt et al. 1999; Gehring et al. 2004). In gynodioecious species in general, the pattern is that hermaphroditic flowers produce more nectar and more concentrated (Eckhart 1999). In *S. vulgaris* and *S. stockenii*, hermaphroditic flowers had greater sugar content than females (Jolls et al. 1994; Talavera et al. 1996). Females of *S. nutans* produced more nectar than hermaphrodites, although less concentrated (Witt et al. 1999).

Other secondary sex characters such as floral fragrance may have important consequences for reproductive success, via pollinator attraction or florivore deterrence (Schaefer and Ruxton 2011). Studies of dioecious species showed that, in most cases, males emitted more volatiles per flower than females (Ashman 2009). Consequently, despite having smaller flowers than females, floral scent emission has been found to be greater in males of *S. latifolia*, which has important consequences on pollinator behavior (Waelti et al. 2009).

On the other hand, biotic interactions mediated by insect visitors (pollinators, predators, or pathogens) may potentially be affected by gender specialization and dimorphism. In several species of *Silene*, a nursery pollination system is found in which noctuids of the genera *Hadena* and *Perizoma* act as pollinators as well as seed predators because they lay eggs on the ovaries of flowers (Jürgens et al. 1996; Kephart et al. 2006; Reynolds et al. 2012). Kephart et al. (2006) found that fruit predation was lower in species with diurnal pollination and hermaphroditic sexual systems compared with nocturnal pollination and dioecious or gynodioecious sexual systems. In this scenario, moths could act as selective agents on sexual systems favoring hermaphroditism. Furthermore, species of *Silene* are frequently infected by the fungal disease *Microbotryum violaceum*, in which pollinators are also potential vectors (Thrall et al. 1993; Hood et al. 2010). This anther-smut pathogen causes plant sterility, producing aborted ovaries and spores instead of pollen. Interestingly, the fungus induces anther development (with fungus spores) in females of dioecious and gynodioecious host species; thus, the fungus is transmitted by male and female plants (Antonovics et al. 2002). Because flower visitors are needed for fertilization and for fungus dispersal, plants may be exposed to conflicting selective forces, attempting to attract pollinators while avoiding fungus transmission (Thrall et al. 1997). Thus, sexual dimorphism in characters related to insect attraction, such as flower number or flower duration, may affect this conflict (Ågren et al. 1999). For example, males of *S. latifolia* and *S. dioica* produce more flowers and are more frequently visited by insects, and therefore a higher number of spores are deposited in male flowers (Alexander 1989; Shykoff and Bucheli 1995; Carlsson-Granér et al. 1998). Similarly, male flowers of *S. latifolia* have a short flower life span compared with females, which reduces the risk of infestation (Kaltz and Shykoff 2001).

7.2 Vegetative Traits

In long-lived species, males are usually more vigorous than females, and the opposite pattern is found in short-lived species (Barrett and Hough 2013). In *S. latifolia*, males were taller and dedicated more biomass to leaves than females; females had shorter and stouter inflorescences, as a consequence of stopping flower production when fruits started to develop (Gehring and Linhart 1993) and had longer leaves (Delph et al. 2002). In *S. dioica* males had more leaves per rosette, although shorter than females (Cox 1981; Van Nigtevech 1966). In spite of these apparently clear differences, it is important to consider the life history of the plant, because sexual dimorphism before the development of inflorescence seems to be very rare (Zluvova et al. 2010). Significant differences can be found before and after flowering, because reproductive costs can influence future resource distribution (Sanchez-Vilas and Pannell 2011; Barrett and Hough 2013). For example, in *S. latifolia*, Meagher (1992) found that male and female plants had equal seed sizes, early establishment, and growth before sexual reproduction. Moreover, in this species, many sexually dimorphic vegetative traits, such as plant height, length of inflorescence branches, and allocation of leaves and branches, were found to be correlated with the number of flowers (Gehring and Linhart 1993; Delph et al. 2002, 2005). Purrington and Schmitt (1998) eliminated age differences by sowing seeds of *S. latifolia* on a single day, finding that females emerged earlier but flowered later. A new step in the analysis of dimorphic sex expression is the analysis of sex-specific genes or gene expression. Recently, Zluvova et al. (2010) found three sex-specifically expressed genes in the rosette stage in *S. latifolia*. Lastly, physiological traits also differ between sexes once flowering has started, but not before. For example, males of *S. latifolia* have higher photosynthetic rates than females (Laporte and Delph 1996), but females live longer than males (Lovett Doust et al. 1987; Carroll and Mulcahy 1993).

8 Conclusions and Future Directions

Studies of sexual systems in plants, and specifically *Silene*, are on the increase. In *Silene*, different aspects of sexual systems such as the above described are mostly studied in 14 species, but almost all literature focused on three (*S. latifolia*, *S. vulgaris*, and *S. acaulis*). The dioecious *S. latifolia* has become a model species for the study of sexual dimorphism (Delph and Herlihy 2012; Barrett and Hough 2013), repetitive DNA (Meagher and Vassiliadis 2005), and sex chromosomes in plants (Charlesworth 2013; Slancarova et al. 2013), among other aspects (Bernasconi et al. 2009). In addition, studies of other species have helped to understand new features of sexual systems in plants; for instance, *S. acaulis* has been used to study the role of environmental factors in the transition from gynodioecy to dioecy (Delph 2003) and *S. nutans* for the genetic basis of male sterility in gynodioecy (Garraud et al. 2011).

The three most frequent sexual systems in *Silene* are, in order, hermaphroditism, gynodioecy, and dioecy; but combinations of these three types are also present. This combination of sexual systems in *Silene* makes the genus particularly engaging for the study of evolutionary transitions. A reliable reconstruction of the evolution of sexual systems in *Silene* remains incomplete. This could be due to the difficulties of building accurate phylogenetic relationships among and within the groups of species due to introgression and complex mutational processes (Erixon and Oxelman 2008; Petri and Oxelman 2011) but also to the lack of clear information about the sexual systems of a relevant number of species (Casimiro-Soriguer et al. 2015). Recently, the evolution of dioecy has been clarified in the two dioecious groups of *Silene* (Marais et al. 2011; Käfer et al. 2013), and gynodioecy seems the most probable pathway for the evolution of dioecy. However, the ancestral sexual system remains unclear (Marais et al. 2011).

Finally, a considerable number of species of *Silene* also showed a variable sexual system within and/or among populations (Jürgens et al. 2002; Casimiro-Soriguer et al. 2015 and references therein), which suggests a plasticity of expression of sexual systems. Particularly interesting are the cases of subdioecious or gynodioecious-gynomonoecious species (Desfeux et al. 1996; Dufay et al. 2010; Casimiro-Soriguer et al. 2013). However, the roles of these sexual systems in the evolution of sexual systems in *Silene* are not yet defined. Experiments designed to clarify the advantage of both sexual systems under different selective scenarios with consideration of functional gender estimates are required.

Acknowledgements We apologize to the authors whose work has not been cited due to space constraints. The authors thank U. Lüttge for the invitation to participate in this issue, two anonymous reviewers for helpful comments on the manuscript, and Anna Crandell for English proofreading. This work was supported by FEDER funds and grants from the Spanish Ministry of Science and Innovation through a Research Personnel Training grant to ICS [BES-2010-031073] and the research projects CGL2009-08257 and CGL2012-37646.

References

Ågren J, Danell K, Elmqvist T, Ericson L, Hjältén J (1999) Sexual dimorphism and biotic interactions. In: Geber MA, Dawson TE, Delph LF (eds) Gender and sexual dimorphism in flowering plants. Springer, Berlin, pp 217–246

Alatalo JM, Molau U (2001) Pollen viability and limitation of seed production in a population of the circumpolar cushion plant, *Silene acaulis* (Caryophyllaceae). Nord J Bot 21:365–372

Alexander HM (1989) An experimental field study of anther-smut disease of *Silene alba* caused by *Ustilago violacea*: genotypic variation and disease incidence. Evolution 43:835–847

Antonovics J, Hood M, Partain J (2002) The ecology and genetics of a host shift: *microbotryum* as a model system. Am Nat 160:S40–S53

Ashman TL (2009) Sniffing out patterns of sexual dimorphism in floral scent. Funct Ecol 23:852–862

Austen EJ, Weis AE (2014) Temporal variation in phenotypic gender and expected functional gender within and among individuals in an annual plant. Ann Bot 114:167–177

Austerlitz F, Gleiser G, Teixeira S, Bernasconi G (2012) The effects of inbreeding, genetic dissimilarity and phenotype on male reproductive success in a dioecious plant. Proc R Soc B 279:91–100

Bailey MF, Delph LF (2007a) A field guide to models of sex-ratio evolution in gynodioecious species. Oikos 116:1609–1617

Bailey MF, Delph LF (2007b) Sex-ratio evolution in nuclear-cytoplasmic gynodioecy when restoration is a threshold trait. Genetics 176:2465–2476

Bailey MF, McCauley DE (2005) Offspring sex ratio under inbreeding and outbreeding in a gynodioecious plant. Evolution 59:287–295

Barrett SCH (2002) The evolution of plant sexual diversity. Nat Rev Genet 3:274–284

Barrett SCH (2013) The evolution of plant reproductive systems: how often are transitions irreversible? Proc R Soc B 280:20130913

Barrett SCH, Harder LD (2006) David G. Lloyd and the evolution of floral biology: from natural history to strategic analysis. In: Harder LD, Barrett SCH (eds) Ecology and evolution of flowers. Oxford University Press, New York, NY, pp 1–21

Barrett SCH, Hough J (2013) Sexual dimorphism in flowering plants. J Exp Bot 64:67–82

Barrett SCH, Yakimowski SB, Field DL, Pickup M (2010) Ecological genetics of sex ratios in plant populations. Philos Trans R Soc Lond B Biol Sci 365:2549–2557

Bateman AJ (1948) Intrasexual selection in *Drosophila*. Heredity 2:349–368

Bawa KS (1980) Evolution of dioecy in flowering plants. Annu Rev Ecol Syst 11:15–39

Bergero R, Charlesworth D (2009) The evolution of restricted recombination in sex chromosomes. Trends Ecol Evol 24:94–102

Bernasconi G, Antonovics J, Biere A, Charlesworth D, Delph LF, Filatov D, Giraud T, Hood ME, Marais GAB, McCauley D, Pannell JR, Shykoff JA, Vyskot B, Wolfe LM, Widmer A (2009) *Silene* as a model system in ecology and evolution. Heredity 103:5–14

Bertin RI, Newman CN (1993) Dichogamy in angiosperms. Bot Rev 59:112–150

Brunet J, Charlesworth D (1995) Floral sex allocation in sequentially blooming plants. Evolution 49:70–79

Carlsson-Granér U, Elmqvist T, Ågren J, Gardfjell H, Ingvarsson P (1998) Floral sex ratios, disease and seed set in dioecious *Silene dioica*. J Ecol 86:79–91

Carroll SB, Delph LF (1996) The effects of gender and plant architecture on allocation to flowers in dioecious *Silene latifolia* (Caryophyllaceae). Int J Plant Sci 157:493–500

Carroll SB, Mulcahy DL (1993) Progeny sex ratios in dioecious *Silene latifolia* (Caryophyllaceae). Am J Bot 80:551–556

Casimiro-Soriguer I (2015) Sistemas sexuales y polimorfismo de color en Silene: una aproximación en la sección Psammophilae. Ph.D. dissertation, Pablo de Olavide University, Seville

Casimiro-Soriguer I, Buide ML, Narbona E (2013) The roles of female and hermaphroditic flowers in the gynodioecious-gynomonoecious *Silene littorea*: insights into the phenology of sex expression. Plant Biol 15:941–947

Casimiro-Soriguer I, Buide ML, Narbona E (2015) Diversity of sexual systems within different lineages of the genus *Silene*. AoB Plants 7:plv037

Charlesworth D (1999) Theories of the evolution of dioecy. In: Geber MA, Dawson TE, Delph LF (eds) Gender and sexual dimorphism in flowering plants. Springer, Berlin, pp 33–60

Charlesworth D (2013) Plant sex chromosome evolution. J Exp Bot 64:405–420

Charlesworth B, Charlesworth D (1978) A model for the evolution of dioecy and gynodioecy. Am Nat 112:975–997

Charlesworth D, Laporte V (1998) The male-sterility polymorphism of *Silene vulgaris*: analysis of genetic data from two populations and comparison with *Thymus vulgaris*. Genetics 150:1267–1282

Cox PA (1981) Niche partitioning between sexes of dioecious plants. Am Nat 117:295–307

Crossman A, Charlesworth D (2014) Breakdown of dioecy: models where males acquire cosexual functions. Evolution 68:426–440

Cruden RW, Hermann-Parker SM (1977) Temporal dioecism: an alternative to dioecism. Evolution 31:863–866

Cruden RW, Lloyd RM (1995) Embryophytes have equivalent sexual phenotypes and breeding systems: why not a common terminology to describe them? Am J Bot 82:816–825

Darwin C (1877) The different forms of flowers on plants of the same species. John Murray, London

Davis SL, Delph LF (2005) Prior selfing and gynomonoecy in *Silene noctiflora* L. (Caryophyllaceae): opportunities for enhanced outcrossing and reproductive assurance. Int J Plant Sci 166:475–480

De Jong TJ, Shmida A, Thuijsman F (2008) Sex allocation in plants and the evolution of monoecy. Evol Ecol Res 10:1087–1109

Delph LF (1999) Sexual dimorphism in life history. In: Geber MA, Dawson TE, Delph LF (eds) Gender and sexual dimorphism in flowering plants. Springer, Berlin, pp 149–173

Delph LF (2003) Sexual dimorphism in gender plasticity and its consequences for breeding system evolution. Evol Dev 5:34–39

Delph LF, Carroll SB (2001) Factors affecting relative seed fitness and female frequency in a gynodioecious species, *Silene acaulis*. Evol Ecol Res 3:487–505

Delph LF, Herlihy CR (2012) Sexual, fecundity, and viability selection on flower size and number in a sexually dimorphic plant. Evolution 66:1154–1166

Delph LF, Meagher TR (1995) Sexual dimorphism masks life history trade-offs in the dioecious plant *Silene latifolia*. Ecology 76:775–785

Delph LF, Wolf DE (2005) Evolutionary consequences of gender plasticity in genetically dimorphic breeding systems. New Phytol 166:119–128

Delph LF, Knapczyk FN, Taylor DR (2002) Among-population variation and correlations in sexually dimorphic traits of *Silene latifolia*. J Evol Biol 15:1011–1020

Delph LF, Gehring JL, Frey FM, Arntz M, Levri M (2004) Genetic constraints on floral evolution in a sexually dimorphic plant revealed by artificial selection. Evolution 58:1936–1946

Delph LF, Gehring JL, Arntz AM, Levri M, Frey FM (2005) Genetic correlations with floral display lead to sexual dimorphism in the cost of reproduction. Am Nat 166:S31–S41

Desfeux C, Maurice S, Henry JP, Lejeune B, Gouyon PH (1996) Evolution of reproductive systems in the genus *Silene*. Proc R Soc Lond 263:409–414

Devlin B, Stephenson AG (1987) Sexual variations among plants of a perfect-flowered species. Am Nat 130:199–218

Dufay M, Billard E (2012) How much better are females? The occurrence of female advantage, its proximal causes and its variation within and among gynodioecious species. Ann Bot 109:505–519

Dufay M, Lahiani E, Brachi B (2010) Gender variation and inbreeding depression in gynodioecious-gynomonoecious *Silene nutans* (Caryophyllaceae). Int J Plant Sci 171:53–62

Dufay M, Champelovier P, Käfer J, Henry JP, Mousset S, Marais GAB (2014) An angiosperm-wide analysis of the gynodioecy-dioecy pathway. Ann Bot 114:539–548

Dykstra AB, Brock MT, Delph LF, Weinig C (2009) Sex-specific trade-offs and responses to foliar shade in the gynodioecious species *Silene vulgaris* (Caryophyllaceae). Int J Plant Sci 170:575–583

Eckhart VM (1999) Sexual dimorphism in flowers and inflorescences. In: Geber MA, Dawson TE, Delph LF (eds) Gender and sexual dimorphism in flowering plants. Springer, Berlin, pp 123–148

Ehlers BK, Thompson JD (2004) Temporal variation in sex allocation in hermaphrodites of gynodioecious *Thymus vulgaris* L. J Ecol 92:15–23

Elle E, Meagher TR (2000) Sex allocation and reproductive success in the andromonoecious perennial *Solanum carolinense* (Solanaceae). II. Paternity and functional gender. Am Nat 156:622–636

Erixon P, Oxelman B (2008) Whole-gene positive selection, elevated synonymous substitution rates, duplication, and indel evolution of the chloroplast *clpP1* gene. PLoS One 1:e1386

Filatov DA (2005) Evolutionary history of *Silene latifolia* sex chromosomes revealed by genetic mapping of four genes. Genetics 170:975–979

Fleming TH, Maurice S, Buchmann SL, Tuttle MD (1994) Reproductive biology and relative male and female fitness in a trioecious cactus, *Pachycereus pringlei* (Cactaceae). Am J Bot 81:858–867

Folke SH, Delph LF (1997) Environmental and physiological effects on pistillate flower production in *Silene noctiflora* L. (Caryophyllaceae). Int J Plant Sci 158:501–509

Garraud C, Brachi B, Dufay M, Touzet P, Shykoff JA (2011) Genetic determination of male sterility in gynodioecious *Silene nutans*. Heredity 106:757–764

Gehring JL, Linhart YB (1993) Sexual dimorphism and response to low resources in the dioecious plant *Silene latifolia* (Caryophyllaceae). Int J Plant Sci 154:152–162

Gehring JL, Scoby J, Parsons M, Delph LF (2004) Whole-plant investment in nectar is greater for males than pollinated females in the dioecious plant *Silene latifolia*. Evol Ecol Res 6:1237–1252

Gleiser G, Verdú M, Segarra-Moragues JG, González-Martínez SC, Pannell JR (2008) Disassortative mating, sexual specialization, and the evolution of gender dimorphism in heterodichogamous *Acer opalus*. Evolution 62:1676–1688

Golenberg EM, West NW (2013) Hormonal interactions and gene regulation can link monoecy and environmental plasticity to the evolution of dioecy in plants. Am J Bot 100:1022–1037

Guitián P, Medrano M (2000) Sex expression and fruit set in *Silene littorea* (Caryophyllaceae): variation among populations. Nord J Bot 20:467–473

Harder LD, Barrett SCH (2006) Ecology and evolution of flowers. Oxford University Press, Oxford, UK

Hood ME, Mena-Alí JI, Gibson AK, Oxelman B, Giraud T, Yockteng R, Arroyo MTK, Conti F, Pedersen AB, Gladieux P, Antonovics J (2010) Distribution of the anther-smut pathogen *Microbotryum* on species of the Caryophyllaceae. New Phytol 187:217–229

Jolls CL, Chenier TC, Hatley CL (1994) Spectrophotometric analysis of nectar production in *Silene vulgaris* (Caryophyllaceae). Am J Bot 81:60–64

Jürgens A, Witt T, Gottsberger G (1996) Reproduction and pollination in Central European populations of *Silene* and *Saponaria* species. Bot Acta 109:316–324

Jürgens A, Witt T, Gottsberger G (2002) Pollen grain numbers, ovule numbers and pollen-ovule ratios in Caryophylloideae: correlation with breeding system, pollination, life form, style number, and sexual system. Sex Plant Reprod 14:279–289

Käfer J, Talianová M, Bigot T, Michu E, Guéguen L, Widmer A, Žlůvová J, Glémin S (2013) Patterns of molecular evolution in dioecious and non-dioecious *Silene*. J Evol Biol 26:335–346

Kaltz O, Shykoff JA (2001) Male and female *Silene latifolia* plants differ in per-contact risk of infection by a sexually transmitted disease. J Ecol 89:99–109

Kay QON, Lack AJ, Bamber FC, Davies CR (1984) Differences between sexes in floral morphology, nectar production and insect visits in a dioecious species, *Silene dioica*. New Phytol 98:515–529

Kephart S, Reynolds RJ, Rutter MT, Fenster CB, Dudash MR (2006) Pollination and seed predation by moths on *Silene* and allied Caryophyllaceae: evaluating a model system to study the evolution of mutualisms. New Phytol 169:667–680

Knuth P (1908) Handbook of flower pollination, vol II (English translation by JR Ainsworth Davis). Clarendon Press, Oxford, UK

Koelewijn HP, Van Damme JMM (1996) Gender variation, partial male sterility and labile sex expression in gynodioecious *Plantago coronopus*. New Phytol 132:67–76

Lafuma L, Maurice S (2006) Reproductive characters in a gynodioecious species, *Silene italica* (Caryophyllaceae), with attention to the gynomonoecious phenotype. Biol J Linn Soc 87:583–591

Laporte MM, Delph LF (1996) Sex-specific physiology and source-sink relations in the dioecious plant *Silene latifolia*. Oecologia 106:63–72

Lloyd DG (1976) The transmission of genes via pollen and ovules in gynodioecious angiosperms. Theor Popul Biol 9:299–316

Lloyd DG (1979) Parental strategies in angiosperms. N Z J Bot 17:595–606

Lloyd DG (1980) Sexual strategies in plants. III. A quantitative method for describing the gender of plants. N Z J Bot 18:103–108

Lloyd DG, Bawa KS (1984) Modification of the gender of seed plants in varying conditions. Evol Biol 17:255–338

Lovett Doust J, O'Brien G, Lovett Doust L (1987) Effect of density on secondary sex characteristics and sex ratio in *Silene alba* (Caryophyllaceae). Am J Bot 74:40–46

Mabberley DJ (2008) Mabberley's plant-book. Cambridge University Press, Cambridge, UK

Marais GAB, Forrest A, Kamau E, Käfer J, Daubin V, Charlesworth D (2011) Multiple nuclear gene phylogenetic analysis of the evolution of dioecy and sex chromosomes in the genus *Silene*. PLoS One 6:e21915

Matsunaga S, Isono E, Kejnovsky E, Vyskot B, Dolezel J, Kawano S, Charlesworth D (2003) Duplicative transfer of a MADS box gene to a plant Y chromosome. Mol Biol Evol 20:1062–1069

Maurice S (1999) Gynomonoecy in *Silene italica* (Caryophyllaceae): sexual phenotypes in natural populations. Plant Biol 1:346–350

Maurice S, Charlesworth D, Desfeux C, Couvet D, Gouyon PH (1993) The evolution of gender in hermaphrodites of gynodioecious populations with nucleo-cytoplasmic male-sterility. Proc R Soc Lond B 251:253–261

Maurice S, Desfeux C, Mignot A, Henry JP (1999) Is *Silene acaulis* (Caryophyllaceae) a trioecious species? Reproductive biology of two subspecies. Can J Bot 76:478–485

Meagher TR (1992) The quantitative genetics of sexual dimorphism in *Silene latifolia* (Caryophyllaceae). I Genetic variation. Evolution 46:445–457

Meagher TR (1999) The quantitative genetics of sexual dimorphism. In: Geber MA, Dawson TE, Delph LF (eds) Gender and sexual dimorphism in flowering plants. Springer, Berlin, pp 275–294

Meagher TR, Costich DE (1994) Sexual dimorphism in nuclear DNA content and floral morphology in populations of *Silene*. Am J Bot 81:1198–1204

Meagher TR, Costich DE (2004) 'Junk' DNA and long-term phenotypic evolution in *Silene* section *Elisanthe* (Caryophyllaceae). Proc R Soc Lond B 271:S493–S497

Meagher TR, Vassiliadis C (2005) Phenotypic impacts of repetitive DNA in flowering plants. New Phytol 168:71–80

Meagher TR, Gillies ACM, Costich DE (2005) Genome size, quantitative genetics and the genomic basis for flower size evolution in *Silene latifolia*. Ann Bot 95:247–254

Méndez M (1998) Modification of phenotypic and functional gender in the monoecious *Arum italicum* (Araceae). Am J Bot 85:225–234

Ming R, Moore PH (2007) Genomics of sex chromosomes. Curr Opin Plant Biol 10:123–130

Mrackova M, Nicolas M, Hobza R, Negrutiu I, Monéger F, Widmer A, Vyskot B, Janousek B (2008) Independent origin of sex chromosomes in two species of the genus *Silene*. Genetics 179:1129–1133

Muenchow GE, Grebus M (1989) The evolution of dioecy from distyly: reevaluation of the hypothesis of the loss of long-tongued pollinators. Am Nat 133:149–156

Müller H (1883) The fertilization of flowers. Macmillan, London, UK

Narbona E, Ortiz PL, Arista M (2005) Dichogamy and sexual dimorphism in floral traits in the andromonoecious *Euphorbia boetica*. Ann Bot 95:779–787

Narbona E, Ortiz PL, Arista M (2011) Linking self-incompatibility, dichogamy and flowering synchrony in two *Euphorbia* species: alternative mechanisms for avoiding self-fertilization? PLoS One 6:e20668

Neal PR, Anderson GJ (2005) Are 'mating systems' 'breeding systems' of inconsistent and confusing terminology in plant reproductive biology? or is it the other way around? Plant Syst Evol 250:173–185

Oxelman B, Lidén M, Turland NJ (2001) Taxonomic and nomenclatural notes on Chinese *Silene* (Caryophyllaceae). Novon 11:322–324

Oxelman B, Rautenberg A, Thollesson M, Larsson A, Frajman B, Eggens F, Petri A, Aydin Z, Töpel M, Brandtberg-Falkman A (2013) Sileneae taxonomy and systematics. http://www.Sileneae.info. Accessed 14 July 2014

Pannell JR (1997) Widespread functional androdioecy in *Mercurialis annua* L. (Euphorbiaceae). Biol J Linn Soc 61:95–116

Pannell JR, Obbard DJ, Buggs RJA (2004) Polyploidy and the sexual system: what can we learn from *Mercurialis annua*? Biol J Linn Soc 82:547–560

Pannell JR, Dorken ME, Pujol B, Berjano R (2008) Gender variation and transitions between sexual systems in *Mercurialis annua* (Euphorbiaceae). Int J Plant Sci 169:129–139

Petri A, Oxelman B (2011) Phylogenetic relationships within *Silene* (Caryophyllaceae) section *Physolychnis*. Taxon 60:953–968

Pettersson MW (1992) Advantage of being a specialist female in the gynodioecious *Silene vulgaris* s.l. (Caryophyllaceae). Am J Bot 79:1389–1395

Popp M, Oxelman B (2004) Evolution of a RNA polymerase gene family in *Silene* (Caryophyllaceae)-incomplete concerted evolution and topological congruence among paralogues. Syst Biol 53:914–932

Primack RB, Lloyd DG (1980) Sexual strategies in plants IV. The distributions of gender in two monomorphic shrub populations. N Z J Bot 18:109–114

Purrington CB, Schmitt J (1998) Consequences of sexually dimorphic timing of emergence and flowering in Silene latifolia. J Ecol 86:397–404

Qiu S, Bergero R, Zeng K, Charlesworth D (2011) Patterns of codon usage bias in *Silene latifolia*. Mol Biol Evol 28:771–780

Rautenberg A, Hathaway L, Oxelman B, Prentice HC (2010) Geographic and phylogenetic patterns in *Silene* section *Melandrium* (Caryophyllaceae) as inferred from chloroplast and nuclear DNA sequences. Mol Phylogenet Evol 57:978–991

Renner SS, Won H (2001) Repeated evolution of dioecy from monoecy in Siparunaceae (Laurales). Syst Biol 50:700–712

Reynolds RJ, Westbrook MJ, Rohde AS, Cridland JM, Fenster CB, Dudash MR (2009) Pollinator specialization and pollination syndromes of three related North American *Silene*. Ecology 90:2077–2087

Reynolds RJ, Kula AAR, Fenster CB, Dudash MR (2012) Variable nursery pollinator importance and its effect on plant reproductive success. Oecologia 168:439–448

Richards AJ (1997) Plant breeding systems. Chapman & Hall, London, UK

Sakai AK, Weller SG (1999) Gender and sexual dimorphism in flowering plants: a review of terminology, biogeographic patterns, ecological correlates, and phylogenetic approaches. In: Geber MA, Dawson TE, Delph LF (eds) Gender and sexual dimorphism in flowering plants. Springer, Berlin, pp 1–31

Sakai AK, Wagner WL, Ferguson DM, Herbst DR (1995) Origins of dioecy in the Hawaiian Flora. Ecology 76:2517–2529

Sanchez-Vilas J, Pannell JR (2011) Sexual dimorphism in resource acquisition and deployment: both size and timing matter. Ann Bot 107:119–126

Sansome FW (1938) Sex determination in *Silene otites* and related species. J Genet 35:387–396

Schaefer HM, Ruxton GD (2011) Plant-animal communication. Oxford University Press, New York, NY

Shykoff JA, Bucheli E (1995) Pollinator visitation patterns, floral rewards and the probability of transmission of *Microbotryum violaceum*, a venereal disease of plants. J Ecol 83:189–198

Slancarova V, Zdanska J, Janousek B, Talianova M, Zschach C, Zluvova J, Siroky J, Kovacova V, Blavet H, Danihelka J, Oxelman B, Widmer A, Vyskot B (2013) Evolution of sex determination systems with heterogametic males and females in *Silene*. Evolution 67:3669–3677

Spigler RB, Ashman TL (2012) Gynodioecy to dioecy: are we there yet? Ann Bot 109:531–543

Talavera S, Arista M, Salgueiro FJ (1996) Population size, pollination and breeding system of *Silene stockenii* Chater (Caryophyllaceae), an annual gynodioecious species of southern Spain. Bot Acta 109:333–339

Taylor DR, Olson MS, McCauley DE (2001) A quantitative genetic analysis of nuclear-cytoplasmic male sterility in structure populations of *Silene vulgaris*. Genetics 158:833–841

Thomson JD, Barrett CH (1981) Temporal variation of gender in *Aralia hispida* Vent. (Araliaceae). Evolution 35:1094–1107

Thrall PH, Biere A, Antonovics J (1993) Plant life-history and disease susceptibility—the occurrence of *Ustilago violacea* on different species within the Caryophyllaceae. J Ecol 81:489–498

Thrall PH, Antonovics J, Bever JD (1997) Sexual transmission of disease and host mating systems: within-season reproductive success. Am Nat 149:485–506

Torices R, Méndez M, Gómez JM (2011) Where do monomorphic sexual systems fit in the evolution of dioecy? Insights from the largest family of angiosperms. New Phytol 190:234–248

Van Nigtevech G (1966) Genetic studies in dioecious *Melandrium*. I. Sex-linked and sex-influenced inheritance in *M. album* and *M. dioicum*. Genetica 37:281–306

Verdú M, Montilla AI, Pannell JR (2004) Paternal effects on functional gender account for cryptic dioecy in a perennial plant. Proc R Soc Lond B 271:2017–2023

Vilas C, San Miguel E, Amaro R, Garcia C (2006) Relative contribution of inbreeding depression and eroded adaptive diversity to extinction risk in small populations of shore campion. Conserv Biol 20:229–238

Waelti MO, Page PA, Widmer A, Schiestl FP (2009) How to be an attractive male: floral dimorphism and attractiveness to pollinators in a dioecious plant. BMC Evol Biol 9:190

Weiblen GD, Oyama RK, Donoghue MJ (2000) Phylogenetic analysis of dioecy in monocotyledons. Am Nat 155:46–58

Weingartner LA, Delph LF (2014) Neo-sex chromosome inheritance across species in *Silene* hybrids. J Evol Biol 27:1491–1499

Willson MF (1994) Sexual selection in plants: perspective and overview. Am Nat 144:S13–S39

Witt T, Jürgens A, Geyer R, Gottsberger G (1999) Nectar dynamics and sugar composition in flowers of *Silene* and *Saponaria* (Caryophyllaceae). Plant Biol 1:334–345

Witt T, Jürgens A, Gottsberger G (2013) Nectar sugar composition of European Caryophylloideae (Caryophyllaceae) in relation to flower length, pollination biology and phylogeny. J Evol Biol 26:2244–2259

Wright JW, Meagher TR (2004) Selection on floral characters in natural Spanish populations of *Silene latifolia*. J Evol Biol 17:382–395

Yampolsky C, Yampolsky H (1922) Distribution of sex forms in the phanerogamic flora. Bibliogr Genet 3:1–62

Zluvova J, Zak J, Janousek B, Vyskot B (2010) Dioecious *Silene latifolia* plants show sexual dimorphism in the vegetative stage. BMC Plant Biol 10:208

Part IV
Ecology

Freezing Stress in Tree Xylem

Stefan Mayr and Thierry Améglio

Contents

Abstract Freezing in plant xylem is a complex process affecting living and dead components. This book chapter gives a brief overview of methods for analyzing freezing dynamics and tissue damage and focuses on the effects of freezing stress in the xylem symplast and apoplast. Survival strategies, such as supercooling, extracellular freezing, or avoidance of critical bubble formation/expansion in conduits are discussed, and insights from experimental and field studies available in the literature summarized. The final part deals with trees at the Alpine timberline, which are exposed to intense freezing as well as extreme drought stress every winter. Timberline trees are thus an interesting model system to study combined

S. Mayr (✉)
Institute of Botany, University of Innsbruck, Sternwartestraße 15, 6020 Innsbruck, Austria
e-mail: stefan.mayr@uibk.ac.at

T. Améglio
INRA, UMR PIAF, 63039 Clermont-Ferrand, France

Clermont University, Université Blaise Pascal, UMR PIAF, 63100 Clermont-Ferrand, France

U. Lüttge et al. (eds.), *Progress in Botany*, Progress in Botany 77,
DOI 10.1007/978-3-319-25688-7_13

effects of drought and freezing stress in tree xylem and respective avoidance, tolerance, and repair strategies of plants.

1 Introduction

1.1 Relevance of Ice Formation in Xylem

The life-form "tree" is based on the formation of wood. Wood, i.e., secondary xylem, fulfills hydraulic, mechanical, as well as storage function and thus is a prerequisite to form high trunks and complex axes systems. The xylem is composed of several cell types and based on an interplay of living and dead cells. Due to the upright habit of trees, their (aboveground) axes system is exposed to the atmosphere and their tissues, including the xylem, influenced by respective temperature conditions (Fig. 1).

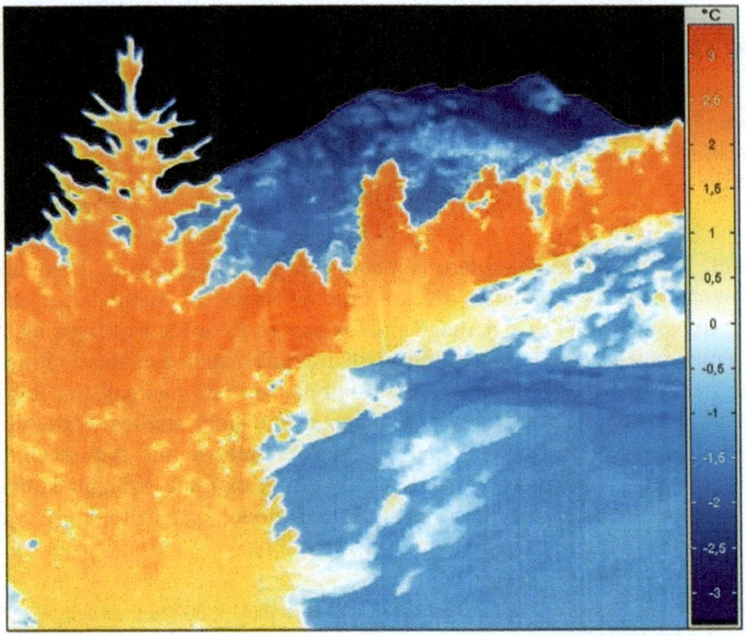

Fig. 1 Infrared-thermography of trees near the timberline cooling down in the evening. The picture was taken around 22 h on November 26, 2014, at 1.800 m in Praxmar, Tyrol. Although it was a cloudy day with slow temperature decrease after sunset, temperature gradients are visible within trees, with coldest parts at the tips and (due to snow fields) at lower crown parts. These trees froze some hours after the picture was taken

In many regions of the world, plants have to cope with freezing stress. In temperate regions with subzero temperatures during one season, plants have to withstand at least one freezing event per year. In most cases, repeated freezing (and intermediate thawing processes) occurs at the onset of the cold season until permanent subzero temperatures are reached. Similarly, multiple frost cycles occur at the end of the cold season. The temperature course in plants and the extremes reached are influenced by the thermal mass and conductance of tissues, exposure, insulation, or spectral properties (determining overheating or radiation cooling). Normally, trees are most affected as they cannot profit from a moderate microclimate near the ground or protection under snow. In contrast, small plants may avoid freezing, critically low temperatures, rapid temperature changes, or frequent freeze–thaw events when protected by a snow cover, by growing in protected microhabitats or by creating mild microclimatic conditions due to special growth forms. Many Alpine growth forms, like cushion plants or dwarf shrubs, follow this strategy (Körner 2003), while trees are exposed to harsh temperature conditions. For instance, more than 100 freeze–thaw events per winter were observed in distal twigs of trees at the Alpine timberline (Mayr et al. 2006b).

In this book chapter, we will deal with timberline trees in the final part (Sect. 5) after briefly discussing the monitoring of freezing and damage in tree xylem (Sect. 2) and the effects of freezing in living (Sect. 3) as well as dead components of the xylem (Sect. 4). Much effort has been made during the last decades to monitor and analyze freezing in plant tissues and to understand the processes causing frost damage. Ice in plant tissues has enormous destructive potential, but plants evolved in an impressive repertoire of strategies to avoid or compensate damage in their symplastic and apoplastic components.

1.2 *Some Physical and Chemical Aspects of Freezing in Xylem*

Freezing is a complex process (Fig. 2) and difficult to analyze in plant xylem. The freezing temperature depends on the sap composition and solute concentrations as well as on the availability of structures for ice nucleation. Ice nucleation points may be bacteria, organic and inorganic debris, or cell wall structures. Ice formation starts at these nucleation points, where first water molecules arrange in ice lattice. As described in Sect. 3, the location of ice nucleation is critical and enables one strategy for plants to avoid freezing damage. When water molecules arrange in the ice lattice, a structure of increased stability is formed. The resulting latent heat of 334 J g^{-1} is released, which can be recorded as an exotherm (Muldrew et al. 2004). Accordingly, thawing is an endothermal process causing a decrease in temperature around melting ice crystals.

From the point of ice nucleation, the ice crystal grows by attraction of water molecules from the surrounding liquid into the ice lattice. This results in a strong

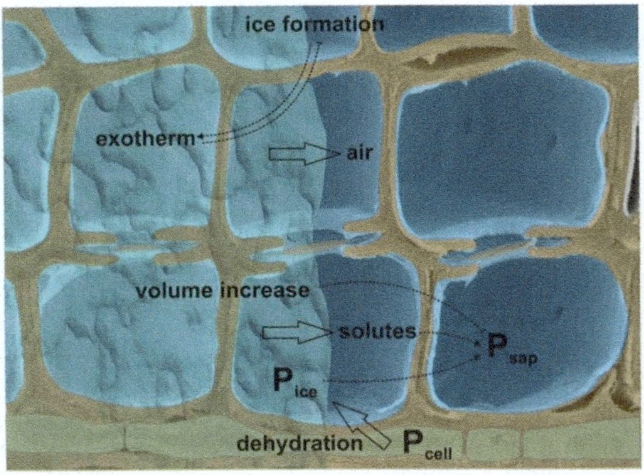

Fig. 2 Some important chemical and physical processes during freezing of xylem. Formation of ice (*left part* of picture) causes a temperature increase (exotherm), which can inhibit subsequent freezing. Ice formation also leads to an increase in volume which reduces tension in the sap (P_{sap}). In contrast, the low water potential of ice (P_{ice}) and up-concentration of solutes increase tensions. Ice formation also leads to gas (air) shifts to the liquid phase and to dehydration of adjacent parenchyma cells (*bottom* of figure) and a decrease of the total water potential in the cell (P_{cell}) as well as an up-concentration of solutes in the cell therefore decreasing the freezing point. The figure is based on a SEM cross-sectional image of *Picea abies*

gradient in water potential (Ψ) toward the ice surface, as the Ψ of ice is temperature dependent. Following Rajashekar and Burke (1982) and Rajashekar et al. (1983), the Ψ of ice changes at about 1.16 MPa K^{-1} decrease in temperature. Thus, the freezing temperature has a dramatic influence on local Ψ, which may be of special relevance in plant tissues (Hansen and Beck 1988; Améglio et al. 2001a; Cavender-Bares 2005; Cavender-Bares et al. 2005). Ice at, e.g., −10 °C, causes a Ψ of less than −10 MPa, which is much lower than Ψ recorded in most plants even under extreme drought (e.g., Choat et al. 2012). When structures for an ice nucleation are lacking (see Sect. 3.2), low temperatures without freezing, and in consequence without exotherm formation and decreasing Ψ, can be reached (for physics of supercooled water, see, e.g., Holten et al. 2012). The ice nucleation temperature decreases with smaller conduits across species (Lintunen et al. 2013), and at a temperature around −40 °C, water molecules spontaneously turn from liquid to ice (Fujikawa and Kuroda 2000). Whenever ice formation has started, ice propagates through plant tissues at rates up to 40 cm s^{-1}, i.e., 1.5 km h^{-1} (e.g., Pearce and Fuller 2001; Neuner et al. 2010) depending on temperature, structural traits, and especially on ice barriers (e.g., Hacker and Neuner 2007; Hacker et al. 2011). Freezing also causes a ca. 9 % increase in volume as the density of ice is 917 kg m^{-3} versus 1000 kg m^{-3} of liquid water (at 0 °C). This causes a local increase in Ψ, while in parallel the up-concentration of osmolytes in the remaining sap between ice crystals decreases Ψ (Sevanto et al. 2012).

Gases are hardly soluble in ice, and thus bubbles are formed during freezing. The size of bubbles depends on the amount of gas dissolved in the sap and the volume of freezing sap within a compartment (Ewers 1985; Pittermann and Sperry 2006). Recent findings indicate that some parts of dissolved gases may be pushed out of stems during freezing, thus reducing the amount of gases entrapped in ice (Lintunen et al. 2014). The stability of these bubbles in the liquid water between ice crystals is determined by their size and Ψ of the sap (Laplace's law, see Sect. 4) and thus also by the physicochemical aspects of ice formation described above.

Overall, freezing in xylem is a process not yet fully understood with various, partly contrary effects: Ice formation causes tension due to the low Ψ of ice, but at the same time, positive pressure due to the expansion in volume is created. Appearing bubbles may buffer pressure changes but cause additional effects due to the surface tension at the air–water interfaces. Several other aspects, such as mentioned changes in solute concentrations, temperature fluctuations due to exotherm formation, or water shifts between the symplast and apoplast, further complicate the process of ice formation. It also has to be considered that this process occurs in a heterogeneous system of dead and living tissue components with a mix of various solid, fluid, and gaseous phases. Last but not least, freezing is followed by thawing, a process with similar complexity and often affecting xylem in combination with freezing during freeze–thaw cycles.

2 Monitoring of Freezing and Damage

2.1 Detection of Freezing and Ice

Many liquids in plants do not freeze at the melting point of the solid phase (Sakai and Larcher 1987). Rasmussen and MacKenzie (1972) demonstrated that pure water supercooled to −38.1 °C provided that no heterogeneous nucleators were present. In nature, supercooling in plant liquids is usually small in amount and therefore important only during mild freezes (−3 °C to −8 °C; Levitt 1980). The ability of the water to remain supercooled varies inversely with the diameter of the capillary where it occurs. Thus, water is much more likely to supercool in smaller than in larger cells or large well-filled vessels (Asahina 1956). As we have described in part in Sect. 1.2, crystallization heat causes an exotherm temperature peak (Muldrew et al. 2004), whereby in most cases water crystallizes around a nucleus, which is often located in the apoplast. Latent heat released by the crystallization of extracellular water can be recorded as a high-temperature exotherm (HTE; see Fig. 3). When the temperature further decreases, a second exotherm corresponding to intracellular ice nucleation appears (low-temperature exotherm, LTE).

Accordingly, important information about freezing in plants is based on exotherm detection (Kuroda et al. 1999; Pearce 2001), e.g., via differential thermal

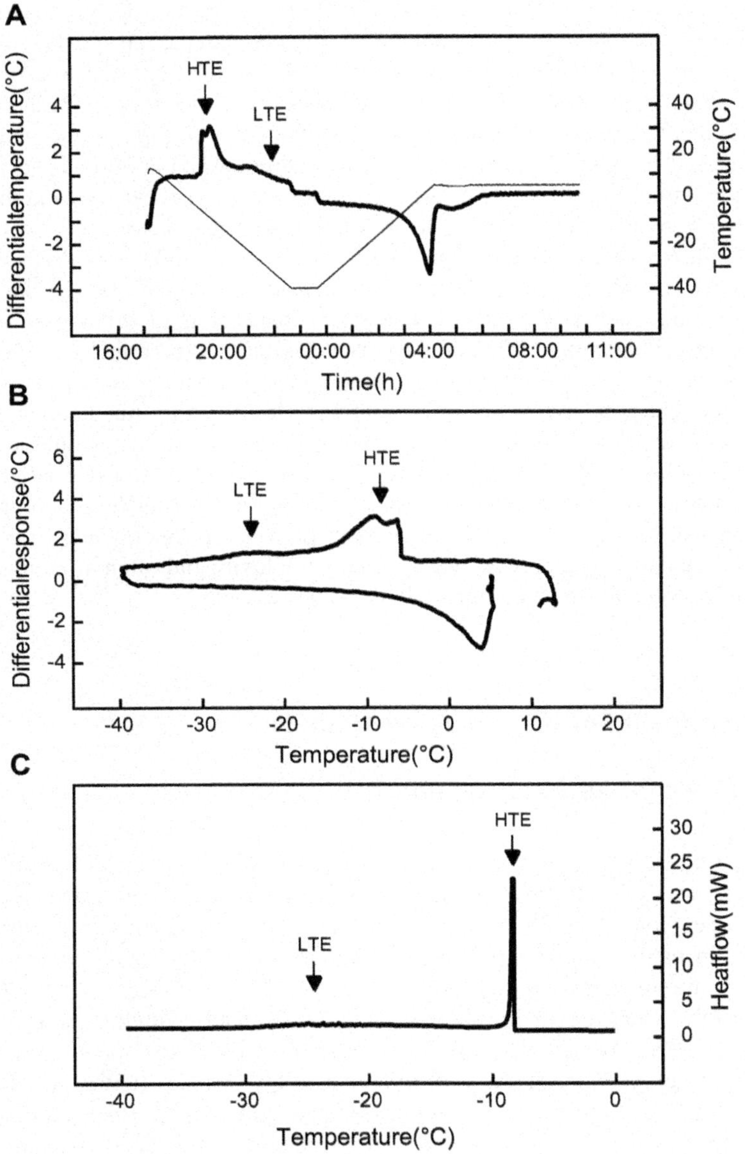

Fig. 3 Effects of one freeze–thaw cycle between 5 °C and −40 °C in walnut twigs. (**a**) Typical thermocouple record of temperature from twigs. Differential temperature between fresh walnut twigs and similar dried twigs (*bold line*) and air temperature in the temperature control chamber (*thin line*). (**b**) Differential temperature between fresh walnut twigs and similar dried twigs during cooling to −40 °C. (**c**) Heat fluxes measurement on walnut xylem with a differential scanning calorimeter during cooling to −40 °C. *HTE* high-temperature exotherm, *LTE* low-temperature exotherm (Améglio and Charrier, unpublished)

analysis (DTA; Burke et al. 1976; Fujikawa et al. 1994; Pramsohler et al. 2012; Fig. 3) or via infrared differential thermal analysis (IDTA; Wisniewski et al. 1997). DTA cannot detect the location of ice nucleation or the path of the initial spread of ice, while IDTA gives a real-time image of the temperature at the plant surface, thus revealing the location at which ice first forms and showing the route and rate of ice growth through the plant (Wisniewski et al. 1997; Hacker and Neuner 2007). In contrast, time–domain reflectometry was used to detect ice fractions in stems (Sparks et al. 2001), and MRI (magnetic resonance imaging; Ishikawa et al. 2009) has sufficient resolution to identify supercooled water in small organs or in tissues such as the xylem. More recently, Charrier et al. (2014a) proposed a new acoustic method to detect freezing events and spatial and temporal dynamics of freezing in plant stems (see Sect. 2.3).

2.2 *Detection of Injury in Living Cells*

Due to the importance of frost resistance for winter survival of perennial plants, there has been considerable interest in developing methods for determining the level of plant cold hardiness. Most of these methods are based on controlled freezing tests followed by determination of freezing injuries and recovery by regrowth tests in spring. However, as exhaustive analyses of cold hardiness in whole plants were only applied on small specimens and necessitated plenty of time and large and expensive cooling chambers, simple screening tests have been developed with exposure of plant parts to controlled freezing (Lindén et al. 2000; Lindén 2002). Thus, using detached plant parts provides detailed information on the level of hardiness in different tissues and organs. For example, the measurement of relative fluorescence (F_v/F_m) was used to estimate the cold hardiness level in leaves of evergreen species (Boorse et al. 1998), while stem diameter changes enabled to estimate the cold hardiness level in bark (Améglio et al. 2003). However, possible artifacts upon excision have to be considered, as the excision zone may be exposed to most extreme temperatures (direct contact of cold air with the xylem, which is thermally protected by the bark *in natura*), and cellular damage may occur due to cutting.

Other methods assess injury caused by freezing via optical microscopy (Levitt 1980), visual observation (Boorse et al. 1998), or vital dye observation (Sutinen et al. 1992). In woody species, an estimation of frost hardiness by microscopic methods was not precise because cell deformation was only observed in the cambial zone, where cell walls are poorly lignified (see Fig. 4).

Electrolyte leakage is the most common method used for frost hardiness monitoring in different tissues (Griffith and McIntyre 1990; Sutinen et al. 1992; Shirazi and Fuchigami 1995; Leinonen 1996; Maldonado et al. 1997; Morin et al. 2007; Poirier et al. 2010; Charra-Vaskou et al. 2012; Charrier and Améglio 2011; Stattin et al. 2012; Charrier et al. 2013b, 2014b). This was also one of the first applied methods in cold hardiness research (Dexter et al. 1930, 1932). This test is based on

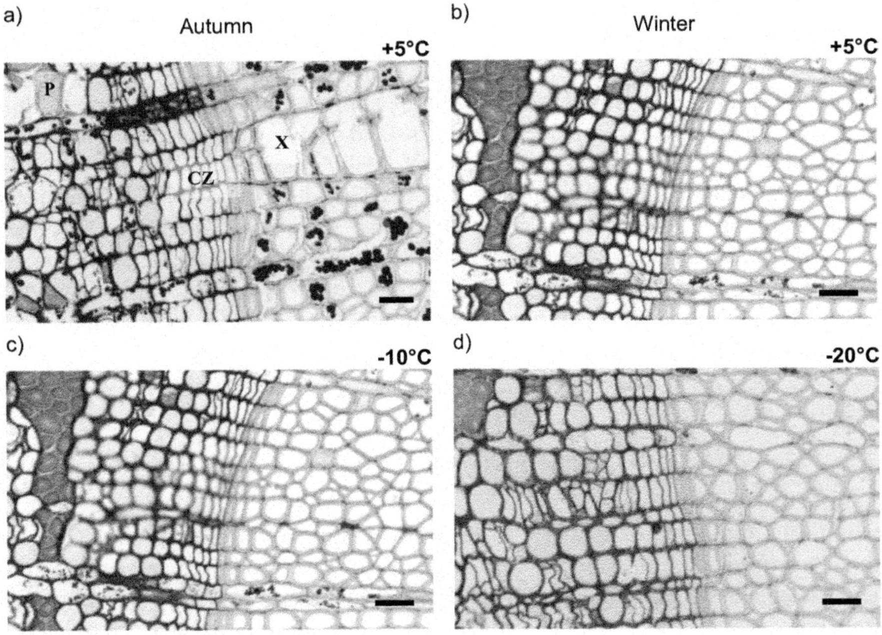

Fig. 4 Transverse sections from 1-year-old twigs of a walnut stem (*Juglans regia* L.) in autumn (**a**, **c**) and in winter (**b**, **d**). On all micrographs, the phloem (P) is located on the *left*, the xylem (X) is situated on the *right*, and the cambial zone (CZ) is in the *center*. In autumn, there are a lot of starch grains in parenchyma cells contrarily to winter time, where all starch grains have disappeared. A deformation of cells due to a freeze–thaw cycle was observed only at the cambial zone level in autumn after one freeze–thaw cycle at −10 °C (**c**). In winter, the cambial zone appeared without damage (*cold hardy*) after one freeze–thaw cycle down to a temperature of −20 °C (**d**). *Scale bar* = 20 μm (Poirier, Brunel, and Améglio unpublished)

the principle that damage to cell membranes results in an enhanced leakage of electrolytes from the cell. Recording the amount of leakage will thus provide an estimate of tissue damage. The test is fairly simple and rapid, yields quantitative data, and requires only small amounts of plant material. However, certain concerns limit the validity of this technique to determine the temperature at 50 % lethality (LT_{50}). Hence, several authors proposed to simultaneously use two or more viability tests and to combine the results of such tests for an estimation of damage intensities (Zhang and Willison 1987; Burr et al. 1990; Lindén 2002). In contrast, the electrolyte leakage method allowed consistent approximations of LT_{50} in some species and permitted to assess dynamics of frost hardiness in trees during the entire winter seasons via identical and, therefore, comparable measurement protocols (Repo et al. 1990; Charrier and Améglio 2011).

2.3 Detection of Damage in the Water Transport System

Hydraulic measurements are important as they allow a direct analysis of effects of freeze stress on the xylem's transport function. The water transport system in plants is composed of dead cells, which form a complex network of tubes interconnected by the pits (e.g., Tyree et al. 1994; Tyree and Zimmermann 2002; Hacke and Sperry 2001). This network provides the hydraulic transport capacity to supply the distal leaves with water, and dysfunction of network components reduces transport capacities. The hydraulic conductivity of stems can readily be analyzed by flow measurements at a given pressure difference and sample length. The specific hydraulic conductivity (k_s) is related to the xylem cross-sectional area and the leaf-specific conductivity (k_l) to the supplied leaf area (e.g., Zimmermann 1978; Tyree and Zimmermann 2002). Reduction in conductivity by embolism (see Sect. 4) can be quantified by the Sperry method (Sperry et al. 1988a), which compares the conductivity of samples before and after repeated high pressure flushes (PLC, percent loss of conductivity). There are numerous systems in use (e.g., Sperry et al. 1988a; Cochard et al. 2000; Vogt 2001; Mayr et al. 2002) both for measurements of native and experimentally induced embolism.

Imaging techniques also enable the analysis of embolism in xylem samples. For Cryo-SEM imaging, samples are rapidly frozen in liquid nitrogen. Consequently, empty conduits can be distinguished from functional conduits (which contained water before shock freezing and ice afterwards; e.g., Canny 1997; Cochard et al. 2000, 2004; Mayr et al. 2007). According to Cochard et al. (2000) and Richter (2001), tension during freezing in liquid nitrogen has to be low enough to avoid artificial embolism. More recently, X-ray tomography systems were used to image xylem structures and water content in intact samples or even intact plants (e.g., Brodersen et al. 2010, 2013; Strullu-Derrien et al. 2014; Cochard et al. 2015). Imaging in intact plants is advantageous as possible artifacts due to sample preparation (e.g., Wheeler et al. 2013; Trifilò et al. 2014; Venturas et al. 2014) can be avoided. On the other hand, X-rays can damage fine structures and warm hit regions within the sample, which might limit studies on freezing. From images, the hydraulic conductivity is estimated based on cross-sectional areas of the functional conduits. It should be mentioned that such assessments cannot account for the pit resistance in the xylem.

Ultrasonic acoustic emission (UAE) analysis has also been used in studies on xylem freezing (e.g., Raschi et al. 1989; Kikuta and Richter 2003; Mayr et al. 2007). Originally, ultrasonic emissions were observed in dehydrating plant stems and used to monitor the formation of drought-induced embolism. UAE analysis methods were refined during the last years (e.g., Mayr and Rosner 2011; Wolkerstorfer et al. 2012; Vergeynst et al. 2015), but the extraction of hydraulically relevant signals from background noise remains challenging. Accordingly, the meaning of ultrasonic signals on freezing of xylem is under debate (see Sect. 4). The origin of UAE on drought as well as freezing are probably cell walls, which are deformed under tension and relax when tensions are released due to cavitation (e.g., Tyree and

Sperry 1989; Mayr and Sperry 2010; Charrier et al. 2014a). For a recent review of hydraulic, imaging, and acoustic methods, see Cochard et al. (2013).

3 Freezing of Living Components

3.1 Cell Damage on Freezing

Following frost, the fate of cells depends on where the formation of ice crystals takes place (Mazur 1966). Ice formation can be either intracellular, which is fatal due to the destruction of membranes (e.g., tonoplast) and causes cell death if the cooling is rapid, or extracellular, which protects the cells, at least temporarily (Rodrigo 2000). Adaptations that allow plants to survive freezing temperatures can be classified in two ways: (1) such plants that exhibit deep supercooling characteristics (Ashworth et al. 1993) and (2) such that exhibits extracellular freezing (Burke et al. 1976). Due to the formation of extracellular ice, on the surface of the cell wall, in lumina of nonliving fibers and vessels, or in the extracellular spaces (Guy 1990), cell dehydration occurs. Liquid water moves out of the cell (Mazur 1969), and the osmotic concentration inside cells increases, thus preventing intracellular freezing. The same phenomenon explains why the living bark of trees shrinks at freezing temperatures (Wiegand 1906; Winget and Kozlowski 1964): Ice formed between cells attracts water from surrounding cells and occupies more and more intercellular spaces. The displacement of gases by ice causes an overall reduction in volume despite the volume increase due to ice formation. Although this mechanism was reported in the nineteenth century (Hoffmann 1857; Sachs 1860; Friedrich 1897), it has received little attention until recently (Loris et al. 1999; Zweifel and Häsler 2000; Améglio et al. 2001a; Mayr et al. 2007). The reversible bark shrinkage of mature subalpine conifers (Zweifel and Häsler 2000) or deciduous angiosperms (Améglio et al. 2001a) represented the transport of water between bark and wood. Swelling during melting indicated a backflow of water from the bark to living cells and consecutive refilling of intercellular spaces with air.

Whenever intracellular ice is formed, it can be detected by an LTE (see Sect. 2.1). LTE differs across species in relation to their geographic origin (Burke et al. 1976; Kaku and Iwaya 1979) and season (Pramsohler and Neuner 2013). Intracellular ice induces low Ψ at the ice–water interface, which interferes with molecular bonds (hydrogen, van der Waals, and hydrophobic bonds), leading to macromolecule denaturation (enzyme and structural protein) or membrane disruption (Uemura et al. 2006; Ruelland et al. 2009). Under natural conditions, intracellular ice formation leads to the death of the cell (Wolfe and Bryant 2001; Gusta et al. 2004; Muldrew et al. 2004).

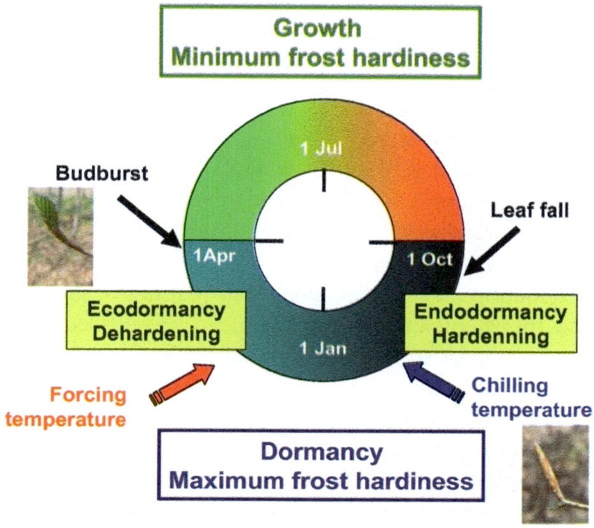

Fig. 5 Annual cycle of tree development and frost hardiness. Minimum frost hardiness is observed in the growing season between budburst and leaf fall and highest frost hardiness during the dormancy period between leaf fall and consecutive budburst (in deciduous trees). This phenological synchronization with environmental conditions is a mechanism of freezing resistance (i.e., avoidance of freezing temperature for the most vulnerable tissue to freeze). The dormancy period is composed of an endodormancy (e.g., suspension of growth in every plant part containing a meristem) and hardening periods. Chilling temperatures are necessary to release endodormancy and induce cold hardening. In winter, at maximum frost hardiness, an ecodormancy (e.g., a dormant state is limited only by environmental factors) and dehardening periods are entered. Forcing temperatures are then necessary to prepare budburst time and growing season

3.2 *Survival Strategies*

As we have seen in Sect. 2.1, the first case of frost avoidance *in natura* is based on supercooling of water and only observed in mild freezes (−3 °C to −8 °C; Levitt 1980). A few extreme cases have been reported for buds of some deciduous trees (Wiegand 1906; Pramsohler and Neuner 2013). Ice formation can also be avoided by the use of antinucleators, such as antifreeze proteins (Pearce 2001), which allow supercooling of the sap. Moreover, structural or thermal ice barriers can block ice propagation within sensitive tissues at different periods (e.g., buds in winter; Dereuddre and Gazeau 1992; Hacker et al. 2011; Pramsohler et al. 2012).

Escape strategies by freezing avoidance may be achieved by ecological distribution (i.e., latitude or altitude), topography, and position (e.g., of dormancy buds belowground/aboveground, under cover/exposed) highlighted by the Raunkiaer classification (Raunkiaer 1934). Phenological synchronization with environmental conditions (Fig. 5), such as leaf fall for deciduous and growth cessation for all species with dormancy periods (Weiser 1970; Lang et al. 1987), is a freezing avoidance mechanism. Thereby, bud growth is blocked by dormancy status which is driven by several environmental factors (Perry 1971), i.e., mainly by the shortening photoperiod at the end of the summer (Moshkov 1935; Welling et al. 1997) or cold nights (Irving and Lanphear 1968; Weiser 1970; Aronsson 1975; Heide and Prestrud 2005). The short-day signal alone is only sufficient to induce a first level of hardiness (Howell and Weiser 1970; Greer and Warrington 1982), while a combination with low temperatures is required to reach full hardiness. For instance, *Juglans* reaches full hardiness at −6 °C (Améglio et al. 2001b) and *Picea sitchensis* at −20 °C (Cannell et al. 1985).

In autumn, endodormancy (the temporary suspension of growth in every plant part containing a meristem; Lang et al. 1987) can be released by chilling (Weinberger 1950; Landsberg 1974; Richardson et al. 1974; Sarvas 1974). The chilling requirements for endodormancy release prevent trees from initiating growth during transient warm events and avoid freezing risks in sensitive tissues (e.g., bud, cambium, and leaves; see Sect. 3.1). Thereafter, the ecodormancy status (a dormant state that is only limited by environmental factors) takes over the endodormancy status. The enlargement and the production of new leaves in the bud with increasing temperatures ("heat requirement"; Lang et al. 1987) and, for some species, the longer photoperiod (Heide 1993a, 1993b) induce budburst.

Differences in chilling and heat requirements are genetic adaptations to environmental conditions. Genotypes with low chilling requirement are able to flush early in cold environmental conditions although the risk of frost damage is high (Scorza and Okie 1990). On the other hand, genotypes with a high chilling requirement may be exposed to insufficient chilling, which triggers delayed or even erratic budburst, limiting the length of the photosynthetic period (Dennis 1987, 1994; Topp et al. 2008). Thus, signals driving dormancy and cold acclimation are identical (Sakai and Larcher 1987). While endodormancy is occurring during winter, trees are acclimating; then, as ecodormancy proceeds, they are deacclimating (Campbell and Sugano 1975; Thomson and Moncrieff 1982). Some authors have studied those two phenomena and shown co-occurrence (Druart et al. 2007; Welling and Palva 2006; Charrier et al. 2011), enabling parallel modeling of both processes in some cases (Fuchigami et al. 1982; Leinonen 1996).

In order to limit freezing damage of tissues, freezing tolerance is another mechanism which is also modulated in an annual cycle. During winter, aboveground parts of trees develop resistance to freezing temperatures by acclimation processes (Aronsson 1975; Christersson 1978) via synthesis of cryoprotectant sugars (Guy et al. 1992; Taji et al. 2002), solutes, or proteins (to inhibit growth and crystallization of ice that would otherwise be fatal; Xing and Rajashekar 2001) as well as tissue dehydration (Tanino et al. 1990; Améglio et al. 2002; Gusta et al. 2004; Charrier et al. 2013b). When environmental conditions are warming, trees are deacclimating in response to warm temperatures (Kalberer et al. 2006). Other resistance mechanisms are based on the avoidance of intracellular ice formation. Increase in intracellular osmotic potential is a strategy shared by different crop genera, such as *Juglans* (Améglio et al. 2004; Charrier et al. 2013b) or *Malus* (Pramsohler and Neuner 2013), and forest genera, such as *Quercus*, *Fagus*, or

Betula (Morin et al. 2007; Charrier et al. 2013a). Osmotic control maintains or stabilizes intracellular structures by use of osmosis, such as mono- and oligosaccharides, polyols, amino acids, lipids, and macromolecules like dehydrins (Yoshida 1984; Khanizadeh et al. 1992; Arora and Wisniewski 1996; Arora et al. 1997, 2004). An increase in cell wall thickness (greater stiffness) and in membrane fluidity at low temperature with unsaturated fatty acids was also observed (Yoshida and Uemura 1986; Uemura and Steponkus 1994). Cryoprotectant molecules help cells to withstand dehydration by expelling water from sensitive areas. Peng et al. (2008) demonstrated that upregulation of aquaporin expression can significantly lower frost resistance by increasing cell membrane permeability. In extreme cell dehydration, remaining water is tightly bound to all structures in a "vitrified state" (Wolfe and Bryant 2001). Membrane stabilization during freeze-induced contraction and thaw-induced expansion is a key process in cell survival (Uemura et al. 2006). The rate of temperature change is therefore critical in damage development, as it determines the time slot available for water to cross the plasma membrane.

4 Freezing of Dead Components

4.1 Damage in the Transport System

Embolism upon freezing stress has been reported from numerous woody species, both angiosperms and conifers (Table 1). According to the "thaw-expansion hypothesis" (or "bubble formation hypothesis," e.g., Sucoff 1969; Ewers 1985; Lo Gullo and Salleo 1993; Davis et al. 1999; Lemoine et al. 1999; Hacke and Sperry 2001; Sperry and Robson 2001; Pittermann and Sperry 2003, 2006), it is the combination of freezing and subsequent thawing which ultimately leads to xylem dysfunction: When the sap freezes, gas bubbles are formed in the conduits because air is hardly soluble in ice. On thawing, these bubbles will expand when the negative pressure of the surrounding sap overcomes the bubble-collapsing force of surface tension. This force is negatively correlated to bubble size (Laplace's law, Pittermann and Sperry 2006) so that larger bubbles expand at lower tensions than small bubbles. Therefore, the risk of embolism increases with the conduit diameter, since wide elements contain large amounts of dissolved gas, which forms large bubbles within the ice (e.g., Davis et al. 1999; Sperry and Robson 2001; Pittermann and Sperry 2003, 2006).

The "thaw-expansion hypothesis" was supported by several studies. Bubbles within the ice of frozen conduits were demonstrated (Sucoff 1969; Ewers 1985; Robson et al. 1988), and conduit size was found to be critical for freeze–thaw-induced embolism. Conifers with small tracheids as well as vessel-less angiosperms were reported to be hardly susceptible to freeze–thaw-induced embolism (Hammel 1967; Sucoff 1969; Sperry and Sullivan 1992; Sperry et al. 1994; Davis et al. 1999;

Table 1 Studies on freezing-induced embolism. Studied species, methodical approaches (H, hydraulic; U, ultrasonic; I, imaging), study type (L, laboratory or experimental study; F, field observation), and citations are given, respectively

Species	Method	Study	References
Conifers			
Abies alba	U	L	Mayr and Zublasing (2010)
Abies grandis	H	F	McCulloh et al. (2011)
Abies lasiocarpa	H	L	Pittermann and Sperry (2003)
Juniperus communis	U	L	Mayr and Zublasing (2010)
Juniperus scopulorum	H	L	Sperry and Sullivan (1992), Pittermann and Sperry (2006), Willson and Jackson (2006)
Juniperus deppeana	H	L	Willson and Jackson (2006)
Juniperus monosperma	H	L	Willson and Jackson (2006)
Juniperus osteosperma	H	L	Willson and Jackson (2006)
Larix decidua	U	L	Mayr and Zublasing (2010)
Larix laricina	H	F	Sperry et al. (1994)
Larix lyallii	H	F	Sparks and Black (2000)
Picea abies	H, U, I	L, F	Mayr et al. (2002, 2003a, b, 2007), Mayr and Zublasing (2010)
Picea glauca	H	F	Sperry et al. (1994)
Pinus albicaulis	H	F	Sparks and Black (2000)
Pinus cembra	H	L, F	Mayr et al. (2003a, b), Mayr and Zublasing (2010)
Pinus contorta	H,U	L, F	Sparks et al. (2001), Pittermann and Sperry (2006), Mayr and Sperry (2010)
Pinus edulis	U	L	Weiser and Wallner (1988)
Pinus mugo	U	L	Mayr and Zublasing (2010)
Pinus ponderosa			
Pinus sylvestris	H, U	L, F	Mayr and Zublasing (2010), Charrier et al. (2013a)
Pseudotsuga menziesii	H	F	McCulloh et al. (2011)
Thuja plicata	H	F	McCulloh et al. (2011)
Tsuga heterophylla	H	F	McCulloh et al. (2011)
Angiosperms, ring porous			
Fraxinus americana	U	L	Weiser and Wallner (1988)
Fraxinus pennsylvanica	U	L	Weiser and Wallner (1988)
Quercus alba	H	L, F	Cochard and Tyree (1990)
Quercus gambelii	H	L, F	Sperry and Sullivan (1992), Sperry et al. (1994)
Quercus robur	H	F	Charrier et al. (2013a, b)
Quercus rubra	H	L, F	Cochard and Tyree (1990)

(continued)

Table 1 (continued)

Species	Method	Study	References
Angiosperms, diffuse porous			
Acacia obtusifolia	H	L	Choat et al. (2011)
Acer pseudoplatanus	H	F	Charrier et al. (2013a, b)
Acer saccharum	H, I	L, F	Sperry et al. (1988b)
Allocasuarina littoralis	H	L	Choat et al. (2011)
Alnus cordata	H	F	Charrier et al. (2013a, b)
Alnus crispa	H	F	Sperry et al. (1994)
Alnus incana	H	F	Sperry et al. (1994)
Atherosperma moschatum	H	F	Feild and Brodribb (2001)
Avicennia germinans	H	L, F	Stuart et al. (2007)
Banksia ericifolia	H	L	Choat et al. (2011)
Banksia spinulosa	H	L	Choat et al. (2011)
Betula alleghaniensis	H	L	Zhu et al. (2001)
Betula occidentalis	H	L, F	Sperry and Sullivan (1992), Sperry et al. (1994), Davis et al. (1999)
Betula papyrifera	H	F	Sperry et al. (1994)
Betula pendula	H, U	L, F	Charrier et al. (2013a, b, 2014a, b)
Betula platyphylla	I	F	Utsumi et al. (1998)
Canella winterana	H	L	Feild et al. (2002)
Ceanothus megacarpus	H	L	Langan et al. (1997)
Cornus sericea	H	L	Davis et al. (1999)
Corylus avellana	H, U	L, F	Charrier et al. (2013a, b, 2014a, b)
Crataegus monogyna	H, U	L	Charrier et al. (2014a, b)
Elaeagnus angustifolia	H	L	Davis et al. (1999)
Eucalyptus burgessiana	H	L	Choat et al. (2011)
Eucalyptus cinerea	U	L	Raschi et al. (1989)
Eucalyptus coccifera	H	F	Feild and Brodribb (2001)

(continued)

Table 1 (continued)

Species	Method	Study	References
Eucalyptus pauciflora	I	L	Ball et al. (2006)
Eucalyptus sieberi	H	L	Choat et al. (2011)
Euonymus kiautschovicus	H	L	Davis et al. (1999)
Euonymus latifolius	U	L	Kikuta and Richter (2003)
Fagus sylvatica	H, U	L, F	Lemoine et al. (1999), Charrier et al. (2013a, b, 2014a, b)
Hedera helix	H	L	Davis et al. (1999)
Larrea tridentata	H	L, F	Pockman and Sperry (1997), Martinez-Vilalta and Pockman (2002)
Juglans regia	H, U	L, F	Améglio et al. (1995, 2002), Kikuta and Richter (2003), Charrier et al. (2014a, b)
Juglans regia x nigra	H	F	Charrier et al. (2013a, b)
Leptospermum rupestre	H	F	Feild and Brodribb (2001)
Malus sp.	U	L	Weiser and Wallner (1988)
Nothofagus cunninghamii	H	F	Feild and Brodribb (2001)
Nothofagus gunnii	H	F	Feild and Brodribb (2001)
Orites revoluta	H	F	Feild and Brodribb (2001)
Ozothamnus rodwayi	H	F	Feild and Brodribb (2001)
Petrophile pulchella	H	L	Choat et al. (2011)
Populus balsamifera	H	F	Sperry et al. (1994)
Populus canadensis	H	L, F	Just and Sauter (1991)
Populus tremuloides	H	L, F	Sperry and Sullivan (1992), Sperry et al. (1994)
Prunus cerasifera	H, U	L, F	Charrier et al. (2013a, b, 2014a, b)
Prunus cerasus	H, U	L	Charrier et al. (2014a, b)
Prunus persica	H, U	L, F	Améglio et al. (2002), Charrier et al. (2014a, b)
Pyrus communis	U	L	Weiser and Wallner (1988)
Quercus gambelii	H	L	Davis et al. (1999)
Quercus ilex	H, U	L, F	Lo Gullo and Salleo (1993), Nardini et al. (2000)
Rhizophora mangle	H	L, F	Stuart et al. (2007)
Rhizophora stylosa	H	L, F	Stuart et al. (2007)

(continued)

Table 1 (continued)

Species	Method	Study	References
Rhododendron maximum	H	F	Lipp and Nilsen (1997)
Rhus aromatica	H	L	Davis et al. (1999)
Rhus laurina	H	L	Langan et al. (1997)
Richea scoparia	H	F	Feild and Brodribb (2001)
Robinia pseudoacacia	H	F	Charrier et al. (2013a, b)
Salix alba	H, U	L	Charrier et al. (2014a, b)
Salix sachalinensis	I	F	Utsumi et al. (1998)
Sorbus aucuparia	H, U	L	Charrier et al. (2014a, b)
Zygogynum balansae	H	L	Feild et al. (2002)
Zygogynum pancheri	H	L	Feild et al. (2002)
Zygogynum queenslandiana	H	L	Feild et al. (2002)

Feild and Brodribb 2001; Feild et al. 2002), while ring-porous species showed extreme vulnerability (Cochard and Tyree 1990; Lo Gullo and Salleo 1993; Sperry et al. 1994; Nardini et al. 2000). Charrier et al. (2014b) compared ten angiosperms and observed a positive correlation between the hydraulic diameter and conductivity losses. It was also demonstrated that freeze–thaw-induced embolism increased with increasing tension in the xylem sap (Sperry and Sullivan 1992; Langan et al. 1997; Davis et al. 1999; Sperry and Robson 2001; Mayr et al. 2003a; Pittermann and Sperry 2006). Even small conduits can embolize when the sap pressure is negative enough to expand the small bubbles formed (Pittermann and Sperry 2006). Accordingly, field studies revealed high conductivity losses upon combinations of drought stress (i.e., high tensions in the xylem sap) and freeze–thaw stress (e.g., Lemoine et al. 1999; Sparks and Black 2000; Sparks et al. 2001; Mayr et al. 2002, 2003b). The "thaw-expansion hypothesis" was also confirmed by a series of centrifuge experiments, which allowed analyzing the role of tension during the freeze and the thaw. As expected, tension during thawing induced embolism while tension during freezing had no effect on resulting embolism (Mayr and Sperry 2010).

In contrast, there are some findings which cannot be explained by this hypothesis: First, in centrifuge experiments (Mayr and Sperry 2010), two freeze–thaw cycles produced more embolism than one cycle. Similarly, conductivity losses increased with the number of freeze–thaw events in field studies (Sperry et al. 1994; Sparks and Black 2000; Sparks et al. 2001; Mayr et al. 2003b) as well as in experiments (Mayr et al. 2003a, 2007; Mayr and Zublasing 2010). It is unclear how conduits may escape from embolism during the first frost cycle although they embolize in consecutive cycles. One explanation might be fatigue effects as described by Christensen-Dalsgaard and Tyree (2014). Second, a

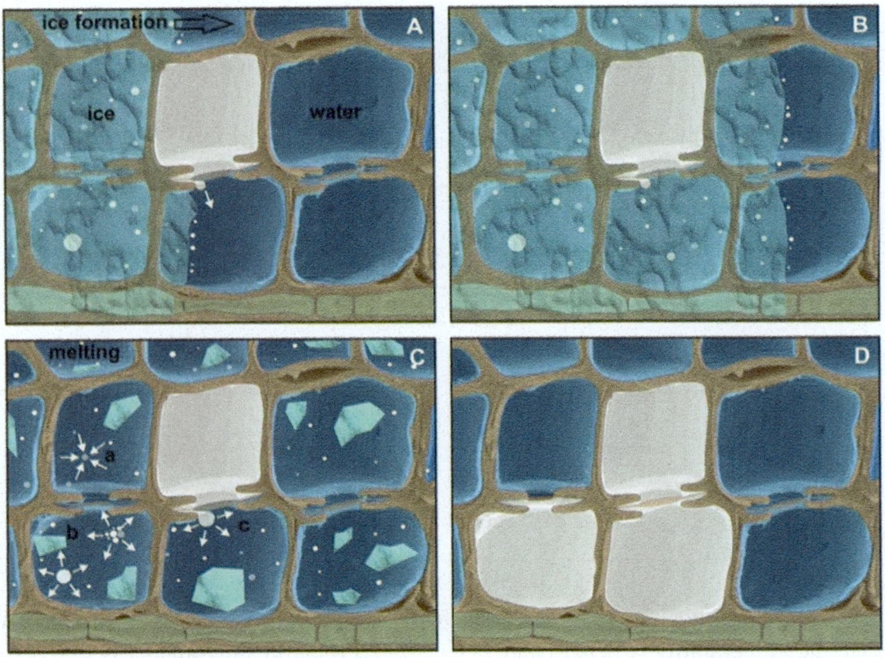

Fig. 6 Proposed mechanism of freeze–thaw-induced embolism according to Charrier et al. (2014b) visualized on a conifer cross section. During ice formation (**A**) air bubbles are formed in the sap as air is hardly soluble in ice. High tensions near the ice front can also induce air seeding, but bubble enlargement is stopped by the growing ice front (**B**). On thawing (**C**), small bubbles collapse (a), while larger bubbles or coalescing bubbles (b) expand depending on the sap tension. Air seeding (c) may also lead to embolism (**D**). The figure is based on a SEM cross-sectional image of *Picea abies*

Cryo-SEM study on *Picea abies* revealed that the widest conduits are not consistently the most vulnerable ones (Mayr et al. 2007), and finally, ultrasonic acoustic emissions were observed on freezing although the "bubble expansion hypothesis" predicts embolism formation during the thaw. This was found in conifers (Mayr et al. 2007; Mayr and Zublasing 2010; Mayr and Sperry 2010) and angiosperms (Weiser and Wallner 1988; Raschi et al. 1989; Kikuta and Richter 2003; Charrier et al. 2014b). While *Pinus contorta* samples cooled down to −15 °C and −25 °C and showed similar loss of conductivity (Mayr and Sperry 2010), Charrier et al. (2014b) demonstrated a clear correlation between the minimum temperature and conductivity losses in angiosperms. PLC (see Sect. 2.3) corresponded to ultrasonic activities, and species with high resistance to drought-induced embolism exhibited the smallest increase in ultrasonic activity with decreasing temperature. These findings indicate that "air seeding" (e.g., Tyree and Zimmermann 2002), which is the process responsible for drought-induced embolism, plays a role also during freezing. According to Charrier et al. (2014b), low Ψ near the ice front may cause air seeding from adjacent conduits and induce ultrasonic acoustic emissions due to relaxation of cell walls (see Sect. 2.3). Immediately afterwards, the small bubbles are entrapped in ice until they can coalescence on thawing and lead to embolism (see Fig. 6).

4.2 Survival Strategies

Many plants escape from freezing stress, e.g., by overwintering under a protective snow layer. But how can plants avoid freezing-induced embolism when their xylem is exposed to subzero temperatures?

Evolution of small conduits increased the xylem's safety to freezing damage. As described above, narrow (and/or short) conduits contain smaller volumes of dissolved air so that formed bubbles are smaller, and the risk of bubble expansion is lower (Pittermann and Sperry 2003). Accordingly, many conifers with small tracheids are very resistant to freezing-induced embolism (e.g., see Sect. 4.1), unless their xylem is exposed to high tensions. As a consequence, avoidance of critical tensions is another important strategy (Pittermann and Sperry 2006). The latter might be based on efficient stomata closure and cuticular shields, efficient water uptake, or internal water storage. For instance, frozen xylem sections can provide large pools of stored water, which release water as soon as the ice melts (e.g., Mayr and Charra-Vaskou 2007).

According to the findings of Charrier et al. (2014b), the extent of freezing-induced embolism might be reduced by a high resistance to air seeding and by avoidance of low minimum temperatures, e.g., via thick bark layers providing sufficient insulation. Interestingly, while low Ψ increases the risk of freezing-induced embolism, freezing can also affect the xylem's resistance to drought-induced embolism ("frost fatigue," Christensen-Dalsgaard and Tyree 2013, 2014).

When trees cannot avoid the formation of freezing-induced embolism, the growth of new xylem may help to overcome transport deficits. Ring-porous species, which lose most of their sapwood's conductivity with the first autumn frost, have to build a new, highly conductive ring of conduits in spring to supply the crown with water. Embolized vessels are closed by tyloses i.e., an outgrowth of the plasma membrane of the vessel-associated cells (VACs; Alves et al. 2001) into the lumen of xylem vessels blocked by embolism (Zimmermann 1983), and heartwood is formed after some years (Cochard and Tyree 1990). Thus, it is likely that the presence of air in the vessels due to embolism has a function and is not without reason with the development of tylosis and ultimately heartwood.

Repair processes are another important possibility to overcome xylem dysfunction caused by freezing-induced embolism. This might be facilitated by positive pressure produced in roots (Ewers et al. 2001) and/or stems (e.g., Sperry et al. 1988b; Améglio and Cruiziat 1992; Améglio et al. 1995, 2001b, 2002, 2004; Hacke and Sauter 1996; Cochard et al. 2001). Stem pressure represents a particular mechanism where xylem parenchyma cells and VAC (see Alves et al. 2001, 2004, 2007) interact with vessels to repair embolism (see Figs. 7 and 8) but also to transport sugars over long distance in the absence of transpiration

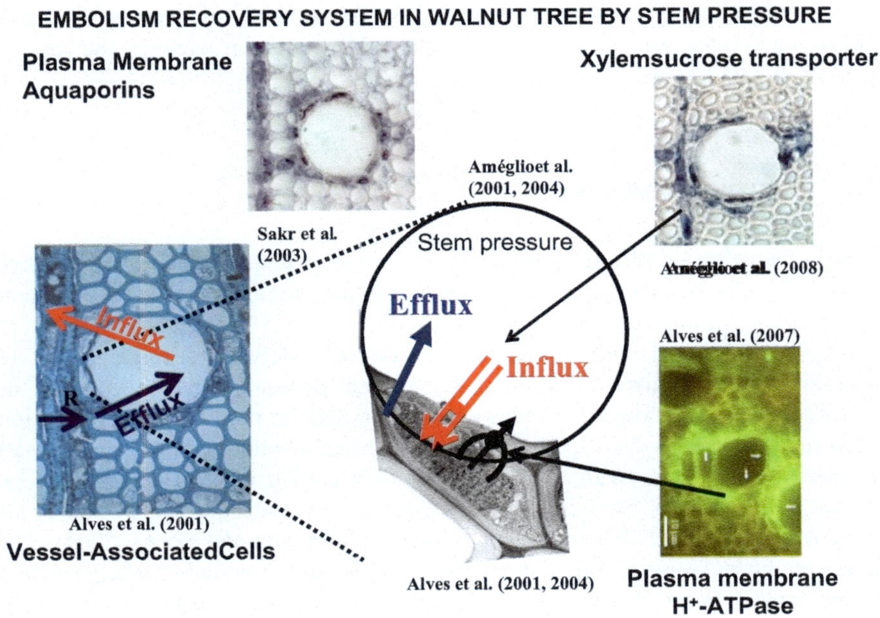

Fig. 7 Schematic of embolism recovery system in a *Juglans* tree by stem pressure. Vessel-associated cells (VACs) around xylem vessel (V) and connected by plasmodesm with xylem parenchyma cells (P) are the principal control points for embolism repair. This complex mechanism involves different elements: plasma membrane aquaporins (see Sakr et al. 2003), xylem sucrose transporter (see Decourteix et al. 2008), plasma membrane H^+-ATPase (see Alves et al. 2001, 2004, 2007), and an induced positive stem pressure in vessels (see Améglio and Cruiziat 1992; Améglio et al. 1995, 2001b, 2002, 2004). All pictures represented a tissue localization of the different protein (i.e., immunolocalization of JrSUT1 for xylem sucrose transporter or immunofluorescence labeling for plasma membrane H^+-ATPase). References indicated the original works for these different elements and methods

(Decourteix et al. 2006; Lacointe et al. 2004; Bonhomme et al. 2010). Repair processes also occur at negative Ψ as demonstrated by several authors (e.g., Zwieniecki and Holbrook 2009; Brodersen et al. 2010; Nardini et al. 2011; Brodersen and McElrone 2013; Zwieniecki et al. 2013). Most of these studies refer to refilling after drought-induced embolism, but there are also indications that some species repair dysfunctional xylem after freeze–thaw stress. For instance, McCulloh et al. (2011) demonstrated recovery from winter embolism in branches of *Abies grandis*, *Pseudotsuga menziesii*, *Thuja plicata*, and *Tsuga heterophylla*, and Sparks et al. (2001) found recovery from winter embolism in *Pinus contorta*. Another remarkable example is conifers growing at the Alpine timberline, which will be dealt with in the following chapter.

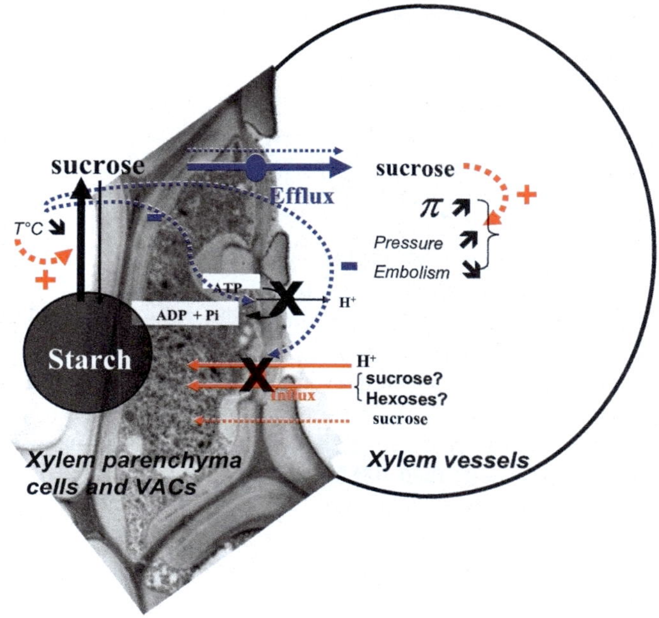

Fig. 8 Schematic of sugar fluxes between parenchyma cells and xylem vessels adapted from Améglio et al. (2004). Vessel-associated cells (VACs) are a control point for sugar fluxes. Efflux of sucrose from VACs into xylem vessels is considered to be a facilitated diffusive flux, putatively mediated by a DEPC-sensitive protein, and influx from vessels into VACs is considered to be an ATPase-dependent active transport. At low temperatures ($-2\,°C < T < 5\,°C$), starch is converted to sucrose. Sucrose accumulates in xylem sap, as a result of facilitated diffusion (efflux) to the apoplast combined with the absence of influx from vessels into VACs. At low temperature, plasma membrane H^+-ATPase is considered inactive, and sugar influx via H^+-sugar symports is blocked. Under these conditions, sap osmolarity (π) and consequently stem pressure increase (P), allowing embolism repair. *Dashed arrows* and *minus signs* (−) indicate negative regulation. *Dotted arrows* and *plus sign* (+) indicate positive regulation. *Question marks* indicate putative mechanisms

5 Timberline: An Example from an Extreme Environment

Trees at the timberline are exposed to harsh conditions unless they are covered by snow. The access to soil water is blocked completely when the soil or the root system is frozen. Even when only upper soil layers are frozen, the respective xylem sections are ice filled and do not allow water transport for months. In parallel, transpirational forces above the snow cover are high due to overheating of branches, which is caused by high radiation during (often frequent and long) sunny winter periods and further amplified by reflection from the snow cover (Turner 1961; Tranquillini 1979) and cold air, which physically is restricted in allowing high absolute humidity. Overheating of conifer needles above air temperature can reach more than 20 K in winter (Mayr et al. 2006b), so that evaporative forces can even be similar to those during summer (Mayr 2007). Deciduous plants have a reduced transpiring surface during winter and are thus generally less prone to transpirational water losses. Nevertheless, peridermal transpiration can also lead to substantial drought stress in some deciduous trees (Richards and Bliss 1986; Tranquillini and Platter 1983). Evergreen species have to keep stomata closed during winter and require a cuticular protection, which sufficiently minimizes water losses as long as water supply is blocked (e.g., Michaelis 1934; Larcher 1972; Tranquillini 1979; Smith et al. 1984). The reduced stomatal conductance thereby corresponds to a reduced photosynthetic capacity (Bauer et al. 1994). Such a frost drought was demonstrated to cause high tensions in the xylem of timberline trees with extreme negative Ψ in late winter (e.g., Mayr et al. 2002, 2003b, 2006a).

Besides drought, low temperature is a main stress for living and dead xylem tissues during winter. Evergreen trees are exposed to numerous frost cycles during winter as strong overheating during sunny days and negative net radiation in clear nights can cause daily frost cycles. During winter season, exposed branches and their xylem can pass more than 100 freeze–thaw cycles (e.g., Groß et al. 1991; Mayr et al. 2006b). Temperature fluctuations in plant parts above the snow cover can be extreme as Mayr et al. (2006b) reported on daily temperature amplitudes of up to 30 K and maximum freezing and thawing rates of 5.4 and 7.0 K h^{-1} in twigs of *Picea abies*.

The combination of drought and freezing stress at the timberline was found to cause dramatic conductivity losses in some conifers. In branch xylem of *Picea abies*, up to 100 % loss of conductivity (PLC; see Sect. 2.3) was observed in late winter (Mayr et al. 2002, 2003b, 2006a). High embolism rates were also found in *Pinus mugo* and *Juniperus communis*, while other species showed moderate or hardly embolism (Sparks and Black 2000; Mayr et al. 2006a). This winter embolism results from the combinatorial stress of frost drought (Fig. 9) and multiple freezing events and occurs nearly each winter (Mayr et al. unpublished data). Freeze–thaw events in the axes system thereby cause complex and dynamic patterns: Ice in the xylem hydraulically separates axes sections and avoids equilibration of water potentials. This causes different drought stress intensities, which in turn influences the vulnerability to freeze–thaw-induced embolism. Due to changing ice barriers, drought and freeze–thaw stress varies spatially and temporarily within the tree crown, and refilling (see below) might further complicate this situation (Mayr and Charra-Vaskou 2007). Like in angiosperms (Charrier et al. 2014b, see Sect. 4.1), minimum temperatures influence the intensity of ultrasonic emissions and thus probably the intensity of embolism formation also in Alpine conifer species (Fig. 10).

But how can trees, which suffer from extreme winter embolism, survive? Several studies indicated that affected trees refill their xylem in late winter and spring (Sperry and Sullivan 1992; Sperry et al. 1994; Mayr et al. 2003b), but the underlying mechanism (e.g., Zwieniecki and Holbrook 2009) is not yet understood. Katz et al. (1989) and Sparks et al. (2001) suggested water uptake by branches or leaves. Mayr et al. (2014) demonstrated that this in fact can substantially contribute to internal water contents. This study also revealed that refilling activities correspond to changes in carbohydrate and aquaporin levels in needles so that an active,

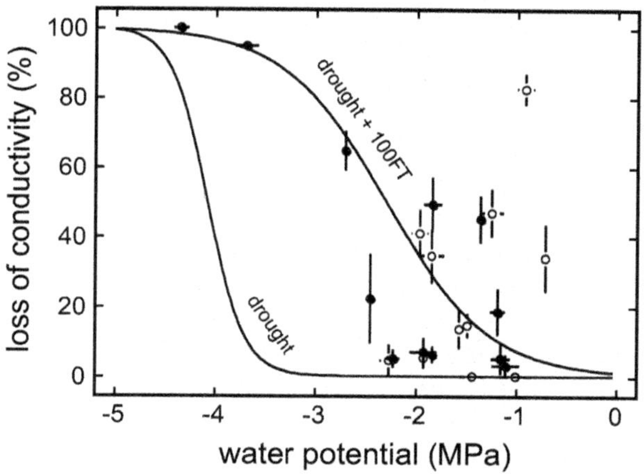

Fig. 9 Combined effects of drought and freezing stress at the timberline. *Curves* show embolism formation on drought or a combination of drought and 100 freeze–thaw cycles (vulnerability curves of potted trees; see Mayr et al. 2003a, b) in *Picea abies*. Points show percent loss of conductivity versus water potentials measured at the timberline during winter 2001/2002 (see Mayr et al. 2003b) and 2003/2004 (see Mayr et al. 2006a) with high stress intensities. *Solid points* indicate samples during embolism formation until midwinter, and *open points* indicate measurements during recovery in late winter and spring. Mean ± SE

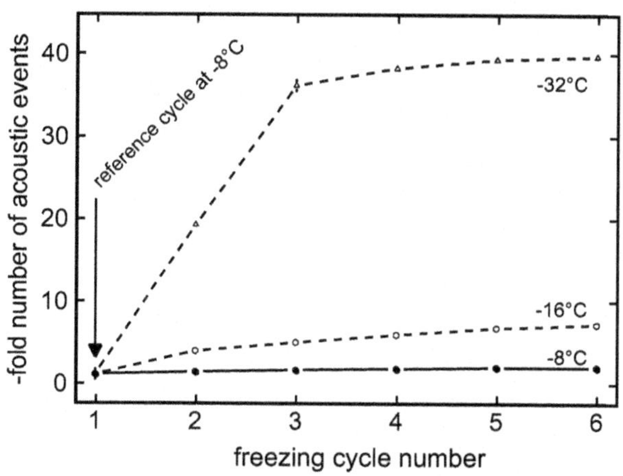

Fig. 10 Effect of the minimum temperature on ultrasonic acoustic emissions in *Picea abies*. The cumulative number of acoustic events after six consecutive freeze–thaw cycles was set to 100 %. Samples were exposed to −8 °C in the first cycle and to −8 °C, −16 °C, or −32 °C in the consecutive five cycles. The increase in cumulative acoustic events was related to the first (reference) cycle. Mean ± SE (Zublasing and Mayr unpublished)

cellular process seems likely. An uptake of water via needles and corresponding aquaporin activity in *Picea glauca* was also described by Laur and Hacke (2014). Surprisingly, yearly embolism-refilling cycles in timberline trees do not seem to cause critical cavitation fatigue (Hacke et al. 2001) in studied conifers (Mayr 2007; Mayr et al. 2014, but also see Christensen-Dalsgaard and Tyree 2013). It was suggested that melting snow might be used by refilling timberline conifers to reduce tensions in the xylem and enable refilling (Mayr et al. 2014).

A recent study of Charrier et al. (2013a) investigated different frost resistance mechanisms, such as the resistance to winter embolism and frost hardiness of living cells (LT_{50}; see Sect. 2.2), in 1-year-old branches of 11 European tree species between leaf fall and budburst. These ecophysiological traits were analyzed according to the potential altitudinal limit, which is highly related to frost exposure. Although seasonal frost hardiness and PLC changes were relatively different across species, maximal PLC observed in winter (PLC_{Max}) was closely correlated to the potential altitudinal limit of species. Moreover, PLC_{Max} was related to the mean hydraulic diameter of vessels (indicating embolism sensitivity) and to osmotic compounds (indicating ability of living cells to refill conducting elements in the xylem). The physiological and anatomical parameters studied enabled to model the potential elevational limit of tree species according to their frost resistance strategies.

According to Körner (2003, 2012), the formation of tree lines is caused by low temperature stress and related growth limitations. Growth limitations may also influence the trees' ability to avoid or repair embolism. Winter embolism thus is another temperature-related factor probably modulating the position of timberlines in temperate regions (but cannot explain timberlines on a global scale). Timberline trees exposed to winter embolism are an interesting model system to study combined effects of drought and freezing stress on tree xylem. Both stress factors occur every year, and stress intensities increase with elevation so that pronounced effects can be observed at the timberline. It also enables insights into avoidance, tolerance, and repair strategies of plants. The underlying processes of xylem recovery are still under research, and timberline conifers are one fascinating example for this probably important survival strategy.

6 Conclusions

The Alpine timberline is only one (although impressive) example of ecosystems, where freezing stress plays a dominant role for plant life and survival. Countless studies enabled insights into the complexity of the freezing process and resulting freezing injury as well as on avoidance and adaptation strategies of plants, but our knowledge still is far from being complete. Regarding plant xylem, many important aspects, like the small-scale patterns and dynamics of water potentials during freezing, the distribution of ice and water phases in tissues after freezing, or the substantial shifts of water and gases on freezing and thawing, remain to be studied. Thereby, the interplay between symplastic and apoplastic components within the xylem deserves closer attention. As demonstrated in this review, plants cope with freezing stress in both living cells and the apoplastic water transport system based on the close connection between these components. For instance, frost tolerance is enabled by water shifts out of living cells toward apoplastic spaces and refilling of embolized xylem by water shifts via living cells.

The interplay between apoplast and symplast may also be the base for acclimation processes, which are of especial relevance in the light of expected climate change. Although temperatures will overall rise, freezing stress may even increase: Extreme weather events are predicted to occur more frequently and with higher intensity, which may be of relevance for frost as well as drought periods. Drought and freezing stress may act in direct combination during winter, while summer drought may indirectly affect cold acclimation by limited carbohydrate reserves. Furthermore, higher mean temperatures may lead to phenological shifts and, in consequence, delayed or insufficient hardening of plant tissues. Xylem tissue, which is formed over years and contains many dead components, is probably limited in its short-term acclimation potential and thus very vulnerable. Knowledge on freezing in plant xylem thus is a prerequisite to understand plant life under present and future conditions.

Acknowledgments This review was prepared in the frame of the project I826-B25 "Acoufreeze" funded by the Austrian and French research agencies (FWF and ANR), respectively. We thank Georg Leitinger, Department of Ecology, University of Innsbruck, for providing the thermographic camera. We thank Prof. Rainer Matyssek and the anonymous reviewer for thoughtful comments and help to improve the manuscript.

References

Alves G, Sauter JJ, Julien JL, Fleurat-Lessard P, Améglio T, Guilliot A, Pétel G, Lacointe A (2001) Plasma membrane H^+-ATPase, succinate and isocitrate dehydrogenases activities of vessel-associated cells in walnut trees. J Plant Physiol 158:1263–1271

Alves G, Améglio T, Guilliot A, Fleurat-Lessard P, Lacointe A, Pétel G, Julien JL (2004) Winter variation of xylem sap pH in walnut tree: implication of the plasma membrane H^+-ATPase of vessel-associated cells. Tree Physiol 24:99–105

Alves G, Decourteix M, Fleurat-Lessard P, Sakr S, Bonhomme M, Améglio T, Lacointe A, Julien JL, Petel G, Guilliot A (2007) Spatial activity and expression of plasma membrane H^+-ATPase in stem xylem of walnut during dormancy and growth resumption. Tree Physiol 27:1471–1480

Améglio T, Cruiziat P (1992) Tension/pressure alternation in walnut xylem sap during winter: the role of winter temperature. C R Acad Sci III 315:429–435

Améglio T, Cruiziat P, Béraud S (1995) Tension-pressure alternation in walnut xylem sap during winter: effect on hydraulic conductivity of twigs. C R Acad Sci III 318:351–357

Améglio T, Cochard H, Ewers F (2001a) Stem diameter variations and cold hardiness in walnut trees. J Exp Bot 52:2135–2142

Améglio T, Ewers FW, Cochard H, Martignac M, Vandame M, Bodet C, Cruiziat P (2001b) Winter stem pressures in walnut trees: effects of carbohydrates, cooling and freezing. Tree Physiol 21:387–394

Améglio T, Bodet C, Lacointe A, Cochard H (2002) Winter embolism, mechanisms of xylem hydraulic conductivity recovery and springtime growth patterns in walnut and peach trees. Tree Physiol 22:1211–1220

Améglio T, Ewers FW, Cochard H (2003) Gelista™: a new tool for testing frost hardiness by stem diameter variations. Acta Horticult 618:509–515

Améglio T, Decourteix M, Alves G, Valentin V, Sakr S, Julien JL, Petel G, Guilliot A, Lacointe A (2004) Temperature effects on xylem sap osmolarity in walnut trees: evidence for a vitalistic model of winter embolism repair. Tree Physiol 24:785–793

Aronsson A (1975) Influence of photo- and thermoperiod on the initial stages of frost hardening and dehardening of phytotron-grown seedlings of scots pine (*Pinus sylvestris* L.) and Norway spruce (*Picea abies* L. Karst.). Stud Forestalia Suecia 128:1–20

Arora R, Wisniewski M (1996) Accumulation of a 60-kD dehydrin protein in peach xylem tissues and its relationship to cold acclimation. Hortscience 31:923–925

Arora R, Rowland LJ, Palta GR (1997) Chill-responsive dehydrins in blueberry: are they associated with cold hardiness or dormancy transitions? Physiol Plant 101:8–16

Arora R, Rowland LJ, Ogden EL, Dhanaraj AL, Marian CO, Ehlenfeldt MK, Vinyard B (2004) Dehardening kinetics, bud development, and dehydrin metabolism in blueberry cultivars during deacclimation at constant, warm temperatures. J Am Soc Hortic Sci 129:667–674

Asahina E (1956) The freezing process of plant cells. Contr Inst Low Temp Sci Hokkaido Univ 10:83–126

Ashworth AN, Malone SR, Ristic Z (1993) Response of woody plant cells to dehydrative stress. Int J Plant Sci 154:90–99

Ball MC, Canny MJ, Huang CX, Egerton JJG, Wolfe J (2006) Freeze/thaw-induced embolism depends on nadir temperature: the heterogeneous hydration hypothesis. Plant Cell Environ 29:729–745

Bauer H, Nagele M, Comploj M, Galler V, Mair M, Unterpertinger E (1994) Photosynthesis in cold acclimated leaves of plants with various degrees of frost tolerance. Physiol Plant 91:403–412

Bonhomme M, Peuch M, Améglio T, Rageau R, Guilliot A, Decourteix M, Alves G, Sakr S, Lacointe A (2010) Carbohydrate uptake from xylem vessels and its distribution among stem tissues and buds in walnut (*Juglans regia* L.). Tree Physiol 30:89–102

Boorse GC, Bosma TL, Meyer AC, Ewers FW, Davis SD (1998) Comparative methods of estimating freezing temperatures and freezing injury in leaves of chaparral shrubs. Int J Plant Sci 159:513–521

Brodersen CR, McElrone AJ (2013) Maintenance of xylem network transport capacity: a review of embolism repair in vascular plants. Front Plant Sci 4:108

Brodersen CR, McElrone AJ, Choat B, Matthews MA, Shackel KA (2010) The dynamics of embolism repair in xylem: in vivo visualizations using high-resolution computed tomography. Plant Physiol 154:1088–1095

Brodersen CR, McElrone AJ, Choat B, Lee EF, Shackel KA, Matthews MA (2013) In vivo visualizations of drought-induced embolism spread in *Vitis vinifera*. Plant Physiol 161:1820–1829

Burke MJ, Gusta LV, Quamme HA, Weiser CJ, Li PH (1976) Freezing and injury in plants. Annu Rev Plant Physiol Plant Mol Biol 27:507–528

Burr KE, Tinus RW, Wallner SJ, King RM (1990) Comparison of 3 cold hardiness tests for conifer seedlings. Tree Physiol 6:351–369

Campbell RK, Sugano AI (1975) Phenology of bud burst in Douglas-fir related to provenance, photoperiod, chilling, and flushing temperature. Bot Gaz 136:290–298

Cannell MGR, Sheppard LJ, Smith RI, Murray MB (1985) Autumn frost damage on young *Picea sitchensis*. Shoot frost hardening and the probability of frost damage in Scotland. Forestry 58:145–166

Canny MJ (1997) Vessel contents of leaves after excision—a test of Scholander's assumption. Am J Bot 84:1217–1222

Cavender-Bares J (2005) Impacts of freezing on long-distance transport in woody plants. In: Holbrook NM, Zwieniecki M, Melcher P (eds) Vascular transport in plants. Elsevier, Oxford

Cavender-Bares J, Cortes P, Rambal S, Joffre R, Miles B, Rocheteau A (2005) Summer and winter sensitivity of leaves and xylem to minimum freezing temperatures: a comparison of co-occurring Mediterranean oaks that differ in leaf lifespan. New Phytol 168:597–612

Charra-Vaskou K, Charrier G, Wortemann R, Beikircher B, Cochard H, Améglio T, Mayr S (2012) Drought and frost resistance of trees: a comparison of four species at different sites and altitudes. Ann For Sci 69:325–333

Charrier G, Améglio T (2011) The timing of leaf fall affects cold acclimation by interactions with air temperature through water and carbohydrate contents. Environ Exp Bot 72:351–357

Charrier G, Bonhomme M, Lacointe A, Améglio T (2011) Are budburst dates, dormancy and cold acclimation in walnut trees (*Juglans regia* L.) under mainly genotypic or environmental control? Int J Biometeorol 55:763–774

Charrier G, Cochard H, Améglio T (2013a) Evaluation of the impact of frost resistances on potential altitudinal limit of trees. Tree Physiol 33:891–902

Charrier G, Poirier M, Bonhomme M, Lacointe A, Améglio T (2013b) Frost acclimation in different organs of walnut trees *Juglans regia* L.: how to link physiology and modelling? Tree Physiol 33:1229–1241

Charrier G, Charra-Vaskou K, Benoit L, Améglio T, Mayr S (2014a) Changes in ultrasound velocity and attenuation indicate freezing of xylem sap. Agric For Meteorol 185:20–25

Charrier G, Charra-Vaskou K, Kasuga J, Cochard H, Mayr S, Améglio T (2014b) Freeze-thaw stress. Effects of temperature on hydraulic conductivity and ultrasonic activity in ten woody angiosperms. Plant Physiol 164:992–998

Choat B, Medek DE, Stuart SA, Pasquet-Kok J, Egerton JJG, Salari H, Sack L, Ball MC (2011) Xylem traits mediate a trade-off between resistance to freeze–thaw-induced embolism and photosynthetic capacity in overwintering evergreens. New Phytol 191:996–1005

Choat B, Jansen S, Brodribb TJ, Cochard H, Delzon S, Bhaskar R, Bucci SJ, Feild TS, Gleason SM, Hacke UG, Jacobsen AL, Lens F, Maherali H, Martinez-Vilalta J, Mayr S, Mencuccini M, Mitchell PJ, Nardini A, Pittermann J, Pratt RB, Sperry JS, Westoby M, Wright IJ, Zanne AE (2012) Global convergence in the vulnerability of forests to drought. Nature 491:752–755

Christensen-Dalsgaard KK, Tyree M (2013) Does freezing and dynamic flexing of frozen branches impact the cavitation resistance of *Malus domestica* and the *Populus* clone Walker? Oecologia 173:665–674

Christensen-Dalsgaard KK, Tyree M (2014) Frost fatigue and spring recovery of xylem vessels in three diffuse-porous trees in situ. Plant Cell Environ 37:1074–1085

Christersson L (1978) The influence of photoperiod and temperature on the development of frost hardiness in seedlings of *Pinus sylvestris* and *Picea abies*. Physiol Plant 44:288–294

Cochard H, Tyree MT (1990) Xylem dysfunction in *Quercus*: vessel sizes, tyloses, cavitation and seasonal changes in embolism. Tree Physiol 6:393–407

Cochard H, Bodet C, Ameglio T, Cruiziat P (2000) Cryo-scanning electron microscopy observations of vessel content during transpiration in walnut petioles. Plant Physiol 124:1191–1202

Cochard H, Lemoine D, Améglio T, Granier A (2001) Mechanisms of xylem recovery from winter embolism in *Fagus sylvatica*. Tree Physiol 21:27–33

Cochard H, Froux F, Mayr S, Coutand C (2004) Xylem wall collapse in water-stressed pine needles. Plant Physiol 134:401–408

Cochard H, Badel E, Herbette S, Delzon S, Choat B, Jansen S (2013) Methods for measuring plant vulnerability to cavitation: a critical review. J Exp Bot 64:4779–4791

Cochard H, Delzon S, Badel E (2015) X-ray microtomography (micro-CT): a reference technology for high-resolution quantification of xylem embolism in trees. Plant Cell Environ 38:201–206. doi:10.1111/pce.12391

Davis SD, Sperry JS, Hacke UG (1999) The relationship between xylem conduit diameter and cavitation caused by freezing. Am J Bot 86:1367–1372

Decourteix M, Alves G, Brunel N, Améglio T, Guilliot A, Lemoine R, Pétel; G, Sakr S (2006) JrSUT1, a putative xylem sucrose transporter, could mediate sucrose influx into xylem parenchyma cells and be up-regulated by freeze–thaw cycles over the autumn–winter period in walnut tree (*Juglans regia* L.). Plant Cell Environ 29:36–47

Decourteix M, Alves G, Bonhomme M, Peuch M, Baaziz KB, Brunel N, Guilliot A, Rageau R, Améglio T, Pételet G, Sakr S (2008) Sucrose (JrSUT1) and hexose (JrHT1 and JrHT2)

transporters in walnut xylem parenchyma cells: their potential role in early events of growth resumption. Tree Physiol 28:215–224
Dennis FG Jr (1987) Producing temperate-zone fruits at low latitudes: an overview. Hortscience 22:1226–1227
Dennis FG Jr (1994) Dormancy. What we know (and don't know). Hortscience 29:1249–1255
Dereuddre J, Gazeau C (1992) Les végétaux et les très basses températures. In: Come D (ed) Les végétaux et le froid. Hermann, Paris, pp 107–175
Dexter ST, Tottingham WE, Graber LF (1930) Preliminary results in measuring the hardiness of plants. Plant Physiol 5:215–223
Dexter ST, Tottingham WE, Graber LF (1932) Investigations of the hardiness of plants by measurement of electrical conductivity. Plant Physiol 7:63–78
Druart N, Johansson A, Baba K, Schrader J, Sjodin A, Bhalerao RR, Resman L, Trygg J, Moritz T, Bhalerao RP (2007) Environmental and hormonal regulation of the activity-dormancy cycle in the cambial meristem involves stage-specific modulation of transcriptional and metabolic networks. Plant J 50:557–573
Ewers FW (1985) Xylem structure and water conduction in conifer trees, dicot trees, and lianas. Int Assoc Wood Anat Bull 6:309–317
Ewers FW, Améglio T, Cochard H, Beaujard F, Martignac M, Vandame M, Bodet C, Cruiziat P (2001) Seasonal variation of xylem pressure in walnut trees: root and stem pressure. Tree Physiol 21:1123–1132
Feild TS, Brodribb T (2001) Stem water transport and freeze-thaw xylem embolism in conifers and angiosperms in a Tasmanian treeline heath. Oecologia 127:314–320
Feild TS, Brodribb T, Holbrook NM (2002) Hardly a relict: freezing and the evolution of vesselless wood in Winteraceae. Evolution 56:464–478
Friedrich J (1897) Über den Einfluss der Witterung auf den Baumzuwachs. Mitt Forstl Versuchswesen Österreichs 22:155
Fuchigami LH, Weiser CJ, Kobayashi K, Timmis R, Gusta LV (1982) A degree growth stage (degree GS) model and cold acclimation in temperate woody plants. In: Li PH, Sakai A (eds) Plant cold hardiness and freezing stress. Mechanisms and crop implications, vol 2. Academic, New York, NY, pp 93–116
Fujikawa S, Kuroda K (2000) Cryo-scanning electron microscopic study on freezing behavior of xylem ray parenchyma cells in hardwood species. Micron 31:669–686
Fujikawa S, Kuroda K, Fukazawa K (1994) Ultrastructural study of deep super-cooling of xylem ray parenchyma cells from *Styrax obassia*. Micron 25:241–252
Greer DH, Warrington IJ (1982) Effect of photoperiod, night temperature, and frost incidence on development of frost hardiness in *Pinus radiata*. Aust J Plant Physiol 9:333–334
Griffith M, McIntyre CH (1990) The effect of photoperiod and temperature on growth and frost resistance of winter rye root systems. Physiol Plant 79:519–525
Groß M, Rainer I, Tranquillini W (1991) Über die Frostresistenz der Fichte mit besonderer Berücksichtigung der Zahl der Gefrierzyklen und der Geschwindigkeit der Temperaturänderung beim Frieren und Auftauen. Forstwiss Centralbl 110:207–217
Gusta LV, Wisniewski M, Nesbitt NT, Gusta ML (2004) The effect of water, sugars, and proteins on the pattern of ice nucleation and propagation in acclimated and non acclimated canola leaves. Plant Physiol 135:1642–1653
Guy CL (1990) Cold acclimation and freezing stress tolerance role of protein metabolism. Annu Rev Plant Physiol Plant Mol Biol 41:187–223
Guy CL, Huber JLA, Huber SC (1992) Sucrose phosphate synthase and sucrose accumulation at low-temperature. Plant Physiol 100:502–508
Hacke U, Sauter JJ (1996) Xylem dysfunction during winter and recovery of hydraulic conductivity in diffuse-porous and ring-porous trees. Oecologia 105:435–439
Hacke UG, Sperry JS (2001) Functional and ecological xylem anatomy. Perspect Plant Ecol Evol Syst 4:97–115

Hacke UG, Stiller V, Sperry JS, Pittermann J, McCulloh KA (2001) Cavitation fatigue. Embolism and refilling cycles can weaken the cavitation resistance of xylem. Plant Physiol 125:779–786
Hacker J, Neuner G (2007) Ice propagation in plants visualized at the tissue level by infrared differential thermal analysis (IDTA). Tree Physiol 27:1661–1670
Hacker J, Ladinig U, Wagner J, Neuner G (2011) Inflorescences of alpine cushion plants freeze autonomously and may survive subzero temperatures by supercooling. Plant Sci 180:149–156
Hammel HT (1967) Freezing of xylem sap without cavitation. Plant Physiol 42:55–66
Hansen J, Beck E (1988) Evidence for ideal and non-ideal equilibrium freezing of leaf water in frost hardy ivy (*Hedera helix*) and winter barley (*Hordeum vulgare*). Bot Acta 101:76–82
Heide OM (1993a) Daylength and thermal time responses of budburst during dormancy release in some northern deciduous trees. Physiol Plant 88:531–540
Heide OM (1993b) Dormancy release in beech buds (*Fagus sylvatica*) requires both chilling and long days. Physiol Plant 89:187–191
Heide OM, Prestrud AK (2005) Low temperature, but not photoperiod controls growth cessation and dormancy induction and release in apple and pear. Tree Physiol 25:109–114
Hoffmann H (1857) Witterung und Wachstum, oder Grundzüge der Pflanzenklimatologie. Leipzig 312–334
Holten V, Bertrand CE, Anisimov MA, Sengers JV (2012) Thermodynamics of supercooled water. J Chem Phys 136:094507
Howell GS, Weiser CJ (1970) The environmental control of cold acclimation in apple. Plant Physiol 45:390–394
Irving RM, Lanphear FO (1968) Regulation of cold hardiness in *Acer negundo*. Plant Physiol 43:9–13
Ishikawa M, Ide H, Price WS, Arata Y, Nakamura T, Kishimoto T (2009) Freezing behaviours in plant tissues: visualization using NMR micro-imaging and biochemical regulatory factors involved. In: Gusta LV, Wisniewski ME, Tanino KK (eds) Plant cold hardiness: from the laboratory to the field. CAB International, Cambridge, pp 19–28
Just J, Sauter JJ (1991) Changes in hydraulic conductivity upon freezing of the xylem of *Populus x canadensis* Moench "robusta". Trees 5:117–121
Kaku S, Iwaya M (1979) Deep supercooling in xylems and ecological distribution in the genera *Ilex*, *Viburnum* and *Quercus* in Japan. Oikos 33:402–411
Kalberer SR, Wisniewski M, Arora R (2006) Deacclimation and reacclimation of cold-hardy plants: current understanding and emerging concepts. Plant Sci 171:3–16
Katz C, Oren R, Schulze E-D, Milburn JA (1989) Uptake of water and solutes through twigs of *Picea abies* (L.) Karst. Trees 3:33–37
Khanizadeh S, Buszard D, Zarkadas CG (1992) Effect of crop load on hardiness, protein and amino acids content of apple flower buds at the wintering stage and the beginning of the growth. J Plant Nutr 15:2441–2455
Kikuta SB, Richter H (2003) Ultrasound acoustic emissions from freezing xylem. Plant Cell Environ 26:383–388
Körner C (2003) Alpine plant life, 2nd edn. Springer, Heidelberg
Körner C (2012) Alpine treelines. Springer, Heidelberg
Kuroda K, Ohtani J, Kubota M, Fujikawa S (1999) Seasonal changes in the freezing behavior of xylem ray parenchyma cells in four boreal hardwood species. Cryobiology 38:81–88
Lacointe A, Deleens E, Améglio T, Saint-Joanis B, Lelarge C, Vandame M, Song GC, Daudet FA (2004) Testing the branch autonomy theory: a $^{13}C/^{14}C$ double-labeling experiment on differentially shaded branches. Plant Cell Environ 27:1159–1168
Landsberg JJ (1974) Apple fruit bud development and growth. Analysis and an empirical model. Ann Bot 38:1013–1023
Lang GA, Early JD, Martin GC, Darnell RL (1987) Endo-, para- and ecodormancy: physiological terminology and classification for dormancy research. Hortscience 22:371–377
Langan SJ, Ewers FW, Davis SD (1997) Xylem dysfunction caused by water stress and freezing in two species of co-occurring chaparral shrubs. Plant Cell Environ 20:425–437

Larcher W (1972) Der Wasserhaushalt immergrüner Pflanzen im Winter. Ber Dtsch Bot Ges 85:315–327

Laur J, Hacke UG (2014) Exploring *Picea glauca* aquaporins in the context of needle water uptake and xylem refilling. New Phytol 2013:388–400

Leinonen I (1996) A simulation model for the annual frost hardiness and freeze damage of Scots Pine. Ann Bot 78:687–693

Lemoine D, Granier A, Cochard H (1999) Mechanism of freeze-induced embolism in *Fagus sylvatica* L. Trees 13:206–210

Levitt J (1980) Responses of plants to environmental stresses. Chilling, freezing, and high temperature stress: physiological ecology series. Academic, New York, NY

Lindén L (2002) Measuring cold hardiness in woody plants. University of Helsinki, Department of Applied Biology, Helsinki, p 57

Lindén L, Palonen P, Lindén M (2000) Relating freeze-induced electrolyte leakage measurements to lethal temperature in red raspberry. J Am Soc Hortic Sci 125:429–435

Lintunen A, Hölttä, T, Kulmala M (2013) Anatomical regulation of ice nucleation and cavitation helps trees to survive freezing and drought stress. Sci Rep 3.2031. doi:10.1038/srep02031

Lintunen A, Lindfors L, Kolari P, Juurola E, Nikinmaa E, Hölttä T (2014) Bursts of CO_2 released during freezing offer a new perspective on avoidance of winter embolism in trees. Ann Bot. 114:1711–1718. doi:10.1093/aob/mcu190

Lipp CC, Nilsen ET (1997) The impact of subcanopy light environment on the hydraulic vulnerability of *Rhododendron maximum* to freeze-thaw cycles and drought. Plant Cell Environ 20:1264–1272

Lo Gullo MA, Salleo S (1993) Different vulnerabilities of *Quercus ilex* L. to freeze- and summer drought-induced xylem embolism: an ecological interpretation. Plant Cell Environ 16:511–519

Loris K, Havranek WM, Wieser G (1999) The ecological significance of thickness changes in stem, branches and twigs of *Pinus cembra* L. during winter. Phyton 39:117–122

Maldonado CA, Zuniga GE, Corcuera LJ, Alberdi M (1997) Effect of water stress on frost resistance of oat leaves. Environ Exp Bot 38:99–107

Martinez-Vilalta J, Pockman W (2002) The vulnerability to freezing-induced xylem cavitation of *Larrea tridentata* (Zygophyllaceae) in the Chihuahuan desert. Am J Bot 89:1916–1924

Mayr S (2007) Limits in water relations. In: Wieser G, Tausz M (eds) Trees at their upper limit. Treelife limitation at the alpine timberline. Springer, Berlin, pp 145–162

Mayr S, Charra-Vaskou K (2007) Winter at the alpine timberline causes complex within-tree patterns of water potential and embolism in *Picea abies*. Physiol Plant 131:131–139

Mayr S, Rosner S (2011) Cavitation in dehydrating xylem of *Picea abies*: energy properties of ultrasonic emissions reflect tracheid dimensions. Tree Physiol 31:59–67

Mayr S, Sperry JS (2010) Freeze-thaw induced embolism in *Pinus contorta*: centrifuge experiments validate the "thaw-expansion hypothesis" but conflict with ultrasonic emission data. New Phytol 185:1016–1024

Mayr S, Zublasing V (2010) Ultrasonic emissions from conifer xylem exposed to repeated freezing. J Plant Physiol 167:34–40

Mayr S, Wolfschwenger M, Bauer H (2002) Winter-drought induced embolism in Norway spruce (*Picea abies*) at the Alpine timberline. Physiol Plant 115:74–80

Mayr S, Gruber A, Bauer H (2003a) Repeated freeze-thaw cycles induce embolism in drought stressed conifers (Norway spruce, stone pine). Planta 217:436–441

Mayr S, Schwienbacher F, Bauer H (2003b) Winter at the Alpine timberline: why does embolism occur in Norway spruce but not in stone pine? Plant Physiol 131:780–792

Mayr S, Hacke U, Schmid P, Schwienbacher F, Gruber A (2006a) Frost drought in conifers at the alpine timberline: xylem dysfunction and adaptations. Ecology 87:3175–3185

Mayr S, Wieser G, Bauer H (2006b) Xylem temperatures during winter in conifers at the alpine timberline. Agric For Meteorol 137:81–88

Mayr S, Cochard H, Améglio T, Kikuta S (2007) Embolism formation during freezing in the wood of *Picea abies*. Plant Physiol 143:60–67

Mayr S, Schmid P, Laur J, Rosner S, Charra-Vaskou K, Daemon B, Hacke UG (2014) Uptake of water via branches helps timberline conifers refill embolized xylem in late winter. Plant Physiol 164:1731–1740

Mazur P (1966) Physical and chemical basis of injury in single-celled microorganisms subjected to freezing and thawing. In: Meryman HT (ed) Cryobiology. Academic, New York, NY, pp 213–315

Mazur P (1969) Freezing injury in plants. Annu Rev Plant Physiol 20:419–448

McCulloh KA, Johnson DM, Meinzer FC, Lachenbruch B (2011) An annual pattern of native embolism in upper branches of four tall conifer species. Am J Bot 98:1007–1015

Michaelis P (1934) Ökologische Studien an der Baumgrenze, IV. Zur Kenntnis des winterlichen Wasserhaushaltes. Jahrb Wiss Bot 80:169–247

Morin X, Améglio T, Ahas R, Besson C, Lanta V, Lebourgeois F, Miglietta F, Chuine I (2007) Variation of cold hardiness and carbohydrate content from dormancy induction to budburst among provenances of three European oak species. Tree Physiol 27:817–825

Moshkov BS (1935) Photoperiodismus und Frosthärte ausdauernder Gewächse. Planta 23:774–803

Muldrew K, Acker JP, Elliott JAW, McGann LE (2004) The water to ice transition: implications for living cells. In: Fuller BJ, Lane N, Benson EE (eds) Life in the Frozen State. CRC Press, Boca Raton

Nardini A, Salleo S, LoGullo MA, Pitt F (2000) Different responses to drought and freeze stress of *Quercus ilex* L. growing along a latitudinal gradient. Plant Ecol 148:139–147

Nardini A, Lo Gullo MA, Salleo S (2011) Refilling embolized xylem conduits: is it a matter of phloem unloading? Plant Sci 180:604–611

Neuner G, Xu B, Hacker J (2010) Velocity and pattern of ice propagation and deep supercooling in woody stems of *Castanea sativa*, *Morus nigra* and *Quercus robur* measured by IDTA. Tree Physiol 30:1037–1045

Pearce RS (2001) Plant freezing and damage. Ann Bot 87:417–424

Pearce RS, Fuller MP (2001) Freezing of barley studied by infrared video thermography. Plant Physiol 125:227–240

Peng Y, Arora R, Li G, Wang X, Fessehaie A (2008) *Rhododendron catawbiense* plasma membrane intrinsic proteins are aquaporins, and their over-expression compromises constitutive freezing tolerance and cold acclimation ability of transgenic *Arabidopsis* plants. Plant Cell Environ 31:1275–1289

Perry TO (1971) Dormancy of trees in winter. Science 171:29–36

Pittermann J, Sperry JS (2003) Tracheid diameter is the key trait determining the extent of freezing-induced embolism in conifers. Tree Physiol 23:907–914

Pittermann J, Sperry JS (2006) Analysis of freeze–thaw embolism in conifers. The interaction between cavitation pressure and tracheid size. Plant Physiol 140:374–382

Pockman WT, Sperry JS (1997) Freezing-induced xylem cavitation and the northern limit of *Larrea tridentata*. Oecologia 109:19–27

Poirier M, Lacointe A, Améglio T (2010) A semi-physiological model of cold hardening and dehardening in walnut stem. Tree Physiol 30:1555–1569

Pramsohler M, Neuner G (2013) Dehydration and osmotic adjustment in apple stem tissue during winter as it relates to the frost resistance of buds. Tree Physiol 33:807–816

Pramsohler M, Hacker J, Neuner G (2012) Freezing pattern and frost killing temperature of apple (*Malus domestica*) wood under controlled conditions and in nature. Tree Physiol 32:819–828

Rajashekar CB, Burke MJ (1982) Liquid water during slow freezing based on cell water relations and limited experimental testing. In: Li PH, Sakai A (eds) Plant cold hardiness and freezing. Academic, New York, NY

Rajashekar CB, Li PH, Caerter JV (1983) Frost injury and heterogenous ice nucleation in leaves of tuber-bearing *Solanum* species. Ice nucleation activity of external source of nucleants. Plant Physiol 71:749–755

Raschi A, Scarascia Mugnozza G, Surace R, Valentini R, Vazzana C (1989) The use of ultrasound technique to monitor freezing and thawing of water in plants. Agric Ecosyst Environ 27:411–418

Rasmussen DH, MacKenzie AP (1972) Effect of solute on ice-solution interfacial free energy; calculation from measured homogeneous nucleation temperatures. In: Water structure at the water-polymer interface. Springer, pp. 126–145

Raunkiaer C (1934) The life forms of plants and statistical plant geography. Claredon Press, Oxford

Repo T, Makela A, Hanninen H (1990) Modeling frost resistance of trees. Silva Carelica 15:61–74

Richards JH, Bliss LC (1986) Winter water relations of a deciduous timberline conifer, *Larix lyallii* Parl. Oecologia 69:16–24

Richardson EA, Seeley SD, Walker DR (1974) A model for estimating the completion of rest for Redhaven and Elberta peach trees. Hortscience 9:331–332

Richter H (2001) The cohesion theory debate continues: the pitfalls of cryobiology. Trends Plant Sci 6:456–457

Robson DJ, McHardy WJ, Petty JA (1988) Freezing in Conifer Xylem II. Pit aspiration and bubble formation. J Exp Bot 39:1617–1621

Rodrigo J (2000) Spring frosts in deciduous fruit trees—morphological damage and flower hardiness. Sci Hortic 85:155–173

Ruelland E, Vaultier MN, Zachowski A, Hurry V (2009) Cold signalling and cold acclimation in plants. Adv Bot Res 49:35–150

Sachs J (1860) Kristallbildung beim Gefrienen und Auftauen saftiger Pflanzenteile, mitgeteilt von W. Hofmeister. Ber Verhandl Sächs Akad Wiss 12:1–50

Sakai A, Larcher W (1987) Frost survival of plants. Responses and adaptation to freezing stress. Ecological studies. Springer, Berlin

Sakr S, Alves G, Morillon RL, Maurel K, Decourteix M, Guilliot A, Fleurat-Lessard P, Julien JL, Chrispeels MJ (2003) Plasma membrane aquaporins are involved in winter embolism recovery in walnut tree. Plant Physiol 133:630–641

Sarvas R (1974) Investigations on the annual cycle of development of forest trees. 2. Autumn dormancy and winter dormancy. Commun Inst For Fenn 84:101

Scorza R, Okie WR (1990) Peaches (Prunus). In: Moore JN, Ballington J (eds) ISHS Acta Horticulturae 290: genetic resources of temperate fruit and nut crops. Wageningen, ISHS, pp 175–231

Sevanto S, Holbrook NM, Ball MC (2012) Freeze/thaw-induced embolism: probability of critical bubble formation depends on speed of ice formation. Front Plant Sci 3:107

Shirazi AM, Fuchigami LH (1995) Effect of near lethal stress on bud dormancy and stem cold hardiness in red-osier dogwood. Tree Physiol 15:275–279

Smith WK, Young DR, Carter GA, Hadley JL, McNaughton GM (1984) Autumn stomatal closure in six conifer species of the Central Rocky Mountains. Oecologia 63:237–242

Sparks JP, Black RA (2000) Winter hydraulic conductivity and xylem cavitation in coniferous trees from upper and lower treeline. Arct Antarct Alp Res 32:397–403

Sparks JP, Campbell GS, Black RA (2001) Water content, hydraulic conductivity, and ice formation in winter stems of *Pinus contorta*: a TDR case study. Oecologia 127:468–475

Sperry JS, Robson DJ (2001) Xylem cavitation and freezing in conifers. In: Bigras FJ, Colombo SJ (eds) Conifer cold hardiness. Kluwer Academic, Dordrecht, pp 121–136

Sperry JS, Sullivan JEM (1992) Xylem embolism in response to freeze-thaw cycles and water stress in ring-porous, diffuse-porous and conifer species. Plant Physiol 100:605–613

Sperry JS, Donnelly JR, Tyree MT (1988a) A method for measuring hydraulic conductivity and embolism in xylem. Plant Cell Environ 11:35–40

Sperry JS, Donnelly JR, Tyree MT (1988b) Seasonal occurrence of xylem embolism in sugar maple (*Acer saccharum*). Am J Bot 75:1212–1218

Sperry JS, Nichols KL, Sullivan JEM, Eastlack SE (1994) Xylem embolism in ring-porous, diffuse-porous, and coniferous trees of northern Utah and interior Alaska. Ecology 75:1736–1752

Stattin E, Verhoef N, Balk P, van Wordragen M, Lindstrom A (2012) Development of a molecular test to determine the vitality status of Norway spruce (*Picea abies*) seedlings during frozen storage. New For 43:665–678

Strullu-Derrien C, Kenrick P, Tafforeau P, Cochard H, Bonnemain JL, Le Herisse A, Lardeux H, Badel E (2014) The earliest wood and its hydraulic properties documented in c. 407-million-year-old fossils using synchrotron microtomography. Bot J Linn Soc 175:423–437

Stuart SA, Choat B, Martin KC, Holbrook NM, Ball MC (2007) The role of freezing in setting the latitudinal limits of mangrove forests. New Phytol 173:576–583

Sucoff E (1969) Freezing of conifer xylem and the cohesion-tension theory. Physiol Plant 22:424–431

Sutinen ML, Palta JP, Reich PB (1992) Seasonal differences in freezing stress resistance of needles of *Pinus nigra* and *Pinus resinosa*: evaluation of the electrolyte leakage method. Tree Physiol 11:241–254

Taji T, Ohsumi C, Iuchi S, Seki M, Kasuga M, Kobayashi M, Yamaguchi-Shinozaki K, Shinozaki K (2002) Important roles of drought- and cold-inducible genes for galactinol synthase in stress tolerance in *Arabidopsis thaliana*. Plant J 294:417–426

Tanino K, Weiser CJ, Fuchigami LH, Chen THH (1990) Water content during abscisic acid induced freezing tolerance in bromegrass cells. Plant Physiol 93:460–464

Thomson AJ, Moncrieff SM (1982) Prediction of bud burst in Douglas-fir by degree-day accumulation. Can J For Res 12:448–452

Topp BL, Sherman WB, Raseira MCB (2008) Low-chill cultivar development. In: Layne DR, Bassi D (eds) The peach: botany, production and uses. CABI, Wallingford, UK, pp 106–138

Tranquillini W (1979) Physiological ecology of the alpine timberline. Tree existence at high altitudes with special reference to the European Alps. Ecological studies, vol 31. Springer, Berlin

Tranquillini W, Platter W (1983) Der winterliche Wasserhaushalt der Lärche (*Larix decidua* Mill.) an der alpinen Waldgrenze. Verh Ges Ökol 9:433–443

Trifiló P, Raimondo F, Lo Gullo MA, Barbera PM, Salleo S, Nardini A (2014) Relax and refill: xylem rehydration prior to hydraulic measurements favours embolism repair in stems and generates artificially low PLC values. Plant Cell Environ 37:2491–2499

Turner H (1961) Jahresgang und biologische Wirkung der Sonnen- und Himmelsstrahlung an der Waldgrenze der Ötztaler Alpen. Wetter Leben 13:93–113

Tyree MT, Sperry JS (1989) Characterization and propagation of acoustic emission signals in woody plants: towards an improved acoustic emission counter. Plant Cell Environ 12:371–382

Tyree MT, Zimmermann MH (2002) Xylem structure and the ascent of sap, 2nd edn. Springer, Berlin

Tyree MT, Davis SD, Cochard H (1994) Biophysical perspectives of xylem evolution: is there a tradeoff of hydraulic efficiency for vulnerability to dysfunction? IAWA J 15:335–360

Uemura M, Steponkus PL (1994) A contrast of the plasma membrane lipid composition of oat and rye leaves in relation to freezing tolerance. Plant Physiol 104:479–496

Uemura M, Tominaga Y, Nakagawara C, Shigematsu S, Minami A, Kawamura Y (2006) Responses of the plasma membrane to low temperatures. Physiol Plant 126:81–89

Utsumi Y, Sano Y, Fujikawa S, Funada R, Ohtani J (1998) Visualization of cavitated vessels in winter and refilled vessels in spring in diffuse-porous trees by cryo-scanning electron microscopy. Plant Physiol 117:1463–1471

Venturas MD, MacKinnon ED, Jacobsen AL, Pratt RB (2014) Excising stem samples underwater at native tension does not induce xylem cavitation. Plant Cell Environ. 38:1060–1068. doi:10.1111/pce.12461

Vergeynst LL, Dierick M, Bogaerts J, Cnudde V, Steppe K (2015) Cavitation: a blessing in disguise? New method to establish vulnerability curves and assess hydraulic capacitance of woody tissues. Tree Physiol 35:400–409. doi:10.1093/treephys/tpu056

Vogt UK (2001) Hydraulic vulnerability, vessel refilling, and seasonal courses of stem water potential of *Sorbus aucuparia* L. and *Sambucus nigra* L. J Exp Bot 52:1527–1536

Weinberger JH (1950) Prolonged dormancy of peaches. Proc Am Soc Hortic Sci 56:129–133

Weiser CJ (1970) Cold resistance and acclimation in woody plants. In: Cold hardiness, dormancy and freeze protection of fruit crops. LRP Publications, Pullman, WA, pp 403–410

Weiser RL, Wallner SJ (1988) Freezing woody plant stems produces acoustic emissions. J Am Soc Hortic Sci 113:636–639

Welling A, Palva ET (2006) Molecular control of cold acclimation in trees. Physiol Plant 127:167–181

Welling A, Kaikuranta P, Rinne P (1997) Photoperiodic induction of dormancy and freezing tolerance in *Betula pubescens*. Involvement of ABA and dehydrins. Physiol Plant 100:119–125

Wheeler JK, Huggett BA, Tofte AN, Rockwell FE, Holbrook NM (2013) Cutting xylem under tension or supersaturated with gas can generate PLC and the appearance of rapid recovery from embolism. Plant Cell Environ 36:1938–1949

Wiegand KM (1906) Pressure and flow of sap in the maple. Am Nat 40:409–453

Willson CJ, Jackson RB (2006) Xylem cavitation caused by drought and freezing stress in four co-occurring *Juniperus* species. Physiol Plant 127:374–382

Winget CH, Kozlowski TT (1964) Winter shrinkage in stems of forest trees. J For 62:335–337

Wisniewski M, Lindow SE, Ashworth EN (1997) Observations of ice nucleation and propagation in plants using infrared video thermography. Plant Physiol 113:327–334

Wolfe J, Bryant G (2001) Cellular cryobiology: thermodynamic and mechanical effects. Int J Refrig 24:438–450

Wolkerstorfer SV, Rosner S, Hietz P (2012) An improved method and data analysis for ultrasound acoustic emissions and xylem vulnerability in conifer wood. Physiol Plant 146:184–191

Xing W, Rajashekar CB (2001) Glycine betaine involvement in freezing tolerance and water stress in *Arabidopsis thaliana*. Environ Exp Bot 46:21–28

Yoshida S (1984) Studies on freezing injury of plant cells. 1. Relation between thermotropic poperties of isolated plasma membrane vesicles and freezing injury. Plant Physiol 75:38–42

Yoshida S, Uemura M (1986) Lipid composition of plasma membrane and tonoplasts isolated from etiolated seedlings of mung bean (*Vigna radiata* L.). Plant Physiol 82:807–812

Zhang MIN, Willison JHM (1987) An improved conductivity method for the measurement of frost hardiness. Can J Bot 65:710–715

Zhu XB, Cox RM, Meng FR, Arp PA (2001) Responses of xylem cavitation, freezing injury and shoot dieback to a simulated winter thaw in yellow birch seedlings growing in different nursery culture regimes. For Ecol Manag 145:243–253

Zimmermann MH (1978) Hydraulic architecture of some diffuse-porous trees. Can J Bot 56:2286–2295

Zimmermann MH (1983) Xylem structure and the ascent of sap. Springer, New York, NY, p 143

Zweifel R, Häsler R (2000) Frost-induced reversible shrinkage of bark of mature, subalpine conifers. Agric For Meteorol 102:213–222

Zwieniecki MA, Holbrook NM (2009) Confronting Maxwell's demon: biophysics of xylem embolism repair. Trends Plant Sci 14:530–534

Zwieniecki MA, Melcher PJ, Ahrens ET (2013) Analysis of spatial and temporal dynamics of xylem refilling in *Acer rubrum* L. using magnetic resonance imaging. Front Plant Sci 4:265

Canary Island Pine (*Pinus canariensis*), an Evergreen Species in a Semiarid Treeline

Gerhard Wieser, Patricia Brito, José R. Lorenzo, Águeda Ma. González-Rodríguez, Domingo Morales, and María S. Jiménez

Contents

Abstract Canary Island pine (*Pinus canariensis*) is an endemic conifer of the Canary Archipelago where it forms the treeline in Tenerife and La Palma at 2,000–2,100 m a.s.l. Due to climatic and edaphic drought and immature soils, the treeline in the Canary Islands is 1,000–1,900 m lower than in continental mountains at similar latitude. This review summarizes the present knowledge on the ecophysiology of *P. canariensis* growing at treeline where the climate is typically semiarid with high winter precipitation and summer drought. Studies on needle anatomy together with specific root patterns, allowing to search for water, suggest that *P. canariensis* is able to withstand climatic and edaphic drought. At the treeline in Tenerife, drought relates to the quantity of winter precipitation. Treeline trees are able to tap water from deep soil water reserves originating from ample winter precipitation prior to a dry summer. Winter precipitation also influences growth and determines whether forests at treeline are carbon sinks or carbon sources. Topsoil

G. Wieser (✉)
Department of Alpine Timberline Ecophysiology, Federal Research and Training Centre for Forests, Natural Hazards and Landscape (BFW), Rennweg 1, 6020 Innsbruck, Austria
e-mail: gerhard.wieser@uibk.ac.at

P. Brito • J.R. Lorenzo • Á.M. González-Rodríguez • D. Morales • M.S. Jiménez
Department of Plant Biology, Universidad de La Laguna (ULL), C/Astrofísico Francisco Sánchez s/n, 38207 La Laguna, Tenerife, Spain

U. Lüttge et al. (eds.), *Progress in Botany*, Progress in Botany 77,
DOI 10.1007/978-3-319-25688-7_14

desiccation, however, impedes seedling establishment, a prerequisite for regeneration and potential treeline migration.

1 Introduction

Canary Island pine (*Pinus canariensis* Chr. Sm. ex DC in Buch) is a paleoendemic three-needle conifer of the Canary Islands after having a much wider distribution along the northern Tethys shore in the Tertiary (Page 1974; Klaus 1989). The closest relative of *P. canariensis* is the Mediterranean species *P. pinea* (Liston et al. 1999). Moreover, phylogenetic studies have demonstrated that *P. canariensis* is also closely genetically related to other Mediterranean pines (*P. brutia*, *P. halepensis*, *P. pinaster*, *P. heldreichii*) and to *P. roxburghii* in the Himalayas (Gernandt et al. 2005). The Canary Archipelago is located in the Atlantic Ocean 100 km southwest of Morocco between 27° 37′–29° 25′ N and 13° 20′–18° 10′ W. These oceanic islands are of volcanic origin, dating back as far as to the pre-Cretaceous era (Danobeita and Canales 2000), without ever having been connected to continental landmasses. Currently, Canary Island pine forests cover approximately 760 km^2 (Arevalo et al. 2010), which is around 10 % of the total surface area of the archipelago. The distribution of *P. canariensis* is limited to the islands of Tenerife, La Palma, Gran Canaria, El Hierro, and La Gomera. The largest and most natural stands are found on La Palma and Tenerife (Arevalo et al. 2010). Canary Island pine forms pure stands under widely different ecological conditions (Blanco et al. 1989) because of the geographical position of each island (from more humid western to dryer eastern islands), because of the importance of exposition to the trade winds in the north-facing slopes that induce changes in humidity, and because of the huge altitudinal range from near the sea level up to 2,000–2,100 m a.s.l. (locally 2,400 m; Fernández-Palacios and de Nicolás 1995). At this upper level, *P. canariensis* forms the actual treeline in Tenerife and La Palma.

High mountains in remote subtropical and warm temperate oceanic islands differ from continental mountains in having considerably lower alpine treelines (Leuschner 1996; Holtmeier 2009). Although mountains in Tenerife and La Palma rise up to 3,718 m (Pico Teide) and 2,424 m a.s.l. (Roque de los Muchachos) respectively, the upper treeline is about 1,000 m lower than in the High Atlas of Morocco (Hermes 1955) and even 1,900 m lower than in the Himalayas of Nepal (Stainton 1972). This may be attributed to climatic and edaphic drought above the trade wind inversion and immature soils (Schwarzbach 1964; Leuschner 1996; Jonsson et al. 2002). Another reason might be human impact (Höllermann 1978) as there is evidence that open stands of dwarfed *Juniperus cedrus* trees had occurred 100–200 m above the present treeline, probably destroyed by the colonizing Spaniards several hundred years ago (cf. Gieger and Leuschner 2004). Such juniper coppices, however, can still be found close to the current treeline in La Palma (Santos Guerra 1983; Köhler et al. 2006).

At present, no other tree species can compete with *P. canariensis* at the semiarid treeline of the Canary Islands. Therefore, this review highlights the currently available knowledge on the ecophysiology of *P. canariensis* growing within the treeline ecotone. The focus, however, will be on Tenerife only, as most of the work has been carried out at the treeline of Las Cañadas (Teide National Park). After examining the treeline environment, we will address the physiological performance of *P. canariensis*. Finally, water and carbon fluxes as well as tree growth will be discussed, emphasizing potential impacts of climate change on *P. canariensis* forests at its upper distribution limit.

2 Treeline Elevation and Structure

The treeline altitude correlates with the geological age of the bedrock (Höllermann 1978; Holtmeier 2009). The treeline is highest (2,400 m) on old phonolithical substrate with advanced soil development on southern slopes of the island, whereas the treeline is rather low (1,800–2,000 m) on western slopes where lithosols prevail. On basaltic substrate of intermediate age along the Cordillera Dorsal, the treeline has a midway position of 2,150–2,250 m a.s.l.

The stands at the treeline are open (<20 % ground cover, Höllermann 1978; Srutek et al. 2002). Wind and snow-shaped forms are generally missing, although some flagged trees occur on exposed ridges near wind gaps (Höllermann 1978). As cripple and krummholz-forms are also missing, a rather sharp line of isolated erected trees with up to 14 m in height (Srutek et al. 2002; Gieger and Leuschner 2004) gives way upslope to an open sclerophyllous scrubland dominated by *Spartocytisus supranubius*, *Descurainia bourgaeana*, *Pterocephalus lasiospermus*, and *Adenocarpus viscosus* (Fernández-Palacios and de Nicolás 1995).

Forest fires are considered to influence the distribution of pine forests in the Canary Islands (Climent et al. 2004; Otto et al. 2010). However, due to stand openness, fire does not seem to be a factor controlling treeline in Tenerife (Höllermann 1978). Due to overheating during the summer and freezing of the soil surface during the winter, natural reproduction outside the protective canopy of the trees is rapidly eliminated (Srutek et al. 2002).

Finally, there is also a cold-induced inverted treeline in the Caldera basin of the Cañadas (Höllermann 1978). While the Caldera base at 2,000–2,250 m a.s.l. is free of trees, isolated trees occasionally grow up to higher elevations in the surrounding rims. Probably this inverted treeline is caused by temperature inversions down to −16 °C during the winter (Burchard 1929), high insulation, and extreme aridity during the summer.

3 Treeline Environment

The main environmental feature at treeline in the Canary Islands arises from the existence of a quasi-permanent temperature inversion layer associated with the trade wind regime. Due to this inversion layer, the convective and orographic rise of humid air masses toward the summits is prevented, thus leading to an accumulation of clouds below the inversion zone on upwind northern slopes. Below the inversion layer, pine forests are able to augment their water supply by fog condensation (Aboal et al. 2000), whereas leeward southern slopes are drier (Fernández-Palacios and de Nicolás 1995; Del-Arco et al. 2006). At treeline, beyond the influence of the trade winds, the air becomes dry, giving way to high insolation, especially in summer. Moreover, in contrast to most continental mountain sites in the temperate zone, rainfall decreases with altitude above 2,000 m a.s.l. (Del-Arco et al. 2006; AEMET 2012).

The climate at treeline is typically semiarid Mediterranean with maximum precipitation during the winter and almost no rain in summer, when temperatures reach their maximum. Figure 1 provides an example for the seasonal changes in thermal conditions and precipitation at the treeline of Las Cañadas. Due to recent climate warming (Fig. 2), mean annual air temperature during the last 9 years (2005–2013) was on average 1.3 °C higher (11.4 °C) as compared to the mean of the previous 30 years (10.1 °C). In high mountain areas of Tenerife (2,000–2,400 m a.s.l.), Martin et al. (2012) also observed a statistically significant ($P < 0.05$) increase in mean air temperature of 0.31 ± 0.12 °C/decade since 1970. In contrast to temperature, total annual precipitation did not change significantly during the last 39 years and varied between 111 mm in 1998 and 1,165 mm in 2006 (Fig. 2). It is also important to consider that snow may fall occasionally between October and April. During the winter, night frosts down to −7.3 °C are common, while during the summer, air temperature can be up to 32.4 °C. Especially during the arid summer, the insolation is high, so that the Izaña Observatory at about 250 m above treeline recorded 3,448.5 sunshine hours per year on average, the highest sum registered in Spain (http://www.izana.org).

Due to intense solar radiation, the soil surface may heat up to almost 40 °C during the winter (Höllermann 1978), whereas in summer, soil surface temperatures can even reach 70 °C (Köhler et al. 2006). Annual mean soil temperature at 10 cm soil depth is around 13.4 °C, with daily mean winter minima falling down to 1.7 °C and mean maxima rising up to 26.9 °C during the summer (Brito et al. 2013a). As winter is the rainy season, water potential in the topsoil (down to 30 cm soil depth) rarely drops to values below −0.02 MPa (approximating soil water contents >0.3 m^3 m^{-3}). During the arid summer, soil water potential remains close to the wilting point (−1.5 MPa) for 3 months or even longer (Höllermann 1978; Brito et al. 2013a).

Up-growing trees at treeline generally experience a cooler climate as compared to nearby low-stature vegetation. Trees generally operate close to air temperature (author's unpublished observations), as their canopy is aerodynamically rougher

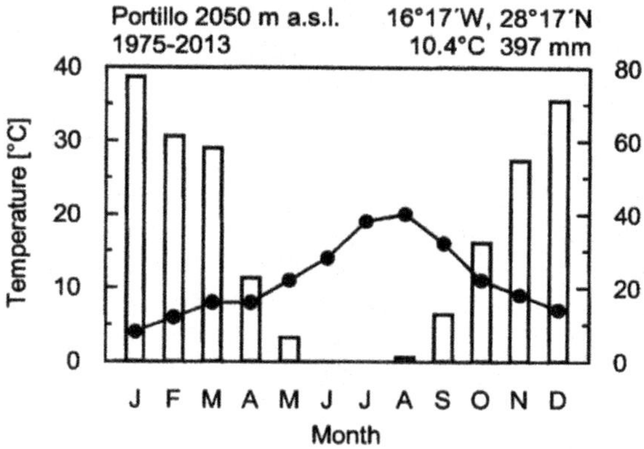

Fig. 1 Climate diagram Tenerife, treeline El Portillo (Las Cañadas), 2,050 m a.s.l. Observation period 1975–2013. Compiled after http://www.iac.es, Brito et al. (2013a), and unpublished data

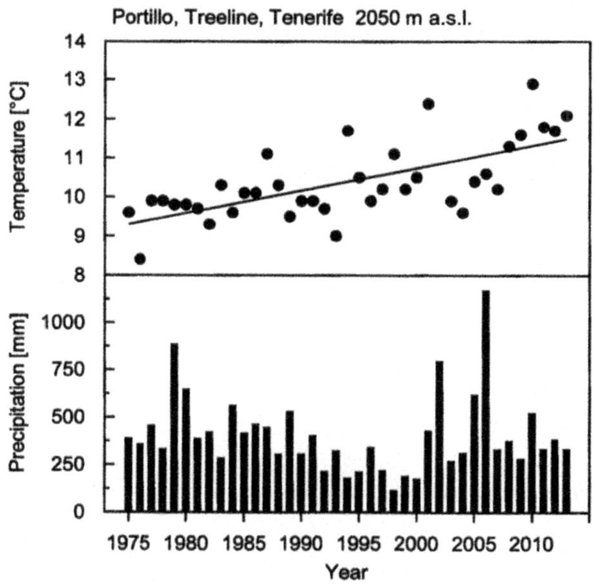

Fig. 2 Mean annual air temperature (*top*) and annual precipitation (*bottom*) El Portillo (Las Cañadas), 2,050 m a.s.l. Observation period 1975–2013. Compiled after http://www.iac.es, Brito et al. (2013a), and unpublished data

and well coupled to the atmosphere. Generally, daily mean leaf-to-air temperature differences (Gieger and Leuschner 2004) and daily mean temperature difference between aboveground woody tissues and the air are rather small (Brito et al. 2010). However, depending on exposure and position, periods may occur when daytime stem temperature in fully sunlit parts of the tree crowns can be up to 3.7 °C warmer than air temperature. Conversely, due to radiative cooling, night-time stem temperature can be up to 4.1 °C lower than air temperature (Brito et al. 2010).

The bedrock is of volcanic origin. As pedogenesis strongly depends on topography, and because soil formation is limited due to severe climatic conditions (Arbelo et al. 2009), the full range of transitions from typical andosols to lithosols and leptosols can occur (Fernández-Caldas and Guerra 1971). The soils at treeline are rich in skeletal material and needle litter can accumulate on the soil surface. The soil texture of the mineral horizon is sandy to sandy clay loam (Höllermann 1978; Brito et al. 2013b). The soils are acidic, possess a low cation exchange capacity, and are poor in organic matter. As low soil temperatures during the winter and soil drought during the summer (Brito et al. 2014, 2015) inhibit mineralization and decomposition, nutrients available for plants become immobilized (Köhler et al. 2006). The upper mineral horizons have a water holding capacity of around 0.40 $m^3\ m^{-3}$ at field capacity (−0.001 MPa) and are very permeable (Höllermann 1978). Conversely, a pumice cover (Höllermann 1978) as well as needle litter on top of the mineral horizon (Brito et al. 2013b) restricts evaporation from deeper soil layers.

4 Water and Carbon Relations

At treeline in Tenerife, it is the limited soil water availability occurring in the dry season that enforces plant adaptations enabling plants to survive drought (Lösch 2000; Körner 2003; Pallardy 2008; Lubczynski 2009). Whole tree acclimation to drought requires sufficient gas exchange while avoiding hydraulic failure. Adaptation mechanisms are related to a high efficiency of water use during periods of photosynthetic carbon gain and seasonal variability of tree transpiration. An efficient water use is achieved by morphological features of the foliage, changes in aboveground carbon allocation between transpiring and conducting tissues, and alterations in belowground hydraulic properties with respect to root patterns allowing search for water.

4.1 Tree Characteristics

Canary Island pine needles are generally xeromorphic (Grill et al. 2004). The stomata are deeply inserted (Jiménez et al. 2000) with an elongated epistomatal chamber. As these epistomatal chambers are covered with epicuticular waxes (Zellnig et al. 2002; Stabentheiner et al. 2004), water loss is restricted proportionally more than CO_2 gas exchange (Jeffree et al. 1971; Riederer 1989). Furthermore, Canary Island pine can adapt anatomically to drought through fostering thickness in needle formation (Grill et al. 2004). Provenances of *P. canariensis* trees from sites experiencing prolonged summer drought, such as at treeline, displayed also thicker needles and thus a higher leaf mass area as compared to provenances originating from the trade wind zone (López et al. 2013). This genetic differentiation in leaf mass area was more pronounced in a xeric common garden ($P = 320$ mm $year^{-1}$) as compared to a more mesic one in the trade wind zone ($P = 795$ mm $year^{-1}$). Moreover, independent of provenance lower leaf-to-sapwood area ratios at the branch level have been found in the xeric than in the mesic common garden (López et al. 2013). The lowered leaf-to-sapwood area ratio at the mesic site resulted from drought-induced needle loss. Adjustment to dryness by reducing foliage reflects a well-known regulation of avoiding severe drought stress in semiarid regions (Lösch 2000; Körner 2003; Pallardy 2008). A reduction in transpiring leaf area per unit of land area has also been observed at the stand level at treeline in Tenerife when summer drought was severe (Brito et al. 2014).

Moreover, Canary Island pine populations from xeric environments are less vulnerable to drought-induced cavitation than those growing at mesic sites. The phenomenon becomes substantiated as the xylem pressure causing a 50 % loss in hydraulic conductance (Breda et al. 2006) dropped below −4.6 and −6.1 MPa in treeline provenances of *P. canariensis* trees when growing under mesic and xeric environmental conditions, respectively (López et al. 2009).

In addition, trees also alter the belowground hydraulic properties by modifying carbon allocation (Steudle 1994; Sperry et al. 1998). In general, *P. canariensis* seedlings allocate more dry matter to roots than to needles and shoots when subjected to drought stress (López et al. 2009). Thus, on the long term, an acclimation in the root-to-shoot ratio to soil water deficits may favor drought tolerance beyond the seedling stage. *P. canariensis* also adapts to soil drought by differentiation between shallow lateral roots and deep groundwater tapping roots. Deep roots extending down to 15 m belowground (Luis et al. 2005; Climent et al. 2007) allow trees to use soil water reserves in deep soil layers when topsoil moisture pools are exhausted.

4.2 Water Relations

Groundwater reserves originate from cold and wet season rainfall (from previous-year October into current-year March), which typically provides more than 90 % of the annual precipitation (Fig. 1). After refilling, surplus of wet season precipitation in the topsoil not used for evapotranspiration percolates via the macroporosity pathway (drainage) to deeper soil layers and is available to deep-rooted plants later during the growing season. Conversely, when October–March rainfall is low and groundwater reserves are depleted, plants may suffer severe drought stress. As treeline drought in Tenerife relates to low wet season precipitation, it is not the dry summer per se that determines the seasonal course of plant water relations (Brito et al. 2015).

Under non-limiting soil water availability including deep soil water reserves, the fine-tuned stomatal response to evaporative demand in terms of leaf-to-air mole fraction difference of water vapor (ΔW) is mirrored in a close linear relationship between transpiration and needle water potential. The greater the transpirational flux, the more the needle water potential is reduced (Fig. 3). On a seasonal time scale, the response of stomatal aperture primarily depends on soil water availability and on ΔW. When soil water availability is non-limiting, stomatal conductance for

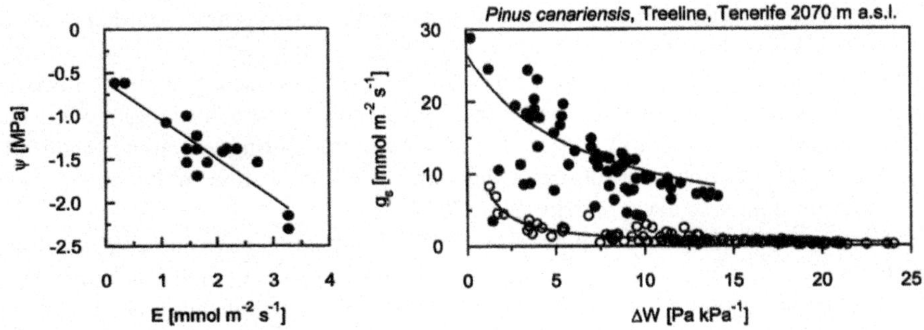

Fig. 3 (*Left*) Relationship between transpiration and leaf water potential in *P. canariensis* at treeline at non-limiting water supply. (*Right*) Daily mean stomatal conductance for water vapor (g_s) of the entire *Pinus canariensis* tree crowns in relation to daily mean leaf/air mole fraction difference of water vapor (ΔW) under conditions of non-limiting soil water availability (*solid symbols*) and conditions of soil drought (*open symbols*). Modified after data from Gieger and Leuschner (2004; *left panel*) and Brito et al. (2014; *right panel*)

Table 1 Maximum leaf conductance for water vapor (g_{max}; mmol m^{-2} s^{-1}), net photosynthetic capacity (A_{max}; μmol m^{-2} s^{-1}), and instantaneous water-use efficiency (WUE, μmol CO_2 mmol^{-1} H_2O), of current-year *Pinus canariensis* needles at treeline (2,070 m a.s.l.) and in a north-facing closed *P. canariensis* forest within the trade wind zone (1,650 m a.s.l.) obtained under conditions of good water supply and under soil drought

	Good water supply		Soil drought	
	Treeline	Trade wind zone	Treeline	Trade wind zone
g_{max}	370.0 ± 72.1	296.7 ± 50.0	95.0 ± 32.1	158.9 ± 25.1
A_{max}	16.48 ± 1.36	16.49 ± 0.50	4.83 ± 0.49	12.44 ± 0.83
WUE	2.34 ± 0.31	3.45 ± 0.50	9.27 ± 2.85	4.17 ± 0.50

Data are related to projected needle surface area (After Brito unpublished treeline and Peters et al. 2008; 1,650 m)

water vapor (g_s) remains high and is solely dependent on ΔW. During periods of water shortage g_s declines gradually to lower values. Notwithstanding, independent of soil water availability, g_s declines exponentially with increasing ΔW (Fig. 3). This commonly observed relationship, however, is more pronounced at treeline than under north-exposed humid forest conditions within the trade wind zone at 1,650 m elevation (Wieser et al. 2002; Peters et al. 2003, 2008). In addition, at non-limiting water supply, maximum g_s increases with elevation (Table 1) as also observed in temperate zone transect studies (Benecke et al. 1981; Körner and Cochrane 1985; Wieser and Havranek 1995). In Tenerife, however, the elevational gradient in g_s becomes reversed during the dry summer (Table 1; Gieger and Leuschner 2004) when soil water availability is limited at the treeline (Brito et al. 2014, 2015) while sufficient in the trade wind zone thereunder (Peters et al. 2008).

Canopy transpiration of a *P. canariensis* forest at treeline has been studied during 2 years, differing in hydrological year precipitation. The hydrological year

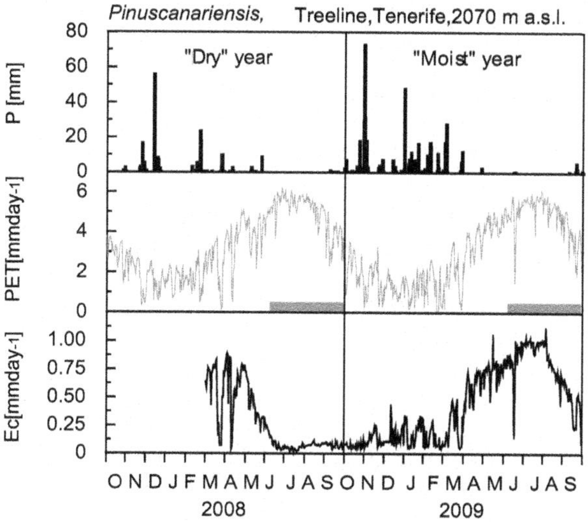

Fig. 4 Seasonal course of daily sums of precipitation (P), potential evaporation (PET), and canopy transpiration (Ec) of *Pinus canariensis* in a "dry" (2008) and a "moist" year (2009). Precipitation from October of the previous year throughout September of the current year was 216 mm in 2008 and 492 mm in 2009. The corresponding PET values were 1,209 and 1,096 mm, respectively. The *gray horizontal bars* highlight the period of topsoil drought that had lasted from June throughout September in both years. Modified after data from Brito et al. (2015)

initiates when rainfall starts to fill up the water reserves in the ground (1st October of the previous year), continues until the evaporation starts to deplete this stored water, and carries on until 1st October of the current year, when it starts to replenish and the cycle begins again. October–March precipitation prior to the dry summer (affecting accumulated soil water including deep soil water pools available to plants) was much lower in 2008 than in 2009 (Fig. 4). As a result, in dry 2008, canopy transpiration was severely reduced during the summer when potential evapotranspiration (PET) was highest (Fig. 4). Such strong reductions in tree and canopy transpiration are typical for Mediterranean forest ecosystems when trees have no access to groundwater (Granier et al. 2007; Infante et al. 2001; Limousin et al. 2009; Martinez-Vilalta et al. 2003; Reichstein et al. 2002) because the primary physiological response to soil water limitation is stomatal closure (Fig. 3; Epron and Dryer 1978; Hinckley et al. 1978; Reich and Hinckley 1989; Borghetti et al. 1998; Damesin and Rambal 1995). In contrast, during moist 2009, trees were able to utilize deep moisture pools caused by ample October–March rainfall (Fig. 4). Hence, the seasonal trends of canopy transpiration and PET resembled with maxima occurring during the dry summer (Fig. 4) when the topsoil was exhausted (Brito et al. 2013a, 2015). The accessibility of deep soil moisture pools originating from ample October to March rainfall prior to a dry summer which are available to deep-rooted treeline plants in Tenerife is well supported by *Spartocytisus supranubius*, a dominant treeline shrub species (Fig. 5). Nevertheless, even when

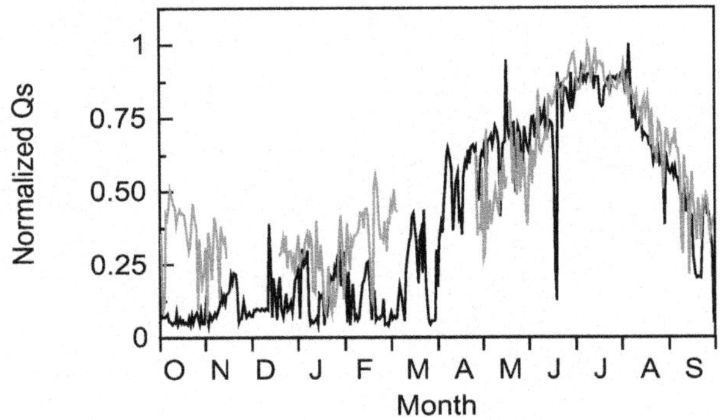

Fig. 5 Seasonal course of normalized sap flow density of *Pinus canariensis* (*black line*) and *Spartocytisus supranubius* (*gray line*) in a moist year at treeline in Tenerife. Compiled after Brito et al. (2015) (*P. canariensis*) and unpublished data (*S. supranubius*)

trees were able to tap water from deep soil layers monthly, mean canopy transpiration amounted only up to 45 % of PET (Brito et al. 2015) indicating an upper limit of tree water loss, even in moist years.

Total annual canopy transpiration varies between 80 and 168 mm year^{-1} in dry and moist years, respectively (Brito et al. 2015). Both values are considerably lower than the value of 252 mm year^{-1} obtained for a *P. canariensis* forest in the trade wind zone at 1650 m elevation by Luis et al. (2005) and the range reported from other Mediterranean forest ecosystems (375–860 mm year^{-1}; Lösch 2000). Annual canopy transpiration at treeline in Tenerife, however, is comparable to levels of drought-affected *P. sylvestris* forests in northeast Germany (82–113 mm year^{-1}; Lüttenschwager et al. 1999) and in a dry valley of Tyrol, Central Alps, Austria (74–110 mm year^{-1}; Wieser et al. 2014a). Evidently, low annual canopy transpiration relates to extreme climatic and/or soil conditions.

4.3 Carbon Relations

Net photosynthetic capacity (sensu Larcher 2001) of *P. canariensis* at treeline resembles that in the trade wind zone when measured under ambient CO_2 partial pressure (Table 1). Stomatal limitation of net photosynthesis occurs under conditions of soil drought and high evaporative demand (Morales et al. 1999; Peters et al. 2003), which is also evident by enhanced instantaneous water-use efficiency (WUE) during the dry summer (Table 1). Due to moderate water use (higher WUE) at treeline as compared to that at lower elevation, the carbon isotope signature ($\delta^{13}C$) of *P. canariensis* needles becomes less negative with elevation, increasing from −24 ‰ at 1,400 m a.s.l. to −21.0 ‰ at treeline (2,100 m a.s.l.; Gieger and Leuschner 2004). Besides decreasing atmospheric CO_2 partial pressure with increasing elevation, the altitude-related increase of $\delta^{13}C$ in *P. canariensis* needles is associated with such in needle thickness (considered as a morphological feature of trees from xeric sites; cf. López et al. 2013). Furthermore, dry mass-related foliar nitrogen levels of treeline trees are comparable to levels at lower elevation (Gieger and Leuschner 2004).

At any given temperature, trees at treeline respire at a higher rate (Brito et al. 2013a) than do trees in the warmer trade wind zone of lower elevation (Wieser et al. 2009). However, when accounting for the average in situ temperature prevailing in the local habitat, carbon release from treeline trees matches those of trees at lower elevation sites. Soil water availability hardly affects respiratory carbon losses of the foliage and aboveground wood tissues, which contrasts with net CO_2 uptake. Foliar dark respiration and aboveground woody tissue CO_2 efflux of *P. canariensis* trees at treeline generally follow seasonal trends in temperature and are highest during the warm and dry summer (Brito et al. 2010, 2013a) as also observed in the trade wind zone (Wieser et al. 2009). Woody tissue CO_2 efflux is also well synchronized with stem growth and is reduced to the level of maintenance respiration when stem growth ceases due to limited soil water availability (Brito et al. 2010). On an annual basis, maintenance respiration accounts for 84 % of total aboveground CO_2 efflux from woody tissues, which is significantly above the range of 30–60 % reported for conifers by Matyssek and Schulze (1988), Ryan (1990), Stockfors and Linder (1998), and Havranek and Matyssek (2005). The estimate of Brito et al. (2010), however, includes an extensively warm and dry period without growth during more than 100 days, which considerably had reduced year round growth respiration. Similarly, at the alpine timberline in the Central Austrian Alps, where stem growth is nil during a 5-month period of dormancy (Gruber et al. 2009), the contribution of maintenance respiration to total annual aboveground woody tissue CO_2 efflux of *Pinus cembra* was 73 % (Wieser and Bahn 2004).

Canary Island pine trees at Tenerife treeline show radial stem growth annually of 0.7–7 mm (Jonsson et al. 2002). On an interannual scale ring width closely follows hydrological year precipitation (Fig. 6), indicating that precipitation and hence soil water availability is the paramount factor limiting tree growth of *P. canariensis* at treeline. Similar findings have also been reported for *Pinus hartwegii* at the topical treeline on Colima Volcano, Mexico (Biondi 2001), and for *Pinus halepensis* subsp. brutia at a low elevation site in Samos, Greece (Sarris et al. 2013).

Table 2 shows the annual carbon balance of a 50–60-year-old *P. canariensis* forest at treeline. Combining annual stem increment data with additive biomass models (Ruiz-Peinado et al. 2011) yields a net primary production (NPP) of 130 g carbon m^{-2} ground surface area $year^{-1}$ in a dry year (2008) and 294 g carbon m^{-2} ground surface area $year^{-1}$ in a moist year (2009), with most of the carbon allocated into the stem fraction. Annual ecosystem respiration (RE) was 550 g carbon m^{-2} ground surface area $year^{-1}$ (average of 2008 and 2009) and comprised the following component fluxes: 33 % from the foliage, 10 % from above ground woody tissue, and 57 % from the soil (Brito et al. 2013a). Interannual differences in ecosystem respiration and its components were <3 % (Table 2). Table 2 also underpins the importance of hydrological year precipitation on NPP and hence also on gross primary production (GPP) as well as net ecosystem production (NEP). Such data have not been available for treeline-associated forest ecosystems in semiarid climates, but can be estimated assuming a contribution of root respiration

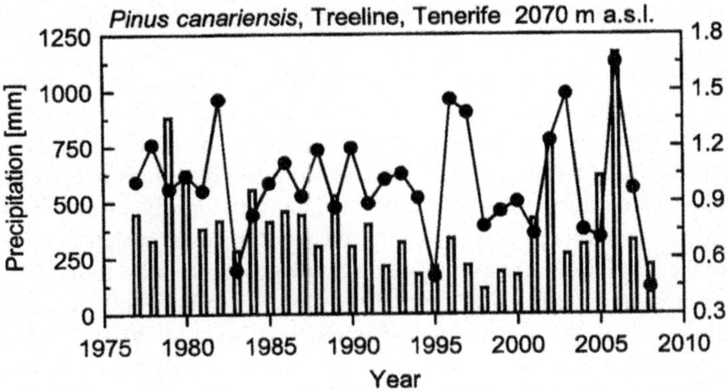

Fig. 6 Hydrological year (October–September) precipitation (*open bars*) and standardized (5-year running mean) tree ring (TRW) width for *P. canariensis* at treeline. The *dotted line* indicates the average (1977–2008) hydrological year precipitation. Compiled after http://www.iac.es, Brito et al. (2013a), and unpublished data

Table 2 Net primary production, respiratory fluxes, gross primary production, and net ecosystem production of a 50–60-year-old *Pinus canariensis* forest at treeline (2070 m a.s.l.) in a hydrological dry (2008) and a moist year (2009)

	Net production (g carbon m^{-2} $year^{-1}$)	
	Dry year (2008)	Moist year (2009)
Net primary production[a]	130	294
Foliage	21	51
Branches	5	11
Stems	64	136
Roots	40	96
Ecosystem respiration[b]	555	554
Foliage	180	185
Above ground woody tissue	53	51
Soil (autotrophic and heterotrophic)	322	308
Total autotrophic respiration[c]	368	365
Gross primary production[c]	498	659
Net ecosystem production[c]	−57	105

[a]Calculated from annual stem increment data (Brito et al. 2010) and biomass models (Ruiz-Peinado et al. 2011); [b]from Brito et al. (2013a). [c]Calculated by assuming root respiration to be 42 % of total soil respiration (Epron 2009). Note that hydrological year precipitation from October of the previous year to September of the current year was 216 mm in 2008 and 492 mm in 2009

to total soil CO_2 efflux of 42 %, corresponding to the overall mean found in forests by Epron (2009). Such a modeling exercise suggests that *P. canariensis* forests at treeline in Tenerife are carbon sinks in moist years but may switch to carbon sources in dry years like 2008 (Table 2) when trees have no access to groundwater reserves (Brito et al. 2015) so that carbon gain becomes severely limited

(Reichstein et al. 2002). Nevertheless, the NEP of 105 g carbon m^{-2} ground surface area $year^{-1}$ estimated for the moist year 2009 falls within the range of 133 ± 47 g carbon m^{-2} $year^{-1}$ reported for temperate semiarid forest ecosystems (Luyssaert et al. 2007).

Another prominent feature of semiarid and Mediterranean-type ecosystems is the effect of episodic precipitation to soil CO_2 efflux during the summer (Rey et al. 2002; Jarvis et al. 2007). After summer rains, two mechanisms produce enhanced CO_2 efflux rates. One is a rapid CO_2 outburst from the soil surface following immediately after a precipitation pulse due to the physical displacement of inorganic CO_2 by water within the soil air spaces (Huxman et al. 2004). The other mechanism is an activation of heterotrophic respiration (Birch 1958). The overall effect of short-term soil CO_2 outbursts occurring in response to summer rain episodes on year round soil CO_2 efflux derived for Canary Island pine forests is <3 % (Wieser et al. 2009; Brito et al. 2013c).

5 Photooxidative Stress and Frost Resistance

Besides summer drought, trees at treeline also experience a combination of high solar radiation with low temperatures and high ground-level ozone (O_3) concentrations as recently reviewed by Jimenez et al. (2005). Such harsh environmental conditions at treeline also cause photooxidative stress in plants and increase the formation of reactive oxygen species (ROS) in the cells (Elstner and Osswald 1994; Polle and Rennenberg 1994). Antioxidant and photoprotective defense systems counteract adverse ROS action, reduce ROS formation, and enable plants to withstand a stress-induced imbalance between carbon fixation and light-driven electron transport in the photosynthetic apparatus, including effects by O_3. Biochemical data obtained along an elevational transect from 550 m a.s.l. up to treeline at 1,950 m a.s.l. (Jimenez et al. 1997, 2005; Tausz et al. 1998) show a tendency of increasing ascorbate concentrations with elevation in *P. canariensis* needles, while contents of glutathione and α-tocopherol do not vary. Chlorophyll contents of *P. canariensis* needles decreased significantly with elevation, while the opposite trend had been observed for the carotenoids/chlorophyll ratio. In addition, the decline in the α/β-carotene ratio with increasing elevation was accompanied by a decline in the ratio of violaxanthin versus the sum of carotenoids (violaxanthin, antheraxanthin, and zeaxanthin) of the xanthophyll cycle.

Presently treeline-associated forests in Tenerife experience O_3 episodes of >75 nl l^{-1} with annual mean values around 45 nl l^{-1} (www.izana.org; Guerra et al. 2004; Oltmans et al. 2006). Such O_3 regimes can adversely affect forest trees (Matyssek and Sandermann 2003; Matyssek et al. 2010), and injury symptoms observed in *P. canariensis* forests have been attributed to photooxidative air pollutants (Arhoun et al. 2000). Then et al. (2009) investigated the effects of O_3 on photosynthesis and biochemical parameters of *P. canariensis* seedlings exposed to free-air O_3 fumigation at Kranzberg Forest, Germany, where ambient O_3 levels were similar to those at high-elevation forest sites in the Canary Islands. After 95 days of exposure, twice-ambient O_3 concentration neither caused any visible injury nor significantly affected the photosynthetic machinery and antioxidant levels in fully developed needles of *P. canariensis* seedlings, the more so, as the stomatal conductance of the well-watered plants did not restrict O_3 uptake at high risk of O_3 injury. Therefore, in the experiment of Then and coworkers, tree metabolism was challenged more intensely by O_3 immediately than to be expected at natural forest sites in the Canary Islands where stomatal closure during the hot and dry summer restricts O_3 uptake (Wieser et al. 2006).

Freezing stress is another important stress at treeline (Körner 2012). In 1-year-old needles of mature Canary Island pine trees at treeline, initial frost damage (LT_{10}, temperature at 10 % frost damage) occurs at −8 to −9 °C, and the LT_{50} (temperature at 50 % frost damage) is −10 °C (Peters et al. 1999). This latter temperature is lower than the reported minimum air temperature of −9.8 °C for the upper distribution limit of *P. canariensis* in Tenerife (Izaña, 2,400 m a.s.l.; http://www.izana.org). Canary Island pine seedlings and saplings with their juvenile growth habit (Klaus 1989), however, might be exposed to frost damage during the winter, as LT_{10} and LT_{50} of current-year-needles are only −3.4 °C and −4.7 °C, respectively (Luis et al. 2007). Nevertheless, even after complete needle loss, sprouting from resting buds on remaining intact shoots made a regrowth possible.

6 Climate Change Perspectives and Conclusions

The semiarid climate at treeline in the Canary Islands with severe summer drought and winter frost differs from most alpine environments on continental mountains in temperate and tropical zonobiomes. Treeline trees show an upright stature, vital habitus, and are not carbon limited (Gieger and Leuschner 2004). Edaphic and climatic drought are suggested to be the paramount factors limiting growth of mature *P. canariensis* trees at treeline to which this species is well adapted. Wind-, frost-, and drought-induced tissue losses do not severely reduce tree growth for trees at treeline (Gieger and Leuschner 2004). Moreover, annual mean air temperatures at treeline (Figs. 1 and 2) are noticeably higher than the mean air temperature range of 5.5–7.0 °C suggested to limit growth in continental treelines worldwide by Körner (2003, 2012).

However, growth of established trees at treeline may not be as important to understand the treeline limit in the Canary Islands as understanding limitations to seedling establishment within the treeline ecotone (c.f. Wieser et al. 2014b, and further references therein). Seedling establishment is a prerequisite for potential treeline migration, demanding for clarification (Smith et al. 2003, 2009). Currently two environmental constraints inhibit the existence of trees above treeline at the seedling stage (Höllermann 1978; Srutek et al. 2002). Topsoil desiccation for about 5 months during the dry summer severely impedes seedling establishment. Seedling establishment also faces frequent night frosts during the winter.

Future climate change might be of paramount importance for *P. canariensis* forests at their upper distribution limit. For the next three decades, climate change and ecophysiological models for Mediterranean ecosystems predict an increase in surface air temperature of 1 °C and a decrease in soil water availability of 15–20 % (Sabaté et al. 2002; IPCC 2013) due to a more than 30 % reduction in precipitation (Giorgi 2006; Somot et al. 2008). An increase in temperature has been observed at Canary Islands, and the temperature increase was highest in high mountain areas above the stratocumulus layer of the trade wind zone (Martin et al. 2012; Luque et al. 2014). Seasonal precipitation patterns by contrast do not display significant changes (De Luque 2011; Cropper and Hanna 2014). Climate warming is expected to increase evapotranspiration and thus making soils dryer. Presently, the topsoil desiccates completely during the warm summer at treeline (Brito et al. 2013a, 2014, p. 215) while topsoil water availability is sufficient at lower elevation sites (Peters et al. 2003, 2008; Luis et al. 2005) where cloud cover is frequent and pine forests are able to augment their water supply from fog condensation (Aboal et al. 2000). Under predicted scenarios of increasing aridity, a shift may occur in the contribution of foliage, stem, and soil CO_2 efflux to ecosystem respiration from presently predominately belowground toward aboveground sources (Brito et al. 2013a). Thus, it is likely that *P. canariensis* forests at their upper distribution limit might be more vulnerable to a future warmer environment as compared to forests at lower elevation sites.

Future climate change may influence treeline advancement (Holtmeier 2009 and further references therein). In Tenerife, a potential shift in the treeline may depend on the future altitude of the trade wind inversion layer. General circulation models for tropical and subtropical mountain are conflicting in proposing both upward (Pounds et al. 1999; Still et al. 1999) and downward shifts (Sperling et al. 2004) of the inversion layer above the trade wind zone where clouds accumulate. Given that the inversion layer may move upward, dragging more humid air masses in a warmer atmosphere, the treeline is expected to advance. At a downward shift of the inversion layer, the treeline may move downward.

In summary, in the semiarid climate at treeline in Tenerife, the timing of the wet and dry season is the major control of gas exchange. As a result, carbon dioxide and water vapor fluxes are out of phase with the seasonal patterns in temperate treelines. In Tenerife, trees at treeline are physiologically active during the winter when soil water availability is sufficient, while the dry summer severely limits gas exchange. To survive, *P. canariensis* has evolved various mechanisms including needle morphology, anatomy, and physiology as well as root systems to endure climatic and edaphic drought at treeline.

Acknowledgments This work was supported by the Spanish Government (CGL2010-21366-C04-04 MCI) and Austrian Science Fund Project (FWF P 22206-B16). P.B. received a fellowship from the Canarian Agency for Research, Innovation and Information Society (ACIISI), cofinanced by FEDER.

References

Aboal JR, Jiménez MS, Morales D, Gil P (2000) Effects of thinning on throughfall in Canary Island pine forest—the role of fog. J Hydrol 238:218–230

AEMET (2012) Climate atlas of the archipelagos of the Canary Islands, Madeira and the Azores. Agencia Estatal de Meteorologia, Ministerio de Agricultura, Alimentacion y medio Ambiente, Gobierno de España. Instituto de Meteorologia Portugal, Lisboa, Portugal

Arbelo CD, Rodrıguez A, Sánchez J, Notario JS, Recatala L, Mora JL, Guerra JA, Armas CM (2009) Caracterizacion en Entorno SIG de los Suelos del Parque Nacional del Teide. Dinamica de Nutrientes y Carbono en los Suelos. Departamento de Edafologıa y Geologıa, Universidad de La Laguna. Proyectos de investigacion en parques nacionales: 2005–2008. www.mma.es/secciones/el_ministerio/organismos/oapn/pdf/oapn_inv-art_0503.pdf

Arevalo JR, Cigala AN, Fernandez-Palacios JM, Fernandez-Lugo S (2010) Ecology and management of natural and reforested Canary Island pine stands. In: Wallace EB (ed) Woodlands: ecology, management and conservation. Nova Publishers, Hauppauge, NY, pp 137–159

Arhoun M, Barreno E, Fos S, Torres-Lapasio JR, Ramis-Ramos G (2000) Injury symptoms and releasing rates of inorganic ions from pine needles as indicators of atmospheric pollution in the Canary Islands forests. Water Air Soil Pollut 117:105–122

Benecke U, Schulze E-D, Matyssek R, Havranek WM (1981) Environmental control of CO_2-assimilation and leaf conductance in *Larix decidua* Mill. I. a comparison of contrasting natural environments. Oecologia 50:54–61

Biondi F (2001) A 400-year tree-ring chronology from the tropical treeline of Northern America. Ambio 30:162–166

Birch HF (1958) The effect of soil drying on humus decomposition and nitrogen availability. Plant Soil 10:9–31

Blanco A, Castroviejo M, Fraile JL, Ganduillo JM, Muñoz LA, Sanchez O (1989) Estudio ecologico del pino canario. ICONA, Instituto nacional para la conservacion de la naturaleza, Madrid

Borghetti M, Cinnirella S, Magani S, Sarracín A (1998) Impact of long-term drought on xylem embolism and growth in *Pinus halepensis*. Trees 12:187–195

Breda N, Huc N, Granier A, Dryer E (2006) Temperate forest trees and stands under severe drought: a review of ecophysiological responses, adaptation processes and long-term consequences. Ann For Sci 63:625–644

Brito P, Morales D, Wieser G, Jimenez MS (2010) Spatial and seasonal variations in stem CO_2 efflux of *Pinus canariensis* at their upper distribution limit. Trees 24:523–531

Brito P, Jimenez MS, Morales D, Wieser G (2013a) Assessment of ecosystem CO_2 efflux and its components in a *Pinus canariensis* forest at treeline. Trees 27:999–1009

Brito P, Trujillo JL, Morales D, Jimenez MS, Wieser G (2013b) Soil moisture overshadows temperature control over soil CO_2 efflux in a *Pinus canariensis* forest at treeline in Tenerife, Canary Islands. Acta Oecol 48:1–6

Brito P, Brito P, Trujillo JL, Morales D, Jimenez MS, Wieser G (2013c) Response of soil CO_2 efflux to simulated precipitation pulsed in a Canary Island pine forest at treeline. Arid Land Res Manag 27:178–187

Brito P, Lorenzo JR, Gonzales-Rodriguez AM, Morales D, Wieser G, Jimenez MS (2014) Canopy transpiration of a *Pinus canariensis* forest at treeline: implications for its distribution under predicted climate warming. Trees 133:491–500

Brito P, Lorenzo JR, Morales D, Jimenez MS, Wieser G (2015) Canopy transpiration of a semi arid *Pinus canariensis* forest at a treeline ecotone in two hydrologically contrasting years. Agric For Meteorol 201:120–127

Burchard O (1929) Beiträge zur Ökologie und Biologie der Kanarenpflanzen. Bibliotheca Botanica H 98, Stuttgart

Climent J, Tapis R, Prados JA, Gil L (2004) Fire adaptations in the Canary Island pine (*Pinus canariensis*). Plant Ecol 171:185–196

Climent J, López R, Gonzalez S, Gil L (2007) El pino canario (*Pinus canariensis*), uns especie singular. Ecosistemas 16:80–89

Cropper TE, Hanna E (2014) An analysis of the climate of Macaronesia, 1865–2012. Int J Climatol 34:604–622

Damesin C, Rambal S (1995) Field study on leaf photosynthetic performance by a Mediterranean deciduous oak (*Quercus pubescens*) during severe summer drought. New Phytol 131:159–167

Danobeita JJ, Canales JP (2000) Magmatic underplating in the Canary Archipelago. J Volcanol Geotherm Res 103:27–41

De Luque AL (2011) Cualificacio´n y homogenizacio´n de las series clima´ticas mensuales de precipitacio´n de Canarias. Estimacio´n de tendencias de la precipitacio´n. Memoria explicativa de resultados. Informe te´cnico proyecto climaimpacto (MAC/3/ C159) del programa de cooperacio´n transnacional Madeira–Azores–Canarias 2007–2013

Del-Arco M, Pérez-de-Paz PL, Acebes JR, González-Mancebo JM, Reyes-Betancort JA, Bermejo JA, De-Armas A, González-González R (2006) Bioclimatology and climatophilous vegetation of Tenerife (Canary Islands). Ann Bot Fenn 43:167–192

Elstner EF, Osswald MW (1994) Mechanisms of oxygen activation during plant stress. Proc R Soc Edinb 102B:131–154

Epron D (2009) Separating autotrophic and heterotrophic components of soil respiration: lessons learned from trenching and related root-exclusion experiments. In: Kutsch W, Bahn M, Heinemeyer A (eds) Soil carbon dynamics. An integrated methodology. Cambridge University Press, Cambridge, UK, pp 157–168

Epron D, Dryer E (1978) Long term effects of drought on photosynthesis of adult oak trees [*Quercus petrea* (Mat.) Liebl. and *Quercus robur* L.] in a natural stand. New Phytol 125:381–389

Fernández-Caldas EF, Guerra A (1971) Condiciones de formación y evolución de los suelos de Tenerife. An Edaf Agrobiol 30:565–610

Fernández-Palacios JM, de Nicolás JP (1995) Altitudinal pattern of vegetation variation on Tenerife. J Veg Sci 6:183–190

Gernandt DS, Lopez GG, García SO, Liston A (2005) Phylogeny and classification of *Pinus*. Taxon 54:29–42

Gieger T, Leuschner C (2004) Altitudinal change in needle water relations of *Pinus canariensis* and possible evidence of a drought-induced alpine timberline on Mt. Teide, Tenerife. Flora 199:100–109

Giorgi F (2006) Climate change hot-spots. Geophys Res Lett 33:L08707. doi:10.1029/ 2006GL025734

Granier A, Reichstein M, Breda N, Janssens IA, Falge E, Ciais P, Grünwald T, Aubinet M, Berbigier P, Bernhofer C, Buchmann N, Facini O, Grassi G, Heinesch B, Ilvesniemi H, Keronen P, Knohl A, Köstner B, Lagergren F, Lindroth A, Longdoz B, Loustau B, Mateus J, Montagnani L, Nys C, Moors E, Papale D, Pfeiffer M, Pilegaard K, Pita G, Pumpanen J, Rambal S, Rebmann C, Rodrigues A, Seufert G, Tenhunen J, Vesala T, Wang Q (2007) Evidence for soil water control on carbon and water dynamics in European forests during the extremely dry year 2003. Agric For Meteorol 143:123–145

Grill D, Tausz M, Pöllinger U, Jiménez MA, Morales D (2004) Effects of drought on needle anatomy of *Pinus canariensis*. Flora 199:85–89

Gruber A, Wieser G, Oberhuber A (2009) Intra-annual dynamics in stem CO_2 efflux in relation to cambial activity and xylem development in *Pinus cembra*. Tree Physiol 29:641–649

Guerra JC, Rodriguez S, Arencibia MT, Garcia MD (2004) Study on the formation and transport of ozone in relation to the air quality management and vegetation protection in Tenerife (Canary Islands). Chemosphere 56:1157–1167

Havranek WM, Matyssek K (2005) The carbon balance of European larch (*Larix decidua*) at the alpine timberline. Phyton 45:213–231

Hermes K (1955) Zur Lage der oberen Waldgrenze in den Gebirgen der Erde und ihr Abstand zur Schneegrenze. Köllner Geogr Arb 5

Hinckley TM, Lassioe JP, Running SW (1978) Temporal and spatial variations in the water status of forest trees. Forest Sci Monogr 20:72
Höllermann PW (1978) Geoecological aspects of the upper timberline in Tenerife, Canary Islands. Arct Alp Res 10:365–382
Holtmeier F-K (2009) Mountain timberlines. Ecology, patchiness, and dynamics. Advances in global change research, vol 36. Springer, Berlin
Huxman TE, Snyder KA, Tissue D, Leffler AJ, Ogle K (2004) Precipitation pulses and carbon fluxes in semiarid and arid ecosystems. Oecologia 141:254–268
Infante JM, Mauchamp A, Fernandez-Ales R, Joffre R, Rambal S (2001) Within-tree variation in transpiration in isolated evergreen oak trees: evidence in support of the pipe model theory. Tree Physiol 21:409–414
IPCC: Intergovernmental Panel on Climate Change (2013) Climate change 2013: the physical science basis. Cambridge University Press, Cambridge
Jarvis P, Rey A, Petsikos C, Wingate L, Rayment M (2007) Drying and wetting of Mediterranean soils stimulates decomposition and carbon dioxide emission: the “Birch effect”. Tree Physiol 27:929–940
Jeffree CE, Johnson RPC, Jarvis PG (1971) Epicuticular wax in the stomatal antechamber of Sitka spruce and its effects on the diffusion of water vapour and carbon dioxide. Planta 98:1–10
Jimenez MS, Tausz M, Zellnig G, Peters J, Grill D, Morales D (1997) Environmental stresses and antioxidative response of *Pinus canariensis* at different field stands in Tenerife. Phyton 37:109–114
Jiménez MS, Zellnig G, Stabentheiner E, Peters J, Morales D, Grill D (2000) Structure and ultrastructure of *Pinus canariensis* needles. Flora 195:228–235
Jimenez MS, Luis VC, Peters J, Gonzales-Rodriguez AM, Morales D (2005) Ecophysiological studies on *Pinus canariensis*. Phyton 45:169–177
Jonsson S, Gunnarson B, Criado C (2002) Drought is the major limiting factor for tree-ring growth of high-altitude Canary Island pines on Tenerife. Geogr Ann 84(A):51–71
Klaus W (1989) Mediterranean pines and their history. Plant Syst Evol 162:133–163
Köhler T, Gieger T, Leuschner C (2006) Altitudinal change in soil and foliar nutrient concentrations and in microclimate across the tree line on the subtropical island mountain Mt. Teide (Canary Islands). Flora 201:202–214
Körner C (2003) Alpine Plant Life—functional plant ecology of high mountain ecosystems, 2nd edn. Springer, Berlin
Körner C (2012) Alpine treelines. Functional ecology of the global high elevation limits. Springer, Basel
Körner C, Cochrane PM (1985) Stomatal responses to water relations in *Eucalyptus pauciflora* in summer along an elevational gradient. Oecologia 66:443–455
Larcher W (2001) Ökophysiologie der Pflanzen: Leben, Leistung und Stressbewältigung der Pflanzen in ihrer Umwelt, 6th edn. Ulmer, Stuttgart
Leuschner C (1996) Timberline and alpine vegetation on the tropical and warm-temperate oceanic islands of the world: elevation, structure and floristics. Vegetatio 123:193–206
Limousin JM, Rambal S, Ourcival JM, Rocheteau A, Joffre R, Rodriguez-Cortina R (2009) Long-term transpiration change with rainfall decline in a Mediterranean *Quercus ilex* forest. Glob Change Biol 15:2163–2175
Liston A, Robinson WA, Piñero D, Alvarez-Buylla ER (1999) Phylogenetics of *Pinus* (Pinaceae based on nuclear ribosomal DNA internal transcribed spacer sequences. Mol Phytogenet Evol 11:95–109
López R, Rodriguez-Calcerrada J, Gil L (2009) Physiological and morphological response to water deficit in seedlings of five provenances of *Pinus canariensis*: potential to detect variation in drought tolerance. Trees 23:509–519
López R, Lopez-de Heredina U, Collada C, Cano FJ, Emerson BC, Cochard H, Gil L (2013) Vulnerability to cavitation, hydraulic efficiency, growth and survival in an insular pine (*Pinus canariensis*). Ann Bot 111:1179–2013

Lösch R (2000) Wasserhaushalt der Pflanzen (Plant water relationships). UTB für Wissenschaft, Quelle und Meyer, Wiebelsheim
Lubczynski MW (2009) The hydrological role of trees in water-limited environments. Hydrogeol J 17:247–259
Luis VC, Jimenez MS, Morales D, Kucera J, Wieser G (2005) Canopy transpiration of a Canary Island pine forest. Agric For Meteorol 135:117–123
Luis VC, Taschler D, Hacker J, Jimenez MS, Wieser G, Neuner G (2007) Ice nucleation and frost resistance of *Pinus canariensis* seedlings bearing needles in three different developmental states. Ann For Sci 64:177–182
Luque A, Martin JL, Dorta P, Mayer P (2014) Temperature trends on Gran Canaria (Canary Islands). An example of global warming over the subtropical Northeastern Atlantic. Atmos Clim Sci 4:20–28
Lüttenschwager D, Wulf M, Rust F, Forkert J, Hüttl RF (1999) Tree canopy and field layer transpiration in Scots pine stands. In: Hüttl RF, Belmann K (eds) Changes of atmospheric chemistry and effects on forest ecosystems. Kluwer, Dordrecht, pp 97–110
Luyssaert S, Inglima I, Jung M, Richardson AD, Reichstein M et al (2007) CO_2 balance of boreal, temperate, and tropical forests derived from a global database. Glob Chang Biol 13:2509–2537
Martin JL, Bethencourt E, Cuevas A (2012) Assessment of global warming on the island of Tenerife, Canary Islands (Spain). Trends in minimum, maximum and mean temperatures since 1944. Clim Chang 144:401–415
Martinez-Vilalta J, Mangiron M, Ogaya R, Sauret M, Serrano L, Penuelas J, Piñol J (2003) Sap flow of three co-occurring Mediterranean woody species under varying atmospheric and soil water conditions. Tree Physiol 23:747–758
Matyssek R, Sandermann H Jr (2003) Impacts of ozone on trees: an ecological perspective. Prog Bot 64:349–404
Matyssek R, Schulze E-D (1988) Carbon uptake and respiration in above-ground parts of a *Larix decidua* x *leptolepis* tree. Trees 2:233–241
Matyssek R, Karnosky DF, Wieser G, Precy K, Oksanen E, Grams TEE, Kubiske M, Hanke D, Pretzsch H (2010) Advances in understanding ozone risk in forest trees: messages from novel phytotron and free-air fumigation studies. Environ Pollut 158:1990–2006
Morales D, Peters J, Jimenez MS, Tausz M, Wonisch A, Grill D (1999) Gas exchange of irrigated and non-irrigated *Pinus canariensis* seedlings growing outdoors in La Laguna, Canary Islands, Spain. Z Naturforsch C 54:693–697
Oltmans SJ, Lefohn AS, Harris JM, Galbally I, Scheel HE, Bodeker G, Brunke E, Claude H, Tarasick D, Johnson BJ, Simmonds P, Shadwick D, Anlauf K, Hyaden K, Schmidlin F, Fujimoto T, Akagi K, Meyer C, Nicjol S, Davies J, Redondas A, Cuevas E (2006) Long-term changes in tropospheric ozone. Atmos Environ 40:3156–3173
Otto R, Garcia-del-Rey E, Munoz PG, Fernandez-Palacios FM (2010) The effect of fire severity on first-year seedlings establishment in a *Pinus canariensis* forest on Tenerife, Canary Islands. Eur J For Res 129:499–508
Page CN (1974) Morphology and affinities of *Pinus canariensis*. Notes R Bot Gard Edinb 33:317–323
Pallardy SG (2008) Physiology of woody plants, 3rd edn. Academic, San Diego
Peters J, Jimenez MS, Morales D (1999) Effect of extreme temperature on quantum yield of fluorescence and membrane leakage of Canarian endemic pine (*Pinus canariensis*). Z Naturforsch C 54:681–687
Peters J, Morales D, Jiménez MS (2003) Gas exchange characteristics of *Pinus canariensis* needles in a forest stand on Tenerife, Canary Islands. Trees 17:492–500
Peters J, Gonzalez-Rodríguez AM, Jiménez MS, Morales D, Wieser G (2008) Influence of canopy position, needle age and season on the foliar gas exchange of *Pinus canariensis*. Eur J For Res 127:293–299

Polle A, Rennenberg H (1994) Photooxidative stress in trees. In: Foyer CH, Mullineaux PM (eds) Causes of photooxidative stress and amelioration of defense systems in plants. CRC Press, Boca Raton, FL, pp 199–218

Pounds JA, Fodgen MPL, Campbell JH (1999) Biological response to climate change on a tropical mountain. Nature 398:411–415

Reich PB, Hinckley TM (1989) Influence of pre-dawn water potential and soil-to leaf hydraulic conductance on the maximum daily diffusive conductance in two oak trees. Funct Ecol 3:719–726

Reichstein M, Tenhunen JD, Roupsard O, Ourcival J-M, Rambal S, Miglietta F, Peresotti A, Pecchiaris M, Tirone G, Valentini R (2002) Severe drought effects on ecosystem CO_2 and H_2O fluxes at three Mediterranean evergreen sites: revision of current hypothesis? Glob Chang Biol 8:999–1017

Rey A, Pegoraro E, Tedeschi V, De Parri I, Jarvis P, Valentini R (2002) Annual variation in soil respiration and its components in a coppice oak forest in central Italy. Glob Chang Biol 8:851–866

Riederer M (1989) The cuticles of conifers: structure, composition and transport properties. In: Schulze ED, Lange OL, Oren R (eds) Forest decline and air pollution. Ecological studies. Springer, Berlin, pp 157–192

Ruiz-Peinado R, del Rio M, Montero G (2011) New models for estimating the carbon sink capacity of Spanish softwood species. For Ecosyst 20:176–188

Ryan MG (1990) Growth and maintenance respiration in stems of *Pinus contorta and Pinus engelmannii*. Can J For Res 20:48–57

Sabaté S, Gracia CA, Sánchez A (2002) Likely effects of climate change on growth of Quercus ilex, Pinus halepensis and Fagus sylvatica forest in the Mediterranean region. For Ecol Manag 162:23–37

Santos Guerra A (1983) Vegetacion y flora de La Palma. Editorial Interinsular Canaria, Santa Cruz, Spain

Sarris D, Siegwolf R, Körner C (2013) Inter-and intra-annual stable carbon and oxygen isotope signals in response to drought in Mediterranean pines. Agric For Meteorol 168:59–68

Schwarzbach M (1964) Edaphisch bedingte Wüsten. Mit Beispielen aus Island, Teneriffa und Hawaii. Z Geomorphol 8:440–452

Smith WK, Germino TE, Hancock TE, Johnson DM (2003) Another perspective on the altitudinal occurrence of alpine tree lines. Tree Physiol 23:1101–1113

Smith WK, Geronimo MJ, Johnson DM, Reinhardt K (2009) The altitude of alpine treeline: a bellwether of climate change effects. Bot Rev 75:163–190

Somot S, Sevault F, Deque M, Crepon M (2008) 21st century climate change scenarios for the Mediterranean using a coupled atmosphere—ocean regional climate model. Glob Planet Chang 63:112–126

Sperling FN, Washington R, Wittaker RJ (2004) Future climate change of the subtropical Atlantic: implications for the cloud forest of Tenerife. Climate Change 65:103–123

Sperry JS, Adler FR, Campbell GS, Comstock JP (1998) Limitation of plant water use by rhizosphere and xylem conductance: results from a model. Plant Cell Environ 21:347–359

Srutek M, Dolezal J, Hara T (2002) Spatial structure and associations in a *Pinus canariensis* population at the treeline, Pico del Teide, Tenerife, Canary Islands. Arct Alp Res 34:201–210

Stabentheiner E, Pfeifhofer HW, Peters J, Jimenez MS, Morales D, Grill D (2004) Different surface characteristics of primary and secondary needles of *Pinus canariensis*. Flora 199:90–99

Stainton JDA (1972) Forests of Nepal. Hafner Publishing Company, New York, NY

Steudle E (1994) Water transport across roots. Plant Soil 167:79–90

Still CJ, Foster PN, Schneider SH (1999) Simulating the effects of climate change on tropical mountain cloud forests. Nature 398:608–610

Stockfors J, Linder S (1998) Effect of nitrogen on the seasonal course of growth and maintenance respiration in stems of Norway spruce trees. Tree Physiol 18:155–166

Tausz M, Peters J, Jimenez MS, Morales D, Grill D (1998) Element contents and stress-physiological characterization of Pinus canariensis needles in Mediterranean type field stands in Tenerife. Chemosphere 36:1019–1023

Then C, Herbinger K, Luis VC, Heerdt C, Matyssek R, Wieser G (2009) Photosynthesis, chloroplast pigments, and antioxidants in Pinus canariensis under free-air ozone exposure. Environ Pollut 157:392–395

Wieser G, Bahn M (2004) Seasonal and spatial variation of woody tissue respiration in a *Pinus cembra* tree at the alpine timberline in the central Austrian Alps. Trees 18:576–580

Wieser G, Havranek WM (1995) Environmental control of ozone uptake in *Larix decidua* Mill.: a comparison between different altitudes. Tree Physiol 15:53–258

Wieser G, Peters J, Luis VC, Morales D, Jiménez MS (2002) Ecophysiological studies on the water relations in a *Pinus canariensis* stand, Tenerife, Canary Islands. Phyton 42:291–304

Wieser G, Luis VC, Cuevas E (2006) Quantification of ozone uptake at the stand level in a Pinus canariensis forest in Tenerife, Canary Islands: an approach based on sap flow measurements. Environ Pollut 140:383–386

Wieser G, Gruber A, Bahn M, Catalá E, Carrillo E, Jiménez MS, Morales D (2009) Respiratory fluxes in a Canary Islands pine forest. Tree Physiol 29:457–466

Wieser G, Leo M, Oberhuber W (2014a) Transpiration and canopy conductance in an inner alpine Scots pine (*Pinus sylvestris* L.) forest. Flora 209:491–498

Wieser G, Holtmeier FK, Smith WK (2014b) Treelines in a changing environment. In: Tausz M, Grulke N (eds) Trees in a changing environment. Plant ecophysiology, vol 9. Springer, Dordrecht, pp 221–263

Zellnig G, Peters J, Jiménez MS, Morales D, Grill D, Perktold A (2002) Three-dimensional reconstruction of the stomatal complex in *Pinus canariensis* needles using serial sections. Plant Biol 4:70–76

The manufacturer's authorised representative in the EU is Springer Nature Customer Service Centre GmbH, Europaplatz 3, 69115 Heidelberg, Germany. If you have any concerns regarding our products, please contact ProductSafety@springernature.com

Printed and bound by CPI Group (UK) Ltd, Croydon, CR0 4YY
01/10/2026
02225869-0006